AQUATIC ENVIRONMENT AND TOXICOLOGY

The Editor

Professor (Dr) Arvind Kumar is the most senior faculty member of Post-Graduate Department of Zoology, SKM University, Dumka, Jharkhand and has 37 years experience as an outstanding teacher and researcher. He served as Pro Vice-Chancellor of SKM University, Dumka, Jharkhand. Professor Kumar also served as Vice-Chancellors of Vinoba Bhave University, Hazaribagh, Jharkhand and Magadh University, Bodhgaya, Bihar.

He has so far published 86 books and more than 250 research papers in peer reviewed journals. Professor Kumar is recipient of several prestigious awards and medals including Prof. W.A. Nizami Gold Medal by the Zoological Society of India, Dr. S.Z. Quasim Gold Medal by SVBP University of Agriculture & Technology, Meerut, ASEA Environment Gold Medal by Gurukul Kangari University, Haridwar, Odum Gold Medal by International Society for Ecological Communications, Poland.

Aquatic Environment and Toxicology

by
Arvind Kumar

2024
Daya Publishing House®
A Division of
Astral International Pvt. Ltd.
New Delhi – 110 002

First Published, 2003

ISBN: 978-81-7035-312-6 (HB)

Publisher's Note:

Every possible effort has been made to ensure that the information contained in this book is accurate at the time of going to press, and the publisher and author cannot accept responsibility for any errors or omissions, however caused. No responsibility for loss or damage occasioned to any person acting, or refraining from action, as a result of the material in this publication can be accepted by the editor, the publisher or the author. The Publisher is not associated with any product or vendor mentioned in the book. The contents of this work are intended to further general scientific research, understanding and discussion only. Readers should consult with a specialist where appropriate.

Every effort has been made to trace the owners of copyright material used in this book, if any. The author and the publisher will be grateful for any omission brought to their notice for acknowledgement in the future editions of the book.

Published by : **Daya Publishing House®**
A Division of
Astral International Pvt. Ltd.
– ISO 9001:2015 Certified Company –
4736/23, Ansari Road, Darya Ganj
New Delhi-110 002
Ph. 011-43549197, 23278134
E-mail: info@astralint.com
Website: www.astralint.com

Digitally Printed at : Replika Press Pvt. Ltd.

Preface

Life originates in water and so water, being the prime necessity, is the soul and hope of nature. From time immemorial, humanity and its civilization has been flourishing in the vicinity of or along with the water resources only. About 70 per cent of the world's surface area is covered with water. Yet, only 0.00192 per cent of the total stock of water is available for human use because 98 per cent water in seas and oceans and 1.998 per cent locked up in arctic regions, glaciers, mountains and clouds is not available. About 85 per cent of the rain falls directly into the sea and never reaches the land. The remaining reaching the land fills the lakes, ponds underground resources and keeps the rivers flowing. The latter constitutes only 0.00008 per cent of the total water. Thus, humanity is left with only one teaspoonful of sweet water for every five litres of total water.

In the world as a whole, only 20 per cent of the population gets clean drinking water, suffering being maximum in rural areas due to wastes discharged from the cities. Nearly 1230 million persons do not have sufficient water for drinking, whereas 1350 million suffer from diseases due to unhygienic water. Roughly, 25000 persons die, maximum being children, each day due to water borne diseases. Although various individuals and organizations have made a lot of hue and cry at different levels for conservation and improvement of aquatic environment, very little has actually been done in the light of toxicology. The need to do something urgently and effectively cannot be ruled out. All those who stress this need are doing a great service to humanity. Information and education are important inputs in the world's efforts to protect or recover the ecological balance. As the Secretary General of Society of Environmental Sciences, Dumka as well as the editor of *Indian Journal of Environment and Ecoplanning*, I have seriously felt that there is almost total ignorance about what an average man needs to know about the aquatic environment and what he can contribute himself towards its improvement and conservation. Therefore, in order to make pollution free aquatic environment, the present endeavour has been undertaken.

In all 50 chapters are written by authors who are acknowledged experts in their respective fields, with the intention to provide sufficient depth of the subjects to satisfy the needs at a level which will be comprehensive and interesting. The book shall be useful to the students, teachers, scientists and researchers in the field of Environmental Sciences and other people with similar interest.

My special thanks and appreciation go to the scientists whose contributions have enriched this volume. I wish to express my sincere gratitude to Professor (Dr.) Indu Dhan, Hon'ble Vice Chancellor, S. K. University, Dumka who has been a source of constant inspiration. I am also grateful to Professor M. C. Dash, Vice Chancellor, Sambalpur University, Professor H. R. Singh, Pro-Vice Chancellor, University of Allahabad, Professor M. P. Singh, Pro-Vice Chancellor, V.B. University, Hazaribag,

Professor N. C. Datta, Calcutta University, Professor Ajit Varma, J. N. U., New Delhi, Professor S. N. Kaul, NEERI, Professor A. K. Mittal, B. H. U., Varanasi, Professor Tanmay Bhattacharya, Midnapore University, Professor S. K. Konar, Kalyani University, Professor P. C. Mishra, Sambalpur University, Professor Sudhendu Mandal, Viswa Bharati University, Professor P. P. Sood, Saurashtra University, Rajkot, Professor B. D. Joshi, G. K. University, Haridwar, Professor U. S. Bagde, Mumbai University, Professor D. B. Tembhare, Nagpur University, Professor B. N. Pandey, Magadh University, Professor J. Ojha, Bhagalpur University, Professor S. P. Roy, Bhagalpur University, Professor M. Raziuddin, V. B. University, Professor M. V. Subba Rao, Andhra University, Professor P. S. Murthy, Bangalore University, Professor P. Natrajan, Kerala University, Professor S. P. Hosmani, Mysore University, Professor R. Ramani Bai, Chennai University and Dr. M. P. Sinha, Ranchi University who have been a source of continuous encouragement and inspiration.

I am especially thankful to Dr. Sikandar Prasad Yadav, Principal, S. P. College, Dumka for his encouragement. I am indebted to respected Shri Tribhuvan Poddar of University Department of Zoology, Bhagalpur University, Bhagalpur for encouragement. I also acknowledge the incentives provided by one of my research scholars, Dr. Chandan Bohra, Head of Zoology, B. S. K. College, Barharwa for helping me actively in bringing out this book.

I also express my deep sense of gratitude to my parents whose blessings have always prompted me to pursue academic activities deeply. I am also thankful to my sweet wife, Professor Kumari Bimla and my two lovely sons, Kumar Pallav Shivashankaran and Kumar Prasun Ramakrishnan whose natural smiles extended to me relief all through this tiresome endeavour.

Last but not the least, I am also thankful to Mr. Anil Mittal, Publisher of Daya Publishing House, Delhi for publishing this book. Finally, I will always remain a debtor to all my well-wishers for their blessings, without which this book would not have come into existence.

Dumka **Professor Arvind Kumar**

Contents

Section 1

AQUATIC ENVIRONMENT

1

Bioindicator as a Tool for Assessment of Aquatic Environment: A Review

Arvind Kumar and Chandan Bohra*

Environmental Biology Research Unit, Post Graduate Department of Zoology, S.K. University, Dumka - 814 101, Bihar
**Pollution Research Laboratory, Department of Zoology, B. S. K. College, Barharwa - 816 101, Jharkhand*

Introduction

Certain living organisms serve the purpose of monitoring the environmental pollution as they are tolerant to adverse environmental conditions. These are termed as Biological Indicators or Bio-indicators and are capable of measuring the actual response of organisms or populations to the environmental quality. The physiological and biological diversity of species allows a wide range of indicator species for various environmental situations. The presence of the organism can be used as a tool to indicate polluted conditions relative to unimpacted reference conditions. Sometimes a set of species or the structure and function of entire biological community may function as a bio-indicator. In assessing the impact of pollution, bio-indicators are frequently used to evaluate the health of an impacted eco-system relative to a reference area or reference conditions. Field based site specific environmental evaluation on the bio-indicator approach generally are complemented with laboratory studies of toxicity testing and bioassay experiments.

The use of individual species or a community structure as bio-indicator involves the identification, classification and quantification of biota in the affected area. While many species are in use, the most widely used biological communities are the benthic macro-invertebrates. These are the sedentary and crawling worm and insect larvae that reside in the bottom sediments of aquatic system. The bottom sediments usually contain most of the pollutants introduced in an aquatic ecosystem since these macro-invertebrates have limited mobility, they are continually exposed to highest concentrations of pollutants in the system. Therefore, the benthic community is an ideal bioindicator: stationary, localized and exposed to maximum pollutant concentrations within a specific location. Examples of indicator organisms can be found among plants, fishes and other biological groups. Giant reed grass a common marsh plant that is typically indicative of disturbed condition in wetlands. Among fish disturbed conditions may be indicated by the disappearance of sensitive species like trout which required clear, cold water to thrive.

Although the two terms are sometimes used interchangeably. Indicator organisms should not be confused with monitor organisms (also called biomonitors) which are organisms that have bio-accumulative toxic chemical substances present in trace amounts in the environment. For example, when it is different to measure directly the low concentrations of a pollutant in water, chemical analysis of shellfish tissues from that location may show higher, easily detected concentrations of that pollutant. In this shellfish is used to monitor the level of the long-term presence of that pollutant in the area.

In the environmental field, bio-indicators are commonly used in field investigation of contaminated sites to document impacts on the biological community. In future bio-indicators may be used more widely as investigative and decision making tools from the initial pollution and impact assessment stage to the remediation and post-remediation monitoring stage.

	Indicator	*Specific Organisms*	*Use*
1.	Faecal coliforms	*E. coli*	Indicator of faecal contamination by warm-blooded animals.
2.	Total coliforms	*E. aerobacter*	Indicator of coliforms originating in soil and intestine of warm blooded animals.
3.	Faecal streptococci	*S. faecalis, S. durans, S. bovis, S. equinus, S. feacium*	Used to judge the probability of human or livestock contaminations source, used in conjunction with faecal coliforms.
4.	Algae	**Polluted Water** *Oscillatoria, Spirogyra, Anabaena,* others	Large accumulation of polluted water farms indicate advancing eutrophication.
		Clean Water *Ulothrix, Calothrix, Cyclotella, Diatom, Chrysococcus*	
5.	Sulfur bacteria	*Thiobacilus* other colourless	Accumulation indicate sulphur bearing water, particularly, containing hydrogen sulphide
6.	Iron bacteria	*Gallionella ferrobacillus* others	Accumulation in natural water indicate iron-bearing waste.
7.	Standard Plate count	Various	Empirical techniques for determining general bacterial density, identification permits judgement concerning relative importance of parasites, pathogens and beneficial saprophytes.

Algae as Marine Pollution Indicators

Pollution at estuaries and of coastal marine water is rapidly becoming a problem at bathing beaches and where the ocean frontage is used for residences, for industry and for recreation. Pollution is influencing the marine fishing and shellfish industries also.

As with fresh water forms, sewage pollution in salt water first tend to reduce the marine algae population to a few of the mare resistant species, but the decomposition products in turn stimulate a vigorous subsequent growth of algae. In marine polluted areas in northern Europe, the sewage pollution is reported to prevent the growth of the brown rock weed Fuscus while algae that are stimulated include *Blindingla maxima, Enteromorpha, Ulva lactuca, Prophyra leucosticta, Erythrothichea carnea, Acrochactium virgatulum, Acrochactium thureti* and *Colothrix confervicola.*

Table 1.1: Algae as Bioindicator in Sewage Fed Pond

Species Names	*Species Names*
Ankistrodesmus falcatus	*Golenkinia radiata*
Chlamydomonas pertusa	*Massartia vorticella*
Chodatella quadriseta	*Ourococcus bicaudatus*
Chromulina vagans	*Planktosphaeria gelatinosa*
Chroomonas caudata	*Polyedriopsis spinulosa*
Closteriopsis brevicula	*Pteromonas angulosa*
Closterium acutum	*Scenedesmus dimorphus*
Cosmarium botrytis	*Schizothrix calcicola*
Cryptomonas cylindrica	*Schroederia setigera*
Diacanthos belenophorus	*Spriulina subtilissima*
Dictyosphaerium ehrenbergianum	*Vacuclario novo-munda*
Elakatothrix gelatinosa	

Table 1.2: Algae as Bioindicator in Polluted Freshwater Bodies

Species Names	*Species Names*
Agmenellum quadriduplicatum	*Tetraedron muticum*
Tenuissima type	*Lepocinclis texta*
Anabaena constricta	*Lyngbya digueti*
Anacystis montana	*Nitzschia palea*
Arthrospira jenneri	*Oscillatoria chlorina (top)*
Carteria multifilis	*Oscillatoria putrida (middle)*
Chlamydomonas reinhardi	*Oscillatoria lauterbomii (bottom)*
Phacus pyrum	*Chlorococcum humicola*
Gomponema parvulum	*Phormidium autumnale*
Chlorogorium euchlorum	*Pyrobotrys stellata*
Chlorella vulgaris	*Spirogyra commumis*
Euglena virdis	*Stigeoclonium temie*

In addition to these large sieved algae microscopic planktonic algae thrive in the polluted areas causing a decrease in the transparency of water. The sea lettuce *Ulva latissima* was stimulated to active growth by sewage in England and India. Industrial pollution in a estuary involving principally iron sulphate stimulate the growth of the pollution algae *Chlorella variegata* and caused the sensitive diatoms *Chaetocerus* and *Skeletonema* to settle out.

Marine and brackish water diatoms are being studied to determine their usefulness indicator of pollution with either sewage or industrial wastes. One apparatus for obtaining the diatoms from the water is a side rack with floats known as diatometer. Its use is being studied in both fresh and salt waters.

Table 1.3: Algae as Bioindicator in Polluted Estuary

Species Names	Species Names
Agardhiella tenera	*Peridinium trochoideum*
Amphidinium fusiforme	*Porphyra atropurpurea*
Asterionella japonica	*Prasiola stipitata*
Chaetoceros decipiens	*Prorocentrum micans*
Chaetomorpha aerea	*Rhodoglossum affine*
Codium fragile	*Scytosiphon lomentaria*
Enteromorpha intenstinalis	*Skeletonema costamum*
Eutreptia virdis	*Spirulina major*
Melosira sulcata	*Stephanoptera gracilis*
Nannochloris atomus	*Stichococcus marinus*
Nitzschia closterium	*Trichodesmium erythraeum*
Pelvetia fastigiata	*Ulva lactuca*

Table 1.4: Algae as Bioindicator in Odorfull Water Body

Species Names	Species Names
Anabaena planctonica	*Nitella gracilis*
Anacystis cyanea	*Pandorina morum*
Aphanizomenon flos-aquae	*Peridinium cinctum*
Asterionella gracillima	*Staruastrum paradoxum*
Ceratium hirundinella	*Synedra uvella*
Dinobryon divergens	*Synura uvella*
Gomphosphaeria lacustris	*Tabellaria fenestrata*
Hydrodictyon reticulatum	*Uroglneopsis americana*
Mallomonas caudata	*Volvox aureus*

Protozoans

There are three primary ways in which protozoans are used to assess the impact of waste discharge:

1. In survey of rivers, streams, lakes and other receiving systems, which are carried out by a field team consisting of variety of specialists working with organisms ranging from bacteria to fish,
2. In laboratory bioassays in which protozoan exposed to various concentration of a particular toxicant to determine the relationship between the concentration and the response and
3. In laboratory micro-ecosystems artificial stream and the like which are designed to fill the gap between the single species laboratory tests and the analysis of the complete systems found in nature.

Many investigators recognize the distinct advantages of protozoans and other microbial species for assessing pollution. A few of more important advantages are

1. They are small easily handled and require small containers as comparison to fish or other larger organisms,
2. They multiply rapidly so that one can test effects of potential pollutants upon reproduction and growth etc.

Sponges

A possible indicator of the presence of toxic wastes has been observed that the megasclear of specimens of *Trochospongilla* growing in iron water pipes exhibited malformation in the axial canal. The malformation is due to a chemical condition in the environment as the tissues of the sponge were strongly marked by iron rust.

When the sponges are treated with sinking agents it had no effects, presumably due to the large particle size of agent employed. Treatment with alone killed the sponges presumably due to clogging of inhalant canal by microscopic sand particles. The effect of oil plus sand were similar to those of oil plus sinking agent except the use of sand as a sinking agent was more deleterious probably due to the additional clogging action of microscopic sand particles.

Annelida

Annelida: Oligochaete

There have been few physiological studies of tubificid in relation to their degree of tolerance of various pollutants so that most of their pollution biology is actually inferred from field studies. The heavy setting of high organic content of silt and associated deoxygenation below organic effluent tend to increase the share of tubificide habitats on the river or lake-bed. Here the number of worm species may decline but the total number of worms may drastically increase in contrast to normal localities, and at the same time the numbers and kind of other benthic species may decline. Oligochaete and some other silf-bodied forms are more tolerant of pesticides than arthropods, but less tolerant of heavy metal ions, and anything that reduces bacterial activity will reduce worm population accordingly. The absolute number of worms may be significant in case of self-evident organic pollution.

Annelida: Hirudinea

Helobdella stagnails and *Helobdella punctata* are commonly associated with many kinds of polluted water and can be considered "Indicator species" only in terms of unusual high densities.

Mollusca: Bivalvia

The presence of usually large population of mussels can be indicative of pollution only in the case of minor degree of organic enrichment. The absence of mussels can logically be an indication of environment disruption only when and where their presence can be demonstrated.

Mollusca: Gastropoda

Analysis of gastropods living in biotopes subjected to these materials can indicate quantitatively the amount of those types of pollutants occurring there. If the chemical and physical characteristic of a body of water appears conducive to a high diversity of organisms, yet only one or two species are present, the stress on that community may be caused by organic pollution.

Anthropods: Insecta

Presence or absence of any species of insects in a stream indicates the presence or absence of pollution. The majority of insect species found in the average eutrophic body of water untolerate a broad enough range of water chemistry and physical conditions to render them unless singly, as indicators of the degree of damage to any body of water. A case study was done in two ponds of Dumka (Jharkhand).

The aquatic insect community samples for the biomonitoring of these two ponds belonged to the following five orders:

1. Odonata
2. Coleoptera
3. Hemiptera
4. Ephemeroptera and
5. Diptera

Only nymphs/larvae were found in case of odonata, ephemeroptera and diptera whereas both nymphs/larvae and adults were present in case of coleoptera and hemiptera. The common and dominant species of aquatic insects of each order were as follows:

1. Odonata

Ischnura senegalensis, Ischnura delicata, Mesogomphus, lineatus and *Potamarcha obscura.*

2. Coleoptera

Berosus sp., *Hydrophilus olivaceous, Canthydrus* sp., and *Amphiops* sp.

3. Hemiptera

Plea frontalis, Gerris fossarum, Anisops sardea, Micronecta scutellaris, Diplonynchus annulatum and *Laccotrephes* sp.

4. Ephemeroptera

Baelis sp. and *Cloeon* sp.

5. Diptera

Chironomus sp., *Clinotanypus* sp., *Monopelopia* sp. and *Conclapelopia* sp.

Variations in species diversity and evenness of species of different orders of aquatic insects calculated for both the ponds are given in Tables 1.5-1.9. The values of species diversity (H) varied from 1.098 to 2.590 bits of odonata, from 1.218 to 2.645 bits of coleoptera, from 1.105 to 2.888 bits of hemiptera, from 0.348 to 2.333 bits of ephemeroptera and from 0.910 to 2.292 bits of diptera in Lakhikundi pond whereas in Singhara pond it ranged from 0.369 to 1.655 bits of odonata, from 0.888 to 2.135 bits of coleoptera, from 0.655 to 1.774 bits of hemiptera, from 0.312 to 1.912 bits of ephemeroptera and from 0.541 to 1.834 of diptera.

Bacteria

The bacteria play a significant role as decomposers and are decisive in determining the biological quality of water bodies. They varied from season to season and pond to pond and are given in Tables 1.10 and 1.11.

Table 1.5: Variation in the Species Diversity Index (H) and Evenness of Species (J) of Odonate Insects from January to December, 2001

Months	*Lakhikundi Pond*		*Singhara Pond*	
	Species Diversity	*Evenness of Species*	*Species Diversity*	*Evenness of Species*
Jan	1.779	0.576	1.525	0.570
Feb	1.685	0.670	1.326	0.568
Mar	1.489	0.584	1.225	0.445
Apr	1.525	0.479	0.825	0.525
May	1.098	0.485	0.689	0.585
Jun	1.569	0.440	0.369	0.675
Jul	1.889	0.842	0.855	0.784
Aug	2.225	0.830	1.528	0.728
Sep	2.590	0.745	1.455	0.625
Oct	2.130	0.872	1.430	0.725
Nov	1.838	0.886	1.658	0.784
Dec	1.925	0.907	1.642	0.805

Table 1.6: Variation in the Species Diversity Index (H) and Evenness of Species (J) of Coleopteran insects from January to December, 2001

Months	*Lakhikundi Pond*		*Singhara Pond*	
	Species Diversity	*Evenness of Species*	*Species Diversity*	*Evenness of Species*
Jan	2.168	0.403	1.625	0.388
Feb	1.825	0.570	1.530	0.348
Mar	1.685	0.546	1.522	0.332
Apr	1.528	0.436	1.335	0.312
May	1.322	0.420	0.888	0.310
Jun	1.218	0.394	0.945	0.380
Jul	1.955	9.513	1.325	0.418
Aug	2.325	0.519	1.889	0.438
Sep	2.645	0.542	2.215	0.488
Oct	2.455	0.540	2.135	0.478
Nov	1.855	0.505	1.455	0.428
Dec	1.926	0.504	1.532	0.482

Total Coliform MPN

The present investigation indicated that there were pronounced fluctuations in density of coliform bacteria in both the ponds. In general, the population was highest in summer season. Their maximum population was $3.68 \times 10^3/100$ ml in Lakhikundi pond and $119.72 \times 10^3/100$ ml in Singhara pond in June. Its population was low during winter and moderate in rainy season. In Lakhikundi pond, its population is lower than the Singhara pond. This clearly reflects low load of civic pollution in Lakhikundi pond.

Faecal Coliform (MPN)

It was maximum during summer season in both the ponds. Its maximum counts was $1.76 \times 10^3/100$ ml in Lakhikundi pond and $80.18 \times 10^3/100$ ml in Singhara pond in June.

Table 1.7: Variation in the Species Diversity Index (H) and Evenness of Species (J) of Hemiptera insects from January to December, 2001

Months	*Lakhikundi Pond*		*Singhara Pond*	
	Species Diversity	*Evenness of Species*	*Species Diversity*	*Evenness of Species*
Jan	2.155	0.855	1.479	0.785
Feb	2.685	0.807	1.689	0.715
Mar	1.225	0.561	1.105	0.505
Apr	1.215	0.355	0.889	0.315
May	1:115	0.468	0.785	0.388
Jun	1.105	0.415	0.655	0.375
Jul	1.888	0.490	0.996	0.393
Aug	1.890	0.770	1.215	0.625
Sep	2.825	0.806	1.682	0.778
Oct	2.885	0.768	1.774	0.755
Nov	2.486	0.755	1.210	0.688
Dec	2.588	0.812	1.386	0.782

Table 1.8: Variation in the Species Diversity Index (H) and Evenness of Species (J) of Ephemeropteran insects from January to December, 2001

Months	*Lakhikundi Pond*		*Singhara Pond*	
	Species Diversity	*Evenness of Species*	*Species Diversity*	*Evenness of Species*
Jan	1.858	0.984	1.628	0.835
Feb	1.541	0.855	1.445	0.754
Mar	0.949	0.399	0.819	0.290
Apr	0.845	0.384	0.729	0.310
May	0.715	0.389	0.625	0.375
Jun	0.348	0.334	0.312	0.330
Jul	0.985	0.448	0.872	0.415
Aug	2.315	0.549	1.912	0.514
Sep	2.333	0.548	1.815	0.498
Oct	1.448	0.560	1.015	0.490
Nov	1.854	0.666	1.325	0.558
Dec	1.989	0.678	1.525	0.569

Table 1.9: Variation in the Species Diversity Index (H) and Evenness of Species (J) of Dipteran insects from January to December, 2001

Months	*Lakhikundi Pond*		*Singhara Pond*	
	Species Diversity	*Evenness of Species*	*Species Diversity*	*Evenness of Species*
Jan	1.836	0.781	1.676	0.625
Feb	1.855	0.805	1.342	0.688
Mar	0.918	0.524	0.633	0.426
Apr	0.910	0.488	0.546	0.376
May	0.954	0.462	0.541	0.410
Jun	1.015	0.681	0.898	0.512
Jul	1.685	0.697	1.113	0.610
Aug	2.992	0.626	1.834	0.581
Sep	2.086	0.825	1.616	0.782
Oct	2.115	0.824	1.726	0.765
Nov	2.041	0.715	1.215	0.705
Dec	1.987	0.725	1.610	0.698

Table 1.10: Variation in the Bacterial Density of Lakhikundi pond water from January to December, 2001

Months	*Total Coliform ($\times 10^3$/100 ml)*	*Faecal Coliform ($\times 10^3$/100 ml)*	*Faecal Streptococci ($\times 10^3$/100 ml)*	*E. coli ($\times 10^3$/100 ml)*	*Total Bacterial Density ($\times 10^7$/L)*	*FC/FS ratio*
Jan	1.80	0.43	0.52	0.46	0.14	0.82
Feb	1.86	0.48	0..56	0.48	0.14	0.85
Mar	2.72	0.54	0.63	0.68	0.16	0.85
Apr	2.89	0.61	0.72	0.82	0.18	0.84
May	3.46	1.64	1.78	1.01	0.19	0.92
Jun	3.68	1.76	1.86	2.12	0.20	0.94
Jul	2.81	1.21	0.84	1.91	0.22	1.44
Aug	2.62	0.81	0.96	1.82	0.18	0.84
Sep	2.81	0.90	0.82	1.34	0.19	1.09
Oct	2.21	0.82	0.66	1.18	0.18	1.24
Nov	1.68	0.49	0.58	1..02	0..18	0.84
Dec	1.58	0.46	0.49	0.61	0.16	0.93

Faecal Streptococci (MPN)

Streptococci are used as indicator faecal habitat. There maximum density was recorded in Singhara pond and minimum in Lakhikundi pond (Table 1.10 and 1.11).

Table 1.11: Variation in the Bacterial Density of Singhara pond water from January to December, 2001

Months	Total Coliform ($\times 10^3$/100 ml)	Faecal Coliform ($\times 10^3$/100 ml)	Faecal Streptococci ($\times 10^3$/100 ml)	E. coli ($\times 10^3$/100 ml)	Total Bacterial Density ($\times 10^7$/L)	FC/FS ratio
Jan	5.12	3.40	0.46	1.81	0.61	07.39
Feb	6.23	9.20	0.55	2.13	0.82	16.72
Mar	18.41	30.00	0.52	2.24	1.12	57.69
Apr	90.12	49.80	1.69	13.12	2.16	29.46
May	114.61	70.10	1.73	44.32	5.61	40.52
Jun	119.72	80.18	2.78	43.48	5.48	28.84
Jul	58.00	76.00	1.89	12.75	3.31	40.21
Aug	66.00	68.00	0.91	49.00	5.12	74.72
Sep	78.00	64.00	9.82	30.01	2.92	78.04
Oct	80.12	22.00	0.68	14.00	2.14	32.35
Nov	16.12	18.00	0.62	7.21	1.12	29.03
Dec	4.82	8.20	0.43	1.68	0.64	19.06

Escherichia coli (MPN)

It showed high MPN in Singhara pond during the period of investigation. Its higher population in Singhara pond is due to intense civic pollution of Dumka town and cattle dumping while their moderate population at Lakhikundi pond is due to bathing activities of man and cattle specially during summer.

Total Bacterial Density

Total bacterial density showed identical pattern of fluctuations in both the ponds. There maximum density was recorded in July (0.22×10^7/L) in Lakhikundi pond but 5.61×10^7/L in Singhara pond in May, whereas minimum count was recorded during winter season in both the ponds. However their density was comparatively higher in Singhara pond.

The species diversity concept explains that the natural communities are typically characterised by the presence of a few species with many individuals and many species with a few individuals. As the species diversity relates to information theory, more amount of information is contained in the natural community than in a polluted community. A complex natural aquatic insect community will have higher diversity and lesser fluctuations in its range. A measure of the pollution of a particular system over a given time can be obtained from the range of values of species diversity calculated within the time period. Wilhm and Dorris (1968) proposed a relationship between species diversity and pollution status of water as below:

$H > 3$ clean water,

$H = 1$ to 3 moderately polluted,

$H < 1$ heavily polluted water.

According to the above scale, both the ponds showed a fluctuating pattern of moderately polluted water. Although the degree of pollution varies from pond to pond and season to season, but both these ponds showed higher pollution load during the summer than the monsoon and winter, because the values of species diversity is very low, particularly in Singhara pond (H < 1). This might be due to low level of water, competition for space, unavailability of food, predation and overcrowding of the species. During summer, the water of the Singhara pond also becomes hypercarbic (Kumar, 1994c) and high abundance of macrophytes (emergent type) are also responsible for low population densities and low species diversity. Benke (1976) and Kumar (1996a) also confirmed that the period of high abundance of macrovegetation as well as moderate pollution status coincide with the period of low abundance of certain aquatic insects. It is also observed that the low range of species diversity of all kinds of aquatic insects during summer showed very poor condition and gross disturbances in the habitat. In this situation, only pollution tolerant aquatic insects will be present and pollution intolerant species decline. The pollution intolerant species decline, than the pollution tolerant species can grow more rapidly without competition for space, nutrients, predation and other extrinsic and intrinsic factors too. This results in heavy dominance of these species leading to the decline in the values of species diversity and also in the evenness of species (Cairns, 1977).

In the present study, it was observed that the species diversity is found to be low during periods of high abundance which might be due to rapid increase and numerical dominance of one or two species in the community (Sagar and Hasler, 1969). Pollutants generally cause adjustments and alterations in species-abundance and community species composition in aquatic ecosystems. A polluted system is simplified and those species that survive and encounter less competition and therefore, may increase in number (Dennis and Patil, 1977, Kumar, 1995a and b).

In the present investigation, the species diversity was higher during monsoon in both the ponds. It is clear from the data that an increase in diversity index occurs when rainfall was maximum (rainy season) and temperature optimum. Rao (1976) also opined that a heavy rainfall in monsoon period increases the diversity while the poor rainfall has adverse effect on the diversity index of hemipteran insects. Julka (1977) has also suggested that in the complexes of interdependent factors governing the seasonal variations in diversity of aquatic bugs, temperature and rainfall appear to be the important factors.

However, it was also observed during the present study that in those months when the total number of individuals was found to be maximum, even then the species diversity as calculated was found slightly lower than the successive months. In these months, evenness of species was also calculated to be minimum. It might be due to the fact that though the total number of individuals were more in a month, the number of species remained more or less same in the successive months. It was observed that increase in the total number of species resulted in the decrease in the values of species diversity and also in the evenness. Therefore, it is obvious that the species diversity index depends upon the number of species in the environment and not on the number of individuals (Kumar, 1994 a & b and 1998).

The bacterial analysis also shows some interesting results regarding the quality assessment of water of both the ponds. Bacterial population was lower in the monsoon, lowest in winter and highest during summer. It is obvious that both the ponds are polluted in summer. Highest values of coliform in summer has greatly lowered in monsoon months. This may be due to an excessive dilution by rain water that drained into the ponds, although rain drain was the important source of bacterial population in the water body (William and Evans, 1972). However, this does not agree with the observations made by Shastry *et al.* (1970) and Aboo (1968) who noted that the maximum bacterial densities occurred during the rainy season, and smaller counts in winter.

The higher values in both the ponds are explainable on the basis of their being used as bathing and washing ghats making them more polluted. Saxena *et al.* (1966) attributed summer high in coliform organisms to the less available dilution in the summer. The low winter counts can be explained on the basis of lower multiplication and poor growth following low temperatures, whereas fluctuations in the number of coliforms in different waters can be attributed to the intensity and age of pollution in addition to the air temperature and run-off waters (Panicker, 1976).

The FC/FS ratio is also used to distinguish whether the suspected contamination derives from human or from animal excreta. Metcalf and Eddy (1979), Gupta (1994) and Singh and Singh (1995) proposed that if the values of FC/FS ratio in lesser than 1, contamination derives from domestic animals and if the value to greater than 4, pollution derives from human excreta. The FC/FS ratio was in range of 0.82 to 1.44 in Lakhikundi pond and of 7.39 to 78.04 in Singhara pond (Tables 1.10 and 1.11). The above ratio reflects that the pollution occurred in Lakhikundi pond is mainly due to cattle and human based while in Singhara pond, the pollution derived mainly from human sources.

The values for both total and faecal coliform in summer in both the ponds were excessive and greater than all recommended standards. This summons towards the dangers of health hazard when using the pond waters for whatsoever purpose. Thus, the values of species diversity of aquatic insects and bacterial density computed within certain time period can be used as bioindicators for the assessment of quality of freshwater ecosystems.

References

Aboo, K.M. (1968). A study of well waters in Bhopal city. Envir. Hlth. 10: 189-201.

Benke, A.C. (1976). Dragonfly turnover and production. *Ecology*. 57: 915-927.

Cairns, J.J. (1977). Indicator species *vs.* the concept of community structure as an index of pollution. *Water Resources Bulletin*. 10: 338-347.

Dennis, B. and Patil, G.P. (1977). The use of community diversity indices for monitoring trends in water pollution impact. *Tropical Ecol*. 18: 36-51.

Environmental Encyclopaedia (1991). (Eds. P. Cunningham, T. H. Cooper, E. G. Malcolm, T. H.). Jaico Publishing House, Mumbai.

Encyclopaedia of Environmental Science (1995). *Man and Environment* (Eds. Trivedi, R. P. and Raj, G.) Vol. 18, Akshdeep Publishing House.

Gupta, A.K. (1994). Impact of sewage and human activities at Veruna river corridor. International conference on recent trends in water pollution and Research, Dobbi, Jaunpur, 19-21 Nov.'94.

Julka, J.M. (1977). On possible fluctuation in the population of aquatic bugs in a fish pond. Oriental insects. II: 139-149.

Kumar, A. (1994a). Periodicity and abundance of rotifers in relation to certain physico-chemical characteristics of two ecologically different pond of Santhal Pargana (Bihar). *Indian J. Ecol*. 21: 54-59.

Kumar, A. (1994b). Role species diversity of aquatic insects in the assessment of pollution in the wetlands of Santhal Pargana. *J. Environ. and Poll.* I: 117-120.

Kumar, A. (1994c). Population density, species diversity and evenness of dragonfly nymphs in a fresh water pond of Santhal Pargana. *J. Freshwat. Biol.* 6: 231-235.

Kumar, A. (1995a). Seasonal abundance and species diversity of Zygopteran larvae in fish pond of Santhal Pargana, *India. J. Ecobiol.*, 7: 35-39.

Kumar, A. (1995b). Population dynamics and species diversity of odonate larvae in the wetlands of Santhal Pargana, India. Proc. Nat. Sci. 65: 265-278.

Kumar, A. (1996). Seasonal variation in the calorific contents of certain predatory aquatic insects in village fish pond of Santhal Pargana (Bihar). *J. Environ. Biol.* 17: 59-62.

Kumar, A. (1998). Biomonitoring of pollution by aquatic insect community and micro-organisms in freshwater ecosystems. In: *Ecotechnology for Pollution Control and Environmental Management* (Eds. A. Kumar and R.K. Trivedy). Enviro Media, Karad. pp. 255-263.

Mervin, C.P. (1980). Algae and Water Pollution, Castle House Publication Ltd.

Metcalf and Eddy (1979). Waste water engineering: Treatment, disposal and reuse. Tata McGraw Hill. Co. Ltd., New Delhi.

Panicker, P.V. (1976). Coliform spectra of raw water sources of Nagpur water supply. *Environ. Hlth.*, 8: 286-296.

Pollution Ecology of Freshwater Invertebrates (1982). (Eds. Hart, C.W. Jr. and Samuel, L.H. Fuller). Academic Press, New York.

Rao, T.K.R. (1976). Bio-ecological studies on some aquatic Hemiptera, Nepidae. *Entomon.*, 1: 123-132.

Sagar, P.E. and Hasler, A.D. (1969). Species diversity in Lacustrine phytoplankton I. The components of the index from Shanon's formula. *Amer. Nat.*, 103: 51-59.

Saxena, K.L., Chakrabarty, R.N., Khan, R.Q., Chattopadhyay, S.N. and Chandra, H. (1966). Pollution studies of Ganges near Kanpur. *Environ. Hlth.*, 8: 270-285.

Shastry, C.A., Aboo, K.M. and Khare, G.M. (1970). Reduction of microorganism at different stages of water treatment. *Environ. Hlth.*, 12: 66-79.

Singh, T.N. and Singh, S.N. (1995). Impact of river Varuna on the river Ganges water quality at Varanasi. *Indian J. Environ. Hlth.*, 37: 272-277.

Wilhm, H. and Dorris, T.C. (1968). Species diversity of benthic macro-invertebrates in stream receiving domestic and oil refinery effluent. Am. Midland. *Naturalist*, 76: 427-449.

William, H.C. and Evans, M.J.C. (1972). A hydrobiological study of the polluted river Lieve (Ghent, Belgium). *Hydrobiologia*, 39: 99-154.

2

Frequency of External Infection in *Liza parsia* from Polluted Harbour Waters of Visakhapatnam

B. Bharatha Lakshmi, Y. Bangaramma and C.M. Rao

Department of Zoology, Andhra University,
Visakhapatnam - 530 003, A.P.

ABSTRACT

The paper deals with a comparative study on the frequency of external infections like E.U.S, descaling, skin rot, fin rot and gill rot in *Liza parsia* from the polluted channel of Vizag harbour (St-I) and relatively unpolluted Gosthany estuary near Bhimili (St-II) was carried out for a period of one year. The percentage of external infection was significantly more in the polluted area compared to that of Gosthany estuary. Statistical analysis for understanding the significance of pollution between those two stations has also been made and the results are presented. .

Key Words: External infection, *Liza parsia*, Pollution.

Introduction

Mortality of fish often occurs in ponds, tanks, estuaries and other aquatic bodies due to environmental stress, pollution and parasitic infections (helminth, crustacean, bacterial, viral, fungal, protozoan etc.). The occurrence and magnitude of infections are mainly related to the prevalent poor sanitary conditions in the water and also to the general health of the fish. Worth mentioning contributions on fish diseases in India and abroad are those of Meyer (1970) and Snieszko (1974) on the effect of environmental stress on outbreaks of infectious diseases of fishes, Burton *et al.* (1972), Starr and Jones (1957) and Skidmore (1970) on the effect of some heavy metals on mucus and gills, Goutam Barua (1994) on the environmental influence on Epizootic ulcerative syndrome in Bangladesh waters.

Other noteworthy contributions are those of Mackenzie and Hall (1976), Tonguthai (1985), Booyaratapallin (1985), Fao (1986), Roberts *et al.* (1992, 1993), Liobrera and Gacurtan (1987), Minchew and Yarbrought, (1977), Dixon and Compher (1977), Skinner (1982), Valdron. *et al.* (1994) and Scheweiger (1957).

Present study deals with the effect of heavy metal pollution through industrial effluents, in the prevalence of external infections like E.U.S., skin rot, descaling and melanin deposition (pigment) above the eye ball, fin rot, gill rot, opercle rot etc., in *L. parsia* of Vizag harbour waters. The concentration of heavy metals, especially Zn, Pb and Cd were significantly more in the harbour waters and sediments as has been revealed by the studies of Ganapathi and Raman (1973) and Rao and Vani (1995). Gosthani estuary near Bhimili, a relatively unpolluted habitat, has been taken as the control station. The degree of prevalence and percentage frequency of occurrence of different external infections in *L. parsia* were compared between the two habitats.

Materials and Methods

Fish were collected from the two stations i.e. the polluted Vizag harbour (St-I) and the relatively unpolluted Gosthani estuary (St-II), at regular monthly intervals (Fig. 2.1). Fish collection was made with the help of local fisherman using castnet. Rainfall data were procured from the Cyclone Warning Centre of Visakhapatnam. The physicochemical parameters of the two stations were recorded using suitable devices.

Fishes were examined grossly for symptoms of each disease. Most of the infections like fin erosion, skin rot, gill rot etc. could be identified with naked eye. All deformities involving skeletal elements were recorded. Portions of acute erosions were examined under microscope for determining the type of infection. Percentage frequency of disease fish in each sample was estimated and subjected to statistical analysis using χ^2 test to find out the significance between the two stations.

Results

The rainfall (Table 2.1, Fig. 2.2) varied between a minimum of 3.95 mm in March and a maximum of 303.55 mm in September. The atmospheric temperature remained slightly higher than that of the surface water temperature throughout the year at both the stations. Atmospheric temperature ranged between a minimum of 22.5°C in December and maximum of 34.5°C in May at St-I and 20.5°C in December and 33.5°C in July at St-II. The surface water temperature ranged between a minimum of 19°c in December and a maximum of 31.5°C in May and September at St-I and a minimum of 17°C in December and a maximum of 30.5°C in July at St-II. pH varied between a minimum of 6.6 in February and a maximum of 7.8 in September at St-I and a minimum of 7 in April and December and a maximum of 8.15 in September at St-II. Salinity varied between a minimum of 2.75 per cent in January and a maximum of 24.75 per cent in May at St-I and a minimum of 3.00 per cent in January and a maximum of 30.96 per cent in May at St-II. Dissolved oxygen varied between a minimum of 3.5 mg/l in January and a maximum of 7.57 mg/l in August at St-I and a minimum of 3.5 mg/l in December and a maximum of 8.5 mg/l in April at St-II.

Coming to the heavy metals (Table 2.1, Fig. 2.3), zinc varied between a minimum of 1.577 mg/l in December and a maximum of 20.580 mg/l in August at St-I and a minimum of 0.022 mg/l in April and maximum of 0.190 mg/l in August at St-II. Lead varied between a minimum of 0.037 mg/l in October and a maximum of 2.190 mg/l in January at St-I and a minimum of 0.0005 mg/l in April and maximum of 0.023 mg/l in February at St-II. Cadmium varied between a minimum of 0.0026 mg/l in July and a maximum of 0.198 mg/l in February at St-I and a minimum of 0.0001 mg/l in December and a maximum

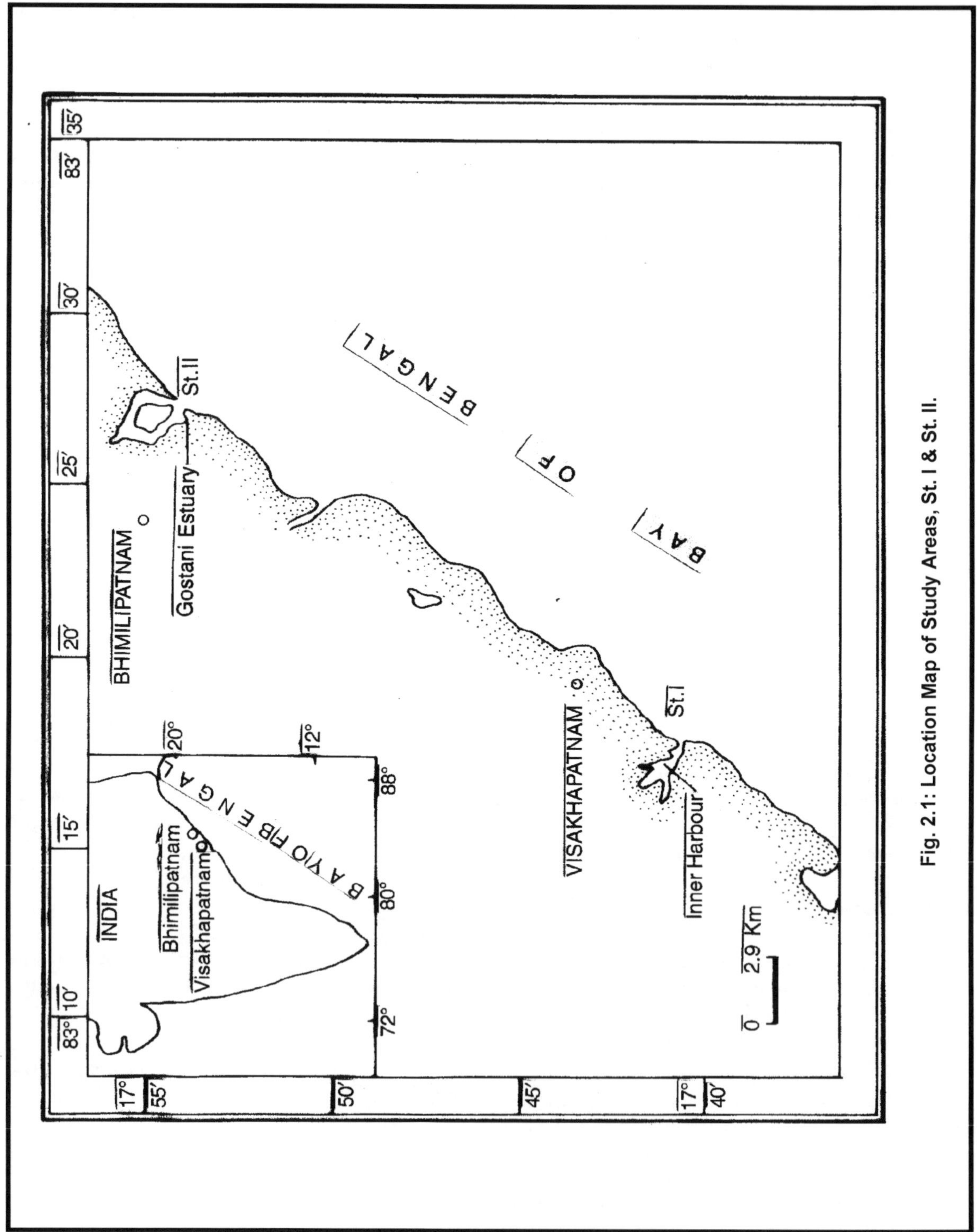

Fig. 2.1: Location Map of Study Areas, St. I & St. II.

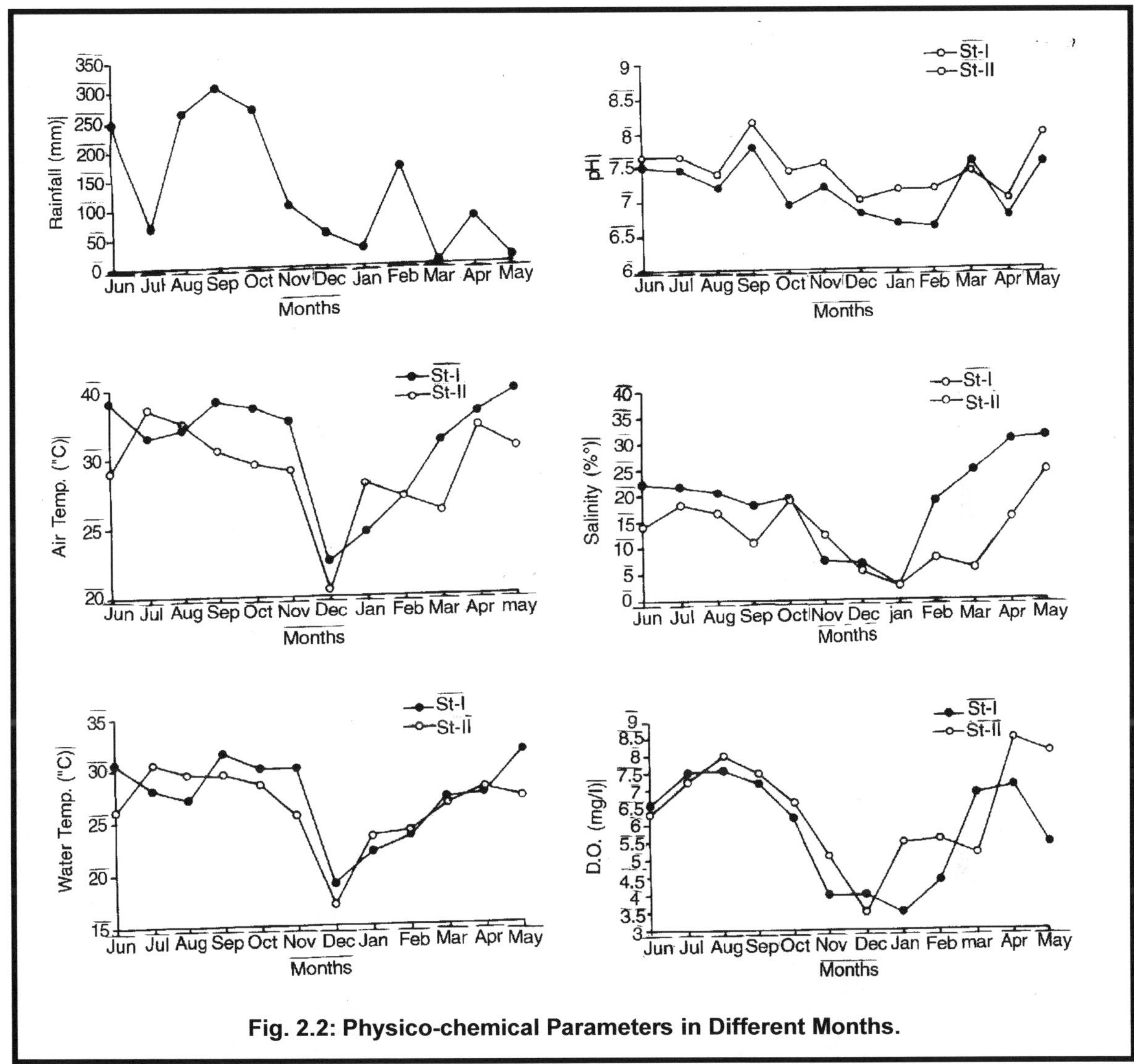

Fig. 2.2: Physico-chemical Parameters in Different Months.

of 0.0048 mg/l in April at St-II.

The overall infection (Table 2.2, Fig. 2.4) was 91.13 per cent (144 out of 158) at St-I and 33.79 per cent (84 out of 249) at St-II. Disease wise infection showed that descaling and black pigment above the eye was 89.87 per cent (142 out of 158) at St-I and 30.12 per cent (75 out of 249) at St-II, E.U.S. and Skin rot was 80.37 per cent (127 out of 158) at St-I and 25.30 per cent (63 out of 249) at St-II, Fin rot was 36.07 per cent (59 out of 158) at St-I, and nil at St-II and Gill rot was 31.64 per cent (50 out of 58) at St-I and nil at St-II.

Month wise details revealed that in *Liza parsia* the overall infection was minimum (70 per cent) in October and maximum (100 per cent) from November to April at St-I, while at St-II it was minimum (26.31 per cent) in November and maximum (80 per cent) in January. Within this, disease-wise the

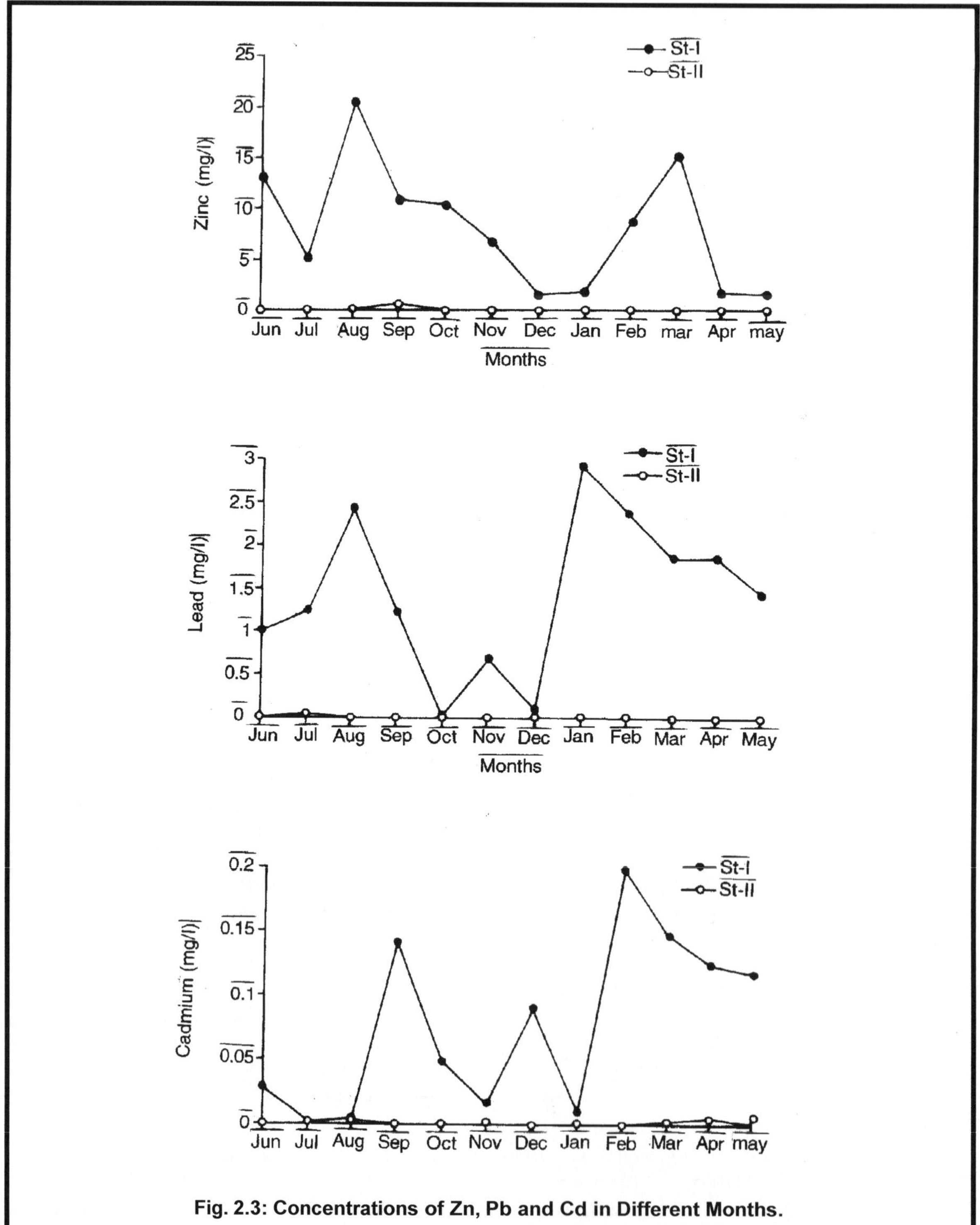

Fig. 2.3: Concentrations of Zn, Pb and Cd in Different Months.

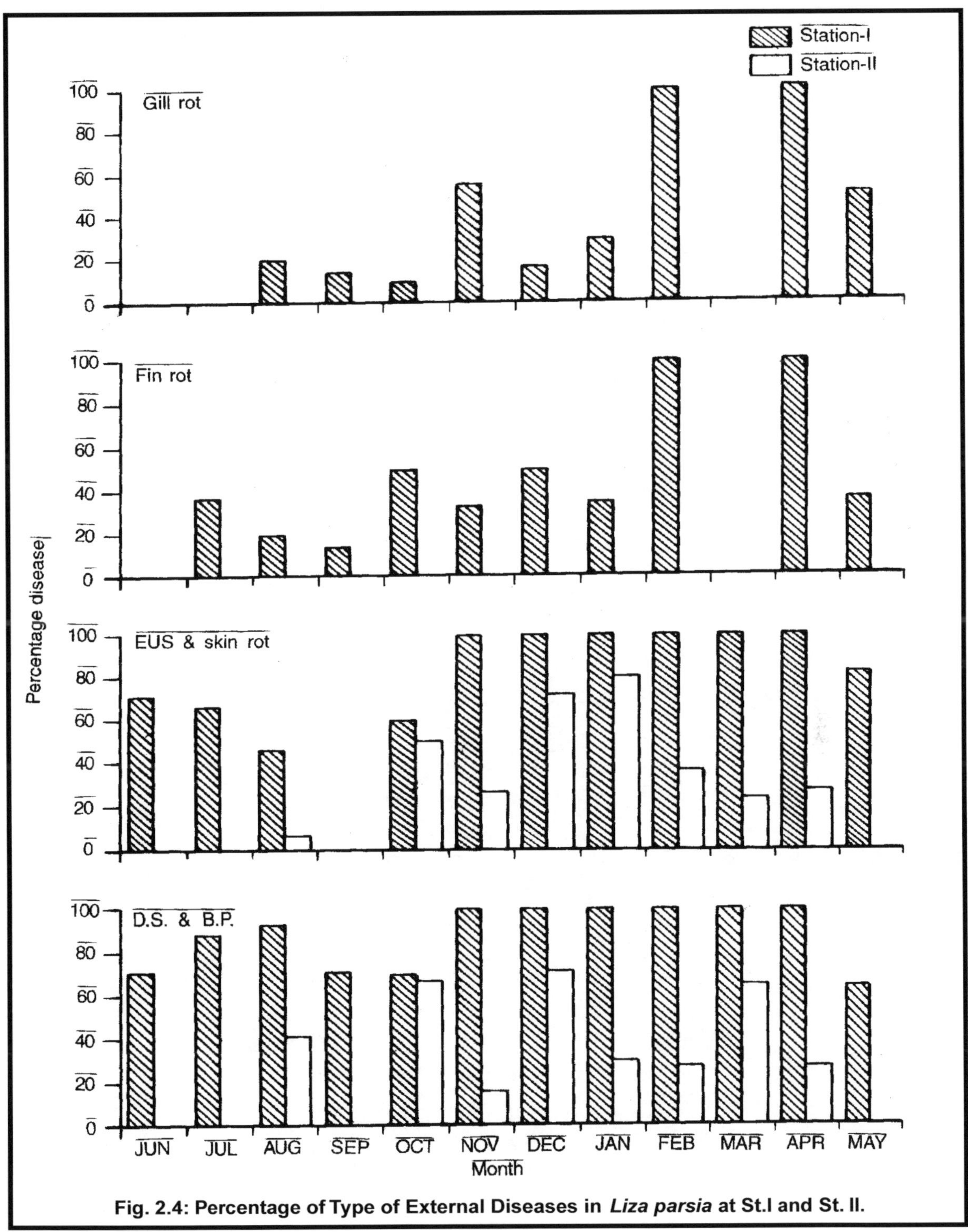

Fig. 2.4: Percentage of Type of External Diseases in *Liza parsia* at St.I and St. II.

Table 2.1: Physico-chemical Parameters at St-I and St-II

Sl.No.	Parameter	Station	Jun	Jul	Aug	Sep	Oct	Nov	Dec	Jan	Feb	Mar	Apr	May
1.	a) Rainfall (mm)		246.25	67.9	262.8	303.55	266.65	102.9	54.25	28.5	163.85	3.95	79.8	12.2
	b) Temperature (°C)													
	Air	I	34	31.5	32	34	33.5	32.5	22.5	24.5	27	31	33	34.5
		II	29	33.5	32.5	30.5	29.5	29.5	20.5	28	27	26	32	30.5
	Water	I	30.5	28	27	31.5	30	30	19	22	23.5	27	27.5	31.5
		II	26.0	30.5	29.5	29.5	28.5	28.5	17	23.5	24	26.5	28	27
	c) pH	I	7.50	7.45	7.20	7.80	6.95	7.20	6.80	6.65	6.60	7.55	6.75	7.52
		II	7.65	7.65	7.40	8.15	7.45	7.55	7.00	7.15	7.15	7.40	7.00	7.95
	d) Salinity (%)	I	14.02	18.11	16.61	11.07	19.50	12.50	5.50	2.75	8.00	6.00	15.50	24.57
		II	22.09	21.59	20.50	18.07	19.65	7.50	7.00	3.00	18.75	24.50	30.50	30.96
	e) D.O. (mg/l)	I	6.55	7.50	7.57	7.20	6.20	4.00	4.00	3.50	4.40	6.90	7.10	5.45
		II	6.30	7.25	8.00	7.50	6.65	5.12	3.50	5.50	5.60	5.20	8.50	8.10
2.	Heavy Metals													
	a) Zinc (mg/l)	I	13.100	5.210	20.580	10.795	10.342	6.792	1.577	1.892	8.720	15.210	1.730	1.610
		II	0.0310	0.1305	0.0190	0.0715	0.0371	0.0270	0.0320	0.0720	0.0995	0.0337	0.2200	0.0307
	b) Lead (mg/l)	I	1.002	1.232	2.423	1.220	0.037	0.672	0.092	2.910	2.370	1.845	1.845	1.425
		II	0.0030	0.0400	0.0018	0.0047	0.0039	0.0029	0.0123	0.0183	0.2340	0.0015	0.0005	0.0076
	c) Cadmium (mg/l)	I	0.0280	0.0026	0.0048	0.1410	0.0480	0.0160	0.0900	0.1000	0.1980	0.1470	0.1240	0.1170
		II	0.0011	0.0020	0.0031	0.0003	0.0007	0.0020	0.0001	0.0019	0.0005	0.0027	0.0048	0.0027

discaling and black pigment above the eye ball was minimum (63.63 per cent) in May and maximum (100 per cent) during November to April at St-I, while at St-II, they were minimum (15.78 per cent) in November and maximum (71.42 per cent) in December. E.U.S. and Skin rot were minimum (46.66 per cent) in August and maximum (100 per cent) during November to April at St-I and they were minimum (6.89 per cent) in August and maximum (80 per cent) in January at St-II. Fin rot was minimum (14.28 per cent) in September and maximum (100 per cent) in February and April at St-I but at St-II it was never encountered. Gill rot was minimum (10 per cent) in October and a maximum (100 per cent) in February and April at St-I but at St-II it was not encountered at all.

Statistical Results

The calculated χ^2 value for the total sample was 129.25. But disease-wise, for loss of scales and presence of black pigment above the eye ball, it was 137.74, for E.U.S. and Skin rot it was 128.39, for fin rot it was 108.75 and for gill rot it was 87.65, while the table value of χ^2 at 5 per cent level is only 5.77.

Discussion

The E.U.S. outbreaks were generally recorded after rainfall, which is well supported by the present study also wherein E.U.S. was observed (Table 2.2 and Fig. 2.3) during the fag end of monsoon period at St-II while at St-I it was encountered considerably in high percentage in almost all the months. Micheal and Phillip (1994) reported strong relationship between seasonal factors, particularly rainfall, and outbreaks of E.U.S., which commonly occur in the drier parts of the year, and the possibility of changes in basic water quality, such as alkalinity, hardness and chloride content, which may lead to the outbreak of E.U.S. According to Das (1988) E.U.S. affected habitats were characteristically of low alkalinity and low hardness, correlated with acidic soils poor in calcium. Mohan and Shankar (1994) reported outbreaks of E.U.S. in Karnataka in the Cauvery river system during August-September 1991, immediately after the floods. Later in November-December 1991 it spread to ponds, major and minor irrigation tanks of Coorg and Mandya districts. During the monsoon months of 1993, when the salinity was very low (< 0.5 ppm) full blown E.U.S. occurred in the estuaries of Dakshina and Uttara Kannada district for the first time. Besides low salinity, the pH and dissolved oxygen were also observed low during December and January at both the stations, when heavy outbreak of E.U.S. and descaling were observed at both the stations. Depletion of oxygen was believed to be the stress factor which resulted in heavy mortalities of thread fin shad, *Dorosoma petense* (Gunther) and American Shad, *Alsa sapidissilna* (Wilson) in Sanjaquin river in Califonrnia (Haley *et al.*, 1967).

Khan (1987), Khan and Tulin (1991) reported that fin rot, skin rot, gill rot and haemorrhages were due to exposure of fish to excess of Polycyclic Aromatic Hydrocarbons (PAHS). Haenesly *et al.* (1982) and Grizzle (1986) provided evidence from natural oil spills that fishes are more prone to a variety of lesions and infections. Khan (1987) investigated effects of Venezulean oil causing severe fin rots and death of *Pseudopleuronectes*. Snieszko (1994) mentioned selected literature to show the coincidence of infectious disease with stress caused by temperature, eutrophication, sewage, metabolic products of fish, industrial pollution and pesticides. In the present study fin rot was very much prevalent only at St-1, which is very close to the HPCL where the unloading of crude oil takes place. Besides this the effluents of HPCL are directly discharged into (Naval Dockyard) the western arm of the harbour which is connected to the Mehadrigedda stream. Thin film of oil upon the surface of water during the study period was noticed many times at this habitat. Haenesly *et al.* (1982) and Khan and Kicenink (1984) reported excessive mucus secretion, capillary dilation, lamellar hyperplasia and fusion of gill filaments in fish exposed to PAHS. These pollutants appear to reduce the respiratory surface (gills) by damaging the capillaries and result in gill impairment. Skinner (1982) reported that low concentration

Table 2.2: External Diseases in *Liza parsia* at St-I and St-II.

Month	*No. of Fish Observed*		*Total No. of Fish Infected*		*% of Disease*		*Descaling and Black Pigment Above the eye balls*				*E.U.S. and skin rot*				*Fin rot*				*Gill rot*			
							Infected Fish No.		*% of Disease*		*Infected Fish No.*		*% of Disease*		*Infected Fish No.*		*% of Disease*		*Infected Fish No.*		*% of Disease*	
	St-I	*St-II*	*St-I*	*St-II*	*St-I*	*St-II*	*St-I*	*St-II*	*St-I*	*St-II*	*St-I*	*St-II*	*St-I*	*St-II*	*St-I*	*St-II*	*St-I*	*St-II*	*St-I*	*St-II*	*St-I*	*St-II*
Jun	14	29	10	A	71.42	34.48	10	A	71.42	A	10	A	71.42	A	A	A	A	A	A	A	A	A
Jul	18	25	16	A	88.88	A	16	A	88.88	A	12	A	66.66	A	5	A	37.34	A	A	A	A	A
Aug	15	29	14	12	93.33	41.37	14	12*	93.33	41.37	7	2*	46.66	6.89	3	A	20	A	3	A	20	A
Sep	7	13	5	A	71.42	A	5	A	71.42	A	A	A	A	A	1	A	14.28	A	1	A	14.28	A
Oct	10	24	7	16	70	66.66	7	16*	70	66.66	6	12*	60	50	5	A	50	A	1	A	10	A
Nov	18	19	18	5	100	26.31	18	3*	100	15.78	18	5*	100	26.31	6	A	33.33	A	10	A	55.55	A
Dec	12	28	12	20	100	71.42	12	20*	100	71.42	12	20*	100	71.42	6		50	A	2	A	16.66	A
Jan	17	10	17	8	100	80	17	3*	100	30	17	8*	100	80	6	A	35.29	A	5	A	29.41	A
Feb	10	22	10	8	100	36.36	10	6*	100	27.27	10	8*	100	36.36	10	A	100	A	10	A	100	A
Mar	13	17	13	11	100	64.70	13	11*	100	64.70	13	4*	100	23.52	A	A	A	A	A	A	A	A
Apr	13	15	13	4	100	26.66	13	4	100	26.66	13	4*	100	26.66	13	A	100	A	13	A	100	A
May	11	20	9	A	81.81	A	7	A	63.63	A	9	A	81.81	A	4	A	36.36	A	5	A	50	A
Total	158	249	144	84	91.13	33.79	142	75	89.87	30.12	127	63	80.37	25.30	59	A	37.07	A	50	A	31.64	A
Calc. χ^2 Value					129.25				137.74				128.30				108.75				87.65	

* Absence of black pigment and skin rot at St-II specimens.
A: Absence of disease
X^2 Table value at 5% level: 5.97

of dissolved oxygen and high temperature increase the toxicity of un-ionised NH_2. Burrows (1964) observed hyperplasia of gill epithelium and epizootic bacterial gill disease in Salmon, exposed to non lethal dosage of Ammonia for prolonged time.

Salts of zinc, copper and other metals cause coagulation and precipitation of mucus and cytological damage to the gills (Burton *et al.*, 1972, Starr and Jones, 1957, Skidmore, 1970) Dixon and Compher (1977) reported inhibition of regeneration in the tailfin of newt by cadmium and have further noted that cadmium treated limbs were also erythromatous due to ruptured capillaries.

The results of the present study show that the percentage frequency of occurrence of E.U.S., descaling, black pigment above the eyeball, skin rot, fin rot and gill rot were very high at St-I. Since St-I receives large amounts of industrial effluents containing heavy metals like zinc, cadmium and lead and some hydrocarbons from zinc smelter, coromandel fertilizers and H.P.C.L., that are situated close to St-I, it is heavily polluted, while St-II is relatively free from such pollution. Low D.O., low pH, high salinity fluctuations and higher temperatures, high concentration of heavy metals like Cd, Zn and Pb have been observed throughout the study period at St-I. This could be attributed to the indiscriminate discharge of industrial effluents in this region. The frequency of infections was high correlating the high concentration of heavy metals at St-I than at St-II. Statistically also St-I was proved to be significantly different from St-II. This clearly indicates that increased concentration of heavy metals in the aquatic environment, perhaps creates a congenial atmosphere for the different parasitic bacteria, viruses, protozoa, copopoda, helminths etc., to in infect the fish occurring in such habitats. This in turn effects the biological condition of the infected fish by way of low growth rate, low fecundity , low survival rate, inferior nutritive value, foul odour, unpalatable taste etc. All the industrial pollutants discharged into the aquatic medium thus hamper the production of the protein rich food in the aquatic environment and cause a great economic loss to the nation. Unless some stringent measures are taken to check the industrial pollutants, it becomes a perennial global problem and is capable of creating both economic and social crises.

Acknowledgement

The authors are thankful to the University authorities and the Head of the Department of Zoology for providing the necessary facilities to carry out the work successfully.

References

Booyaratapalln, S. (1985). Fish disease out break in Burma, FAO special report TCP /BUR/ 4402.

Burrows, R.E. (1964). Effects of accumulated excretory products on hatchery reared salmonids. U.S. *Fish Wildl. Serv. Res. Rep.* 12: p. 66.

Burton, D.T., Jones, A.H., Cairns, J.Jr. (1972). Acute zinc toxicity to rainbow trout *(Salmo gairdneri).* Confirmation of the hypothesis that death is related to tissue hypoxia. *J. Fish. Res. Bd. Con.,* 29: 1463-1466.

Das, M.K. (1988). The disease epizootic ulcerative syndrome an overview souvenir, Inland fisheries society of India, Barrackpore, West Bengal: pp. 25-30.

Dixon, C. and Compher, K. (1977). Protective action of Zinc against the deleterious effects of cadmium in the regenerating forelimb of the adult Newt. *Notopthalmus viridescens. Grolvth* 41: 95-103.

Food and Agricultural Organisation (1986). Field and laboratory investigations into UDS. Creative fish disease in the Asia-Pacific region FAO. Tech. Rep. TCP/RAS, pp. 4508-4509.

Goutam Barua (1994). The Environmental influence on Epizootic ulcerative syndrome in Bangladesh. In: R.J. Roberts, B. Campbell, I.H. MacRae (Eds.), ODA *Regional Seminar on E.U.S.,* The aquatic animal health research institute Bangkok, Thailand, 25-27 January, 1994: 137-141.

Grizzle, J.M. (1986). Lesions in fishes captured near drilling platforms in Gulf of Mexico. *Mar. Env. Res.,* 13: 267-276.

Haensely, W.E., Neff, J.M., Sharp, J.R., Morris, A.C., Bedgood, M.F. and Beom, P.D. (1982). Histopathology of *Pleuronectes platessa* L. from Aber Wrach and Aber Benoit, Brittany, France: Long term effects of the Amoco codiz crude oil spill. *J. Fish. Dis.,* 5: 365-391.

Haley, R., Davis, S.P. and Hyde, J.M. (1967). Environmental stress and *Aeromonas liquesfaciens* in American and Threadfin shad mortalities. *Progre. Fish Cult.* 29: p. 193.

Khan, R.A. (1987). Crude oil and parasites of fish. *Parsitology Today,* 3: 99-100.

Khan, R.A. and Thulin, J. (1991). Influence of pollution on parasites of aquatic animals. *Adv. Parasitol.,* 30: 201-238.

Liobera, A.J. and Gacutan, R.Q. (1987). Aeromonas Hpydrophila associated with ulcerative disease epizootic in Leguna De Bay Philippines. *Aquaculture,* 67: 273-278.

MacKenzie, R.A. and Hall, W.T.K. (1976). Dermal ulceration of mullet *(Mugil cephallls). Aust. Vet. J.,* 52: 230-231.

Meyer, F.P. (1970). Seasonal fluctuations in the incidence of disease on fish forms. *Am. Fish. Soc. Symp. Special Publication,* 5: 21-29.

Michael, K. and Phillips (1994). Relationship between epizootic ulcerative syndrome (EUS) and the environment. In: R.J. Roberts, B. Campbell, I.H. MacRae (Eds.), ODA Regional seminar on E.U.S. General features based on findings of the NACA regional research programme. (Division of food and Agricultural Engineering, Asian Institute of Technology Bangkok) Thailand. The aquatic animal health research institute Bangkok, Thailand: 25-27 January, 1994: 109-117.

Minchew, C.D. and Yarbrought, J.D. (1977). The occurrence of finrot in mullet *(M. cephalus)* associated with oil contamination in an estuarine pond ecosystem. *J. Fish Biol.,* 10: 319-325.

Mohan, C.V. and Shankar, K.M. (1994). Epidemiological analysis of epizootic ulcerative syndrome of fresh and brackish water fishes of Karnataka, India. *Curr. Sci.* 66 (9): 659-668.

Roberts, R.J., Willoughby, L.G. and Chinabut, S. (1993). Mycotic aspects of epizootic ulcerative syndrome EUS of Asian fishes. *J. Fish,. Dis.,* 16: 169-183.

Roberts, R.J.N., Frerichs, N. and Millar, S.D. (1992). Epizootic ulcerative syndrome, the current position. In: I.M. Shariff, R.P. Subasinghe and J.R. Arthur (Eds)., Disease in *Asian Aquaculture.* Fish Health Section, Manila. *Asian Fish. Soc.,* pp. 431-436.

Schweiger, G. (1957). The toxic action of heavy metal salts on fish and organisms on which fish feed. *Arch. Fish. Wiss.,* 8: 54-78.

Skidmore, J.F. (1970). Respiration and Osmoregulation in rain bow trout with gills damaged by zinc sulphate. *J. Exp. Biol.* 52: 481-494.

Skinner, R.H. (1982). The interrelation of water quality, gill parasites and gill pathology of some fishes from south Biscayne Bay, Florida. *Fish. Bull.* 80: 269-280.

Snieszko, S.F. (1974). The effects of environmental stress on outbreaks of infectious diseases of fishes. *J. Fish. Biol.* 6: 197-208.

Starr, T.J. and Jones, M.E. (1957). The effect of copper on the growth of bacteria isolated from marine environments. *Limnol. Oceanogr.*, 2: 33-36.

Tonguthai, K. (1985). A preliminary account of ulcerative disease in the Indo-Pacific region (A comprehensive study based on that experience). National Inland Fisheries Institute, Bangkok, FAO/TCP /RAS/ 4508.

Valdron, K.A., Cone, D.K. and Cusack, R. (1994). Persistent nodular lesions of bacterial origin on the gills of gasterosteid fish inhabiting an estuarine bay in Nova Scotia, Canada. *J. Fish. Dis.*, 17: 673-679.

Wedemeyer, G. (1970). The role of stress in disease resistance of fishes. In: A Symposium on Diseases of Fishes and Shell Fishes. S.F. Snieszko (ED.), Washington. *Am. Fish. Soc.*, pp. 30-35.

3

Cyanobacterial Diversity in Polluted Lakes of Jalgaon District of North Maharashtra

S.N. Nandan and S.R. Mahajan*

P.G. Department of Botany
S.S.V.P.S's L.K. Dr. P.R. Ghogrey Science College, Dhule-424 005 (M.S.)
**Department of Botany, Bhusawal Arts, Science and P.O.N. Commerce College, Bhusawal-425 201, Dist. Jalgaon (M.S.)*

ABSTRACT

Cyanobacteria is an extremely diverse group of prokayotic organisms which make valuable contributions to soil fertility by fixing atmospheric nitrogen and play an important role in all types of water bodies whether they being heterocystous or non-heterocystous. Polluted waters harbour characteristic type of blue-green algae. Considering above their distributional pattern in polluted and mildly polluted lakes of Jalgaon has been worked out. Their role as indicators of water quality and eutrophication has been summarized. Its highest population has been recorded during summer while lowest during monsoon months. 154 species belonging to 35 genera were identified. Of these 39 species were heterocystous and 115 species were non-heterocystous members.

Key Words: *Cyanobateria, Heterocystous, Non-heterocystous, Lake.*

Introduction

During the recent past, studies on Cyanobacterial have emphasized their important role in ecosystems. Cyanobacteria algae grow at any place and in any environments where moisture and sunlight are available. However, specific algae grow in specific environment and therefore, their distributional pattern, ecology, periodicity, qualitative and quantitative occurrence differ widely. Now a days people are facing many problems and water pollution is one of them. Industrial wastes, domestic wastes and sewage of the area either accumulated in the form of ditches of ponds or are being let into nearby rivers. Some nuisance algae grow inside these ditches and ponds which develop bloom and ultimately water becomes unpotable even for animals and not fit for irrigation puposes as well.

To know the role of Cyanobacterial in polluted Hartala and mildly polluted Velhala lake of Jalgaon, present study has been carried out. Their role as indicators of water quality and eutrophication has been investigated. A list of Cyanobacterial collected from these lakes has also been given.

Material and Methods

Two different lakes of Jalgaon (India) have been selected for the present study and work has been carried out from October, 1998 to September, 2000. The Hartala lake is located near the village Hartala of Tal-Muktainagar lies on small tributary of river Tapti. Geographically, Hartala lake is located at latitude 21°.00′ and longitude 76°.00′ East. The drainage area of this lake is 6.61 square miles. The lake has a capacity of 140 millions of cubic feet. Three stations were selected for the collection of water and algal samples. H-I station is very near to village, at this station during the study period, the water utilized for clothes washing and bathing by villagers and also cattle wadding was seen. At this station water was much contaminated due to addition of domestic wastes etc. H-II station is near to Chakradhar Swamy temple nearly 1 k.m. away from the H-I station. At this station during the study period, the water was utilized for bathing and washing clothes by villagers. At this station fishing and cattle wadding were observed in the study period. H-II station remained away from human activities and was free from much contamination.

Another lake selected for study is Velhala lake situated near village Velhala of Bhusawal taluka. It lies at 20°.55′ North latitude and 75°.50′ East longitude. The net available storage of lake is 74 millions of cubic feet. Three stations were selected for the collection of water and algal samples. V-I station of Velhala lake received ash water from the leakage of pipe lines of waste water of Thermal Power station, Deepnagar. During the study period, at number of places through the leakage of pipelines of water, ash water leaked out into rivulets of catchment area and finally get contaminated into lake water at this station. V-II station is near to village Velhala. The water was utilized for bathing and washing clothes by villagers. Cattle wadding were also observed during the study period. V-III station is nearly 3 km away from the village. It was free from much contamination and human activities.

Monthly collection of water and algae were made for two years from October, 1998 to September, 2000. The algal samples were collected in the acid washed plastic collection bottles and immediately preserved in 4 per cent formalin for microscopic observations. Line drawings of different forms of Cyanobacteria were made by camera lucida. Cyanobacteria were identified by relevant monographs and recent available literature. (Desikachary, 1959, Dixit, 1936, Gonzalves and Kamat, 1960, Kamat, 1963, Mahajan,1983, Nandan and Patel, 1985a, Rao, 1937). The characterization and identification of Cyanobacterial taxa was based on the taxonomic keys given by Desikachary (1959). Systematic enumeration of Cyanobacteria, their habitat and month of collection have been recorded.

Systematic Enumeration

In all 154 cyanobacterial forms were identified belonging to families Chroococcaceae, Entophysalidaceae, Cyanidiaceae, Chamaesiphonaceae, Pleurocapsaceae, Oscillatoriaceae, Nostocaceae, Scytonemataceae, Rivulariaceae, Mastigocladaceae and Stigonemataceae. Out of these 39 were heterocystous and 115 were non-heterocystous.

Cyanophyceae

Order - Chroococcales

Family - Chroococcaceae

***Microcystis aeruginosa* Kuetz. (Plate 3.1, F.1)**

Desikachary, 1959, P.93, Pl.17, F.L2,6 and Pl.18, F.10

Cells 3.33μ m in diam., spherical, generally with gas vacuoles.

Habiatat: Hartala lake, H-III, June, 1999.

Velhala lake, V-I, November, 1998, February, 1999.

***Microcystis flox-aquae* (Wittr.) Kirchner. (Plate 3.1, F.2)**

Desikachary, 1959, P.94. Pl.17., F.11.

Cells 3.3μ in diameter, spherical, nannocytes present.

Habitat: Velhala lake, V-I, October, 1999, July, 1999, September, 1999.

***Microcystis marginata* (Menegh.) Kuetz. (Plate 3.1, F.3)**

Desikachary, 1959, P.87, Pl.17, F.3-5

Cells 5.5μ in diam., closely arranged, with gas vacuoles.

Habitat: Velhala lake, V-I, July, 2000.

***Microcystis robusta* (Clark) Nygaard (Plate 3.1, F.4)**

Desikachary, 1959, P.85, Pl.17, F.7-10

Cells 6.1-6.6μ diameter, spherical, without gas-vacuoles.

Habitat: Velhala lake, V-I, March, 1999, V-III, July, 1999, January, 1999.

***Microcystis viridis* (A.Br.) Lemm. (Plate 3.1, F.5)**

Desikachary, 1959, P.87, Pl.18, F.1-6

Cells 3.3μ in diameter, spherical with gas vacuoles.

Habitat: Velhala lake, V-I, May, 1999, V-III, January, 1999.

***Chroococcus cohaerens* (Breb.) Nag. (Plate 3.1, F.6)**

Desikachary, 1959, P.111, Pl.26, F.3-9

Cells single or up to 2-8 in groups, without envelop 2.2μ diameter, and with sheath 2.5μ diameter, sheath thin, colourless, unlamellated.

Habitat: Hartala lake, H-I, March, 1999.

***Chroococcus hansgirgi* Schmidle (Plate 3.1, F.7)**

Desikachary, 1959, P.85, Pl.105

Cells violet-blue, single spherical or oblong 7.2μ broad, 8.3μ long sheath thin hyaline, adhering, cell contents homogeneous.

Habitat: Velhala lake, V-I, April, 2000, V-II, January, 1999.

***Chroococcus minimus* (Keissler) Lamm. (Plate 3.1, F.8)**

Desikachary, 1959, P.106

Cells spherical or ellipsoid, generally in two or many in spherical colonies, without sheath 2.2μ diameter, with sheath 3.8μ in diameter.

Habitat: Velhala lake,V-I, April, 1999, V-II, December, 1998, February, 1999.

***Chroococcus minutus* (Kuetz.) Nag. (Plate 3.1, F.9)**

Desikachary, 1959, P.103, Pl.24, F.4

Cells with sheath 6.1μ diameter and without sheath 4.4μ diameter, sheath colourless.

Habitat: Hartala lake, H-II, June, 2000; Velhala lake, V-III, December, 1998.

***Chroococcus montanus* Hansgirg (Plate 3.1, F.10)**

Desikachary, 1959, P.108, Pl.26, F.12

Thallus slimy, gelatinous, cells 4.9μ broad, 8.8μ long, single or 2-4, colonial sheath lamellose.

Habitat: Hartala lake, H-II, September, 2000.

***Chroococcus varius* A.Br. (Plate 3.1, F.11)**

Desikachary, 1959, P.107, Pl.24, F .5

Cells globular, without sheath 4μ, with sheath 5.95μ diam. sheath apparently thick, indistinctly lamellated, colourless.

Habitat: Velhala lake, V-II, June, 1999, July, 1999, July, 2000.

***Gloeocapsa calcarea* Tilden (Plate 3.1, F.12)**

Desikachary, 1959, P.115, Pl.24, F.6

Cells with or without individual sheath, 6.6μ diameter.

Habitat: Hartala lake, H-II, October, 1999.

***Gloeocapsa kuetzingiana* Nag. (Plate 3.1, F.13)**

Desikachary, 1959, P.118, Pl.23, F.4

Cells without sheath 2.7-3.3μ diameter, with sheath upto 7.7μ diameter.

Habitat: Hartala lake, H-II, July, 1999; Velhala lake, V-II, November, 1998, January, 1999.

***Gloeocapsa crepidium* Thuret. (Plate 3.1, F.14)**

Desikachary, 1959, P.117, Pl.27, F.4,5

Cells in groups of 2-4, 23.8μ diameter

Habitat: Hartala lake, H-III, August, 2000.

***Gloeocapsa punctata* Nag. (Plate 3.1, F.15)**

Desikachary, 1959, P.115, Pl.23, F.2

Cells without sheath 1.6μ diameter with sheath 4.9μ diameter.

Habitat: Hartala lake, H-II, May, 1999.

Velhala lake, V-II, March, 1999, April, 1999, May, 1999.

V-III, May, 1999, August, 2000.

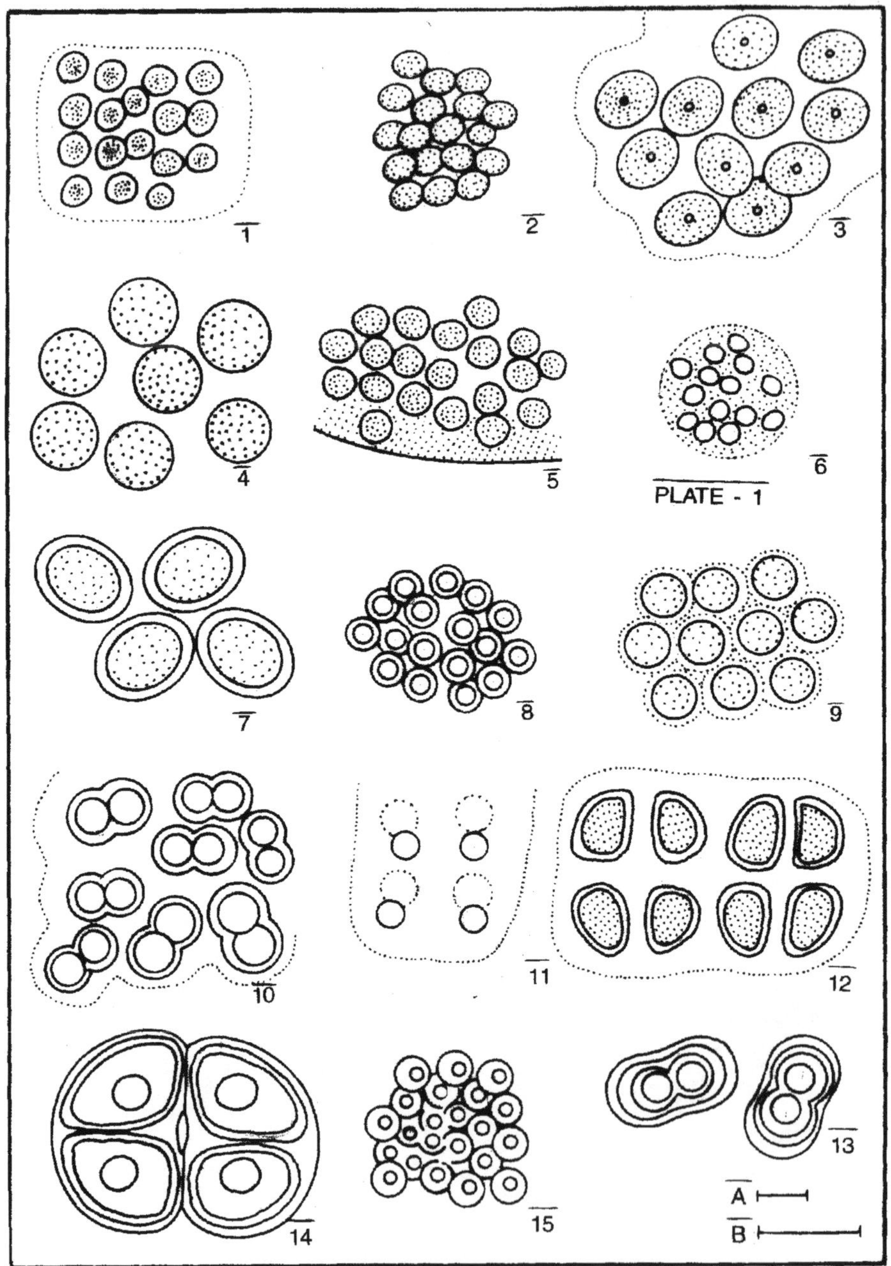

Plate 3.1: (1) *Microcystis aeruginosa* Kuetz., (2) *M. flox-aquae* (Wittr). Kirchn., (3) *M. marginata* (Mengegh.) Kuetz., (4) *M. robusta* (Clark) Nygaard, (5) *M. viridis* (A. Br.) Lamm., (6) *Chroococus coharens* (Breb.) Nag., (7) *C. hansgirgi* Schmidl, (8) *C. minimus* (Keissl.) Lamm., (9) *C. minutus* (Kuetz.) Nag., (10) *C. montanus* Hansg., (11) *C. varius* A. Br., (12) *Gloeocapsa calcarea* Tilden, (13) *G. kuetzingiana* Nag., (14) *G. crepidinum* Thuret, (15) *G. punctata* Nag.

Scale Bars A & B: 10µm Scale A: Fig. 11, Scale B. Fig. 1, 2, 3, 4, 5, 6, 7, 8, 9, 10, 12, 13, 14, 15.

Gloeocapsa stegophila **(Itzings.) Rabenh. v.** ***crassa*** **Rao, C.B. (Plate 3.2, F.1)**

Desikachary, 1959, P.119, Pl.25, F.3

Cells without sheath 3.8μ broad, 5.5μ long, with sheath 4.4μ broad, 6.6μ long, 2-4 cells in globose groups.

Habitat: Velhala lake, V-I, January, 1999, V-III, August, 2000.

Gloeothece rupestris **(Lyngb.) Bornet v.** ***maxima*** **West (Plate 3.2, F.2)**

Desikachary, 1959, P.127, Pl.25, F.4

Cells cylindrical with rounded ends, 9μ broad and 14.2μ long, with large granules, mucilage sheath thick, cells with sheath 11.9μ broad, 22.6μ long.

Habitat: Velhala lake, V-III, January, 1999.

Gloeothece samoensis **Wille (Plate 3.2, F.3)**

Desikachary, 1959, P.108, Pl.23, F.3

Cells ellipsoidal, without sheath 5.5μ broad, 8.3μ long, cells bluish green, in rounded colonies.

Habitat: Velhala lake, V-I, April, 2000.

Gloeothece samoensis **Wille v .*major*** **Wille. (Plate 3.2, F.4)**

Desikachary, 1959, P.128, Pl.23, F.6

Cells ellipsoidal, without sheath 5.5μ broad, 8.8μ long, with envelope 11.6μ broad, 16.6V long, sheath hyaline.

Habitat: Velhala lake, V-II, April, 1999.

Aphanocapsa banaresensis **Bharadwaja (Plate 3.2, F.5)**

Desikachary, 1959, P.133

Cells oval or almost spherical 4.4μ in diameter.

Habitat: Hartala lake, H-I, September, 2000.

Aphanocapsa crassa **Ghose (Plate 3.2, F.6)**

Cells spherical 8.3μ diameter.

Habitat: Hartala lake, H-I, March, 2000.

Aphanocapsa grevillei **(Mass.) Rabenh. (Plate 3.2, F.7)**

Desikachary, 1959, P.134

Cells spherical 5μ diameter closely arranged.

Habitat: Hartala lake, H-I, September, 2000.

Aphanocapsa koordersi **Strom. (Plate 3.2, F.8)**

Desikachary, 1959, P.132

Colony spherical 2-3 mm in diameter, cells loosely arranged or in groups of four, spherical 2.2-2.7μ in diameter.

Habitat: Hartala lake, H-I, October, 1998, December, 1998, May, 1999, June, 1999, July, 1999, H-III, July, 1999.

***Aphanocapsa montana* Cramer (Plate 3.2, F.9)**

Desikachary, 1959, P.135, Pl.20, F.8

Thallus of no definite shape, gelatinous, cells 2.77-3.33μ diameter, spherical, light blue-green, single or in pair, mucilage colourless, diffluent.

Habitat: Hartala lake, H-II, May, 1999, H-III, October, 1998, November, 1998, December, 1998, January, 1999, February, 1999.

***Aphanocapsa roseana* De Bary (Plate 3.2, F.I0)**

Desikachary, 1959, P.131

Cells 5.1-8.3μ diameter.

Habitat: Hartala lake, H-III, October, 1999, March, 2000.

***Aphanothece castagnei* (Breb.) Rabenh. (Plate 3.2, F.11)**

Desikachary, 1959, P.140, Pl.21, F.8

Cells ellipsoidal to cylindrical, 2.2μ broad, 4.4μ long.

Habitat: Velhala lake, V-I, October, 1999.

***Aphanothece pallida* (Kuetz.) Rabenh. (Plate 3.2, F.12)**

Desikachary, 1959, P.140, Pl.22, F.3

Cells oblong, elliptical or cylindrical, 5.5μ broad and 8.3μ long.

Habitat: Velhala lake, V-III, December, 1998, January, 1999, December, 1999, November, 1999.

***Synechococcus aeruginosus* Nag. (Plate 3.2, F.13)**

Desikachary, 1959, P.143, Pl.25, F.6,12

Cells cylindrical, 13.3μ broad, 30.5μ long single, pale blue-green.

Habitat: Velhala lake, V-I, September, 2000.

***Synechocystis aquatilis* Sauv. (Plate 3.2, F.14)**

Desikachary, 1959, P.144, Pl.25, F.9

Cells spherical, single or in rows, 5.5-6.6μ diameter.

Habitat: Hartala lake, H-I, January, 1999, March, 1999, June, 1999. H-II, October, 1998, January, 1999.

Velhala lake: V-II, January, 1999, V-III, October, (1998).

***Synechocystis pevalekii* Ercegovic (Plate 3.2, F.15)**

Desikachary, 1959, P.145, Pl.25, F.11

Cells spherical, 2.7-3.3μ broad, single or two together.

Habitat: Hartala lake, H-I, February, 1999, May, 1999, July, 1999.

***Coelosphaerium kuetzingianum* Nag. (Plate 3.3, F.1)**

Desikachary, 1959, P.148, Pl.28, F.7,8

Cells spherical or sub-spherical, 2.7μ broad, cells loosely arranged.

Habitat: Hartala lake, H-I, October, 1998.

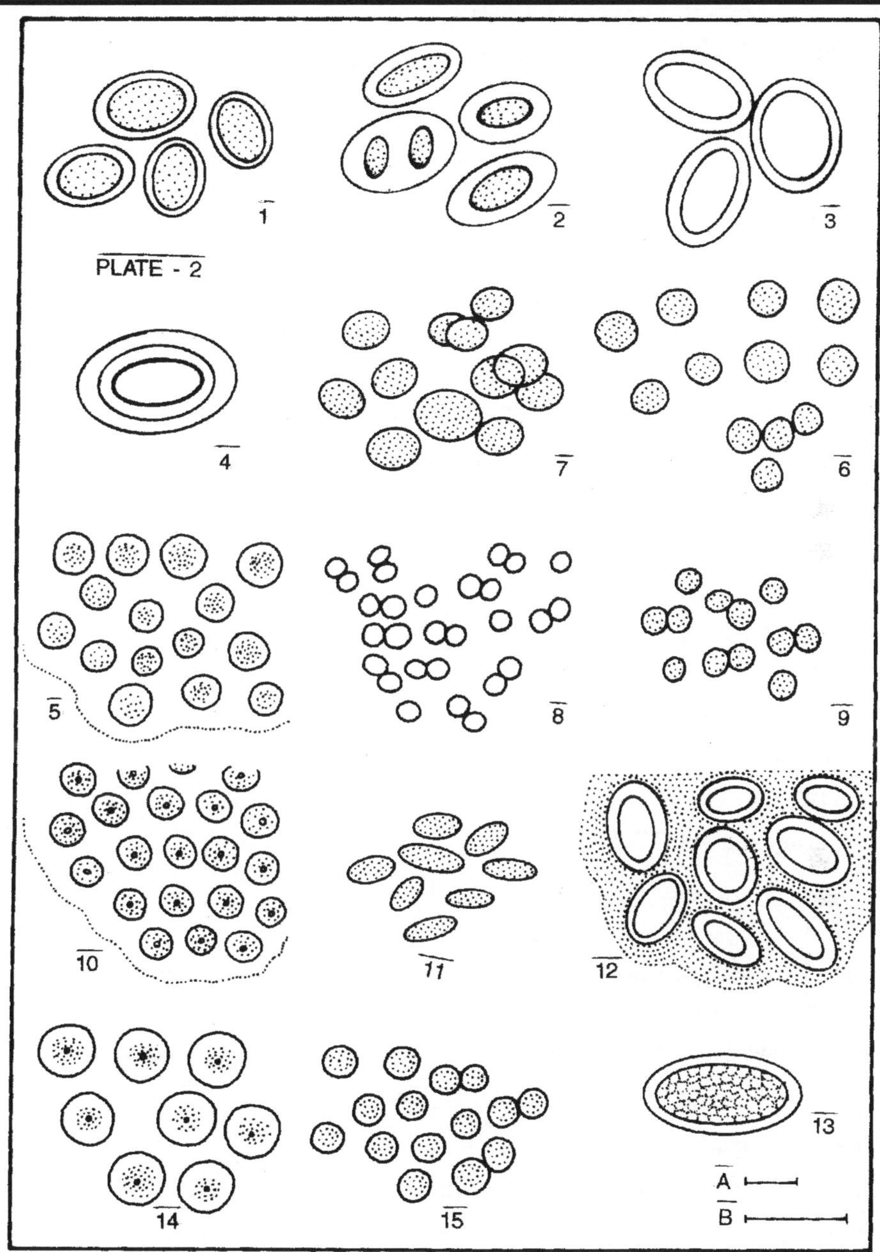

Plate 3.2: (1) *Gloeocapsa stegophila* (Itzings.), Rabenh. *v. crassa* Rao, C.B., (2) *G. rupestris* (Lyngb.) Borb *v. maxima* West, (3) *G. samoensis* Wille, (4) *G. samoensis* Wille *v. major* Wille, (5) *Aphansocapsa banaresensis* Bharadwaj, (6) *A. crassa Ghose*, (7) *A. grevillie* (Hass.) Rebenh., (8) *A. kooredersi* Storm, (9) *A. montana Cramer*, (10) *A. roseana* De Bary, (11) *A. castagnei* (Breb.) Rabenh., (12) *A. pallida* (Kuetz.) Rebenh, (13) *Synechococcus aeruginosus* Nag., (14) *Synechosystis aqualitis* Sauv., (15) *S. pevalekii* Erceg.

Scale Bars A & B: 10µm Scale A: Fig. 2, 6, 10, 13, Scale B. Fig. 1, 3, 4, 5, 7, 8, 9, 11, 12, 14, 15.

***Merismopedia convoluta* Breb. (Plate 3.3, F.2)**

Desikachary, 1959, P.152, Pl.29, F.8,12,13

Cells spherical to oblong, 4.7μ broad, 9.5μ long, forming a large 1-4mm long and broad, flat often leaf like convolute colonies, blue-green.

Habitat: Velhala lake, V-II, October, 1998, November, 1998, January, 1999, February, 1999, January, 2000.

***Merismopedia minima* Beck (Plate 3.3, F.3)**

Desikachary, 1959, P.154, Pl.29, F.11

Cells pale blue-green 4 to many in small colonies, 0.5μ broad, free swimming, groups of four cells 2-3×3μ.

Habitat: Hartala lake, H-I, March, 2000, H-II, August, 2000.

***Merismopedia punctata* Mayen (Plate 3.3, F.4)**

Desikachary, 1959, P.155, Pl.23, F.5

Colonies small, 4-64 cells about 60μ broad, cells not closely packed, spherical to ovoid, 3.8μ broad, pale blue-green.

Habitat: Hartala lake, H-I, March, 2000.

Velhala lake, V-II, May, 2000.

***Merismopedia tenuissima* Lemm. (Plate 3.3, F.5)**

Desikachary, 1959, P.154, Pl.29, F.7

Cells 1.6μ broad.

Habitat: Hartala lake-H-I, September, 1999, August, 2000.

***Dactylococcopsis raphidioids* Hansg. (Plate 3.3, F.6)**

Desikachary, 1959, P.158, Pl.29, F.1,2

Cells spindle-shaped, sigmoid or lunulately bent, 2.7-3.9μ broad, upto 29μ long, light blue-green, single to a few together in a little mucilage.

Habitat: Hartala lake, H-I, June, 1999.

Velhala lake, V-II, July, 1999, April, 2000.

Family: Entophysalidaceae

***Johannesbaptistia pellucida* (Dickie) Taylor and Dr. (Plate 3.3, F.7)**

Desikachary, 1959, P.165, Pl.32, F.14-19.

Cells 5.8μ broad and 2.9μ long, mucilage homogenous.

Habitat: Velhala lake, V-III, March, 1999.

Order - Chamaesiphonales

Family - Cyanidiaceae

***Chroococcidiopis indica* Desikachari (Plate 3.3, F.8)**

Desikachary, 1959, P.167, Pl.31, F.28

Cells solitary, spherical 6.1μ diameter, with a firm wall, endospores 8-12 formed by successive division.

Habitat: Velhala lake, V-II, June, 1999.

Family: Chamaesiphonaceae

***Chamaesiphon rostaffinskii* (Rostaf.) Hansgirg (Plate 3.3, F.9)**

Desikachary, 1959, P.171, Pl.32, F.7,8

Sporangia single, long or short, club-shaped or cylindrical, 2.7μ broad, at the apex-3.3μ broad, 21μ long.

Habitat: Velhala lake, V-III, December, 1999.

***Chamaesiphon sideriphilus* Starmach v. *glabra* Rao, C.B. (Plate 3.3, F.10)**

Desikachary, 1959, P.169, Pl.33, F.3

Sporangia single, club-shaped, slightly bent, 2.7-3.3μ broad, and 4-14μ long, with pseudovagina 3.3-3.8μ and 4.8-17.2μ long.

Habitat: Velhala lake, V-II, December, 1999, V-III, January, 2000.

***Stichosiphon sansibaricus* (Hieron) Drouet and Daily (Plate 3.3, F.11)**

Desikachary, 1959, P .177, Pl.32, F.9-13

Plants microscopic, blue-green, about 400μ high.

Habitat: Hartala lake, H-I, February, 2000, March, 2000, H-II, September, 1999, January, 2000, H-III, July, 2000.

Velhala lake, V-I, January, 1999, V-III, October, 1998, January, 1999, May, 1999, July, 1999, July, 2000, September, 2000.

Order - Pleurocapsales

Family - Pleurocapsaceae

***Myxosarcina burmensis* Skuja (Plate 3.3, F.12)**

Desikachary, 1959, P.178, Pl.32, F.20-22

Plant aquatic, minute rounded, sarcinoid, cells 2-3μ diameter.

Habitat: Hartala lake, H-I, July, 2000, H-III, March, 1999, September, 1999, H-III, July-2000.

Order - Nostocales

Family - Oscillatoriaceae

***Spirulina major* Kuetz. ex Gomont (Plate 3.3, F.13)**

Desikachary, 1959, P.196, Pl.36, F.13

Trichome 1.6μ broad, regularly spirally coiled, blue-green, spirals 3.3μ broad and 4.4μ distant.

Habitat: Hartala lake, H-II, September, 2000.

***Spirulina meneghiniana* Zanard. ex Gomont (Plate 3.3, F.14)**

Desikachary, 1959, P.195, Pl.36, F.8

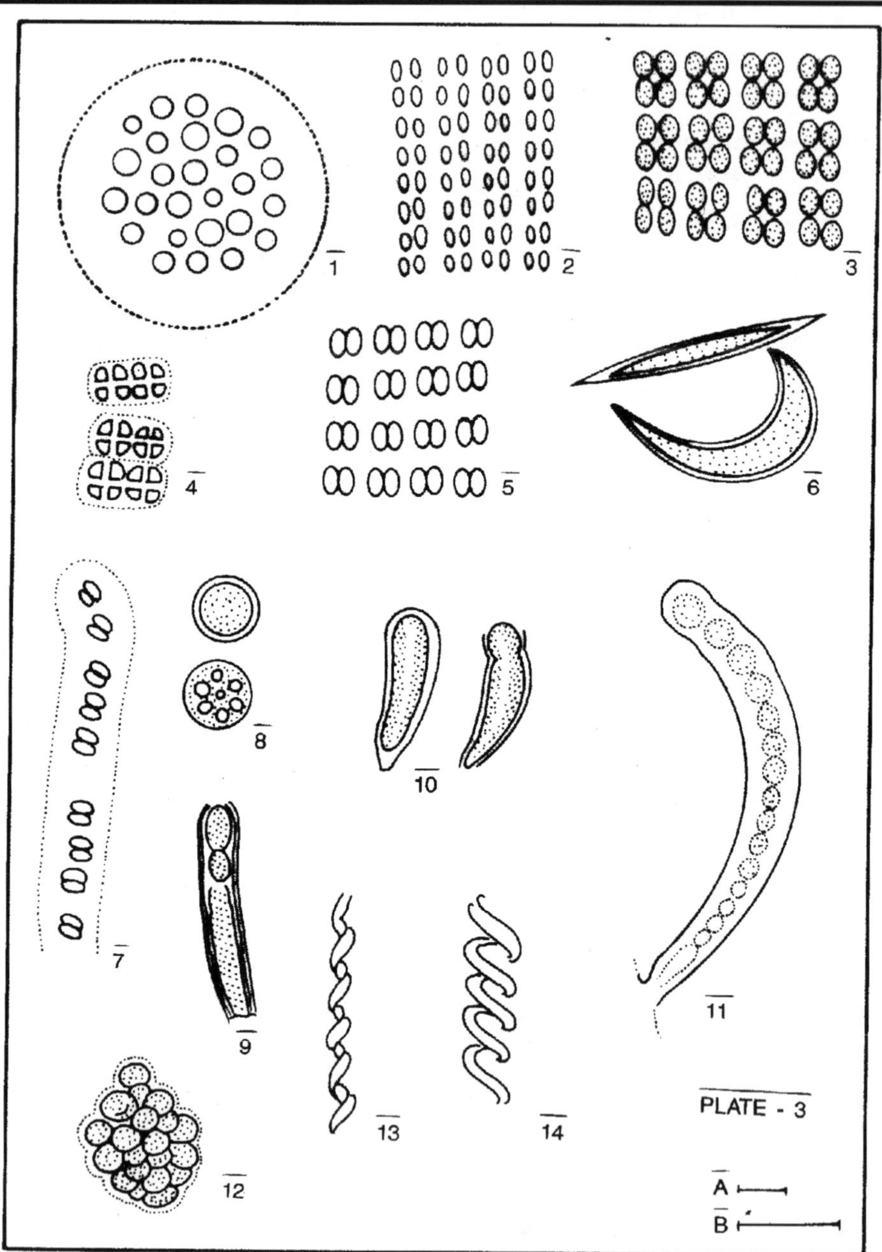

Plate 3.3: (1) *Coelosphaerium kuetzingianum* Nag., (2) *Merismopedia convoluta* Breb, (3) *M. minima* Beck, (4) *M. punctata* Mayen, (5) *M. tenuissima* Lemm., (6) *Dactylococcopis raphidiodes* Hansg., (7) *Johannesbaptistia pellucida* (Dickie) Taylor and Dr. (8) *Chroococcidiopsis indica* Desikachari, (9) *Chamaesiphon rostaffinskii* (Rostaf.) Hansg., (10) *C. sideriphilus* Starmach *v. glabra* Rao, C.B., (11) *Stichosiphon sansibaricus* (Hieron.) Dr. and D., (12) *Myxosarcina burmensis* Skuja, (13) *Spirulina major* Kuetz. ex Gom., (14) *S. meneghiniana* Zanard. ex. Gomont.

Scale Bars A & B: 10μm Scale A: Fig. 2, 3, 4, 7, 11, Scale B: Figs. 1, 5, 6, 8, 9, 10, 12, 13, 14.

Trichomes 1.6µ broad, flexible, spirally coiled, bright blue-green, spirals 4.9µ broad and 3.8µ distant from each other.

Habitat: Hartala lake, H-I, August, 2000.

***Oscillatoria brevis* (Kuetz.) Gom. (Plate 3.4, F.l)**

Desikachary, 1959, P.241.

Cells 6.1µ broad, 3.8µ long, end cell rounded.

Habitat: Hartala lake, H-I, August, 2000.

***Oscillatoria calcuttensis* Biswas (Plate 3.4, F.2)**

Desikachary, 1959, P.237, Pl.42, F.20, 21

Cells 6.6µ broad, cells 6.6µ long, cross-walls with 3 granules.

Habitat: Velhala lake, V-II, April, 2000.

***Oscillatoria chilkensis* Biswas (Plate 3.4, F.3)**

Desikachary, 1959, P.215, Pl.39, F.1

Cells shorter than the diameter, about 2.2µ in length.

Habitat: Velhala lake, V-II, May, 2000.

***Oscillatoria curviceps* Ag. *v.angusta* Ghose (Plate 3.4, F.4)**

Desikachary, 1959, P.209, Pl.38, F.2

Trichomes bent at the end, not attenuated, not constricted at the cross walls, 5.5µ broad 2.7µ long, end cells flat rounded, not capitate.

Habitat: Hartala lake, H-I, August, 2000, H-II, April, 2000.

***Oscillatoria chalybea* (Mertens) Gom (Plate 3.4, F.5)**

Desikachary, 1959, P.218, Pl.38, F.3

Cells 8.9µ broad, 3.5µ long, end wall obtuse, not capitate.

Habitat: Hartal lake, H-I, January, 2000, H-II, August, 2000.

***Oscillatoria formosa* Bory f. *loktakensis* Bruhl and Biswas (Plate 3.4, F.6)**

Desikachary, 1959, P.232, Pl.40, F.15

Trichomes 4.4µ broad, bent, flexible, constricted at the cross-walls, ends slightly attenuated, capitate, cells 2-7µ long.

Habitat: Hartala lake, H-III, October, 1998.

***Oscillatoria hamelli* Fremy (Plate 3.4, F.7)**

Desikachary, 1959, P.225

Cells 4.9µ broad, 7.2µ long.

Habitat: Velhala lake, V-I, May, 2000.

Oscillatoria irrigua **Kuetz. (Plate 3.4, F.8)**

Desikachary, 1959, P.224, Pl.42, F.7,9

Cells nearly quadrate, 4.4μ long.

Habitat: Hartala lake, H-III, January, 1999.

Oscillatoria jasorvenisis **Vouk (Plate 3.4, F.9)**

Desikachary, 1959, P.221

Trichome straight, 2.7μ broad, cells as long as broad.

Habitat: Velhala lake, V-II, July, 2000.

Oscillatoria laete-virens **Gom. (Plate 3.4, F.I0)**

Desikachary, 1959, P.213, Pl.29, F.2,3

Cells 4.7μ broad, apices attenuated, 2.7μ long.

Habitat: Hartala lake, H-II, March, 2000, July, 2000.

Velhala lake, V-II, February, 1999.

Oscillatoria limnetica **Lamm (Plate 3.4, F.11)**

Desikachary,1959, P.226, Pl.37, F.3

Trichome slightly bent, 1.6μ broad, not attenuated, cells 6.6μ long.

Habitat: Hartala lake, H-II, October, 1998.

Velhala lake, V-III, January, 1999.

Oscillatoria limosa **Ag. (Plate 3.4, F.12)**

Desikachary, 1959, P.206, Pl.42, F.11

Trichome more or less straight, 15.4μ broad, cells 1/3 - 1/6 as long as broad, 4.7μ long.

Habitat: Hartala lake, H-I, August, 2000, H-II, October, 1999.

Oscillatoria margaretifera **Kuetz. (Plate 3.4, F.13)**

Desikachary, 1959, P.202, Pl.42, F.8

Cells 1/3 - 1/7 as long as broad, 4.7-5.9μ long.

Habitat: Velhala lake, V-II, September, 1998, V-III, November, 1999.

Oscillatoria nigro-virdis **Thwaites (Plate 3.4, F.14)**

Desikachary, 1959, P.202, Pl.42, F.6

Cells 1/2 - 1/4 as long as broad, 4.4μ long.

Habitat: Hartala lake, H-II, November, 1999, May, 2000.

Oscillatoria ornata **Kuetz. (Plate 3.4, F.15)**

Desikachary, 1959, P.206, Pl.37, F.12

Cells 1/2 - 1/6 as long as broad, 4.7μ long.

Habitat: Hartala lake, H-III, August, 2000.

***Oscillatoria peronata* Skuja (Plate 3.4, F.16)**

Desikachary,1959, P.205, Pl.41, F.8,9,14.

Cells commonly 1/2 - 1/5 as long as broad, 5.3μ 1ong.

Habitat: Hartala lake, H-I, August, 1999, January, 2000, H-Ill, February, 2000.

***Oscillatoria peronata* Skuja v.attenuata Skuja (Plate 3.4, F.17)**

Desikachary, 1959, P.205

Cells 1/2 - 1/3 as long as broad.

Habitat: Hartala lake, H-I, May, 2000.

***Oscillatoria quadripunctulata* Bruhl and Biswas (Plate 3.4, F.18)**

Desikachary, 1959, P.227, Pl.37, F.5

Cells 4.4μ long, 1.5μ in diameter.

Habitat: Hartala lake, H-I, January, 2000.

***Oscillatoria princeps* Voucher ex Gomont (Plate 3.5, F.1)**

Desikachary, 1959, P.210, Pl.37, F.1,10,11

Cells ½ - ¼ as long as broad, 3.6μ long.

Habitat: Hartala lake, H-II, November, 1998.

***Oscillatoria pseudogeminata* Schmid (Plate 3.5, F.2)**

Desikachary, 1959, P.228, Pl.41, F.10

Cells 2.2μ broad, cells as long as broad.

Habitat: Hartala lake, H-I, November, 1999.

***Oscillatoria pseudogeminata* Schmid *v.unigranulata* Biswas (Plate 3.5, F.3)**

Desikachary, 1959, P.228, Pl.39, F.19

Cells 3.8μ in length, cell-wall thick, distinct with one large granule situated at the centre of the partition walls on either side.

Habitat: Hartala lake, H-III, November, 1998.

***Oscillatoria raoi* De Toni, J. (Plate 3.5, F.4)**

Desikachary, 1959, P.223, Pl.42, F.16-19.

Cells 4.4μ long, 6.6μ broad.

Habitat: Hartala lake, H-I, June, 1999.

***Oscillatoria raciborskii* Wolosz (Plate 3.5, F.5)**

Desikachary, 1959, P.236, Pl.37, F.15

Cells ½ times as long as broad, 8.3μ broad.

Habitat: Hartala lake, H-I, September, 1999, H-III, April, 2000.

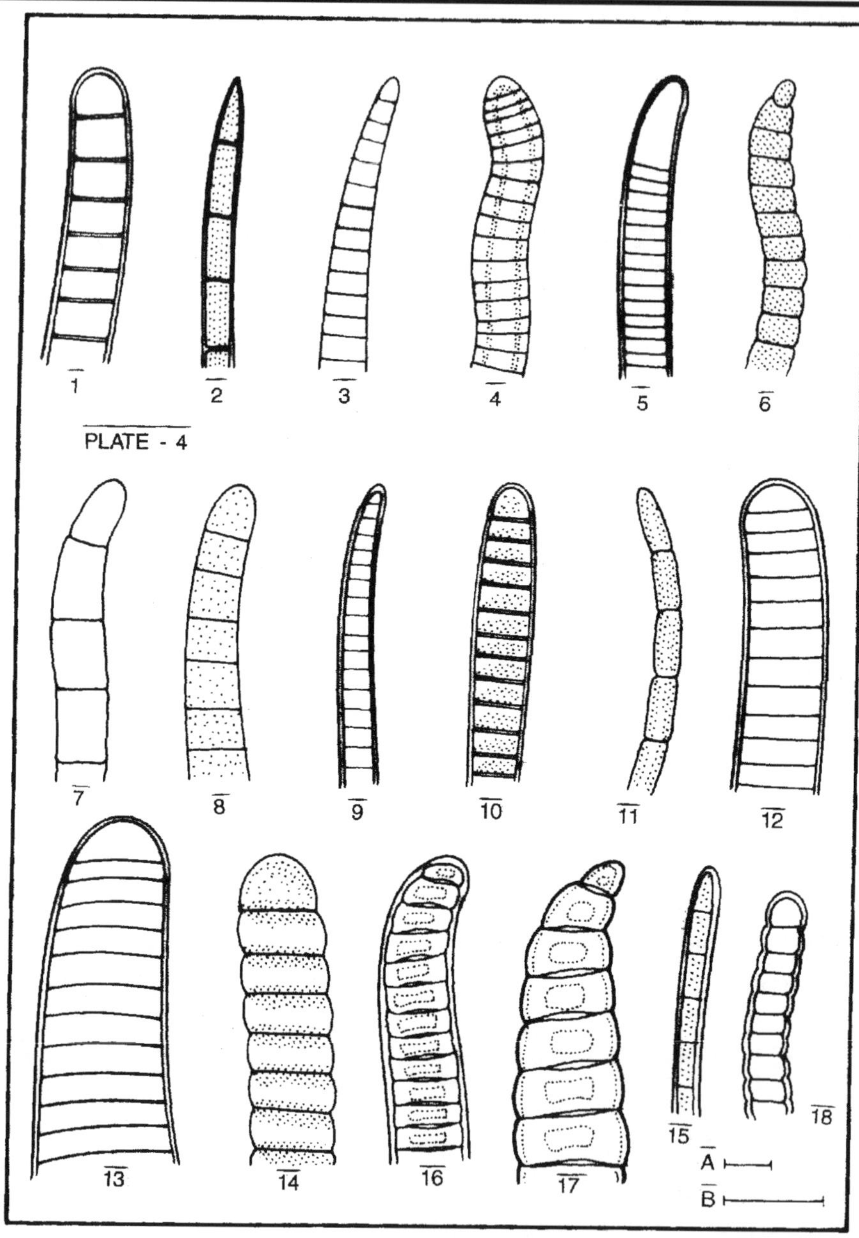

Plate 3.4: (1) *Oscillatoria brevis* (Kuetz.) Gom., (2) *O. calcuttensis* Biswas, (3) *O. chilkensis* Biswas, (4) *O. curviceps* Ag. *v. angusta* Ghose, (5) *O. chalybea* (Mertens) Gom., (6) *O. formosa* Bory f. *loktakensis* Bruhl and Biswas, (7) *O. hamelli* Fremy, (8) *O. irrigua* Kuetz., (9) *O. jasorvenisis,* (10) *O. laete-virens* Gom., (11) *O. limnetica* Lamm., (12) *O. limosa* Ag., (13) *O. margaretifera* Kuetz, (14) *O. nigro-virdis* Thwaites, (15) *O. ornata* Kuetz., (16) *O. perornata* Skuja, (17) *O. perornata* Skuja *v. attenuata* Skuja, (18) *O. quadripanctulata* Bruhl and Biswas.

Scale Bars A & B: 10μm Scale A: Fig. 5, 12, 13, 15, 16, Scale B: Figs. 1, 2, 3, 4, 6, 7, 8, 9, 10, 11, 14, 17, 18.

***Oscillatoria rubescens* D.C. ex Gomont (Plate 3.5, F.6)**

Desikachary, 1959, P.235, Pl.42, F.12

Cells 1/2 - 1/3 as long as broad 3.5μ long.

Habitat: Hartala lake, H-II, August, 2000.

***Oscillatoria subtillissima* Kuetz. (Plate 3.5, F.7)**

Desikachary, 1959, P.215

Trichome, 1.5μ broad, curved, septa indistint, without gas vocuoles.

Habitat: Hartala lake, H-II, July, 2000.

Velhala lake, V-II, December, 1998, January, 1999, January, 2000.

***Oscillatoria terebriformis* Ag. (Plate 3.5, F.8)**

Desikachary, 1959, P.217, Pl.38, F.16

Trichomes straight, 5.9μ broad, 2.3-2.9μ long, end cell rounded.

Habitat: Hartala lake, H-II, March, 1999.

Velhala lake, V-I, March, 2000.

***Oscillatoria tenuis* Ag. (Plate 3.5, F.9)**

Desikachary, 1959, P.222

Trichome straight, 4.9μ broad, bent at ends, 5.5μ long.

Habitat: Hartala lake, H-II, September, 2000.

***Oscillatoria tenuis* Ag. v. *tergestina* Rabenh. (Plate 3.5, F.10)**

Desikachary, 1959, P.222, Pl.42, F.15

Cells as long as broad, 4.4μ long.

Habitat: Hartala lake, H-I, October, 1999.

Velhala lake, V-I, March, 1999, March, 2000, April, 2000.

***Oscillatoria vizagapatensis* Rao C.B. (Plate 3.5, F.11)**

Desikachary, 1959, P.205, Pl.39, F.16,18

Cells much shorter than broad, 1.6μ long.

Habitat: Velhala lake, V-III, April, 1999.

***Oscillatoria willei* Gardner (Plate 3.5, F.12)**

Desikachary, 1959, P.217, Pl.38, F.4,5

Trichome 2.7μ broad, cells 3.8μ long.

Habitat: Hartala lake, H-II, June, 2000.

***Phormidium ambigum* Gom. (Plate 3.5, F.13)**

Desikachary, 1959, P.266, Pl.44, F.16

Trichomes, 4.3-5.8μ broad, cells shorter than broad, 1.4-2.1μ long.

Habitat: Hartala lake, H-II, March, 1999.

Velhala lake, V-I, December, 1999.

Phormidium angustissimum **West and West (Plate 3.5, F.14)**

Desikachary, 1959, P.253.

Trichomes bent, 0.8μ broad, 2μ long.

Habitat: Hartala lake, H-II, December, 1999.

Phormidium autumnale **(Ag.) Gom. (Plate 3.5, F.15)**

Desikachary, 1959, P.276, Pl.44, F.24,25

Trichomes 4.3μ broad cells 2.9μ long.

Habitat: Hartala lake, H-III, December, 1999.

Phormidium bigranulatum **Gardner (Plate 3.5, F.16)**

Desikachary, 1959, P.261

Filaments 0.8μ broad, cells 9μ long, at each end with one granule.

Habitat: Velhala lake, V-II, May, 2000.

Phormidium calcicola **Gardner (Plate 3.5, F.17)**

Desikachary, 1959, P.267, Pl.43, F.4,5.

Trichomes 5.8μ broad, cells slightly longer than broad.

Habitat: Hartala lake, H-III, August, 1999.

Phormidium corium **(Ag.) Gom. v. *capitatum* Gardner (Plate 3.5, F.18)**

Desikachary, 1959, P.269, Pl.43, F.11,12

Trichomes 5.9μ broad, not constricted at the cross-walls.

Habitat: Hartala lake H-III, March, 2000.

Phormidium fragile **(Manegh.) Gom. (Plate 3.5, F.19)**

Desikachary, 1959, P.253, Pl.44, F.1-3

Trichomes 2.2μ broad.

Habitat: Hartala lake, H-II, January, 1999, H-III, February, 2000.

Phormidium incrustatum **(Nag.) Gom. (Plate 3.5, F.20)**

Desikachary, 1959, P.269, Pl.44, F.12

Trichome 4.7μ broad, cells nearly quadrate, 3.5μ long.

Habitat: Hartala lake, H-II, August-1999.

Phormidium inudatum **Kuetz. (Plate 3.5, F.21)**

Desikachary, 1959, P.271

Trichome straight, 5μ broad, 7.1μ long.

Habitat: Hartala lake, H-III, August, 1999, March, 2000.

Velhala lake, V-II, November, 1998, December, (1998).

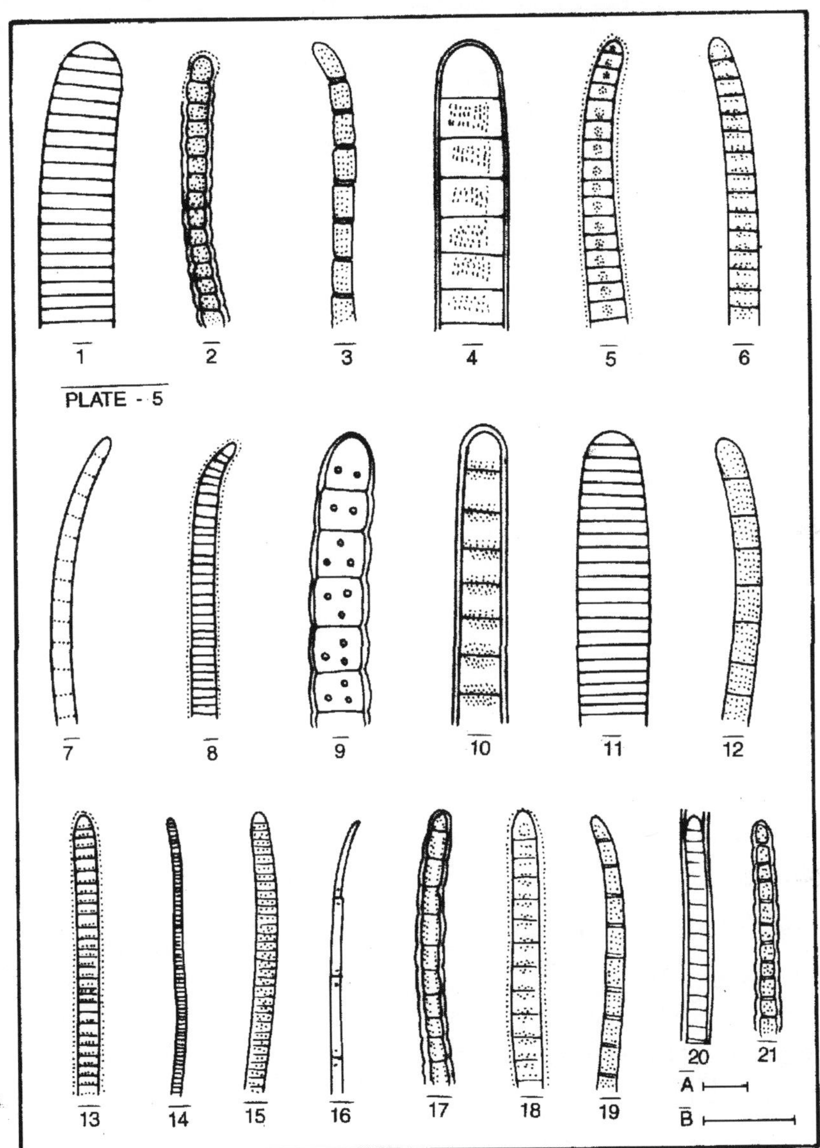

Plate 3.5: (1) *Oscillatoria princeps* Voucher ex Gom, (2) *O. pseudogeminata* Schmid., (3) *O. pseudogeminata* Schmid. *v. unigranulata* Biswas, (4) *O. raoi* De Toni, (5) *O. raciborskii* Wolosz, (6) *O. rubescens* D.C., (7) *O. subtillissima* Kuetz., (8) *O. terebriformis* Ag., (9) *O. tenuis* Ag., (10) *O. tenuis* Ag. *v. tergestina* Rabenh, (11) *O. vizagapatensis* Rao, (12) *O. willei* Gardner, (13) *Phormidium ambigum* Gom., (14) *P. angustissimum* West and West, (15) *P. autumnale* (Ag.) Gom., (16) *P. bigranulatum* Gardner, (17) *P. calcicola* Gardner, (18) *P. corium* (Ag.) Gom. *v. capitatum* Gardner, (19) *P. fragile* (Manegh.) Gom., (20) *P. incrustatum* (Nag) Gom., (21) *P. inudatum* Kuetz.

Scale Bars A & B: 10µm Scale A: Fig. 1, 5, 6, 8, 15, 17, 18, 20, 21, Scale B: Figs. 2, 3, 4, 7, 9, 10, 11, 12, 14, 16, 19.

Phormidium jadianianum **Gomont. (Plate 3.6, F.l)**

Desikachary, 1959, P.256, Pl.55, F.9

Trichome with straight long acuminate ends 4.3μ broad, 2.9μ long.

Habitat: Hartala lake, H-III, May, 1999.

Phormidium lucidium **(Kuetz.) Gomont (Plate 3.6, F.2)**

Desikachary, 1959, P.275, Pl.44, F.17,18

Trichome 7.1μ broad, cells 2.3μ long.

Habitat: Hartala lake H-III, May, 1999.

Phormidium mucosum **Gardner v *.arvense* Rao, C.B. (Plate 3.6, F.3)**

Desikachary, 1959, P.265, Pl.43, F.6,7

Trichome 2.2μ broad, cells 3.3μ long.

Habitat: Hartala lake, H-I, January, 2000.

Phormidium pachydermaticum **Fremy (Plate 3.6, F.4)**

Desikachary, 1959, P.267, Pl.43, F.8-10.

Trichome 4.4μ broad, cells nearly quadrate, 7.7μ long.

Habitat: Hartala lake, H-III, September, 1999.

Phormidium papyraceum **(Ag). Gom.(Plate 3.6, F.5)**

Desikachary, 1959, P.271

Trichomes 4.3μ broad, Cells 3.6μ long.

Habitat: Hartala lake, H-III, December, 1999.

Velhala lake, V-II, October, 1998,1999, V-III, October, 1999.

Phormidium purpurascens **(Kuetz.) Gom. (Plate 3.6, F.6)**

Desikachary, 1959, P.262, Pl.44, F.4

Trichome 1.6μ broad cells 2.2μ long, cross walls marked by two granuals on either side.

Habitat: Hartala lake, H-II, December, 1999.

Velhala lake, V-III, August, 1999.

Phormidium retzii **(Ag.) Gom. (Plate 3.6, F.7)**

Desikachary, 1959, P.268, Pl.44, F.13,15

Trichomes 8.3μ broad, cells 7.1μ long.

Habitat: Hartala lake, H-I, March, 2000, H-III, September, 1999.

Velhala lake, V-III, September, 1999.

Phormidium rotheanum **Itzigs. (Plate 3.6, F.8)**

Desikachary, 1959, P.258, Pl.45, F.14,16

Trichome 10.1μ broad, cells much shorter than broad, 4.3μ long.

Habitat: Hartala lake, H-III, May, 1999.

Phormidium subincrustatum **Fritch and Rich. (Plate 3.6, F.9)**

Desikachary, 1959, P.267.

Trichomes 5.9μ broad.

Habitat: Hartala lake, H-I, August, 1999, March, 2000.

Phormidium tenue **(Meneg.) Gom. (Plate 3.6, F.I0)**

Desikachary, 1959, P.259, Pl.43, F.13,15

Trichome straight, 1.9μ broad, cells up to 3 times longer than broad, 4.9μ long.

Habitat: Hartala lake, H-I, July, 2000.

Phormidium usterii **Schmidle (Plate 3.6, F.11)**

Desikachary, 1959, P.257

Trichomes 3.3μ broad, cells shorter than broad, often rectangular.

Habitat: Hartala lake, H-II, December, 1999.

Lyngbya aestuarii **Liemb. ex Gomont (Plate 3.6, F.12)**

Desikachary, 1959, P.305, Pl.52, F.8

Filaments single straight, cells 8-18μ broad, 2.7-4.7μ long.

Habitat: Hartala lake, H-I, March, 2000, September, 2000.

Velhala lake, V-II, May, 1999, V-III, May, 1999, April, 2000.

Lyngbya allorgei **Fremy (Plate 3.6, F.13)**

Desikachary, 1959, P.313, Pl.54, F.6

Filaments solitary and elongate, end cell round.

Habitat: Velhala lake, V-I, April, 2000, V-II, October, 1998, January, 1999, October, 1999, V-III, September, 2000.

Lyngbya borgertii **Lemm. (Plate 3.6, F.14)**

Desikachary, 1959, P.293, Pl.53, F.12

Filaments single, 2.6-3μ broad, cells 2.2μ broad, 3.8μ long.

Habitat: Hartala lake, H-II, August, 1999.

Velhala lake, V-II, November, 1998, November, 1999, V-III, October, 1998.

Lyngbya ceylanica **Wille (Plate 3.6, F.15)**

Desikachary, 1959, P.299, Pl.54, F.4

Filaments 10-14μ broad, cells 5.9μ long, end cell round.

Habitat: Hartala lake, H-III, July, 1999, September, 1999.

Lyngbya cryptovaginata **Schkorbatow (Plate 3.6, F.16)**

Desikachary, 1959, P.297, P1.50, F.6

Filaments straight, 7.1μ broad, cells nearly quadrate, 5.9μ long.

Habitat: Hartala lake, H-II, March, 2000.

***Lyngbya digueti* Gom. (Plate 3.6, F.17)**

Desikachary, 1959, P.310, Pl.53, F.8

Filaments 2.2μ broad, cells nearly quadrate, 2.2μ long.

Habitat: Hartala lake, H-I, August, 2000.

***Lyngbya epiphytica* Hieron (Plate 3.6, F.18)**

Desikachary, 1959, P.284, Pl.53, F.7

Filaments 1.5-2.2μ broad, cell 1.6μ broad, 1.6μ long.

Habitat: Hartala lake, H-II, February, 1999.

***Lyngbya hieronymusii* Lemm. (Plate 3.6, F.19)**

Desikachary, 1959, P.297, Pl.48, F.4.

Filaments 12-14μ broad, cells 11μ broad, 2.7μ long.

Habitat: Hartala lake, H-III, January, 1999.

***Lyngbya infixa* Fremy (Plate 3.6, F.20)**

Desikachary, 1959, P.282

Filaments 2.7μ broad, cells commonly shorter than long, 2.2μ long.

Habitat: Hartala lake, H-I, March, 2000.

***Lyngbya hieronymusii* Lemm. *v.crassivaginata* Ghose (Plate 3.6, F.21)**

Desikachary, 1959, P.297, Pl.55, F.8

Filaments 15-20μ broad, cells 11.9μ broad, upto 10μ long.

Habitat: Hartala lake, H-II, October, 1999.

***Lyngya laxespiralis* Skuja. (Plate 3.7, F.1)**

Desikachary, 1959, P .289, Pl.50, F.4.

Trichomes 9.5μ broad, cells shorter than broad, 5.9μ long.

Habitat: Hartala lake, H-III, July, 2000.

***Lyngbya limmnetica* Lemm. (Plate 3.7, F.2)**

Desikachary, 1959, P.294, Pl.50, F.11

Filaments 1.2μ broad cells 1.5μ broad, 3μ long.

Habitat: Hartala lake, H-III, July, 1999.

Velhala lake, V-II, February, 1999, May, 2000.

***Lyngbya martensiana* Manegh. ex Gomont (Plate 3.7, F.3)**

Desikachary, 1959, P.318, Pl.52, F.6

Trichome 6-10μ broad, cells 1.75-3.3μ in length.

Habitat: Hartala lake, H-II, July, 1999, March, 2000., H-III, May, 2000.

Velhala lake, V-I, July, 1999, June, 2000, V-II, November, 1998, April, 1999, November, 1999, V-III, January,1999.

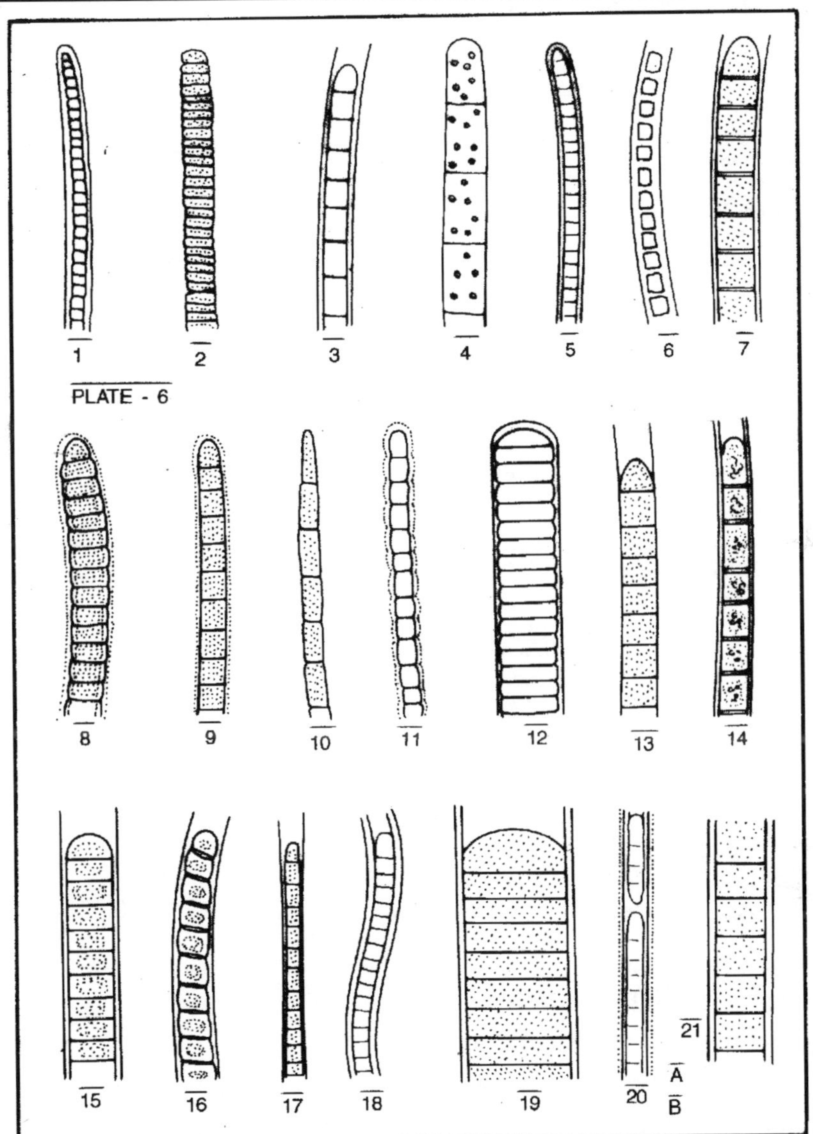

Plate 3.6: (1) *Phromidium jadianianum* Gom., (2) *P. lucidium* (Kuetz.) Gom., (3) *P. mucosum* Gardner *v. arvense* Rao, C.B., (4) *P. pachydermaticum* Fremy, (5) *P. papyraceum* (Ag.) Gom., (6) *P. purpurascens* (Kuetz.) Gom., (7) *P. retzii* (Ag.) Gom., (8) *P. rotheanum* Itzigs., (9) *P. subincrustatum* Fritch and Rich, (10) *P. tenue* (Meneg.) Gom., (11) *P. usterii* Schmidle, (12) *Lyngbya aestuarii* Liemb ex Gomont, (13) *L. allorgei* Fremy, (14) *L. borgertii* Lemm., (15) *L. ceylanica* Wille, (16) *L. cryptovaginata* Schkorbatow, (17) *L. digueti* Gom., (18) *L. epiphytica* Hieron, (19) *L. hieronymusii* Lemm., (20) *L. infixa* Framy, (21) *L. hieronymusii* Lemm. *v. crassivaginata* Ghose.

Scale Bars A & B: 10μm Scale A: Fig. 1, 2, 3, 5, 7, 8, 9, 12, 15, 16, 21, Scale B: Figs. 4, 6, 10, 11, 13, 14, 17, 18, 19, 20.

***Lyngbya mesotricha* Skuja (Plate 3.7, F.4)**

Desikachary, 1959, P.282, Pl.50, F.1,2

Filaments 3-5μ broad, upto 1mm long, trichomes 2.5-3μ broad cell 1-2 times longer than broad, 4.6μ long.

Habitat: Hartala lake, H-II, August, 1999.

***Lyngbya nordgardhii* Wille (Plate 3.7, F.5)**

Desikachary, 1959, P.287, Pl.54, F.1-3

Trichome 2.2μ broad, not attenuated, cells 1-½ as long as broad, 3.3μ long.

Habitat: Hartala lake, H-III, April, 2000, September, 1999.

***Lyngbya perelegans* Lemm. (Plate 3.7, F.6)**

Desikachary, 1959, P.309, Pl.48, F.8

Trichomes 1.6μ broad, cells 7.7μ long.

Habitat: Hartala lake, H-II, November, 1999.

***Lyngbya putealis* Mont. ex Gomont (Plate 3.7, F.7)**

Desikachary, 1959, P.317, Pl.52, F.12

Trichome 7.5-13μ broad, cells 8.7μ long, end cell rounded.

Habitat: Hartala lake, H-III, November, 1999.

Velhala lake, V-II, December, (1998).

***Schizothrix mulleri* Nag. (Plate 3.7, F.8)**

Desikachary, 1959, P.330, Pl.57, F.9,10

Trichome slightly constricted at the cross-walls, 7.2μ broad, cells 4.4μ long.

Habitat: Velhala lake, V-I, October, 1998, V-II, September, 1999, V-III, October, 1999, September, 2000.

***Symploca cartilaginea* (Mont.) Gom. (Plate 3.7, F.9)**

Desikachary, 1959, P .339, Pl.59, F.1,2

Cells 2-3μ broad, mostly longer than the broad, 3-5.8μ long.

Habitat: Hartala lake, H-II, May, 1999.

Velhala lake, V-II, October, 1998, September-October, 1998, V-III, October, December, 1998, September-October, 1999, September, 2000.

***Symploca muscorum* (Ag.) Gom. (Plate 3.7, F.I0)**

Desikachary, 1959, P.337, Pl.59, F.3,4

Cells 5-8μ broad, 5-11μ long.

Habitat: Hartala lake, H-II, June, 1999, November, 1999.

***Microcoleus lacustris* (Rabenh.) Farlow (Plate 3.7, F.l1)**

Desikachary, 1959, P.345, Pl.60, F.4,5

Cell cylindrical, 1-3 times as long as broad, 8.3μ long.

Habitat: Velhala lake, V-II, July, 1999, April, 1999, July, 2000.

Family: Nostocaceae

***Cylindrospermum indicum* Rao, C.B. (Plate 3.7, F.12)**

Desikachary, 1959, P.369, Pl.64, F.4,11

Cells quadrate, 3-4.5μ long, heterocysts spherical, 2.8-5.8μ broad and 3-7.6μ long, spores 10-12μ broad and 18-22μ long.

Habitat: Velhala lake, V-I, February, 2000, January, 1999.

***Nostoc calcicola* Breb. (Plate 3.7, F.13)**

Desikachary, 1959, P.384, Pl.68, F.1

Trichome 2.7μ broad, heterocysts 5.5μ broad, spores 4-5μ broad.

Habitat: Hartala lake, H-III, March, 2000.

Velhala lake, V-II, April-May, 2000.

***Nostoc coeruleum* Lyngb. (Plate 3.7, F.14)**

Desikachary, 1959, P.388

Trichome 7μ broad, heterocysts 10.7μ broad, spores 8.3μ diameter.

Habitat: Hartala lake, H-III, August, 1999, December, 1999.

Velhala lake, V-I, February, 1999, February, 2000, V-II, October-December, 1998, January, 1999, V-III, October, 1998, December, 1998, December, 1999.

***Nostoc ellipsosporum* (Desm.) Rabenh. (Plate 3.7, F.15)**

Desikachary, 1959, P.383, Pl.69, F.5

Trichome 4μ broad-cells 6-14μ long, heterocysts 6-7μ broad, 6- 14μ long, spores 6-8μ broad.

Habitat: Hartala lake, H-III, November, 1998, October, 1999, December, 1999.

Velhala lake, V-I, January-March-April, 2000.

***Nostoc linckia* (Roth) B. and F. *v.arvense* Rao, C.B. (Plate 3.7, F.16)**

Desikachary, 1959, P.377, Pl.67, F.1

Trichomes 4-5.6μ broad, cells spherical 6.4μ long, heterocysts spherical 7.2μ broad.

Habitat: Hartala lake, H-II, September, 1999.

Velhala lake, V-I, July, 1999, June, 2000.

***Nostoc muscorum* Ag. (Plate 3.7, F.17)**

Desikachary, 1959, P.385, Pl.70, F.2

Trichome 3-4 (-5)μ broad, cells short heterocysts spherical, 6.6μ broad, spores oblong, 4-8μ broad, 8-12μ long.

Habitat: Hartala lake, H-III, October, (1998).

Velhala lake, V-II, June-July, 1999, June, 2000.

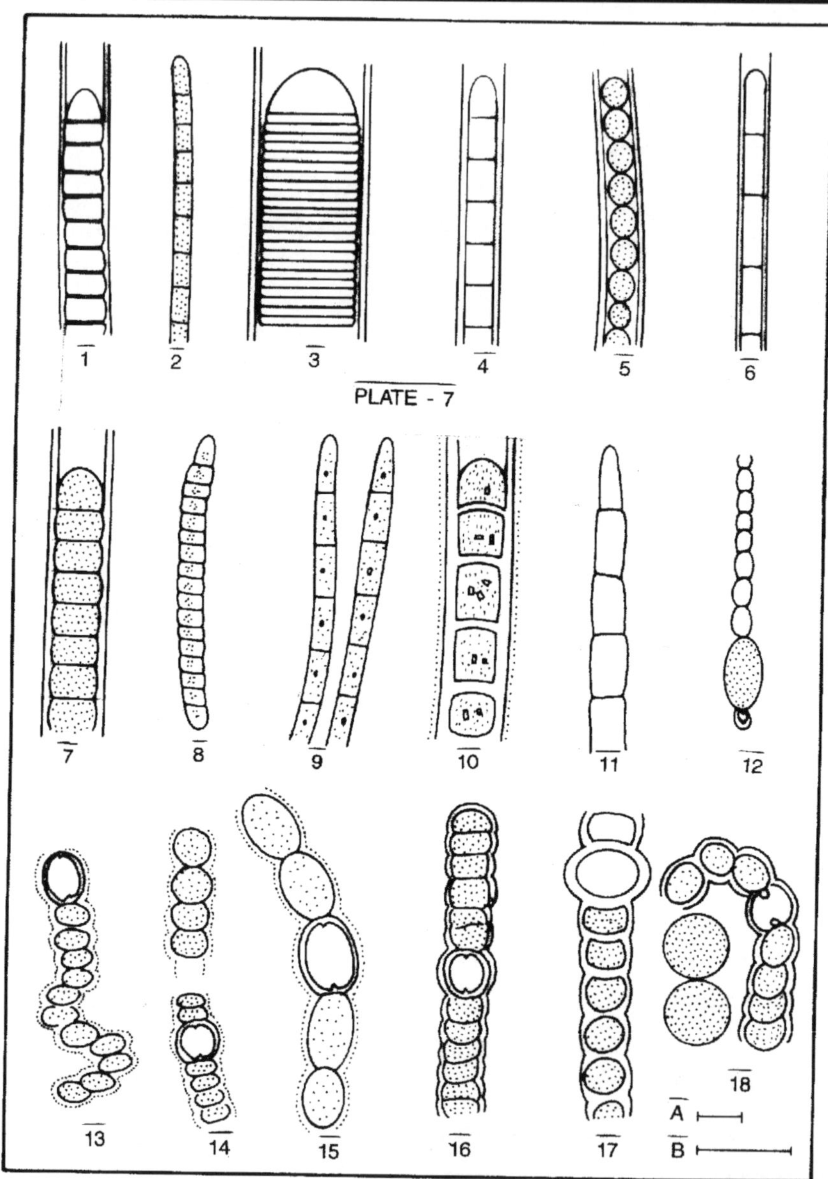

Plate 3.7: (1) *Lyngya laxespiralis* Skuja., (2) *L. limmnetica* Lemm., (3) *L. martensiana* Manegh, (4) *L. mesotricha* Skuja, (5) *L. nordgardhii* Wille, (6) *L. perelegans* Lemm., (7) *L. putealis* Mont., (8) *Schizothrix mulleri* Nag., (9) *Symploca cartilaginea* (Mont.) Gom., (10) *S. muscorum* (Ag.) Gom., (11) *Microcoleus lacustris* (Rabenh.) Farlow, (12) *Cylindrospermum indicum* Rao, C.B., (13) *Nostoc calcicola* Breb., (14) *N. coeruleum* Lyngb., (15) *N. ellipsosporum* (Desm.) Rabenh, (16) *N. linckia* (Rogh.) B. and F. *v. arvense* Rao, C.B., (17) *N. muscorum* Ag., (18) *N. piscinale* Kuetz.

Scale Bars A & B: 10µm Scale A: Fig. 1, 7, 8, 12, 14, 16, Scale B: Figs. 2, 3, 4, 5, 6, 9, 10, 11, 13, 15, 17, 18.

Nostoc piscinale **Kuetz. (Plate 3.7, F.18)**

Desikachary, 1959, P.377, Pl.69, F.3

Trichome 3-5μ broad, heterocysts 4.5-6μ broad spores globose, 6-7μ broad, in long chains.

Habitat: Hartala lake, H-I, September, 2000, H-III, September, 1999.

Velhala lake, V-I, January-February-March-May, 1999, February-March-May, 2000, V-II, March-April, 1999, June, 2000, V-III, September, 1999, March, 2000.

Nostoc pruniformae **(L.) Ag. (Plate 3.8, F.l)**

Desikachary, 1959, P.390

Trichome 4-6μ broad, cells short barrel shaped heterocysts 6-7μ broad spores spherical upto 10μ.

Habitat: Velhala lake, V-I, March-September, 1999, March-September, 2000.

Nostoc punctiformae **(Kuetz.) Hariot (Plate 3.8, F.2)**

Desikachary, 1959, P.374, Pl.69, F.1

Trichome 3-4μ broad, heterocysts 4.4μ broad spores 5-6μ broad and 5-8μ long.

Habitat: Hartala lake, H-III, February, 1999.

Velhala lake, V-I, June, 2000.

Nostoc sphaericum **Vaucher (Plate 3.8, F.3)**

Desikachary, 1959, P.390

Trichome 4-5μ broad, cells spherical, heretocysts 5-6μ broad, spherical, spores oval 5μ broad, 7μ long.

Habitat: Hartala lake, H-I, June, 2000, H-III, November, 1999.

Velhala lake, V-I, April, 1999, April, 2000.

Nostoc sphongiaeformae **Ag. *v.varians* Rao, C.B. (Plate 3.8, F.4)**

Desikachary, 1959, P.380, Pl.68, F.2

Trichomes 3-3.5μ broad, heterocysts trichomes, 4.4μ broad and 6.6μ long, spores in long chains 4.8-6μ diameter.

Habitat: Velhala lake, V-I, April, 2000, V-II, November-December, 1998, December, 1999.

Anabaena ambigua **Rao, C.B. (Plate 3.8, F.5)**

Desikachary, 1959, P.400, Pl.76, F.2

Cells 5.5μ broad, 4.4μ long, heterocysts spherical, 7.7μ in diameter, spores 8.3μ broad and 11.6μ long.

Habitat: Velhala lake, V-I, July, 1999, June, 2000.

Anabaena fertilisimma **Rao, C.B. (Plate 3.8, F.6)**

Desikachary, 1959, P.398, Pl.74, F.1

Cells barrel shaped 5.5μ broad, 4.4μ long, heterocysts 6.6μ broad, spores in long chains 6.6μ in diameter.

Habitat: Hartala lake, H-III, December, 1998, August, 2000.

Velhala lake, V-III, October, 1999.

***Anabaena khannae* Skuja (Plate 3.8, F.7)**

Desikachary, 1959, P.396, Pl.75, F.2

Trichomes 2.5-4μ broad, heterocysts , 3-3.5μ broad 3.5-6μ long, spores, globose 6.6μ in diameter.

Habitat: Velhala lake, V-I, March, 2000.

***Anabaena spinosa* Tiwari (Plate 3.8, F.8)**

Trichome 3.8μ broad, cells 6.6μ long, heterocysts 5μ broad, 8.8μ long, surrounded by spiny papillae, spores 3.8μ broad, 6.6μ long.

Habitat: Velhala lake, V-I, June, 2000.

***Raphidiopsis indica* Singh, R.N. (Plate 3.8, F.9)**

Desikachary, 1959, P.422, Pl.79, F.1,5

Cells 2.7μ broad, cells 6-8μ times as long as broad, with gas vacuoles.

Habitat: Hartala lake, H-I, January, 2000.

Velhala lake, October, 1998, June, 1999, V-III, July, 1999.

***Raphidiopsis curvata* Fritsch and Rich (Plate 3.8, F.10)**

Desikachary, 1959, P.422, Pl.65, F.5,7

Cells 5μ broad, 4-5 times as long as broad, spores 10-19.5μ long, 5μ broad, in the middle of the trichome.

Habitat: Velhala lake, V-II, July, 1999, V-III, February, 1999, May, 1999, July, 2000.

Family - Scytonemataceae

***Plectonema notatum* Schmidle (Plate 3.8, F.11)**

Desikachary, 1959, P.440, Pl.83, F.5

Filaments 2.2μ broad, cells cylindrical, 2-3 times as long as broad.

Habitat: Velhala lake, V-II, May, 1999.

***Scytonema coactile* Mont. (Plate 3.9, F.10)**

Desikachary, 1959, P.455, Pl.90, F.2

Filament 18-24μ broad, trichome 12-18μ broad, cells subquadrate, heterocysts sparse.

Habitat: Velhala lake, V-I, September, 2000.

Family: Rivulariaceae

***Homoethrix balaerica* (Born. et. Flab.) Lamm. (Plate 3.8, F.12)**

Gonzalves and Kamat, 1960, P.4, Plate 3.4, F.36 a,b.

Filaments 8.3μ broad and up to 150μ long, trichome 3.6μ broad cells 3.5μ broad and 1.5μ long.

Habitat: Hartala lake, H-III, Sept. 1999.

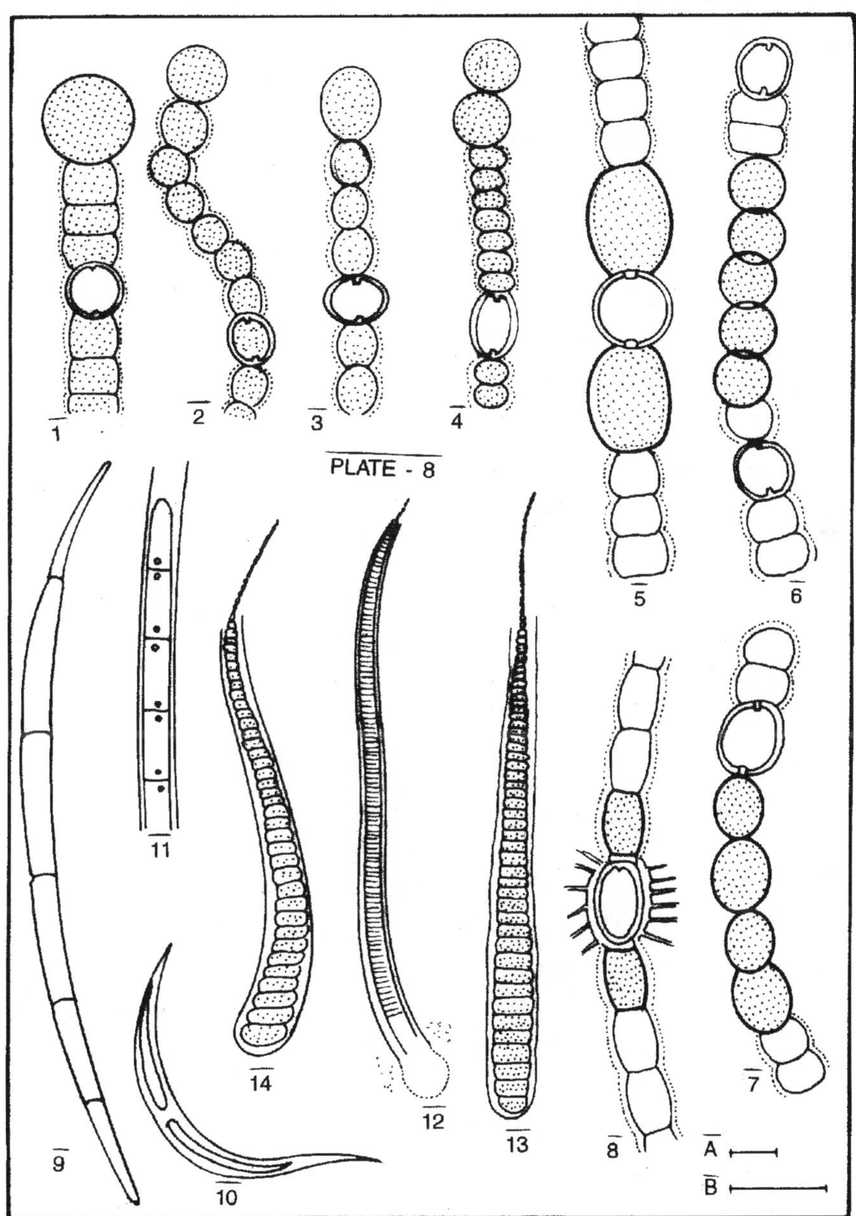

Plate 3.8: (1) *Nostoc pruniformae* (L.) Ag., (2) *N. punctiformae* (Kuetz.) Hariot, (3) *N. sphaericum* Vaucher, (4) *N. sphongiaeformae* Ag. *v. varians* Rao, C.B., (5) *Anabaena ambigua* Rao, C.B., (6) *A. fertilisimma* Rao, C.B., (7) *A. khannae* Skuja, (8) *A. spinosa* Tiwari., (9) *Raphidiopsis indica* Singh, R.N., (10) *R. curvata* Fritsch and Rich, (11) *Plectonema notatum* Schmidle, (12) *Homoethrix balearica* (Born. et Flab.) Lamm., (13) *H. hansgirigi* (Schmidle) Lemm. *v. constricta* v. nov., (14) *H. juliana* (Manegh.) Kirchn.

Scale Bars A & B: 10μm Scale A: Fig. 10, 12, 14, Scale B: Figs. 1, 2, 3, 4, 5, 6, 7, 8, 9, 11, 13.

Homoeothrix hansgirigi **(Schmidle) Lemm.** ***v.constricta*** **v.nov. (Plate 3.8, F.13)**

Filaments 4μ broad, 66.6μ long, cross walls constricted.

Habitat: Hartala lake, H-III, November, 1999.

Homoeothrix juliana **(Manegh.) Kirchn. (Plate 3.8, F.14)**

Desikachary, 1959, P.519, Pl.107, F.7

Filaments 10-15μ broad, upto 2 mm long, trichomes 9.5-12.5μ broad, cells discoid, upto 1/3 - 1/1 as long as broad, 4.7-6μ long.

Habitat: Hartala lake, H-III, November, 1999.

Velhala lake, V-I, November, 1998, November, 1999.

Calothrix brevissima **West, G.S. (Plate 3.9, F.1)**

Desikachary, 1959, P .533, Pl.114, F.1

Filaments epiphytic 65.6μ long, 7.1μ broad, trichomes 58.3μ long, heterocysts basal, single rounded hemispherical.

Habitat: Hartala lake, H-III, July, 2000.

Calothrix brevissima **West, G.S.** ***v.chakradharensis*** **v. nov. (Plate 3.9, F.2)**

Filaments epiphytic, 39.4μ long, 6.6μ broad, trichome, 37.7μ long, Heterocysts basal, single, rounded, hemispherical.

Habitat: Hartala lake, H-III, August, 2000.

Calothrix castellii **(Massal.) B and F.** ***v.somastipurense*** **Rao, C.S. (Plate 3.9, F.3)**

Desikachary, 1959, P.529, Pl.114, F.7

Filaments 10-20μ broad, upto 232μ long cells 8.3μ long at the base, hair about 12μ long, heterocysts basal, single, hemispherical, 10.7μ broad and 5.9μ long.

Habitat: Velhala lake, V-I, April, 2000, V-II, November, 1998, November, 1999, September, 2000.

Calothrix clavata **West, G.S. (Plate 3.9, F.4)**

Desikachary, 1959, P.542, Pl.114, F.2

Filaments 60μ long, straight, prominently swollen at the base, 6.6μ broad, cells discoid at the base, heterocysts basal, single hemispherical.

Habitat: Velhala lake, V-II, September, 1999.

Calothrix javanic **De Wilde (Plate 3.9, F.5)**

Desikachary, 1959, P.525, Pl.106, F.1,2,5-7

Trichomes 3.8μ broad, ending in a thin hair, heterocysts basal 5.5μ broad.

Habitat: Velhala lake, V-II, June, 2000.

Calothrix marchica **Lamm.** ***v.intermedia*** **Rao, C.B. (Plate 3.9, F.6)**

Desikachary, 1959, P .543, Pl.114, F.4

Flaments epiphytic 8.3μ broad and upto 70μ long, trichomes 7.4μ broad, heterocysts single, basal, 7.1μ broad.

Habitat: Hartala lake, H-III, July, 2000.

Velhala lake, V-I, July, September, October, December, 1999, June, 2000.

***Calothrix membranicea* Schmidle (Plate 3.9, F.7)**

Desikachary, 1959, P.542, Pl.106, F.10

Filaments 9.5μ broad at the base and 55μ long, trichome 8.3μ broad, cells 7.1μ broad, 9.5μ long, heterocyst single, basal, spherical, 9.5μ broad.

Habitat: Velhala lake, August, 2000.

***Rivularia aquatica* De Wilde (Plate 3.9, F.8)**

Desikachary, 1959, P.552

Trichome 9.5μ broad, ending in a long thin hair, heterocysts basal, solitary, spherical, 13μ in diameter.

Habitat: Velhala lake, V-I, June-September, 2000, V-II, December, 1998, January, 1999, V-III, December, 1998, January, 1999.

***Rivularia beccariana* (De. Wilde) Geitler (Plate 3.9, F.9)**

Desikachary, 1959, P.551, Pl.106, F.8,9

Trichome 8.3μ broad, cells at the base longer than broad, 13μ long, heterocyst, basal, solitary, spherical, 9.5μ in diameter.

Habitat: Velhala lake, V-I, June, 2000.

***Gloeotrichia indica* Schmidle (Plate 3.10, F.1)**

Desikachary, 1959, P.558, Pl.117, F.1,7,8

Filaments radiating 200-300μ long, trichomes made of 3 cells, spores 60-70μ long.

Habitat: Velhala lake, V-II, November, 1998, January, 1999.

***Gloeotrichia pilgeri* Schmidle (Pl.3.10, F.2)**

Desikachary, 1959, P.558, Pl.116, F.6,7

Filaments 150-190μ long, base with 2-3 heterocysts, trichome 8.3μ broad, spores 12-16μ broad and 29-50μ. long.

Habitat: Velhala lake, V-II, September, 2000, V-III, August, 2000.

***Gloeothrichia raciborskii* Wolosz (Pl.3.10, F.3)**

Desikachary, 1959, P.562, Pl.118, F.14

Trichome 7-8μ broad, ending in a long hair, with hair upto 800μ long, heterocysts spherical 5.9μ broad, spores 15-25μ broad, upto 70μ long.

Habitat: Velhala lake, V-III, December, 1999.

***Gloeothrichia raciborskii* Wolosz *v.conica* Dixit (Pl.3.10, F.4)**

Desikachary, 1959, P.562, Pl.117, F.10-12

Filaments upto 370μ long, trichomes 9.5μ broad, heterocysts upto 10μ. broad, spores 8.5-11.5μ broad and upto 50μ long.

Habitat: Velhala lake, V-II, August, 2000, V-III, December, 1999.

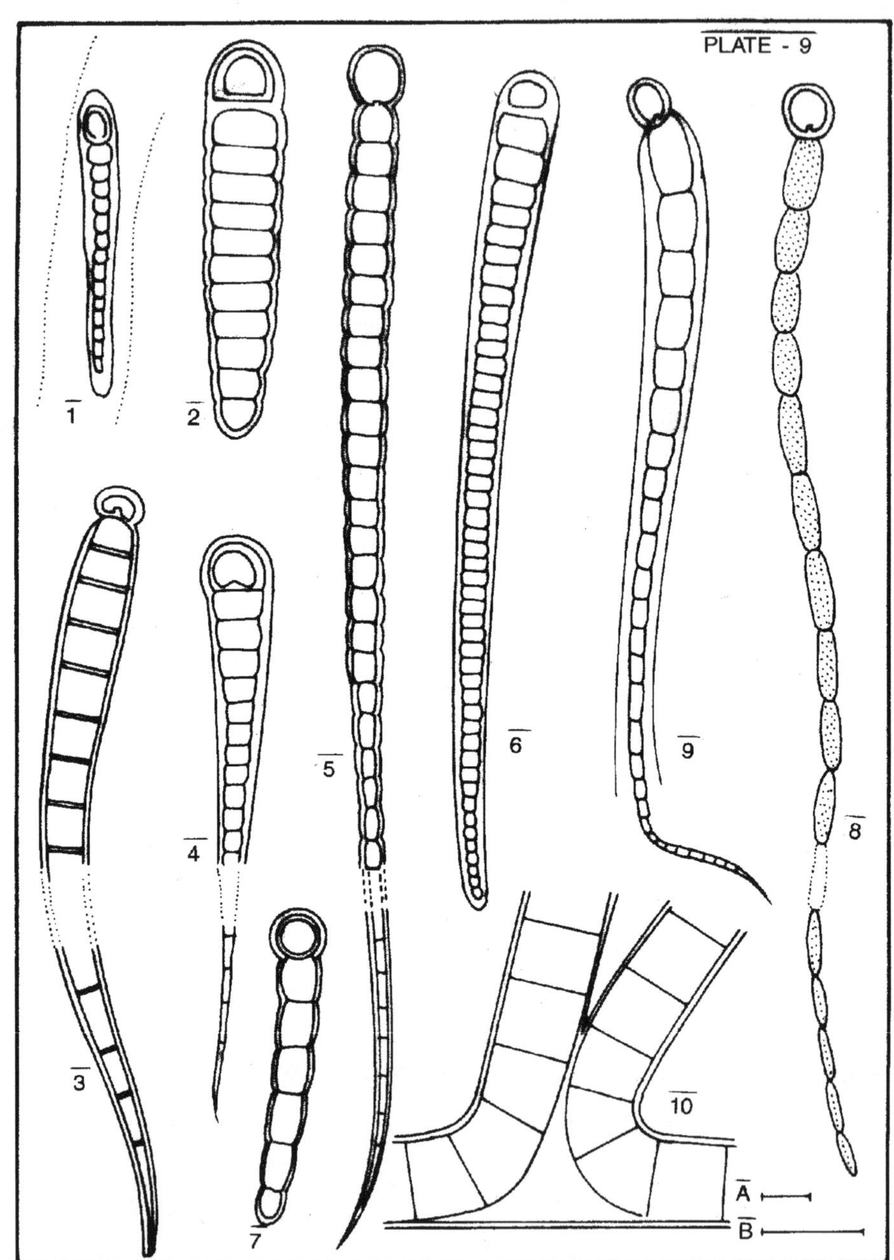

Plate 3.9: (1) *Calothrix brevissima* West G.S., (2) *C. brevissima* West, G.S. *v. chakradharensis* v. nov., (3) *C. castellii* (Massal.) B and F., (4) *C. clavata* West, G.S., (5) *C. javanic* De Wilde, (6) *C. marchica* Lamm., (7) *C. membranicea* Schmidle, (8) *Rivularia aquatica* de Wilde, (9) *Rivularia beccariana* (De. wilded) Geitler, (10) *Schytonema coactile* Mont. Scale Bars A & B: 10µm Scale A: Fig. 1, 3, 6, 7, 8, 9, 10, Scale B: Figs. 2, 4, 5.

Gloeotrichia raciborskii **Wolosz v.kashiense Rao, C.S. (Plate 3.10, F.5)**

Desikachary, 1959, P .562, Pl.117, F.2-6

Trichomes with constrictions at the joints, 8.4-10μ broad at the base, at the apex upto 3.5μ broad, heterocysts single, ellipsoidal, 11μ broad and 16.6μ long, spores 13.8μ broad, 34.4μ long.

Habitat: Velhala lake, V-II, September, 2000.

Gloeothrichia raciborskii **Wolosz v *.longispora* Rao, C.B. (Pl.10, F.6)**

Desikachary, 1959, P.562, Pl.118, F.4-6

Trichomes constricted at the joints at the base 9.5μ broad, at the apex 2.3μ broad, cells at the base barrel shaped and quadratic, 8-12μ long, heterocyst single, spherical 12μ broad, spores 10-12μ broad, 83μ long.

Habitat: Velhala lake, V-II, August, 2000, September, 2000.

Family: Mastigocladaceae

Mastigocladus laminosus **Cohn. *v.indicus* Desik. (Pl.3.10, F.7)**

Desikachary, 1959, P.581, Pl.126, F.1-4.

Filaments interwoven, 6.6-9.2μ broad, cells barrel shaped, heterocysts 10.7μ broad, single.

Habitat: Hartala lake, H-I, December, 1999.

Family: Stigonemataceae

Stigonema hormoides **(Kuetz.) Bom. et Flab. (Pl.3.10, F.8)**

Desikachary, 1959, P.664, Pl.134, F.1-4

Filaments 7-15μ broad, irregularly branched, cells sub-spherical heterocysts sparse intercalary.

Habitat: Hartala lake, H-I, November, 1998.

Observations and Discussion

Cyanobacterial showed maximum population during summer and a minimum during monsoon in both the lakes. Maximum blue-green algal population was observed during rainy and minimum during summer in Suraha lake. (Singh and Swarup, 1979) While Zutshi *et al.* (1984) observed their highest population during January. In the present study, comparatively a higher percentage of blue-greens was recorded at H-I station of Hartala lake. This was supported the fact that sewage contaminated water harbours more blue-green algae has also observed by Parmasivam and Sreenivasan (1981).

Blue-greens was represented by 153 species belonging to 35 genera of which 102 species occurred in Hartala lake and 87 in Velhala lake. *Oscillatoria princeps* and *Lyngbya allorgei, L. martensiana* showed a regular occurrence throughout the study period in the sewage polluted Hartala lake. This tallied with the observations made by Pandey and Tripathi (1988).

In India, *Microcystis aeruginosa* has been reported as a permanent bloom found commonly throughout the country. Singh (1953) studied the factors affecting glooms in some inland waters of India and the best single indicator of pollution, as indicated by him, is *M. aeruginosa.* During the present study *M. aeruginosa* was found commonly growing in the lake particularly during late summer in association of other algae and develop into a bloom.

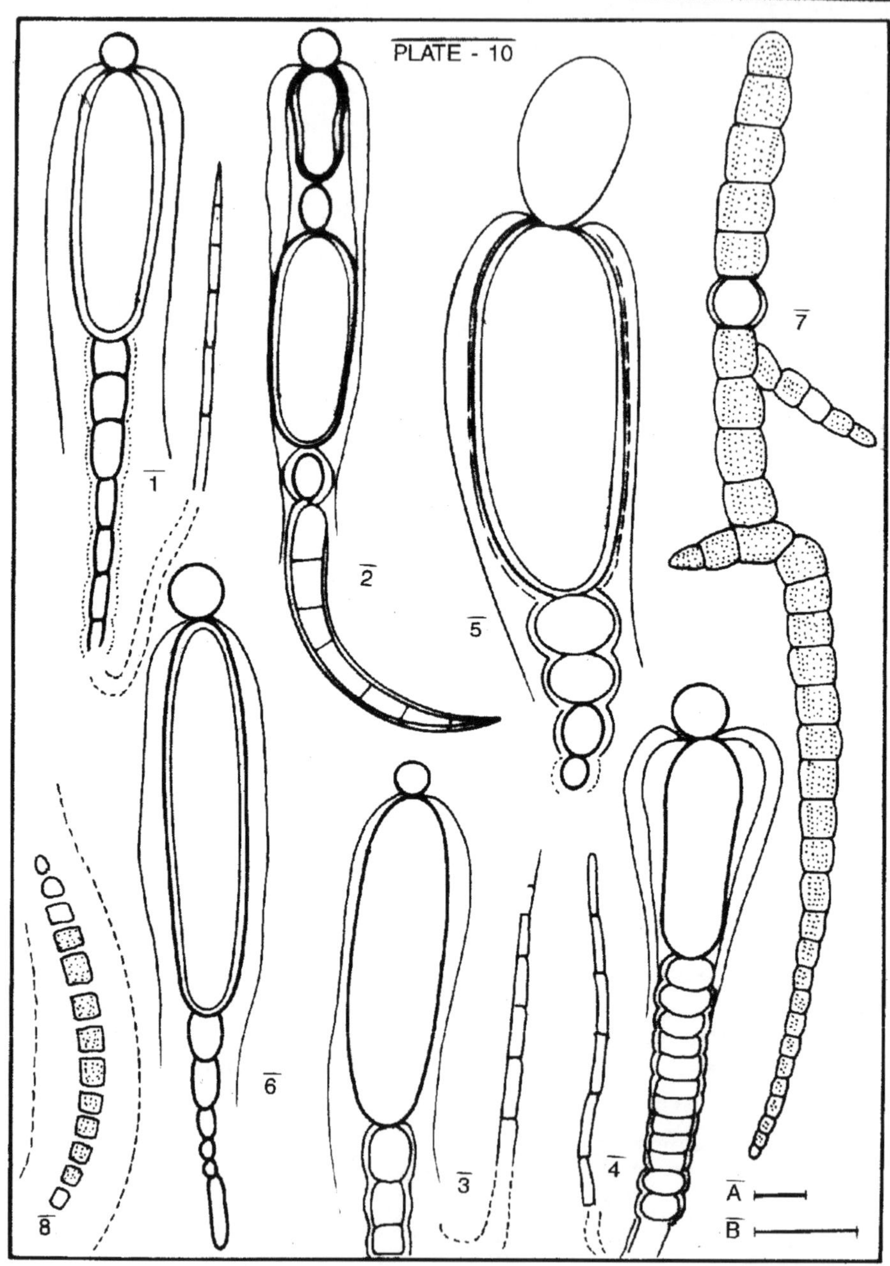

Plate 3.10: (1) *Gloeotrichia indica* Schmidle, (2) *G. pilgeri* Schmidle, (3) *G. raciborskii* Wolosz., (4) *G. raciborskii* Wolosz. *v. conica* Dixit, (5) *G. raciborskii* Wolosz. *v. kashiense* Rao, C.S., (6) *G. raciborskii* Wolosz. *v. longispora* Rao, (7) *Mastigocladus laminosus* Cohn. *v. indicus* Desikachary, (8) *Stigonem hormoides* (Kuetz.) B and F.

Scale Bars A & B: 10μm Scale A: Figs. 1, 2, 3, 4, 5, 6, 7, 8, Scale B: Fig. 5.

Rai and Kumar (1976) observed non-heterocystous forms in the polluted waters rich in nitrogen. Present investigation also showed dominance of Cyanobacterial in these lakes with 38 heterocystous forms indicating eutrophication of the lakes.

Genus *Oscillatoria* has been found to be very tolerant to pollution which frequently inhabits the polluted waters (Rai and Kumar, 1976). Present study confirmed their observations as *Oscillatoria* was found dominating in both the lakes represented by 30 species followed by Phormidium with 20 species. In the present study the dominance of *Oscillatoria* was indicating pollutants of biological origin which agreed with the observations of Rai and Kumar (1976).

Palmer (1980) listed 14 species of Cyanobacterial as pollution tolerant algae which are common in organically enriched water bodies. During the present study 7 species from his list namely *Oscillatoria limosa, O. ChaJybea, O. brevis, O. formosa, O. princeps* and *O. tenuis, Phomidium autumale* have been observed in the lakes.

Following 12 species of Cyanobacterial which showed a wide range of adaptability can be regarded as "pollution tolerant species": *Chlorococcus minutus, Johannesbaptistia pellucida, Myxosarcina burmensis, Lyngbya allorgei, L.martensiana, Oscillatoria limosa, O. princeps, O. tenuis, Phormidium autumale, P. retzii, Stichosiphon sansibaricus, Synechocystis aquatilis.*

Acknowledgement

Authors are thankful to Mr. B.M. Patil, The Principal of S.S.V.P.S's Late Dr. P.R.Ghoghrey Science College, Dhule for providing facilities. We are also thankful to Dr. V.C. Kolhe, The Principal of Bhusawal Arts, Science and P.O. Nahata Commerce College, for encouragement during this work.

References

Desikachary, T.V. (1959). *Cyanophyta* ICAR Monographs on Algae. New Delhi, p. 686.

Dixit, S.C. (1936). The Myxophyceae of the Bombay Presidency, India. I. *Proc. Indian Acad. Sci.* 3: 93-106.

Gonzalves, E.A. and Kamat, N.D. (1960). The Myxophyceae of Karnataka III *J. Bombay Univ:* 28: 28-41.

Kamat, N.D. (1963). The Algae of Kolphapur, India. *Hydrobiologia.* 22 (3-4): 209-305.

Mahajan, A.D. (1983). *Study of algal flora of Khaira district Gujarat.* Ph.D. thesis, Sardar Patel Univ., Anand.

Nandan, S.N. and Patel, R.J. (1985a). Species diversities of algal flora in Vishwamitri, river Baroda *Pol. Arch. Hydrobiology.* Poland 32 (1): 1-6.

Palmer, C.M. (1980). *Algae and water pollution.* Castle House Publications Ltd., England. p.123.

Pandey, S.N. and Tripathi, A.K. (1988). Studies on algae of polluted ponds of Kanpur (India) VIII. Role of blue-green algae, *Phykos:* 27: 38-43.

Parmasivam, M. and Sreenivasan, A. (1981). Changes in algal flora due to pollution in Cauvery river, *Indian J. Environ.* Hlth. 23(3): 222-238.

Rai, L.C. and Kumar, H.D. (1976). Algal growth as a means of evaluation of nutrient status of the effluent of a fertilizer factory near Shahpuri, Varanasi, *Trop. Ecol.* 17(1): 50-56.

Rao, C.B. (1937). The Myxophycea of the Madras Presidency India. *Journal Indian Botanical Society.* 17: 81-96.

Singh, R.N. (1953). Limnological relations of Indian inland waters with special references to water blooms, *Verh. Int. Var. Rher. Anew. Limnol.* 12: 831-836.

Singh, S.R. and Swarup, K. (1979). Limnological studies of Suraha lake (Ballia). II The periodicity of phytoplankton, *J. Indian Bot. Soc.* 58: 319-329.

Zutshi, D.P., Vishin, N. and Subla, B.A. (1984). Nutrient status and plankton dynamics of a perennial pond, *Proc. Indian Natl. Sci. Acad.* 50: 577-581.

4

Studies on Soil Fertility, Productivity and Ecosystem as Influenced by Introduction of Fish and Prawn in Rice-fish Integration System

Rajeeb K. Mohanty

Scientist, Water Technology Centre for Eastern Region (ICAR), Bhubaneswar -751 023

ABSTRACT

This experiment was aimed at to study soil fertility, productivity and ecosystem as influenced by introduction of fish and prawn in rice field. In soil, percentage increase in available N, P and organic carbon in rice-fish culture field was 58.3 per cent, 94 per cent and 54.5 per cent against control respectively while, the unit weight of top soil was 2.3 per cent lighter than the control and porosity was higher by 1.2 per cent. The benthic biomass, indicator of soil productivity, was higher in rice-fish culture field (0.44 ± 0.04 gm^{-2}) than control (0.21 ± 0.05 gm^{-2}). Faster growth rate of *Catla catla* and bottom dwellers were attributed to effective utilization of ecological niches and rich detrital food web. Highest average grain yield was recorded in rice-fish culture field (3595 kg ha^{-1}) which was 20.3 per cent higher against control (2988 kg ha^{-1}). To ensure better productivity, higher planting density must be avoided to provide greater ventilation and illumination, that helps in enhancing both plant and plankton growth which ultimately helps fish growth and yield.

Key Words: *Rice-fish integration, Ecosystem, Water quality, Soil fertility, Productivity.*

Introduction

Being a traditionally old practice, rice-fish culture is very much popular in north-eastern parts of India and south-east Asian countries. In India, about 42 million ha of land is under rice cultivation, out of which 20 million is suitable for fish integration (Rao and Ram Singh, 1998). Unfortunately, the carrying capacity of these suitable lands in India have not been utilized to the fullest extent. However, if these lands are brought under integrated rice-fish system, would help to compensate the economic losses in rice production brought about by natural calamities. This will also optimize the water and land use without bringing about environmental degradation. As paddy fields are complex ecosystems where primary producers and consumers at different levels compete with rice for material and energy, decrease overall productivity. However, fish culture in paddy fields can turn and recycle the available material and energy in to fish production, accelerate the productivity of paddy fields and its yield for the benefit of human beings (Brahmanand and Mohanty, 1999). Although much works have been carried out on different aspects of rice-fish integration system (Likangmin, 1988, Huazhu, 1994 and Rao and Ram Singh, 1998), very little information are available on the mechanism of rice-fish culture and the relationship among rice, fish and other factors such as soil, water, fertilization etc. Keeping these in view, an attempt has been made to evaluate rice-fish integration system with special reference to soil fertility, productivity and ecosystem as influenced by introduction of fish and prawn in rice field.

Material and Methods

The present study was carried out at research farm of WTCER, Bhubaneswar, India (Lat. 20°30′ N and Long. 87°48′10″ E) during 2000-2001 for two successive years. Three plots of 30 × 10 m size were selected for the proposed study with refuge (approximately 10.2 per cent of field area) of 1.75 m deep and peripheral trench of 0.5 m wide and 0.3 m deep. Dyke height of rice fields were maintained at 20 cm. The excess water from the refuge was drained out through a surface drainage system provided with fine-meshed net to prevent escape of fish. Three plots of similar size and dyke height were kept as control with out refuge. No fish and prawn was introduced in control fields. *Swarna* variety of rice was tried in each plot of the present study including control at a spacing of 20 × 10 cm (between rows and plants) during second week of July each year. The seed and fertilizer application rate was 50 kg ha^{-1} and 80:40:40 (N:P:K) ha^{-1} respectively. 50 per cent of N and full dose of P and K were given as basal at the time of transplanting. The rest of Nitrogen was given in two equal splits during tillering (20 days after transplanting) and panicle initiation stage (45 days after transplanting). Crop growth and yield parameters were recorded at regular intervals. However, no pesticide was used during the experiment.

After proper refuge preparation, liming @ 2000 kg ha^{-1}, manuring with raw cattle dung @ 5000 kg ha^{-1} and fertilization (Urea + SSP, 1:1) @ 3 ppm was carried out prior to stocking. Seven days after refuge preparation, fry of *Catla catla, Labeo rohita, Cirrhinus mrigala, Cyprinus carpio* and post larvae (PL_{30}) of freshwater prawn *Macrobachium rosenbergii* were stocked in the refuge with a species composition of 30:30:15:15:10 respectively. Stocking density of fish fry and post larvae of prawn was 25000 ha^{-1} and rearing was continued for 120 days. Supplemental feed (rice bran + groundnut oil cake, 1:1) @ 10 per cent, 8 per cent, 6 per cent and 4 per cent of mean body weight (MBW) was given twice a day, during 1st, 2nd, 3rd and 4th month to harvesting respectively. Periodic manuring @ 500 kg ha^{-1} and liming @ 200 kg ha^{-1} were carried out at every 15 days interval to maintain plankton population in the ecosystem. Plankton estimation, weekly observation on soil and water quality were recorded using standard methods (APHA, 1989 and Biswas, 1993). Field test instruments were also in use to analyze *in situ* water pH (Checker-1, HANNA, USA), Soil pH (DM-13, Japan), and dissolved oxygen (YSI-55, USA). Crop performance, fish/prawn growth parameters, condition factor of fish and prawn, feed conversion ratio were estimated using standard methods (Mohanty, 1999). To estimate the food preference and feed intake pattern of cultured species, gut content analysis, indices of electivity of different food components (Ivlev, 1961), feed abundance/frequency (Mohanty *el al.*, 2001) were carried out. Every year, during the experiments, 12 numbers of each species were sacrificed for this purpose.

Results and Discussion

In this experiment, various hydro-biological parameters did not show any distinct trend between the treatment and control except in the cases of dissolved oxygen, total alkalinity, total suspended solid and total plankton count (2.4×10^2 - 9.1×10^3). Most of the parameters were within/ nearly optimum ranges through out the culture period (Table 4.1). The dissolved oxygen content in rice-fish culture field showed a decreasing trend with the advancement of rearing period, attributed to gradual increase in biomass, resulting in higher oxygen consumption (Mohanty, 1995). Concentration of ammonia also showed an increasing trend as the days of culture increased, probably due to split application of nitrogen to paddy, higher metabolic deposition and organic load (Mohanty, 1999). Many times water in the rice fields were slightly acidic probably due to decomposition of weed and rice stalks. The available nutrients observed (Table 4.1) did not result in higher phytoplankton production and moreover it was also affected by shading and competition for nutrient with rice and weed. Density of zooplankton was however, not high (Fig. 4.1) in comparison to composite fish culture ponds. In general, upper limit of planting density is usually practiced for rice. However, this practice must be avoided in rice-fish integration system as it would result in stunted plant growth, poor ventilation and illumination that adversely affect plankton population, a major source of natural food for fish. In contrast, at relatively lower density, plants grow fast, strong and become more resistant to lodging. In soil, percentage increase in available N, P and organic carbon in rice-fish culture field was 58.3 per cent, 94 per cent and 54.5 per cent against control respectively (Table 4.2). This was probably due to leftover supplemental feed, faecal matter of cultured species and stirring movement of fish/ prawn that aerate the soil (Li Kangmin, 1988) and help in transforming insoluble nutrients in soil in to a soluble state. In rice-fish culture fields, the unit weight of top soil was 2.3 per cent lighter than the control and porosity (vol. of pore space/vol. of soil × 100) was higher by 1.2 per cent. Although, no pesticide was used, there was no substantial decrease in yield probably due to the fact that fish and prawn act as a biological controller of insect/pest and pathogen (Spiller, 1985). The benthic population, indicators of soil productivity, in both rice-fish culture field and control was consisted mainly of *Oligochaetes* and *Chironomids,* while their biomass was higher in rice-fish culture field (0.44 ± 0.04 gm^{-2}) than control (0.21 ± 0.05 gm^{-2}). This was probably due to periodic manuring, faecal matter of cultured species and increased organic load. (Table 4.1, 4.2)

Table 4.1: Minimum and Maximum Average Range of Water Quality Parameters

Parameter	*Rice-fish Culture Plot*		*Control Plot*	
	Min	*Max*	*Min*	*Max*
pH	6.9	8.3	6.7	7.6
Dissolved oxygen (ppm)	3.3	7.6	2.1	4.6
Temp. (°C)	27.5	30.1	27.6	30.2
Total alkalinity (ppm)	56	109	41	87
Nitrite-N (ppm)	0.006	0.071	0.006	0.058
Nitrate-N (ppm)	0.06	0.52	0.06	0.48
Amonnia (ppm)	0.01	0.41	0.02	0.39
TSS (ppm)	116	319	105	211
Phosphate-P (ppm)	0.07	0.33	0.07	0.3

Table 4.2: Changes in Soil Quality Due to Introduction of Fish and Prawn in Rice Field

Parameters	*Rice-fish culture plot*		*Control plot*	
	1st week	*15th week*	*1st week*	*15th week*
Soil pH	6.8	7.0	6.7	6.8
Available - N in soil (mg $100g^{-1}$)	8.8	10.7	8.4	9.6
Available - P in soil (mg $100g^{-1}$)	0.32	0.67	0.36	0.54
Organic carbon in soil (per cent)	0.19	0.53	0.17	0.39

Phytoplankton and zooplankton was most preferred food item for C. *catla* and *L. rohita,* while mud and detritus was highly preferred by *C. mrigala, C. carpio* and *M. rosenbergii* in rice fish integration system (Table 4.3). However, quantity-wise most consumed food items were artificial supplemental feed of rice bran and groundnut oil cake. Among bottom dwellers *(C. mrigala, C. carpio* and *M. rosenbergii),* phytoplankton and benthos were preferred more by *M. rosenbergii* while zooplankton and detritus by C. *Carpio* and *C. mrigala* respectively. Omnivorous feeding behaviour was observed in case of each species except *Catla catla.* Positive indices of electivity (0.02-0.31) were observed for phytoplankton in monsoon-winter, while it was negative for zooplankton during the same period. Negative indices of electivity for zooplankton (-0.14 to -0.46) in case of all species was recorded during monsoon-winter (August-November) and improved thereafter. This was probably due to rich detrital food web in the initial phase of rearing where raw cattle dung was applied @ 5000 kg ha^{-1} for refuge preparation prior to stocking. (Table 4.3)

Table 4.3: Average Percentage of Analyzed Fish and Prawn in which mentioned Food Component was found (Frequency) and Percentage of Individual Gut Content Volume (Abundance)

Food Component	*Frequency (per cent)*					*Abundance (per cent)*				
	1	*2*	*3*	*4*	*5*	*1*	*2*	*3*	*4*	*5*
Phytoplankton	94.4	83.3	55.6	66.6	72.2	<11.2	<5.1	<2.3	<2.7	<4.3
Zooplankton	88.8	83.3	44.4	72.2	44.4	<5.9	<4.3	<1.4	<1.9	<1.6
Mud+Detritus	11.1	22.2	94.4	88.9	77.8	<5.6	<15.4	>29.1	>32.1	<21.0
Benthos/insect	-	5.5	44.5	55.6	61.1	-	<1.0	<12.2	<12.2	<16.4
Artificial feed	72.2	77.8	83.3	88.8	77.8	>56.7	>49.3	>45.8	>46.1	>61.7

1-*Catla catla,* 2-*Labeo rohita,* 3-*Cirrhinus mrigala,* 4-*Cyprinus calpio,* 5-*Macrobrachium rosenbergii*

Faster growth performance under rice-fish culture system was recorded for *Catla catla* followed by *C. carpio, C. mrigala, L. rohita* and *M. rosenbergii* during 120 days of culture at stocking density of 25,000 ha^{-1} (Table 4.4). *C. carpio* and *C. mrigala* performed better growth rate against that of *L. rohita* probably due to the fact that being surface and column dweller, *L. rohita* is more sensitive to oxygen depletion, while being bottom dwellers *C. carpio* and *C. mrigala* are more tolerant to fluctuation of oxygen concentration (Vijayan and Verghese, 1986). Moreover, faster growth rate of *Catla catla* and bottom dwellers were attributed to effective utilization of ecological niches and rich detrital food web that was maintained through periodic manuring, liming and initial fertilization, which agrees to the findings of Mohanty (1995). Observations on feed conversion ratio (FCR) also supports the conclusion of effective utilization of ecological niches, as minimum and maximum FCR ranged between 0.94-1.13. Comparative daily growth performance (Table 4.4) was moderate to good as periodic organic manuring

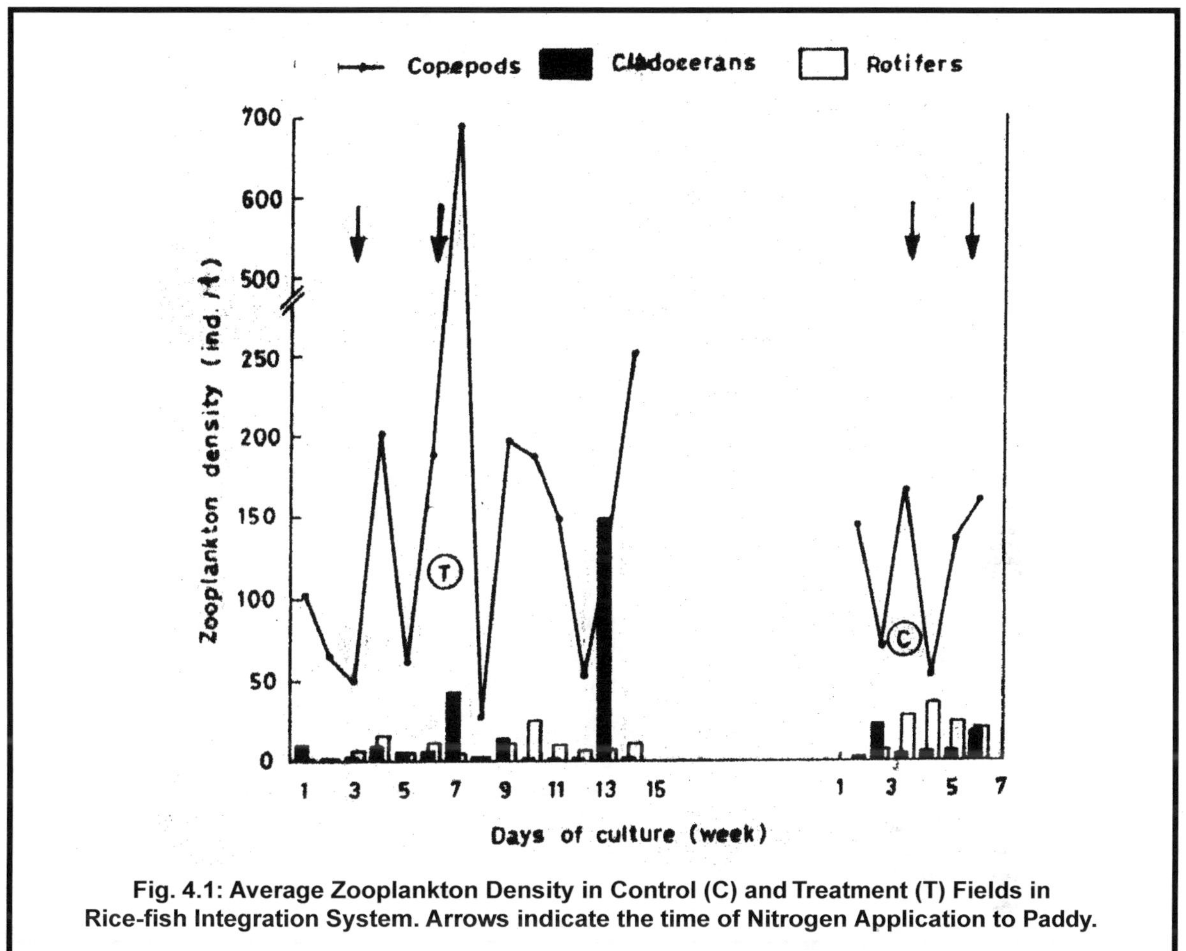

Fig. 4.1: Average Zooplankton Density in Control (C) and Treatment (T) Fields in Rice-fish Integration System. Arrows indicate the time of Nitrogen Application to Paddy.

and liming was a regular practice instead of inorganic fertilization, that improved primary productivity (508.3 ± 71.8 mg Cm^{-2} h^{-1}), which enhanced the growth rate. Similar observation was also made by Noriega-Curtis (1979). Encouraging growth rate of fish/prawn was also attributed by avoidance of periodic inorganic fertilization, that depress growth due to high ammonia concentration in water (Colt and Tehobonoglous, 1978). Condition factor (Ponderal index) of fish and prawn was less than 1.0 at the initial three weeks of rearing and improved thereafter with gradual improvement in water quality. Roy *et al.* (1990) reported that, in traditional deepwater rice-fish system in India, yield of rice in a season range between 1.0-1.5 t ha^{-1} and fish production between 50-200 kg ha^{-1}, while application of cow-dung enhance the productivity of rice and fish to 3.1 and 0.67 t ha^{-1} respectively. However, in the present study, an average productivity of 1156.5 kg ha^{-1} of fish and prawn has been achieved with application of cow-dung @ 5000 kg ha^{-1}, which was much higher than the earlier recorded productivity in a season. This high productivity of fish/prawn under rice-fish integration system was however, low in comparison to composite fish culture, probably due to shorter rearing duration and comparably less favourable environment as that in ponds (Rao and Ram Singh, 1998) (Table 4.4, 4.5).

Table 4.4: Growth Performance in Rearing of Fry to Advanced Fingerlings of Fish and Prawn in Rice-fish Integration System at Stocking Density of 25000 ha $^{-1}$

Species Stocked	*Initial MBW (g)*	*Final MBW (g)*	*ADG (g)*	*Min.-max. (K_n)*	*SR (%)*	*Productivity (Kg ha^{-1} 120 DOC^{-1})*
C.catla	0.82 ± 0.03	98.2 ± 6.1	0.81	0.93-1.22	57.6	
L.rohita	0.88 ± 0.03	76.2 ± 4.9	0.63	0.84-1.17	61.5	
C.mrigala	0.92 ± 0.06	87.7 ± 6.3	0.72	0.97-1.27	61.5	1156.5
C.carpio	0.88 ± 0.09	90.2 ± 6.8	0.74	0.93-1.2	46.1	
M. rosenbergii	0.04 ± 0.01	22.8 ± 0.8	0.19	0.99-1.33	50.0	

MBW-mean body weight, ADG-average daily growth, SR-survival rate, DOC-days of culture

Highest average grain yield (Table 4.5) was recorded in rice-fish culture field (3595 kg ha $^{-1}$) which was 20.3 per cent higher against control (2988 kg ha^{-1}), contributed by higher number of panicles/m^2 (233.2) and number of filled grains/panicle (118.0). This was probably due to introduction of fish and prawn in rice field where, frequent locomotary movement of fish and prawn improves dissolved oxygen level and soil organic matter/nutrient status by adding faecal matter, controls plankton population/macro and micro aquatic insects/bacteria/organic detritus that compete with rice for material and energy. Hora and Pillay (1962) also reported that, introduction of fish has increased paddy yield by 15 per cent in the Indo-Pacific countries due to better aeration of water, greater tillering effect and additional supply of fertilizer in form of leftover feed and fish excreta.

Table 4.5: Average Rice Yield and Yield Attributes as Influenced by Fish and Prawn in Rice Field

Type of Culture	*Nos. of Panicles/m^2*	*Nos. of Grain/panicle*	*Test wt. (g)*	*Grain Yield (kg ha^{-1})*	*Straw Yield (kg ha^{-1})*	*Increase in Grain Yield*
Rice+fish	233.2	118.0	19.9	3595.0	4303.0	20.3%
Rice (control)	204.2	110.2	19.5	2988.0	3572.0	

In brief, as discussed above, it is concluded that rice-fish culture is a feasible and efficient way for optimum utilization of land and water resources that improve soil fertility and productivity. Rice and fish when grow together, are not only compatible but also mutually beneficial. To ensure better productivity, higher planting density must be avoided to provide greater ventilation and illumination, that helps in enhancing both plant and plankton growth, which ultimately helps fish growth and yield.

Acknowledgement

I warmly thank Dr. H.N. Verma, Director, WTCER (ICAR), for providing me the necessary facilities for this experiment and Dr. S.K. Mohanty, Fisheries Consultant, CDA, for reviewing the manuscript in detail.

References

APHA, (1989). Standard methods for examination of water and waste water. *American Public Health Association*. Washington, D.C., U.S.A, 1647 p.

Biswas, K.P. (1993). *Fisheries manual*. Bhagabat Press and Publishers, Orissa, India, 436 p.

Brahmanand, P.S. and Rajeeb K. Mohanty, (1999). Rice-fish Integration. *Yojana*, 43 (9): 11-12.

Colt, J. and Tehobonoglous (1978). Chronic exposure of channel cat fish *(Ictalurus punctatus)* to ammonia: effects on growth and survival. *Aquaculture*, 15: 353-372.

Hora, S.L. and Pillay, T.V.R. (1962). *Handbook on fish culture in the Indo-Pacific region*. *Fish.Biol.Tech.Pap.* 14, FAO, Rome.

Huazhu Yang (1994). Integrated rice-fish culture. In: *Developments in aquaculture and fisheries science*. Elsevier Publication Vol. 28 (1994): 368-385.

Ivlev, V.S. (1961). *Experimental ecology of the feeding of fish*. Yale University Press, New Haven.

Likangmin (1988). Rice-fish culture in China: a review. *Aquaculture*, 71 (1988) 173-186.

Mohanty, U.K. (1995). Comparative evaluation of growth and survival of Indian Major Carp fry in aerated vis-à-vis non-aerated ponds under different stocking densities. M.F.Sc. thesis. Orissa University of Agriculture and Technology, Bhubaneswar, India, 123 p.

Mohanty, Rajeeb K. (1999). Growth performance of *Penaeus monodon* at different stocking density *J. Inland fish Soc. India*, 31 (1): 53-59.

Mohanty, R.K., Verma, H.N. and Brahmanand, P.S. (2001). Rice-fish integration for enhancing land and water productivity. *Annual report*, 2000-2001, WTCER (ICAR), India: 70-74.

Noriega-Curtis, P. (1979). Primary productivity and related fish yield in intensely manured fish pond. *Aquaculture*, 17: 335-344.

Rao, A.P. and Ram Singh, (1998). Rice-fish farming system. In: S.H. Ahmad (ed.), *Advances in Fisheries and Fish Production*. Hindustan Publishing Corporation, NewDelhi, India, 309 p.

Roy, B., Das, D.N. and Mukhopadhyay, P.K. (1990). Rice-fish-vegetable integrated farming: towards a sustainable ecosystem. *Naga*, ICLARM, October, 1990: 17-18.

Spiller, G. (1985). *Rice cum fish culture-environmental aspects of rice and fish production in Asia*. Consultancy report, FAO, Bangkok, Thailand, 60 pp.

Vijayan, M.M. and Verghese, T.J. (1986). Effect of artificial aeration on growth and survival of Indian Major Carps. *Proc. Indian Acad. Sci.* (Anim.sci.), 95 (4): 371-378.

5

Induced Breeding of *Labeo rohita* (Hamilton-Buchanan) Using Single Application of Ovaprim in a Controlled Environment Carp Hatchery

Anita Jhajhria

Department of Zoology, Government P.G. College, Jalore - 334 001, Rajasthan

ABSTRACT

Five induced breeding exercises of *Labeo rohita* (Hamilton-buchanan) were conducted in the modified CIFE-D81 Hatchery from 6.7.1999 to 25.7.2000 using the synthetic fish hormone ovaprim @ 0.2 ml (male) and 0.5 ml (female)/kg. body weight of brooder. The optimum water temperature (24.3-27.6°C) was observed during breeding of rohu. The eggs (1.00-2.59 lakhs) were produced by varying the injection time (7.15 a.m-3.30 p.m.). No hatching occurred at 32.7°C temperature. The fertilization rate (50-85 per cent) as well as hatching success (80-89 per cent) indicates that ovaprim is a convenient ovulating agent for induced breeding.

Key Words: Ovaprim, Induced breeding, *Labeo rohita* (Hamilton-buchanan)

Introduction

During the recent years, intensive aquaculture of carps has emerged as a thrust area and because of this reason, farmer's requirement for seed is increasing (Bhowmick *et. at.*, 1997, Kesavanath and Radhakrishna, 1990, Tripathi, 1990; Bromage and Roberts, 1995). The seed production of cultivable carps is mainly based on the induced breeding and the inducing agents such as carp pituitary extracts (CPE) (Mahanta *et al.*, 1998) or synthetic hormonal preparations of the gonadotropin releasing hormone (GnRH) (Nandeesha *et al.*, 1990b; Peter *et al.*, 1993; Lakra *et al.*, 1996; Patino, 1997; Barlinsky *et al.*, 1997 and Garg *et al.*, 2002).

This paper is an attempt to work out the efficacy of ovaprim for reliable spawning and fingerling production as in Western Rajasthan it is rather difficult to meet the demands of seed supply through the natural breeding of carps.

Materials and Methods

By operating cast and drag net in the local ponds and lakes the healthy brooders of *Labeo rohita* were collected in the happa. These were then brought to the modified CIFE-D81 Hatchery and were fed on oilcake + rice bran (1:1) for seven days.

The CIFE-D81 is a portable carp hatchery made up of low density translucent polyethylene material and consists of a breeding and a hatching unit. The breeding unit comprises of a cooling tower, large pools with spray shower and a water circulating system. The hatching unit comprises of vertical jars, each of 40 litres capacity.

After their acclimatization in the laboratory conditions ripe males with milt oozing freely and females with soft bulging rounded abdomen with reddish vent were selected for the breeding purposes. Male carp can be easily distinguished from the female during the breeding season by the roughness of the dorsal surface of its pectoral fins compared to the smooth surfaces of the female. The male and female brooders were administered ovaprim a synthetic fish hormone continuing calibrated amount of SGnRH (Salmon gonadotropin releasing hormone) and doperidone, a dopamine antagonist dissolved in an organic solvent. Ovaprim @ 0.5 ml/kg body weight of female brooder and @ 0.2 ml/kg body weight of male brooder was injected intramuscularly at an angle of 45° to the body of fish in the region of caudal peduncle by using a 2 ml hypodermic syringe fitted with B.D.H. no.22 needle. Two male and one female brooder were exposed in the breeding unit of the modified CIFE-D81 Hatchery unit after giving the injection.

After successful breeding brooders were removed from the breeding tank. The brooders were weighed before and after induced spawning to evaluate the weight loss. The fertilization rate was assessed by counting the number of fertilized and unfertilized eggs. The eggs of rohu are of non-adhesive type and are red in colour having a mean diameter of 5 mm. These eggs were transferred to the hatching bucket of the hatching unit. Hatching success was evaluated by counting the number of spawn hatched out from the fertilized eggs. Some physico-chemical parameters were studied by following the methods of Welch (1948), APHA (1985), Strickland and Parson (1972), Murphy and Riley (1962) and Mullin and Riley (1956), in the breeding and the hatching units of the modified CIFE-D81 Hatchery.

Results and Discussion

The results of five induced breeding exercises on *Labeo rohita* are summarised in the Table 5.1. The limnological parameters were also recorded in these breeding exercises (Table 5.2).

There are optimum temperature ranges for induced breeding of cultivated fishes and critical temperature limits, above and below which fish will not reproduce. In this study, the optimum water temperature range (24.3-27.6°C) during breeding of rohu was observed. Water temperature plays a crucial role in the sexual maturation and breeding of carps.

Dwivedi and Reddy (1986) reported that in the controlled conditions, the temperature range (26-28.5°C) is suitable for breeding Indian major carps. Chaudhuri and Singh (1984) observed the optimum range of temperature for Indian and exotic carps to be between 24-31°C and they failed to spawn if the temperature is raised above 31°C. In the present work, also no spawn was produced at 32.7°C. Spawing

Table 5.1: Results of Induced Fish Breeding of Rohu from 6.7.1999 to 25.7.2000

Date of breeding	Time of ovaprim Injection	Water temp ↓ Temp. Decr	Brooder Weight in Kg.						Time & date of egg laying	Spawning time in (Hrs.)	No. of eggs laid (Lakhs)	Fertili-zation (%)	Hatching time (Hrs.)	Hatching (%)	No. of spawn produced (Lakhs)
			Male			Female									
			Before breed-ing	After breed-ing	Weight loss	Before breed-ing	After breed-ing	Weight loss							
06.07.1999	9.00 am	27.6	2.700	2.398	0.302	2.500	1.960	0.540	7.10 pm 06.07.1999	10.10	2.63	72	42	80	1.51488
25.07.1999	7.15 am	29.2 ↓ 24.3	2.100	1.996	0.104	1.900	1.700	0.200	7.00 pm 25.07.1999	12.15	1.10	50	39	89	0.48955
28.07.1999	10.30 am	32.7	2.400	2.151	0.249	2.000	1.700	0.300	11.50 pm 28.07.1999	13.20	1.00	12	-	-	-
16.07.2000	10.15 am	33.0 ↓ 26.4	2.900	2.600	0.300	2.500	2.000	0.500	8.00 pm 16.07.2000	10.15	2.72	85	30	84	1.94208
25.07.2000	3.30 pm	26.2	3.200	3.000	0.200	3.000	2.670	0.330	5.40 am 26.07.2000	14.10	2.59	79	37	80	1.63688

Table 5.2: Physico-chemical Parameters Studied During the Breeding and Hatching of Rohu

Date of Injection	During Breeding								During Hatching							
	Water Temp. (°C)	pH	Dissolved Oxygen (ml/l)	Free CO_2 (ppm)	CO_3 (pm)	HCO_3 (ppm)	Nitrate (µg/l)	Phosphate (µg/l)	Water Temp. (°C)	pH	Dissolved Oxygen (ml/l)	Free CO_2 (ppm)	CO_3 (pm)	HCO_3 (ppm)	Nitrate (µg/l)	Phosphate (µg/l)
06.07.1999	27.6 ±0.4	7.2 ±0.2	6.00 ±0.1	2.1 ±1.1	-	98 ±1.1	12.42 ±0.13	0.318 ±0.005	29.8 ±0.3	7.3 ±0.2	4.50 ±0.5	5.6 ±0.5	-	97 ±1.2	11 ±0.12	0.48 ±0.003
25.07.1999	54.3 ±0.3	7.2 ±0.3	5.96 ±0.2	1.9 ±0.3	-	99 ±1.0	8.02 ±0.15	0.312 ±0.001	25.5 ±0.4	7.4 ±0.4	4.40 ±0.2	5.2 ±0.1	-	94 ±1.0	13 ±0.10	0.46 ±0.001
28.07.1999	32.7 ±0.2	7.2 ±0.6	5.40 ±0.4	1.2 ±0.2	-	101 ±1.3	11.0 ±0.11	0.405 ±0.002	32.9 ±0.2	7.5 ±0.2	3.50 ±0.5	8.1 ±0.3	-	95 ±1.0	14 ±0.15	0.52 ±0.004
16.07.2000	26.4 ±0.4	7.2 ±0.3	5.60 ±0.1	1.6 ±0.5	-	104 ±1.1	6.60 ±0.14	0.307 ±0.002	26.6 ±0.1	7.3 ±0.1	4.28 ±0.3	6.8 ±0.4	-	95 ±1.1	10 ±0.17	0.55 ±0.001
25.07.2000	26.2 ±0.2	7.2 ±0.1	5.50 ±0.2	1.5 ±0.1	-	100 ±1.1	9.14 ±0.13	0.401 ±0.001	24.8 ±0.3	7.6 ±0.5	3.62 ±0.1	7.5 ±0.3	-	95 ±1.2	13 ±0.16	0.49 ±0.001

time for carps was observed by Pandey and Singh (1997) between 14-20 hours. Pandey *et al.* (2002) recorded spawning time to be between 10-12 hours after administration of carp pituitary extract (CPE) and between 7-8 hours after injecting ovatide.

Nandeesha *et al.* (1990b) observed after application of ovaprim the fertilization and hatching per cent for rohu (98-99.6 per cent and 90-95 per cent). Lakra *et al.* (1996) used ovaprim and reported fertilization and hatching per cent (96.2 per cent and 94.8 per cent) for rohu. Pandey *et al.* (2002) after injecting ovatide observed rate of fertilization (95-100 per cent) and hatching success (90-98 per cent) and by CPE recorded fertilization rate (75-85 per cent) and hatching per cent (70-78 per cent) in *Labeo rohita.* In the present work, fertilization rate (50-85 per cent) and hatching per cent (80-89 per cent) occurred after injecting ovaprim.

Thus, the results of the present study clearly demonstrate possibility of using the synthetic fish hormone preparation "ovaprim" for effective induced spawning and seed production in *Labeo rohita* (Hamilton-buchanan), in Western Rajasthan.

Acknowledgement

The author is grateful to Dr. Devendra Mohan, Associate Professor, Department of Zoology, Jai Narain Vyas University, Jodhpur for providing valuable suggestions and facilities to carry out the studies.

References

APHA (1985). *American water works association and water environment federation, standard methods for the examination of water and waste water*. 19th edition.

Barlinsky, D.L., William, K.V., Hodson, R.G. and Sullivan, C.V. (1997). Hormone induced spawning of summer flounder, *Paralichthys dentatus.* J. World Aquacult. Soc. 28: 79-86.

Bhowmick, R.M., Kowtal, G.V., Jana, R.K. and Gupta, S.D. (1997). Experiments on second spawning of Indian major carps in the same season by hypophysation. *Aquaculture.* 12: 149-155.

Bromage, N. and Roberts, R.J. (1995). Broodstock management and egg and larval quality. Blackwell Science Publications, Oxford, U.K.

Chaudhuri, H. and Singh, S.B. (1984). *Induced breeding of carps*. ICAR Publication.

Dwivedi, S.N. and Reddy, A.K. (1986). Fish breeding in a controlled environment carp hatchery CIFE-D81. *Aquaculture.* 54: 27-36.

Garg, S.K., Bhatnagar, A., Kalla, A. and Johal, M. S. (2002). Experimental Ichthyology. CBS Publishers and Distributors, New Delhi. pp.140-141.

Kesavanath, P. and Radhakrishnan, K.V. (1990). Carp seed production technology, special publication No.2, Asian fisheries society, Indian branch, Mangalore.

Lakra, W.S., Mishra, A., Dayal, R. and Pandey, A.K. (1996). Breeding of Indian major carps with the synthetic hormone drug ovaprim in Uttar Pradesh. *J. Adv. Zool.* 17: 105-109.

Mahanta, P.C., Rao, K.G., Pandey, G.C. and Pandey, A.K. (1998). Induced double spawing of an Indian major carp *Labeo rohita,* in the agro climatic conditions of Assam. *J. Adv. Zool.* 19: 99-104.

Mullin, J.B. and Riley, J.P. (1956). The spectrophotometer determination of nitrate in natural water. *Anal. Chem. Acta.* p. 464.

Murphy, J. and Riley, J.P. (1962). A modified single solution method for determination of phosphate in natural water. *Anal. Chem. Acta.* 27: 31-36.

Nandeesha, M.C., Rao, K.G., Jayanna, R., Parker, N.C., Varghese, T.J., Keshavanath, P. and Shetty, H.P.C. (1990b). Induced spawning of Indian major carps through single application of ovaprim c. In: The second Asian Fisheries Forum Proceedings. Asian Fisheries Society, Manila, Philippines. (Eds. Hirano, R. and Hanyo, I.). pp. 581-585.

Pandey, A.C. and Singh, R.N. (1997). Breeding of *Catla catla, Labeo rohita* and *Cirrhinus mrigala* by ovaprim injection for quality seed production. *J. Adv. Zool.* 18: 38-41.

Pandey, A.K., Mahapatra, C.T., Kanungo, G., Sarkar, M., Sahoo, G.C. and Singh, B.N. (2002). Ovatide induced spawning in the Indian major carp, *Labeo rohita* (Hamilton-buchanan), *Aquacult.* 3: 1-4.

Patino, R. (1997). Manipulations of the reproductive system of fishes by means of exogenous chemicals. *Prog. Fish. Cult.* 59: 118-128.

Peter, R.E., Lin, H.R., Kraak, G.V. and Little, M. (1993). Releasing hormones, dopamine antagonists and induced spawning In: *Recent advances in aquaculture* (Eds. Muir, J.F. and Roberts, R.J.). Blackwell Science Publications, London. pp.25-30.

Strickland, J.D.R. and Parson, T.R. (1972). *A practical handbook of seawater analysis*. Fish Research Board of Canada publications, Ottawa.

Triphati, S.D. (1990). Aquaculture in India. In: Aquaculture in Asia. (Ed. Joseph, M.M.). Asian Fisheries Society, Indian Branch, College of Fisheries, Mangalore. pp. 191-222.

Welch, P.S. (1948). *Limnological methods*. McGraw Hill Book. Co.

6

Evaluation of Growth Patterns of Two Indian Major Carps - Rohu (*Labeo rohita*) and Mrigal (*Cirrhinus mrigala*) with Tilapia (*Sarotherodon mossambicus*) under Mixed Rearing Conditions

T.A. Sethuramalingam and P. Ramakrishnan

Centre for Aquafeed and Nutrition (CAFeN)
Research Department of Zoology, St. Xavier's College (Autonomous)
Palayankottai - 627 002

ABSTRACT

The food utilization and growth pattern of rohu, mrigal with tilapia fingerlings under mixed rearing conditions were studied. When reared individually mrigal exhibited a reduction in growth but recovered with a marginal increase of weight when combined with rohu and tilapia. The growth rate of rohu decreased significantly ($p<0.05$) from 406.57 J/g fish/day in individual rearing to 132.39 J/g fish/day when reared with mrigal. But the growth rate increased significantly to 34I.24 J/g fish/day when reared in combination with mrigal and tilapia. Rohu consumed 1981.47 J/g fish/day in individual rearing but exhibited a feeding rate of 1488.63 J/g fish/day in combination with tilapia. Despite of this feeding rate, rohu did not show reduction in growth rate because of a significant increase ($p<0.05$) in the food conversion efficiency from 16.52 per cent in individual rearing to 24.14 per cent in combination with tilapia. The interaction between mrigal and rohu positively influenced the energy budget of tilapia. Eventhough the feeding rate of ti1apia decreased from 2793.91 J/g fish/day in individual rearing to 1807.22 J/g fish/day when reared with mrigal, the growth rate increased significantly from 388.97 J/g fish/day to 523.31 J/g fish/day. The increased growth rate even at low level of feeding is attributed to the decreased metabolic rate.

Introduction

In recent years aquaculture practice has been efficiently increased to enhance fish production by exploiting all the niches through composite and polyculture techniques. The pinnacle of composite fish culture is to reduce the competition among a cultured species to a minimum as they exploit all available area of the pond. Increase in growth of some fishes when reared with other species is often reported to have a great benefit for fish farmers (Jena *et al.*, 1998). Enhancement in the number and biomass of brown trout *Salmo trutta* was influenced by the introduction of rainbow trout. Growth of catla and grass carp have shown enhanced growth in mixed culture than cultured individually. (Basavaraju *et a1.*,1987). Williams *et al.*, (1987) have reported that the mean weight of channal cat fish, *Ictalurus punctatus* with blue tilapia, *Oreocromis aurea* reared together was significantly higher than that of channal cat fish in the cages grown without blue tilapia. Raising Nile tilapia with pig or other fishes in integrated culture increases the fish biomass (Woynarovich, 1979, Schroeder and Hapher, 1979, Chiayvareeaajja *et al.*,1989, Nitithamyong *et al.*, 1991). Similar findings are also reported by Mork, (1982), Holm, (1989), Tripathi, (1990), Nortvedt and Holm, (1991), Elangoven and Lethi, (1994), Mohanty, (1995) .The positive interaction between and among the fish species cultured has become one of the important features contributing to the success of polyculture (Murty *et al.*, 1978; Sen and Chakraberty, 1979). Moreover any alteration in food consumption and utilization in the species could alter the production of fishes. So it is important to study the cause of change in the growth of fish pattern during species interaction in mixed rearing conditions. Hence the present study was aimed to find out the growth pattern of carps (rohu and mrigal) under mixed rearing conditions along with tilapia. Tilapia was preferred in the study because of its proliferic breeding habit and also the combination implies a thorough understanding of the interaction of tilapia with carps in composite culture.

Materials and Methods

Fingerlings of rohu and mrigal were procured from Fish Farmers Development Agency (FFDA), Manimuthar, Tirunelveli District and transported to Centre for Aquafeed and Nutrition (CAFeN) laboratory, St. Xavier's College, Palayankotta1 and acclimatized in the round trough (40 lit. capacity) for 15 days. Tilapia were collected from nearby ponds of Potal near Palayankottai. Fish of 2.4 ± 0.27g were recruited from stock and divided into three series. In the first series, five fish were reared individually, in the second series combination of two fishes such as mrigal and rohu, rohu and tilapia and mrigal and tilapia were maintained. The third series contain all the three fishes (mrigal, rohu and tilapia). All the three series were maintained atleast with triplicate and were held in round plastic troughs (20 lit. capacity) interconnected with pipes supplied with aerated water with a flow rate of 4 lit/hr. The experimental fishes were fed *ad libitum* with formulated pelleted diet which were specially prepared for carp with 40 per cent crude protein level (Sethuramalingam, 2002). In the second series of experiments a wire mesh was placed inside each trough between the two types of fishes during the course of feeding to find out the individual food consumption. The collection of unfed and faeces were made prior to the feeding and were oven dried and packed. The experiments were run for 49 days which were carried out at temperature of 26 ± 1°C, DO 6.2 0.23 ml 0/lit and 12:12 LL:DD cycles. After the experimental period, fishes from each series were sacrificed (Maynard and Loosli, 1962) and wet and dry weight were taken. The energy content of the test fishes were also estimated using a semi microbomb calorimeter (Jauncey, 1982).

The scheme of energy budget was followed by IBP (Petrusewicz and MacFadyen, 1970)

$$C = P + R + U + F$$

C = Consumption, P = Growth, R = Metabolism, F = Faeces, U = Urine.

Absorption (A) was calculated by substracting F from C (A = C- F), P = Weight difference or animals at the beginning and at the end of the experiments. Total metabolism was deducted as the difference between absorption and growth. (R = A -P). Urine is not considered in this work.

Rate of feeding (FR)/ absorption (AR)/ Conversion (CR) = Food consumed/ absorbed/ converted(g)/wet weight of the fish(g) × time (days),

Absorption efficiency (AE) % = Food absorbed (g)/ food consumed(g) × 100

Gross conversion efficiency (GCE-K1%) = Growth(g)/ food consumed(g) × 100

Net conversion efficiency (NCE-K2%) = Growth(g)/ food absorbed(g) × 100.

Absorption of individual fish in combination could not be calculated because the faeces collection was impossible in that series hence it was assumed that absorption efficiency of each fish in any combination was same as in the individual rearing.

One way ANOVA was performed to examine the level of significance among the various series and correlation coefficient was employed to find out the relationship between and among each series/ combinations (Snedecor and Cochran, 1967).

Results and Discussion

The food consumption and utilization or fishes have been greatly influenced by the inter-specific interaction of rohu, mrigal and tilapia. The bottom feeder mrigal consumed as low as 1219.91 J/g fish/ day when reared individually. However the increase in the body weight was evident significantly ($p<0.05$), amidst rohu and tilapia because of increase in food intake (Table 6.1). Elangoven and Lethi (l994) reported a weight gain of cat1a when combined with rohu or tilapia over monoculture of catla.

Upon individual rearing the feeding and growth rates of rohu declined significantly ($p<0.05$) from 1981.47 to 1387.86 J/g fish/day and 406.57 to 132.39 J/g fish/day respectively than reared with mrigal (Table 6.1b). However in combination with tilapia and mrigal, the reduction in weight gain by this interaction between mrigal and rohu was counter balanced and the growth rate of rohu increased significantly ($p<0.05$) to 341.24 J/g fish/day. This report was well supported by Elizabeth (1986) in which Channa had a negative effect on rohu's growth, but when mrigal was introduced with rohu and Channa combinations, the negative growth effect on rohu was nullified. The same result was accepted by Beem and Gebhert, (1988) with a weight gain in *Ictalurus punctatus* when cultured with *Salmo graidneri* at different age groups. The reduction in the growth of rohu in the presence of mrigal may be attributed to the cumulative effect of reduced food consumption and increased metabolic rate. The interaction between rohu and tilapia was found to be positive and influenced the growth of rohu very well. Though the feeding rate of rohu declined during mixed rearing with tilapia combination, there was no decline in growth rate when reared individually. This may be due to lesser fraction of energy utilized by rohu for metabolism which In turn increased the growth rate and conversion efficiency.

Rearing of rohum mrigal with tilapia has a positive effect on the energy budget of tilapia. Even though the feeding rate of tilapia decreased drastically when reared with mrigal, the growth rate increased and the metabolic rate decreased. This was evident by the concomitant increase in conversion efficiency of tilapia as tilapia (*Oreochromis mossambicus*) was reared with Catla and rohu, the growth

Table 6.1a: Energy Budget of *Mrigal* Reared Individually and in Combination with *Rohu* and *Tilapia*

Rearing Conditions	*Feeding Rate (FR)*	*Absorption Rate (AR)*	*Conversion Rate (CR)*	*Metabolic Rate (MR)*	*Absorption Efficiency (AE)(%)*	*Gross Conversion Efficiency (GCE)(%)*	*Net Conversion Efficiency (NCE)(%)*
Mrigal & Combination Individual	1219.91 ± 81.28	1069.93 ± 24.66	121.73 ± 18.04	921.33 ± 18.62	85.34 ± 3.67	10.43 ± 1.83	11.43 ± 1.07
Mrigal + Rohu	1551.86 ± 33.68	1352.7 ± 18.03	220.9 ± 12.27	1142.9 ± 12.41	87.72 ± 2.91	14.28 ± 2.64	16.37 ± 1.34
Mrigal+Tilapia	1495.82 ± 53.22	903.3 ± 28.86	154.92 ± 12.63	747.36 ± 32.18	61.64 ± 2.88	10.52 ± 2.07	17.05 ± 1.92
Mrigal+Rohu+Tilapia	-	-	169.88 ± 16.71				

Table 6.1b

Rohu & Combination Individual	1981.47 ± 39.81	1653.31 ± 19.78	406.57 ± 11.73	1233.83 ± 31.54	82.84 ± 3.62	16.52 ± 2.16	24.76 ± 2.86
Mrigal + Rohu	1387.86 ± 17.61	1013.86 ± 40.93	132.39 ± 18.47	881.89 ± 26.47	73.04 ± 2.87	9.52 ± 1.37	14.97 ± 2.11
Mrigal + Tilapia	1488.63 ± 43.58	1234.55 ± 32.18	392.38 ± 19.63	846.49 ± 38.18	82.93 ± 3.66	24.14 ± 1.09	31.18 ± 1.99
Mrigal + Rohu + Tilapia	-	-	341.24 ± 18.47	-	-	-	-

Table 6.1c

Rohu and Combination Individual	2793.91 ± 29.36	2468.82 ± 18.33	88.97 ± 10.36	2079.03 ± 14.62	85.27 ± 2.11	17.46 ± 2.13	19.84 ± 1.49
Mrigal + Rohu	1807.22 ± 49.87	1461.39 ± 32.48	523.31 ± 40.43	938.64 ± 47.32	80.85 ± 2.27	27.96 ± 3.62	36.96 ± 3.87
Mrigal + Tilapia	1784.38 ± 18.93	1498.31 ± 45.36	508.47 ± 37.21	991.84 ± 37.28	83.97 ± 2.58	28.54 ± 3.29	33.96 ± 3.72
Mrigal + Rohu + Tilapia	-	-	511.83 ± 43.52	-	-	-	-

rate of tilapia increased significantly under mixed rearing condition (Elangoven and Lethi, 1994). The same trend was noticeable when tilapia was stocked with *Cyprinus carpio*, the growth of fishes were higher in combined rearing rather than in monoculture (Bardach *et al.*, 1972). When catla was grown with *Ctenopharyngodon idella* an increase in the weight of catla was noticed compared to its monoculture (Basavaraju *et al.*, 1987). Stocking tilapia with predatory fishes decreased tilapia production only due to reduced number of recruits but increased the average size/mass of tilapia (Manzano, 1990). Under combined culture improved growth performance was noticed in combination of Indian major carps and exotic carps than individual rearing (Jena *et al.*, 1998).

Under mixed rearing conditions tilapia cultured with rohu and mrigal showed an elevated mean growth rate but these increase was not highly significant ($P>0.05$) over rohu+tilapia or mrigal+tilapia combinations. A similar trend was also reported by Williams *et al.*, (1987), in channal cat fish and tilapia combination and Elangoven and Lethi, (1994) in catla+tilapia combinations.

The present study indicated that in combined rearing conditions one of the main reasons for the better growth even at low feeding level of fishes was due to the reduction in metabolic rate. Tilapia could be included in culture pond along with other carps provided its density and size should be considered and maintained in such a way that it does not exploit the culture pond. Monosex of tilapia with preferred density could also be practiced with carps which could nullify the proliferation of tilapia. Therefore inclusion of tilapia with rohu and mrigal under mixed rearing condition is economically viable.

References

Bardach, H.E., Ryther, J.H., and William, O.M. (1972). *Aquaculture: The Farming and Husbandry of Fresh Water and Marine Organisms*. John Wiley and Sons, New York. pp 354-369.

Basavaraju, Y., Devaraj, K.V., and Keshave (1987). Culture of grass carp and catla in combination. Paper presented at the *First Indian Fisheries Forum*. Mangalore. Dec.4-8, 1987.

Beem, M.D., and Gebhart, G.E. (1988). Winter polyculture of channel cat fish and rainbow trout in cages. *Prog. Fish. Cult.*, 50: 49-51.

Chiayvareesajja, S.C., Wongwit, A., Cronin, K., Supamataya, C., Tantikitu and Tansakul, R. (1989). Utilization of aquatic weed mixture pellet as feed for Nile Tilapia (*Oreochromis niloticus*) and pig pp 143-147. In S.De Silva (ed.), *Fish Nutrition Research in Asia*, Asian Fisheries Society, Manila, Phillippines.

Elangoven, A., and Lethi., C.D. (1994). Growth rate comparisons of two Indian major carps - catla (*Catla catla*) and rohu (*Labeo rohita*) with tilapia (*Oreochromis mossambicus*) *J. Aqua. Trop.* 9: 241-246.

Elizabeth, A.R. (1986). Physiological studies on the composite culture of carp and murrel. M. Phil dissertation, Bharathidasan University, Tiruchirapalli.

Holm, J.C. (1989). Mono and duoculture of Juvenile Atlantic salmon (*Salmo salar*) and Arctic charr (*Salvelinus alphinus*) *Can. J.fish. Aquat*. Sci. 46: 697-704.

Jauncey, K., (1982b). Carp (*Cyprinus carpio*) nutrition. A review in J.F. Muir and R.J. Roberts (eds.), *Recent Advances in Aquaculture*. Croom Helen. London. pp 215-263.

Jena, J.K., Aravindakshan, P.K., Suresh Chandra, Muduli, H.K., Ayyappan, S. (1998). Comparative Evaluation of Growth and survival of Indian major carps and exotic carps in raising fingerlings. *J. Aqua. Trap*. 13(2): 143-150.

Maynard, A.L., and Loosli, J.K. (1962). *Animal Nutrition*. McGraw Hill, New York. pp. 553.

Manzano, V.B. (1990). Polyculture systems using grouper (*Epinephelus tauvina*) and Tilapia (*Oreochromis mossambicus*) in brackishwater ponds pp. 213-216. In R. Hirano and I. Hanyu (eds.), *The Second Fisheries Forum*. Asian Fisheries Society, Manila, Philippines.

Mohanty, U.K. (1995). Comparative evaluation of growth and survival of Indian major carp fry in aerated vis-à-vis non-aerated ponds under different stocking densities. M.F.Sc. thesis. Orissa, University of Agriculture and Technology. Bhubaneswar, p 123.

Mork, O.I. (1982). Growth of three salmonid species in mono and double culture (*Salmo salar, S. trutta* and *S. gairdneri*) *Aquaculture*, 27: 141-147.

Murty, D.S., Dey, B.K., and Reddy, P.V.G.K. (1978). Experiments on rearing exotic carp fingerlings in composite fish culture in India. *Aquaculture*, 13(4): 331-337.

Nitithamyong, C., Chiayvareesajja, J., Chiayvareesajja, S., Wangwit, C., and Tansakul, R. (1991). Production of Nile Tilapia (*Oreochromis niloticus*) in different culture and Harvesting systems. In Fish Nutrition Research in Asia (Ed.) S.S. De Silva. *Proceedings of IVth Asian Fish Nutrition Workshop*. pp. 169-174.

Nortvedt, R., and Holm, J.C. (1991). Atlantic salmon in duoculture with Arctic charr. Decreased aggression enhances growth and stocking density potential. *Aquaculture*. 98: 355-361.

Petrusezicz, K., and MacFadyen, A. (1970). *Productivity of Terrestrial Animals*. Oxford Blackwell Scientific Publication (IBP Handbook).

Schroeder, G., and Hepher, B. (1979). Use of Agricultural and Urban wastes in fish culture. pp. 487-489. In TVR Pillay and W.M.Dill (eds.), *Advances in Aquaculture Fishing*. News Book. Surrey, England.

Sen, P.R., and Chakraborty, R.D. (1979). Rearing of major carp fry to fingerlings in freshwater ponds. Proc. Sem. Inland Aqua. CIFRI, Barrackpore. p 43.

Sethuramalingam, T.A., and Haniffa, M.A. (2002). Effect of formulated diets on digestive enzyme of *Labeo rohita* (Ham.). Indian Journal of Experimental Bio1ogy. 40: 83-88.

Snedecor, G.W. and Cochran, W.G. (1967). *Statistical method*. Oxford and IBH Publishing Co. New Delhi. p. 135.

Tripathi, S.D. (1990). Freshwater aquaculture in India. In: *Aquaculture in Asia*. M. Joseph. (ed.), pp. 191-222. Asian Fisheries Society, Indian Branch, Mangalore.

Williams, K., Gebhart, G.F., and Manghan, O.E. (1987). Enhanced growth of cage cultured channel cat fish through polyculture with blue tilapia. *Aquaculture*, 62: 207-214.

Woynarovich, E. (1979). The feasibility of combining animal husbandry with fish farming with special reference to duck and pig production. pp. 203-208. In TVR Pillay and W.M. Dill (eds.), *Advances in Aquaculture Fishing*. News Book, Surrey, England.

7

Observations on Three New Species of *Myxobolus* bütschli, 1882 from Hybrid Carps of West Bengal

Saugata Basu* and D.P. Haldar**

** Department of Biology, Uttarpara Govt. High School, Uttarpara, Hooghly - 712258, West Bengal*
***Protozoology Laboratory, Department of Zoology, University of Kalyani, Kalyani - 741 235, West Bengal*

ABSTRACT

Three new species of myxozoan parasites (Myxozoa: Bivalvulida) - *Myxobolus rocatlae* sp.n., *Myxobolus manoramae* sp.n. and *Myxobolus shantipuri* sp.n. are described from different organs of hybrid Catla-Rohu host fish. All three species are very much peculiar in having two different spore populations (one with equal and the other with unequal polar capsules) from the same plasmodium of the same host. The detailed structures and measurements of these three species are given in this communication.

Key Words: *Myxobolus rocatlae* sp.n, *Myxobolus manoramae* sp.n., *Myxobolus shantipuri* sp.n., Myxozoa Bivalvuida, Hybrid Carps, India

Abbreviations Used

SP = Spore
LS = Length of spore
BS = Breadth of spore
LPC = Larger polar capsule
SPC = Smaller polar capsule
LLPC = Length of larger polar capsule
BLPC = Breadth of larger polar capsule
LSPC = Length of smaller polar capsule
BSPC = Breadth of smaller polar capsule
DIV = Diameter of iodinophilous vacuole
DSPN = Diameter of sporoplasmic nuclei

Introduction

The genus *Myxobolus* Bütschli, 1882 is a member of the class Myxosporea in the Phylum Myxozoa, which has traditionally been placed in the Kingdom Protista, but molecular genetic evidence indicates that the Myxozoa may be metazoans (Smothers *et al.*, 1994). About 1100 myxosporean species have been described infecting fish (Lom and Arthur, 1989) of pure breed. But in a recent paper we (Basu and Haldar, 1998) also reported the presence of myxosporean parasite infection in hybrid carps.

This communication records three new species of *Myxobolus,* viz., *Myxobolus rocatlae* sp.n., *Myxobolus manoramae* sp.n. and *Myxobolus shantipuri* sp.n. from hybrid Catla-Rohu host fish. The preparation of species description of the present myxozoans has been done in accordance with the guideline of Lom and Arthur (1989) and Lom and Dyková (1992).

Materials and Methods

Hybrid Catla-Rohu host fish [Male parent fish *Catla catla* (Hamilton-Buchanan) and Female parent fish *Labeo rohita* (Hamilton-Buchanan)], weighing 27.7 gm, and measuring 13.9 cm. in the average, were collected from different privately maintained ponds at Ranaghat and Shantipur areas in the district of Nadia, West Bengal, in the months from July, 1997 to April, 1998. The host fishes were brought to the laboratory from their sites of collection as quickly as possible and were immediately examined thoroughly for their myxozoan parasites in their body surfaces, fin and fin rays, gill filaments and other internal organs. Sporogonic plasmodia of the myxozoans, when found attached to the different body parts of the infected fishes, were carefully removed with the help of a sterile forceps, smeared on clean grease-free slides with drops of 0.5 per cent NaCl solution, covered with thin cover-glasses and properly sealed for examination under the oil immersion lens of Olympus CH-2 phase contrast microscope. Some of the fish smears were treated with various concentrations (2-10 per cent) of KOH solution for the extrusion of polar filaments. The India ink method of Lom and Vavra (1963) was employed for observing the mucus envelope of spores. For permanent preparations, air-dried smears were stained with Giemsa after fixation in acetone-free absolute methanol or wet smears fixed in Schaudinn's fluid were stained with Heidenhain's iron alum haematoxylin. Measurements (based on twenty fresh spores treated with Lugol's iodine) were taken with the aid of a calibrated ocular micrometer. All measurements are presented in micrometers as mean ± SD followed in parentheses by the range. Drawings were made on fresh/stained materials with the aid of a Camera Lucida (Mirror type) and computer programme Corel Draw 9.0.

Observations

Myxobolus rocatlae sp.n. (Figures 7.1-7.7)

Plasmodia

Sporogonic plasmodia, appearing as "cysts", encased within hosts' cells and are found attached to the gills and gut wall of infected host fishes. These are oval, 1100 × 560 or horse-shoe shaped, 440 × 218 and light-yellowish in colour. They contain some late developmental stages (Fig. 7.1) as well as many mature spores.

Spore

The most striking feature of the species is the presence of two distinctly separate populations namely Population I (Figs. 7.2-7.4) and Population II (Figs. 7.5-7.7). They occur simultaneously within the single plasmodium of the same host fish. Since they vary in structure and morphometry they are described separately below.

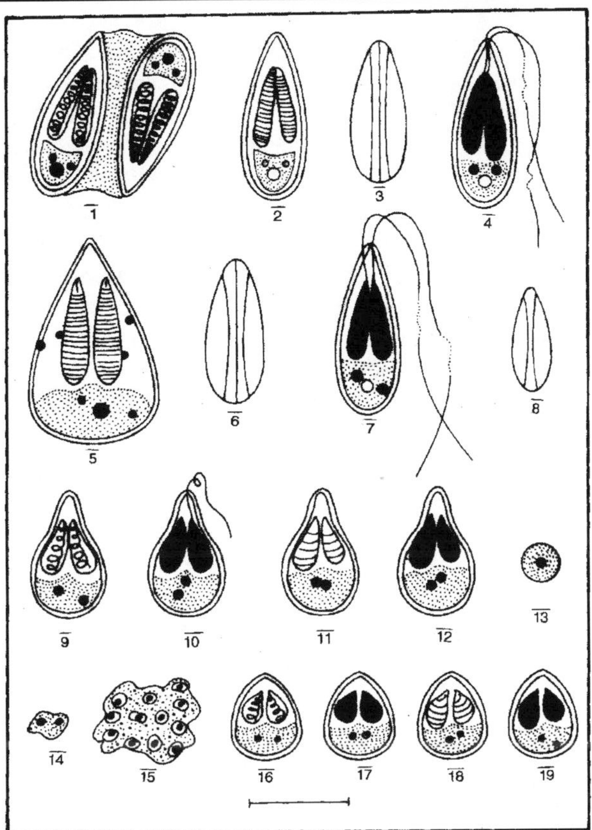

Figs. 7.1-7.7: Camera Lucida Drawings of Plasmodia and Spores of *Myxobolus rocatlae* sp.n. Fig. 7.1: Late Developing Stage - Lugol's Iodine, Figs. 7.2-7.4: Mature Spores of *M. rocatlae* sp.n. of Population I, Fig. 7.2: Fresh Spore in Valvular View - Lugol's Iodine, Fig. 7.3: Fresh Spore in Sutural View - Lugol's Iodine, Fig. 7.4: Fixed Spore in Valvular View with Extruded Polar Filaments - Giemsa, Figs. 7.5-7.7: Immature and Mature Spores of *M. roctlae* sp. n. of Population II, Fig. 7.5: Immature Fresh Spore in Valvular View Showing Six Different Nuclei - Lugol's Iodine, Fig. 7.6: Fresh Mature Spore in Sutural View - Lugol's Iodine, Fig. 7.7: Fixed Mature Spore in Vulvular View with Extruded Polar Filaments - Giemsa, Figs. 7.8-7.12: Camera Lucida Drawings of Mature Spores of *Myxobolus manoramae* sp. n. Figs. 7.8-7.10: Spores of *M. manoramae* sp. n. of Population I, Fig. 7.8: Fresh Spore in Sutural View - Lugol's Iodine, Fig. 7.9: Fixed Mature Spore in Vulvular View - Lugol's Iodine, Fig. 7.10: Fixed Spore in Vulvular with Extruded Polar Filaments from One Polar Capsule Giemsa, Figs. 7.11-7.12: Spores of *M. manoramae* sp. n. of Population II, Fig. 7.11: Fresh Spore in Valvular View - Lugol's Iodine, Fig. 7.12: Fixed Spore in Vulvular View - Giemsa, Figs. 7.13-7.19: Camera Lucida Drawings of Plasmodai and Spores of *Myxobolus shantipuri* sp. n. Figs. 7.13-7.15: Plasmodia Containing Different Types of Pansporoblasts, Fig. 7.13: Uninucleate Pansporoblast - Lugol's Iodine, Fig. 7.14: Binucleate Pansporoblast - Lugol's Iodine, Fig. 7.15: Multinucleate Pansporoblast - Lugol's Iodine, Figs. 7.16-7.17: Mature Spores of *M. shantipuri* sp. n. of Population I, Fig 7.16: Fresh Spore in Vulvular View - Lugol's Iodine, Fig. 7.17: Fixed Spore in Valvular View - Giemsa, Figs. 7.18-7.19: Mature Spores of *M. shantipuri* sp. n. of Population II, Fig 7.18: Fresh Spore in Vulvular View - Lugol's Iodine, Fig. 7.19: Fixed Spore in Valvular View - Giemsa Scale Bar = 10.0 µm.

Population I (Figure 7.2-7.4)

The spores are histozoic, elongated and pyriform in valvular view (Figs. 7.2, 7.4) and lenticular with straight thin sutural line and ridge in sutural view (Fig. 7.3). The spore is bluntly pointed anteriorly with rounded posterior end and 18.5 ± 0.41 (17.5-19.3) × 5.9 ± 0.15 (5.6-6.2) in dimensions. Two symmetrical shell-valves are smooth and thin-walled. Polar capsules are pyriform, two in number, situated anteriorly and cover almost two-third of the spore cavity. The tip of the polar capsules converge at the anterior end (Fig. 7.4). One polar capsule is slightly larger, 12.9 ± 0.53 (11.8-13.7) × 2.85 ± 0.14 (2.5-3.0) than the other which is 11.3 ± 0.61 (10.1-12.2) long with a breadth of 2.2 ± 0.14 (2.0-2.4) (Figs. 7.2, 7.4). The polar filament makes 17-19 spiral coils in the larger and 15-18 tight coils in the smaller polar capsules (Fig. 2). When extruded the larger polar filament reaches a length of 102.0 with a mean of 82.4 ± 8.5, the length of smaller polar filament averages 79.0 ± 3.8 (69.5-83.7) (Fig. 7.4). The posterior one third of the spore cavity is filled with finely granular crescentic mass of sporoplasm which contains two spherical sporoplasmic nuclei of 1.7 ± 0.21 (1.4-2.1) diameter.

The fresh spore, when treated with lugol's iodine, shows the presence of a spherical, 2.0 ± 0.16 (1.8-2.5) iodinophilous vacuole (Fig. 7.2). India ink method of Lom and Vavra (1963) shows negative result as mucus envelopes around spores are absent.

Population II (Figures 7.5-7.7)

The elongated pyriform spores of 18.3 ± 0.57 (16.9-19.3) × 6.0 ± 0.30 (5.6-6.9) dimensions are identical with the spores of Population I. The two shell valves are smooth but slightly thicker and meet along a conspicuous sutural ridge (Fig. 7.6). Some immature spores with larger breadth have also been noticed in which all the six nuclei are present (Fig. 7.5). Two anteriorly situated polar capsules are equal in size unlike those of Population I and 12.6 ± 0.79 (10.4-13.7) × 2.8 ± 0.22 (2.2-3.1) in morphometric measurements (Fig. 7.7). Polar filament forms 18-20 slightly loose coils inside each polar capsule (Fig. 7.5). The length of extruded polar filament is 88.1 ± 5.7 (74.9-96.4) (Fig. 7.7). Finely granular homogenous mass of sporoplasm with two spherical sporoplasmic nuclei. [DSPN.: 1.8 ± 0.13 (1.5-2.0)] and round to oval iodinophilous vacuole [DIV. 2.1 ± 0.20 (1.8-2.5)] covers the entire extra-capsular space (Fig. 7.5). Mucus envelope around spore is absent.

Spore Index

Population I

LS : BS	=	1 : 0.32
LLPC : LSPC	=	1 : 0.88
BLPC : BSPC	=	1 : 0.78
LLPC : BLPC	=	1 : 0.22
LSPC : BSPC	=	1 : 0.19

Population II

LS : BS	=	1 : 0.33
LPC : BPC	=	1 : 0.22
LS : LPC	=	1 : 0.69
BS : BPC	=	1 : 0.46

Affinities

The present myxosporidan resembles *M. angustus* Kudo, 1934 reported from the gill filaments of *Cliola vigilax*, *M. Calbasui* Chakravarty, 1939 reported from the gall-bladder of *Labeo calbasu* (Ham.), *L. rohita* (Ham.) and *Cirrhinus mrigala* (Ham.), *M. punctatus* Roychaudhuri and Chakravarty, 1970 from spleen and pharyngeal epithelium of *Ophicephalus punctatus* (Bloch), *M. bhadrensis* Seenappa and Manohar, 1981 from the muscles of *L. rohita* (Ham.), *M. hosadurgensis* Seenappa and Manohar, 1981 from the gills and muscles of *Cirrhinus mrigala* (Ham.) *M. vedavatiensis* Seenappa and Manohar, 1981 from the gills of *Cirrhinus mrigala* (Ham.) and *M. iranicus* Molnar, Masoumian and Abasi, 1996 from the spleen of *Barbus luteus* (Heckel). The former species, however, differs from all the latter species by the larger length and smaller breadth of spore [spore dimensions in *M. angustus* 14 - 15 × 7 - 8, *M. calbasui* 12.4 - 15.0 × 8.2 - 10.0, *M. punctatus* 12.2 - 15.0 × 5.7 - 7.8, *M. bhadrensis* 8.0 - 11.0 × 7.0 - 8.0, *M. hosadurgensis* 9.0 - 11.0 × 5.0 - 8.0, *M. vedavatiensis* 13.0 - 15.0 × 8.0 - 10.0 and in *M. iranicus* 13.2 - 14.0 × 7.5 - 9.2 respectively]

Further, the myxosporidan in study shows some similarities with *Myxobolus (= Myxosoma) andhrae* (Lalitha Kumari, 1969) Lom and Noble, 1984 reported from the intestinal wall of *Opicephalus punctatus* and *Myxobolus (=Myxosoma) indiae* (Lalitha Kumari, 1969) Lom and Noble, 1984 reported from the gill filaments of *Barbus sarana*. But the smaller spores of *M. (=M.) andhrae* [LS: 12.1 - 15.7 (13.5)] with parietal folds in the shell valves are distinctly different from the present species as in the present species shell valves are smooth and devoid of any parietal folds. Moreover, the ellipsoidal spores of *M = (M.) indiae* with 6- 8 valvular ridges and 2-3 unequal or subequal polar capsules are distinctly different from the present species under study.

The spores of *M. inaequus* Kent and Hoffman, 1984 reported from the brain tissue of *Eigemannia virescens* (V.) are nearly similar in shape with the present myxozoan species. However, in the former species the polar capsules are unequal and the dimensions of two polar capsules are much more different (in *M. inaequus* LPC: 11.8 × 3.6, SPC: 4.8 × 3.6).

However, the present species has two populations from the same plasmodium of the same host one of which has spores with unequal polar capsules and the other with equal polar capsules. These two populations cover almost 50 per cent each of a single plasmodium.

Variation in spore structure is also found in a recently created new myxozoan species - *M. multivaderis* Mukhopadhyay and Haldar, 1998 described from diverse host organs as gills, intestinal wall, brain and palate of juvenile major carp *Catla catla*. According to the authors, such variations are possible due to invasion of the spores in different tissues having different environment. However, variations in the spore population of the present species are recorded from the single plasmodium of a particular host organ.

Finally, the host of the present species is a hybrid carp from which no such species are described earlier. Taking into consideration all the above mentioned criteria the present myxozoan warrants the status of a new species for which the name *Myxobolus rocatlae* sp. n. is given.

Taxonomic Summary

Type Host	:	Hybrid Catla-Rohu host fish
Site of infection	:	Gills and Gut wall
Type locality	:	Ranaghat in the district of Nadia, West Bengal, India

Type Specimens

Holotype : On slide no. CR 2/1 obtained from the smears of plasmodium attached with the gut wall of hybrid Catla-Rohu host fish collected at Ranaghat in the district of Nadia, West Bengal, India on 4 March (1998).

Paratype : On the above-numbered slide as well as on other slides, other particulars are the same as for the holotype material.

Type materials have presently been deposited at the Department of Zoology, University of Kalyani and will finally be deposited at the National Collection of the Zoological Survey of India, Kolkata, West Bengal, India.

Prevalence and intensity of infection: 27/59 (45.76 per cent)

Etymology : The species is named after its type host Hybrid Catla-Rohu host fish.

Myxobolus manoramae sp.n. (Figures 7.8 - 7.12)

Plasmodia

The sporogonic plasmodia encased within host cells are found attached to the tail fins of infected host fishes. These are small, round, creamy white in colour, measuring 200-310 in diameter and contain only the mature spores.

Spore

Two different types of mature, histozoic spores occur simultaneously in the same plasmodium of the same host fish which are described hereunder as Population I (Figs. 7.8-7.10) and Population II (Figs. 7.11-7.12).

Population I (Figures 7.8-7.10)

Spores are pyriform in valvular view, 11.8 ± 10.82 (10.7-13.1) × 5.6 ± 0.55 (4.8-6.3), narrowing toward anterior bluntly pointed end and rounded at posterior extremity (Figs. 7.9-7.10). Shell valves are moderately thick and the sutural ridge is distinct in Lugol's iodine preparation (Fig. 7.8). Two pyriform polar capsules arc equal, 6.2 ± 0.25 (5.5-6.6) × 2.2 ± 0.13 (2.0-2.5), occupy more than half of the spore cavity (Figs. 7.9-7.10). In most of the cases polar filament makes 5-6 spiral coils (Fig. 7.9) inside each polar capsule and in very few cases the polar filament is extruded through the anterior extremity of the polar capsule (Fig. 7.10) and attains a maximum length of 10.2. The cup-shaped, crescentic extra-capsular space is filled with granular homogenous mass of sporoplasm which contains two nuclei of 1.7 ± 0.20 (1.5-2.1) diameter (Fig. 7.10). The iodinophilous vacuole and mucus envelope around the spores are absent.

Population II (Figures 7.11-7.12)

Pyriform spores with bluntly pointed anterior and rounded posterior ends are 11.5 ± 0.56 (10.6-12.5) × 5.7 ± 0.44 (4.8-6.3) in dimensions. Unlike the spores of Population I, the pyriform polar capsules are unequal with slightly elongated neck at their anterior extremities (Figs. 7.11-7.12). The larger popular capsule is 6.1 ± 0.21 (5.8-6.5) × 2.17 ± 0.15 (2.0-2.5) and the smaller one is 5.2 ± 0.44 (4.5-5.9) × 2.0 ± 0.05 (1.9 -2.1) in their dimensions. Polar filament forms 5-6 and 4-5 coils in the larger and smaller capsules respectively (Fig 7.11). But the extrusion of polar filament has not been occurred although all the conventional methods for the extrusion of polar filament have been performed. The extra-capsular space is filled with binucleate homogenous mass of sporoplasm (Figs. 7.11-7.12). Other characteristics are almost identical with the spores of Population I.

Spore Index

Population I

LS : BS	=	1 : 0.48
LPC : BPC	=	1 : 0.36
LS : LPC	=	1 : 0.53
BS : BPC	=	1 : 0.40

Population II

LS : US	=	1 : 0.50
LLPC : LSPC	=	1 : 0.85
BLPC : BSPC	=	1 : 0.94
LLPC : BLPC	=	1 : 0.35
LSPC : BSPC	=	1 : 0.39

Affinities

Thelohan, 1892 in the early days of discovering Myxosporea considered the mere presence or absence of the "iodinophilous vacuole" a reason sufficient to distinguish two separate, but otherwise identical genera *Myxobolus* and *Myxosoma*, and Poche in 1913 used this Character to establish a separate family Myxosomatidae. Walliker (1968), however, concluded that the iodinophilous vacuole, which consists mostly of glycogen, was not a stable mark and he synonymized the two genera. Although the Russian authors, among them Donec and Shulman (1984) still make a distinction between the two genera, other authors (Mitchell 1977, Lom and Noble 1984, Landsberg and Lom 1991, Lom and Dyková 1992) regard them as synonyms, and use the presence or absence of the vacuole only as one of the marks for characterizing the spore. Thus, the present myxozoan under consideration, although has no iodinophilous vacuole, is placed in the genus *Myxobolus* Bütschli, 1882 based on the spore characters given by Kudo 1920, 1933.

The present species resembles *M. koi* Kudo, 1919 described from the connective tissue of gill lamellae and subcutaneous cellular tissue of head of *Cyprinus carpio* and *Hypopthalmichthys molitrix*, *M. cheisini* Shulman, 1962 from gill lamellae of *Ophiocephalus argus*, *M. ophicephali* Bhatt and Siddique, 1964 from the accessory respiratory membrane of *Ophicephalus punctatus*, *M. rohitae* Ilaldar, Das and Sharma, 1983 from scales of *Laheo rohita*, *M. mugilii* Haldar, Samal and Mukhopadhyay, 1996 from gills of *Mugil cephalus* (Linn.) and *M. coelii* Haldar, Samal and Mukhopadhyay, 1996 from gall bladder of *Chanos chanos* (Forskal). However, spore dimensions of *M. koi*, *M. cheisini* and *M. ophicephali* are larger than the present species (spore dimensions in *M. koi*, *M. cheisini* and *M. ophicephali* are 14.0-16.0 × 7.0-9.0, 12.5-3.6 × 7.0-8.5, 11.6-13.3 × 4.6-6.3 respectively). Further, the ovoidal and rounded spores of *M. rohitae* with triangular notch in spore valves are distinctly different from the pyriform spores with smooth shell valves of the present species.

Although the dimensions of spores of *M, mugilii* [SP: 8.1-6.3 (11.7) × 4.0-7.3 (5.5)] and *M. coelii* [SP: 8.3-14.9 (10.6) × 4.1- 8.4 (5.49)] are closely similar to the present species under study but the ovoid spores of *M. mugilii* with equal polar capsules are different from the present species. Moreover, the pyriform spores of *M. coelii* with lodinophilous vacuole and unequal polar capsules lack sutural line which is very much prominent in the present species.

Further, in the present myxozoan species a single plasmodium from the same host contains two different types of spores - one with unequal polar capsules (40 per cent) another with equal polar capsules (60 per cent).

Such characteristic feature is also encountered in the spores of *M. rocatlae* sp. n., described in the present context. But the dimensions of spore and polar capsules in the latter species are much more than the former species under study.

From the above discussion it is evident that the present species is undoubtedly new to science for which it is given separate specific status and is named *Myxobolus manoramae* sp. n.

Taxonomic Summary

Type Host : Hybrid Catla-Rohu host fish

Site of infection : Tail fins

Type locality : Ranaghat in the district of Nadia, West Bengal, India

Type Specimens

Holotype : On slide no. CR 1/2 obtained from the smears of tail fins of hybrid Catla-Rohu host fish collected at Ranaghat in the district of Nadia, West Bengal, India on 4 September (1997).

Paratype : On the above-numbered slide as well as on other slides, other particulars are the same as for the holotype material.

Slides containing the holotype and paratype materials have presently been deposited at the Department of Zoology, University of Kalyani and will finally be deposited at the National Collection of the Zoological Survey of India, Kolkata, West Bengal, India.

Prevalence and intensity of infection : 17/43 (39.53 per cent)

Etymology : The species name *manoramae* is proposed after Late Manorama Basu, the grandmother of the senior author, who had been a constant source of encouragement for him.

Myxobolus shantipuri sp.n. (Figures 7.13-7.19)

Plasmodia

Spherical (diameter 140) or oval (100 × 90), white coloured sporogonic plasmodia are found to attach to the gill filaments of infected host fishes. Smears of cysts reveal the presence or mature spores and, as well as, very few uni-, bi- and multinucleate pansporoblasts (Figs. 7.13-7.15).

Spore

Mature histozoic spores are ovoidal or tear-shaped in valvular view with rounded posterior and bluntly pointed or rounded anterior ends. Spores are of two types which share the same plasmodium of the same host at a ratio of 60 per cent (Population I) (Figs. 7.16-7.17) to 40 per cent (Population II) (Figs. 7.18-7.19) respectively.

Population I (Figures 7.16-7.17)

Mature spores are 7.3 ± 0.59 (6.3-8.2) long with a breadth of 5.8 ± 0.31 (5.2-6.1). Spore valves are symmetrical without any parietal folds and valvular striations. Sutural line and inter-capsular ridge are absent. Ovoidal to pyriform polar capsules with elongated necks are equal in shape and 4.0 ± 0.16 (3.5-4.1) × 2.4 ± 0.29 (2.0-2.8) in dimensions. Polar filament makes 4-5 tight coils, wound spirally within each polar capsule and is not extruded by the usual methods (Fig. 7.16). Finely granular sporoplasm containing two obliquely placed round to oval nuclei of 1.3 ± 0.09 (1.2-1.4) diameter fills the posterior part of the extra-capsular spore cavity (Fig. 7.17). Iodinophilous vacuole and mucus envelope around spores are absent.

Population II (Figures 7.18-7.19)

Spores with dimensions of 7.4 ± 0.59 (6.3-8.1) × 5.8 ± 0.35 (5.1-6.2) possess two polar capsules of unequal shape and sizes (Figs. 7.18-7.19). Larger ovoidal to pyriform polar capsule with slightly elongated neck is 3.9 ± 0.28 (3.2-4.2) × 2.2 ± 0.36 (2.0-3.0) in morphometric measurements. The smaller one is 3.4 ± 0.52 (2.5-4.0) × 2.2 ± 0.25 (1.9-28), ovoidal and without the elongated neck. Number of coiled polar filaments and other characteristics of the spores are almost similar to that of Population I.

Spore Index

Population I

LS : BS	=	1 : 0.79
LPC : BPC	=	1 : 0.61
LS : LPC	=	1 : 0.55
BS : BPC	=	1 : 0.43

Population II

LS : BS	=	1 : 0.78
LLPC : BLPC	=	1 : 0.57
LSPC : BSPC	=	1 : 0.65
LLPC : L.SPC	=	1 : 0.86
BLPC : BSPC	=	1 : 0.97

Affinities

The present species is placed in the genus *Myxobolus* Bütschli, 1882 on the basis of spore structure and it resembles *M. dispar* Thélohan, 1895, *M. mrigalae* Chakravarty, 1939, *M. bengolensis* Chakravarty and Basu, 1948, *M. indicum* Tripathi, 1952, *M. branchialis* Tripathi, 1952, *M. koli* Lalitha Kumari, 1969, *M. potaili* Lalitha Kumari, 1969, *M. psilorhynchi* Lalitha Kumari, 1969, *M. carnaticus* Seenappa and Manahar 1980a, *M.* (=*M.*) *magaudii* (Bajpai *et al.*, 1981) Lom and Noble, 1984 and *M. molae* Sarkar, 1993 either in shape or in morphometry of the spores. However, the spores of *M. dispar*, *M. mrigalae. M. indicum*, *M. branchialis*, *M. koli*, *M. carnaticus* and *M. mogaudii* possess a pair of unequal polar capsules and hence are different from the myxozoan species under study. The spores of *M. potaili* has an anterior knob and striated shell valves which are not seen in the present species. Moreover, the dimensions of spores are also smaller in the present form. In *M. psilorhynchi*, spores are spherical with unequal polar capsules and, therefore, the present myxozoan is distinct from *M. psilorhynchi*, *M. bengalensis* and *M. molae* also possess spores having a similar shape but the overall dimensions of both the spores are significantly different from the present species under study.

Spores of *M. curmucae* Seenappa and Manohar, 1980a described from scales of *Puntius curmuca* (Ham.) and *M. vanivilasae* Seenappa and Manohar, 1980b described from *Cirrhinus mrigala* (Ham.) possess two types of spores, one with equal polar capsules another with unequal polar capsules, show similarity with the present myxozoan as 60 per cent of its spores bear equal polar capsules and 40 per cent have unequal polar capsules. But the oval spores of *M. curmucae* with inter-capsular ridge and sutural ridge (both of which arc absent in the present form) are larger in dimensions (SP in *M. curmucae* is 9.8 × 7.6) than the present species under consideration.

Moreover, ovate spores of *M. vanivilasae* is different from the present species not only in the presence of iodinophilous vacuole but also in the morphometry of spore and polar capsule (in *M. vanivilasae* SP: 8-10 × 7-9, Polar capsule (equal): 3.1 × 2.3, Polar capsule (unequal): 3.5 × 2.5 and 2.9 × 2.5).

Furthermore, the present species is also compared with *M. rocatlae* sp. n. and *M. manoramae* sp. n. described in the present dissertation. But the sites of infestation, dimensions of pyriform spore and polar capsules, and type locality in the latter two species are absolutely different from the tear-shaped spore of the species under consideration.

Taking into account all the above-mentioned criteria, the present myxozoan has, therefore, been considered as a new species for which the name *Myxobolus shantipuri* sp. n. is given.

The morphometric comparisons of Population I and Population II of all three species referred to above, viz *M. rocatlae* sp.n., *M. manoramae* sp.n. and *M. shantipuri* sp.n. have been compiled and presented in Table 7.1.

Taxonomic Summary

Type Host : Hybrid Catla-Rohu host fish

Site of infection : Gill lamellae

Type locality : Shantipur in the district of Nadia, West Bengal, India

Type Specimens

Holotype : On slide no. CR 3/3 obtained from the gill smears of hybrid Catla-Rohu host fish collected at Shantipur in the district of Nadia, West Bengal, India on 4 September (1997).

Paratype : On the above numbered slide as well as on other slides, other particulars are the same as for the holotype material.

Slides containing the holotype and paratype materials have presently been deposited at the Department of Zoology, University of Kalyani and will finally be deposited at the National Collection of the Zoological Survey of India, Kolkata, West Bengal, India.

Prevalence and intensity of infection: 19/56 (33.92 per cent)

Etymology : The species is named after its type locality Shantipur.

Discussion

Fishes are parasitized by different groups of arthropods, helminths and protozoans. Protozoans are probably the most important groups of animal parasites affecting fishes (Rogers and Gains, 1975). Bauer *et al.*, (1981) were of the view that during the past few years new methods of fish cultivation has been elaborated and widely practiced in many countries in addition to traditional systems such as

Table 7.1: Showing the Morphometric Comparison of Population I and Population II of Three Described Species of *Myxobolus* Bütschli, 1882

Species	*Myxobolus rocatlae* sp.n.		*Myxobolus manoramae* sp.n.		*Myxobolus shantipuri* sp.n.	
Characters	*Population I*	*Population II*	*Population I*	*Population II*	*Population I*	*Population II*
Host	Hybrid *Labeo rohita (Ham.)* × *Catla catla (Ham.)* carps		Hybrid *Labeo rohita (Ham.)* × *Catla catla (Ham.)* carps		Hybrid *Labeo rohita (Ham.)* × *Catla catla* (Ham.) carps	
Site of Infection	*Gills and gut wall*		*Tail find*		*Gill lamellae*	
Locality	*Ranghat, Nadia, West Bengal, India*		*Ranghat, Nadia, West Bengal, India*		*Ranghat, Nadia, West Bengal, India*	
Reference	*Present study*		*Present study*		*Present study*	
Spore						
Length	18.5 ± 0.41 (17.5-19.3)	18.3 ± 0.57 (16.9-19.3)	11.8 ± 0.82 (10.7-13.1)	11.5 ± 0.56 (10.6-12.5)	7.3 ± 0.59 (6.3-8.2)	7.4 ± 0.59 (6.3-8.1)
Breadth	5.9 ± 0.15 (5.6-6.2)	6.0 ± 0.30 (5.6-6.9)	5.6 ± 0.55 (4.8-6.3)	5.7 ± 0.44 (4.8-6.3)	5.8 ± 0.31 (5.2-6.1)	5.8 ± 0.35 (5.1-6.2)
Polar Capsule (Equal)						
Length	-	12.6 ± 0.79 (10.4-13.7)	6.2 ± 0.25 (5.5-6.6)	-	4.0 ± 0.16 (3.5-4.1)	-
Breadth	-	2.8 ± 0.22 (2.2-3.1)	2.2 ± 0.13 (2.0-2.5)	-	2.4 ± 0.29 (2.0-2.8)	-
Polar Capsule (Unequal)						
LPC						
Length	12.9 ± 0.53 (11.8-13.7)	-	-	6.1 ± 2.1 (5.8-6.5)	-	3.9 ± 0.28 (3.2-4.2)
Breadth (2.5-3.0)	2.8 ± 0.14	-	- (2.0-2.5)	2.1 ± 0.15	- (2.0-3.0)	2.2 ± 0.36
SPC						
Length	11.3 ± 0.61 (10.1-12.2)	-	-	5.2 ± 0.44 (4.5-5.9)	-	3.4 ± 0.52 (2.5-4.0)
Breadth	2.2 ± 0.14 (2.0-2.4)	-	-	2.0 ± 0.05 (1.9-2.1)	-	2.2 ± 0.25 (1.9-2.8)
Diameter of						
Sporoplasmic nuclei	1.7 ± 0.21 (1.4-2.1)	1.8 ± 0.13 (1.5-2.0)	1.7 ± 0.20 (1.5-2.1)	1.7 ± 0.23 (1.4-2.0)	1.3 ± 0.09 (1.2-1.4)	1.3 ± 0.08 (1.2-1.4)
Iodinophilous vacuole	2.0 ± 0.16 (1.8-2.5)	2.1 ± 0.20 (1.8-2.5)	-	-	-	-

pond rearing, Polyculture of different combinations of fishes has been elaborated and all have influenced the parasitological situation of fish-rearing farms and hatcheries resulting in origination of new diseases caused by new parasites or those which were already known but not described as pathogens.

Hybridization technique is one of the most important techniques which is followed in all places including India. Polyculture of pure and hybrid fishes has influenced the parasitological situation of fish rearing ponds resulting in eruption of infections and diseases in hybrid carps which have been already known for the pure carps.

In a recent paper we (Basu and Haldar, 1998) reported the higher susceptibility of protozoan parasites to the hybrid carps. In many instances, these protozoans prove to be highly pathogenic leading to heavy mortality.

In this communication three new species of *Myxobolus* Bütschli, 1882 have been reported from various organs of hybrid Catla-Rohu host fish. All the three species possess two different types of spores from the same plasmodium of the same host, one bearing polar capsules of equal dimensions another with polar capsules of unequal dimensions - a very peculiar characteristic feature which is not a very common phenomenon in the genus *Myxobolus.* However, extrusion of polar filament has not been achieved in most of the spores, although all the conventional chemical reagents recommended for this purpose were used. This is perhaps due to very thick valves of the spores that prevents desiccation which seems to initiate the required changes leading to extrusion or the filaments (Lom, 1964).

However, from the present state of our knowledge it is difficult to speculate as to whether this type of spore structure have any relation with the development in a hybrid carp. Future studies may perhaps give some answer to this very important question.

Acknowledgements

The senior author is thankful to the Head of the Department of Zoology, University of Kalyani, for providing laboratory facilities. Thanks are also due to the Director, West Bengal Subordinate Education Service, Government of West Bengal and the Headmaster, Uttarpara Government High School for giving permission to continue the research work. Sincere thanks are also expressed to Mr. Amlan Kumar Mitra, Assistant Teacher, Kalna High School, Kalna, Burdwan, for supplying the host fishes from Ranaghat and Shantipur, Nadia, West Bengal, India.

References

Bhatt, V.S. and Siddiqui, W.A. (1964). Four new species of Myxosporidia from the Indian fresh water fish *Ophicephalus punctatus* Bloch. *J. Protozool.,* 11: 314-316.

Bajpai, R.R.N., Kundu, T.K. and Haldar, D.P. (1981). Observations on *Myxosoma magauddi* n.sp. (Myxozoa: Myxosomatidae), parasitic on the gill filaments of *Trichogaster fasciatus* Bl. Schn. *Riv. Parasitol.,* 42(3): 343-350.

Basu, S. and Haldar, D.P. (1998). Comparative study on prevalence of protozoan parasites in pure and hybrid carps *Environ. and Ecol.,* 16(3): 584-587.

Bauer, O.N., Egusa, S., and Hoffman, G.L. (1981). Parasitic infections of economic importance in fishes pp. 425-443. In W. Slusarski, (Ed), *Review of Advances of Parasitology*. Polish Acad. Sci. Committee for Parasitology, PWN, Polish Scientific Publ., Warszawa, Poland.

Chakravarty, M.M. (1939). Studies on Myxosporidia from the fishes of Bengal, with a note on the myxosporidian infection in aquaria fishes, *Arch. Protistendkd.,* 92: 169-178.

Chakravarty, M.M. and Basu, S.P. (1948). Observations on some myxosporidians parasitic in fishes, with an account of nuclear cycles in one of them. *Proc. Zool. Soc. Beng.* 1: 23-33.

Donec, J.S. and Shulman, S.S. (1984). Knidosporidii. In: O.N. BAUER (ed.) opredelitel' parazitov presnovodnikh rib fauni SSSR. 1: 88-251. [Cnidosporidia. In: O. N. BAUER (ed.), Key to parasites or freshwater fishes of USSR Fauna] (in Russian).

Haldar, D.P., Das, M.K. and Sharma, B.K. (1983). Studies on protozoan parasites from fishes. Four new species of the genera *Henneguya* Thelohan, 1892, *Thelohanellud* Kudo, 1933 and *Myxobolus* Bütschli, 1892. *Arch. Prostistenkd.*, 127: 283-296.

Haldar, D.P., Samal, K.K. and Mukhopadhyay, D. (1996). Studies on the protozoan parasites of fishes in Orissa: eight species of *Myxobolus* Bütschli (Myxozoa: Bivalvulida). *J. Beng. Nat. Hist. Soc.* (New series), 16(1): 3-24.

Kent, M.L. and Hoffman, G.L. (1984). Two new species of Myxozoa, *Myxobolus inaequus* sp.n. and *Henneguya theca* sp.n. from the brain of a South American Knife fish-*Eigemannia virescens* (L). *J. Protozool.*, 31: 91-94.

Kudo, R.R. (1920). Studies on Myxosporidia, *Illinois Biol. Monogr.*, 5: 245.

Kudo, R.R. (1933). A taxonomic consideration of Myxosporidia. *Tr. Am. Micr. Soc.*, 52: 195.

Kudo, R.R. (1934). Studies on some protozoan parasites of fishes of Illinois, *Illinois Biol. Monogr.*, 13: 1.

Lalitha Kumari, P.S. (1969). Studies on parasitic protozoa (Myxosporidia) of fresh water fishes of Andhra Pradesh, India. *Riv. Parasitol.*, XXX(3): 153-226.

Landsberg. J.H. and Lom. J. (1991). Taxonomy of the genera *Myxobolus/Myxosoma* group (Myxobolidae: Myxosporea): current listing of species and revision of synonyms. *Syst. Parasitol.*, 18: 165-186.

Lom, J. (1964). Notes on the extrusion and some other features of myxosporidian spores. *Acta Protozool.*, 2: 321-327.

Lom, J. and Arthur, J.R. (1989). A guideline for the preparation of species descriptions in Myxosporea, *J. Fish Dis.*, 12: 151-156.

Lom, J. and Dyková, I. (1992). Protozoan parasites of fishes Development in Aquaculture and Fisheries Sciences. Elsevier Publ., Amsterdam.

Lom, J. and Noble, E.R. (1984). Revised classification of the Class Myxosporea Bütschli, 1881. *Fol. Parasitol.* (Praha), 31: 193-205.

Lom, J. and Vavrá, J. (1963). Mucus envelopes of spores of the subphylum Cnidospora (Doflein, 1901). *Vestn. Cs. Spol. Zool.*, 27: 4-6.

Mitchell, L.G. (1977). Myxosporidia. In: KREIER, J.F. (ed.), Parasitic Protozoa, Vol. 4, pp. 115-154. New York.

Molnár, K., Masaumian, M. and Abasi, S. (1996). Four new *Myxobolus* spp. (Myxosporea: Myxobolidae) from Iranian Barboid Fishes. *Arch. Protistenkd.*, 147: 115-123.

Mukhopadhyay, D. and Haldar, D.P. (1998). Spore variation in a piscine myxozoan parasite (Myxozoa: Myxosporea) from the juvenile major carp *Catla catla*. *Trans. Zool. Soc. India.* 2(1): 46-52.

Poche, F. (1913). Das System der Protozoa. *Arch. Protistenkd.*, 30: 125-321.

Ray Chaudhuri, S. and Chakravarty, M.M. (1970). Studies on Myxosporidia (Protozoa: Sporozoa) from the food fishes of Bengal, I. Three new species from *Ophicephalus punctatus* Bloch. *Acta Protozool.*, VIII: 167-173.

Rogers, W.A. and Gains, Jr. J.L. (1975). Lesions of Protozoan diseases in fish. The Pathology of fishes. Edited by William E. Riberllin and George Migaki, Wisconsin.

Sarkar, N.K. (1993). On two new species of *Myxoholus* Bütschli 1882 (Myxozoa: Myxosporea) from some freshwater fish of West Bengal, India. *Proc. Zool. Soc.*, 46(1): 61-66.

Seenappa, D. and Manohar, L. (1980a). Two new species of *Myxobolus* (Myxosporidea: Protozoa) parasitic on *Cirrhina mrigala* (Hamilton) and *Puntius curmaca* (Hamilton). *Curr. Sci.*, 49(5): 204-206.

Seenappa, D. and Manohar, L. (1980b). *Myxobolus vanivilasae* n. sp. parasitic in *Cirrhina mrigala* (Hamilton). *Proc. Ind. Aca. Sci.*, 89(5): 485-491.

Seenappa, D. and Manohar, L. (1981). Five new species of *Myxobolus* (Myxosporea: Protozoa), parasitic in *Cirrhina mrigala* (Hamilton) and *Laheo rohita* (Hamilton), with a note on a new host record for *M. curmucae* Seenappa and Manohar, 1980. *J. Protozool.*, 28(3): 358-360.

Shulman, S.S. (1962). Key to myxosporidian fauna of the USSR, No.80. *In* Bykhovskaya-Pavlovskaya, I.E. *et al.*, (ed): *Key to Parasites of Fresh Water Fishes of the USSR, Nauka, Moscow-Lelingrad*, 55-155 (Eng. Trans.).

Smothers, J.F., Von Dohlen, C.D., Smith Jr., L.H. and Small, R.D. (1994). Molecular evidence that the Myxozoan protists are metazoans. *Science*, 265: 1719-1721.

Thélohan, P. (1892). Observation sur les myxosporidies et éssai de classification de ces organismes. *Bull. Soc. Philom.* 4: 165-178.

Thélohan, P. (1895). Recherches sur les myxosporidies *Bull. Sc. Fr. Belg.* 26: 100.

Tripathi, Y.R. (1952). Studies on parasites of Indian fishes, I, Protozoa: Myxosporidia together with check list of parasitic Protozoa described from Indian fishes *Rec. Indian Mus.*, 50: 63-88.

Walliker, D. (1968). The nature of the vacuole of myxosporidian spores and a proposal to synonymise the genus *Myxosoma* Thélohan, 1892 with the genus *Myxobolus* Bütschli 1882 *Protozoology*, 15: 571-575.

8

Allelopathic Potential of Three Aquatic Macrophytes on Growth of Water Hyacinth

Jyoti Gupta and Manjula K. Saxena

Department of Botany, University of Rajasthan, Jaipur - 302 004.

ABSTRACT

The two emergent plants like *P. karka* (Retz.) Trin, Ex. Steud. (Poaceae) *Polygonum glabrum* Wild (Polygonaceae) and one rooted floating plant *Nymphaea stellata* were screened for their allelopathic impact on water hyacinth (Eichhornia crassipes, Mart.) Solms. The water hyacinth was allowed to grow in experimental pots containing 3 per cent shoot and root leachate (w/v) of *P. karka, P. glabrum* and *N. stellata* respectively. About 97.42 and 94.78 per cent inhibition of water hyacinth over *control* was observed in aqueous shoot leachate of *P. karka and P. glabrum*. In comparison to roots, shoot leachates of *P. karka* and *N. stellata* were found more toxic to the growth of water hyacinth. The study highlights the weedicidal properties of two emergent plants viz., *P. karka* and *P. glabrum*.

Key Words: Biocontrol, Allelopathy, Allelochemicals, Water hyacinth, *Polygonum glabrum*, *Phragmites karka* and *Nymphaea stellata*.

Introduction

Allelopathy refers to biochemical interaction among plants through the production and release of allelochemicals (Rice, 1984). Allelopathy can be also used as a means for biocontrol of aquatic weeds (Szczepanski, 1977). Suppressed growth of water hyacinth was observed in the presence of planktonic algae such as *Scendesm bifugatus, Chlorella pyrenoidosa* (Sharma, 1985). A number of studies on a duckweed (*Spirodela polyrhiza* L.) and seed germination and seedling growth bioassay to reveal the allelopathic nature of *Phragmites karka* (Kulshreshtha, 1981) and *Polygonum glabrum* (Gupta, 1998, Gogoi, 1992, Singh, 1992, Alsaadawi and Rice, 1982 and Datta and Chatterjee,1980). *P. karka* has been found to inhibit the growth of *S. polyrrhiza,* wheat, barley, pea and chickpea seedlings when grown in 1 per cent extract.

This study was conducted to investigate the allelopathic potential of these two macrophytes along with *Nymphaea stellata (Wild) on water hyacinth.* The water hyacinth [*Eichhornia crassipes* Mart.) Solms. Family Pontederiaceae] is a floating aquatic weed, native to the Amazon River basin, Brazil but now growing in 50 countries on five continents (Kulshreshtha, 1981) as a troublesome weed. Herbicides and biological control are most commonly used to control water hyacinth. However, being harmful to other useful organism in the lake the former has limited scope. Out of more than 100 species tested for biocontrol, only three are reported to be suitable as biocontrol agents viz., water hyacinth weevils *Neochatina eichhornia, N. crassipes* (Julien and Griffiths, 1998) *N. bruchi* [Coleoptera: Curculionidaa (Jimenez *et al.*, 2001)] and a water hyacinth moth *Niphograpta albiguttalis* (Jimenez *et al.*, 2001).

Management of the obnoxious growth of some weed species through allelopathy has attracted considerable attention in recent years (Saxena, 2000, Gopal and Goel, 1993, Rice, 1984). To investigate the role of allelopathy in biocontrol, some aquatic plants were screened for their allelopathic potential on water hyacinth. This study was conducted to determine the effects of aqueous leachates (3 per cent w/v) of three macrophytes namely *Polygonum glabrum* Wild. (Polygonaceae), *Phragmites karka* (Retz.) Trin. Ex. Steud. (Poaceae) and *Nymphaea stellata* (Wild). (Nymphaeaceae) on water hyacinth.

Materials and Methods

The fresh shoots of two emergent plants *P. karka* (Retz.) Trin. Ex. Steud. (Poaceae), *P. glabrum* Wild. (Polygonaceae), and one rooted floating plant *N. stellata* Wild. (Nymphaeaceae) were collected from the Botanical Garden, University campus, Jaipur, India. The aboveground (shoots) and below ground parts (roots and rhizomes) of *P. karka* and *N. stellata* were separated and cut into 5 cm pieces. Only the above ground parts of *P. glabrum* (shoots with inflorescence) were used for these experiments. Three hundred gram fresh shoot and roots were separately soaked in 10 liters tap water (3g plant material per 100 ml of tap water) at room temperature (28-40°C) for 24 hr. Tap water was used instead of distilled water in order to maintain more natural conditions of pond water. After soaking, the mixture was passed twice through the two layers of cheesecloth and the filtrate was permitted to settle for 8h. The supernatant was decanted and used for these experiment. *E. crassipes* was grown in this leachate. The treatments consisted of 3 per cent aqueous leachate (shoots and roots) of the aquatic weeds *P. karka, P. glabrum* and *N. stellata.*

The simple growth experiment was conducted in plastic pots filled with 400 ml per pot of shoot leachate, roots leachate and tap water (control) separately. The pots of treatments and control were replicated five times. Two plants of water hyacinth were placed in each pot. Each plant had a total of 4 leaves and 2 leaf buds. Initial dry weight of the shoot and root of water hyacinth was observed to be 0.735 ± 0.008g, 0.30 ± 0.01g respectively. The pots were kept in an experimental chamber. The light intensity ranged 1000-2500 Lux measured by Lux meter (Aplab ML 4420) and temperature 28-40°C). The water level was maintained by adding 100 ml of water daily in each experimental pot with the aim to see the impact of added plant leachates of three plants on *E. crassipes.* It did not affect the concentration of the leachate, however, the amount absorbed by the plant may affect the concentration but in the field, dilution factor is also very common. The plants were harvested after 21 days. The leaves were counted. After being washed the shoots and roots were separated and dried in hot air oven at 80°C and dry weight recorded. The relative growth rate (RGR) was calculated as (Hillman, 1961):

$$\text{RGR} = \frac{\log(\text{Final frond}) - \log(\text{Initial frond})}{\text{days (d)}}$$

Results and Discussion

The effect of shoot and root leachate of different species on the growth of water hyacinth is shown in Table 8.1. Growth was significantly reduced in all treated sets both in terms of number of leaves and dry weight after 21 days of the experiment. The toxicity of these plant residues was also evidenced from the negative relative growth rate values of *E. crassipes*. About 97.42 and 94.78 per cent inhibition of water hyacinth over control was recorded in aqueous shoot leachate of *P. karka* and *P. glabrum* after 21 days. The present study highlights the allelopathic potential of *P. karka* and *P. glabrum* to suppress the growth of water hyacinth.

Table 8.1: Effect of Aqueous Leachate (3 per cent) of Aquatic Plants on the Growth of Water Hyacinth

Treatments Macrophytes	*No. of Leaves Initial*	*21d*	*RGR[1] 21d*	*Dry weight (g) AG*	*After 21 days BG*
Control	4.0	6.33[a]	09.49	1.23[b]	0.89[a]
P. karka (AG)	4.0	0.16[f] (-97.47)	-113.70	0.01[f] (-99.99)	0.60[b] (-32.58)
P. karka (BG)	4.0	2.50[d] (-64.45)	-21.00	0.12[d] (-79.77)	0.43[e] (-51.6)
N. stellata (AG)	4.0	4.30[b] (-32.06)	-02.80	0.86[c] (-18.69)	0.46[d] (-48.31)
N. stellata (BG)	4.0	3.83[c] (-39.49)	-27.85	1.33[a] (8.13)	0.50[c] (-43.83)
P. glabrum (AG)	4.0	0.33[c] (-94.78)	-68.80	0.04[c] (-96.74)	0.17[f] (-80.89)

1. Relative growth rate in terms of number of leaves,
2. Means within columns followed by different letters are significantly different (0.05) according to Duncan's Multiple Range Test. AG: Above ground part, BG: Below ground part, Data in parentheses indicate per cent reduction over control.

The present study reveals that above ground parts and underground parts of *P. karka* and *N. stellata* have shown varying degree of toxicity. In comparison to roots, the shoots of *P. karka* and *N. stellata* were found more toxic to the growth of water hyacinth. Different parts of plants have been shown to exhibit varying degree of toxicity (Horsle, 1977). In general, leaves have been found to contain a higher concentration of the toxic material than other plant parts (Gupta, 1998 and Fietcher and Renney, 1963). In our earlier study on the effect of 1 per cent extract of the leaves, stem, and roots of *P. karka*, only leaf extract was found to be inhibitory (44 per cent reduction in the number of fronds) to the growth of duckweed *(Spirodela polyrrhiza)*. The toxicity of different plant parts of *P. karka* has been reported in the following order - Leaves > stem > roots > rhizomes using duckweed bioassay.

This study highlights the bioweedicidal properties of two emergent plants *P. karka* and *P. glabrium* which were found highly toxic to the growth of water hyacinth. These plants can be used integrated material in biocontrol programmes for water hyacinth and need to be investigated.

Acknowledgement

We gratefully acknowledge DST, New Delhi for providing financial assistance.

References

Alsaadawi, I.S., Rice, E.L. (1982). Allelopathic effects of *Polygonum aviculare* L. I Vegetational patterning. *J. Chem. Ecol.* 8: 993-1009.

Datta, S.C., Chatterjee, A.K. (1980). Allelopathy in *Polygonum orientate:* Inhibition of seed germination and seedling growth of mustard. *Comp. Physiol. Ecol.* 5: 54-59.

Fietcher, R.A., Renney, A.J. (1963). A growth inhibiter found in Centauraea spp. *Can. J. Plant Sci.* 43: 475- 481.

Gogoi, A.K. (1992). Effect of weed extracts and fibrous organic matters of weeds on emergence and growth of *Micania micrantha.* In: First National Symposium. 'Allelopathy in agroeco-systems Agriculture and Forestry.' 12-14 Feb. (1992). Abstract. p.80.

Gopal, B., Goel, U. (1993). Competition and allelopathy in aquatic plant communities. *Botan. Review.* 59(3): 155-210.

Gupta, J. (1998). Studies on allelopathic potential of some terrestrial and aquatic weeds. 128 pp. Ph.D. Thesis. University of Rajasthan, Jaipur.

Hillman, W.S. (1961). The lemnaceae or duckweeds: A review of the descriptive and experimental literature. *Botan. Rev.* 27: 221-287.

Horsley, S.B. (1977). Allelopathic interferences among plants. II. Physiological modes of actions. In: Proceedings of the fourth North American Forest, Biology, Workshop (Wilcox, H.E. and Hamer, A.F. (Eds.). pp.93-136.

Jimenez, M.M., Lopez, E.G., Delgadillo, R.H. and Franco, E.R. (2001). Importation, rearing, release and establishment of *Neochatina bruchi (Coleopttera Curculioniidae)* for the biological control of water hyacinth in Mexico. *J. Aqua. Pl. Manage.* 39(7): 140-148.

Julien, M.H., Griffiths, M.W. (eds.). (1998). Biological control of weeds: a world catalogue of agents and their target weeds. 4th Ed., CABI Publishing, UK.

Kulshreshtha, M. (1981). Allelopathic influence of some aquatic macrophytes. Acta Limnologia Indica. 1: 35-37.

Rice, E.L. (1984). Allelopathy. 2nd Edition. Academic Press, New York, 421 p.

Saxena, M. (2000). Aqueoua leachate of Lantana camara kills water hyacinth. *J. Chem. Ecol.* 26 (10): 2435-2447.

Sharma, K.P. (1985). Allelopathic influence of algae on the growth of *Eichhornia crassipes* (Mart.) Solms. Aquatic Bot. 22: 71-78.

Singh, M. (1992). Studies on growth competition and allelopathic interactions among aquatic macrophytes. Ph.D. Thesis, University of Rajasthan, Jaipur.

Szczepanski, A.J. (1977). Allelopathy as a means of biological control of water weeds. *Aquatic Bot.* 3: 193-197.

9

Food and Feeding Habit of the Cyprinid Fish *Semiplotus manipurensis* from Manipur

Laishram Kosygin and Waikhom Vishwanath*

Loktak Development Authority, Manipur, Lamphel - 795004
** Department of Life Sciences, Manipur University, Canchipur -795003*

ABSTRACT

Semiplotus manipurensis feeds on phytoplankton, zooplanktons, crustaceans, insects, detritus and sands. The gut content reveals that 70.5-74.0 per cent of food items encountered are algae and diatoms. Alimentary canal of the fish is described and discussed. Feeding intensity was high during November and low during March. The nature of gut length and content, structure of mouth, and nature of gill rakers suggests that *Semiplotus manipurensis* is an omnivorous bottom feeder, feeding predominantly on planktonic materials.

Key Words: *Semiplotus manipurensis*, Food and Feeding Habit, Manipur.

Introduction

A detailed knowledge on the biology of fish is useful in fishery management conservation and selecting the fish of high productivity for fish culture. Now-a-days emphases are being laid more and more on studies of vital statistics of fishes for development and conservation of fishery. This aspect is scarcely touched for many of the riverine fishes of India, many of which incidentally are also favourite fishes among pisciculturists.

Semiplotus manipurensis Vish. and Kosygin, 2002 is a delicious carp which has a high demand for its good taste. This fish is known only from Ukhrul district of Manipur. Due to difficult hill terrain and lack of proper means of communication from the fishing areas to the fish landing centres, fishermen smoke the fish and bring to Ukhrul market where it fetches high prices. However, in this region, fishing operations are usually carried out during the late winter and early summer months (November to May) as fishing is practically difficult during other months. This is because of the following reasons: (i) rough weather conditions and huge volume of water with whirlpools and great velocity during rainy seasons, (ii) hill streams are shallow and do not hold enough water except in the localised deep pools during late pre-monsoon, (iii) use of chemicals and explosives which causes over-exploitation during the pre-monsoon period which result in scarcity of fish. This commercially important fish belongs to a rare genus *Semiplotus* that distributed only in South East Asia and very little is known about the fishes of this genus (Vishwanath & Kosygin, 2000). Menon (1989) included the type species of the genus *S. semiplotus* in the list of endangered freshwater fishes of India. Similarly, Talwar & Jhingran (1991) also remarked that *S. modestus* was a rare species which distributes only in Mayanmar. However, there is no information on the biology of these fishes. Thus a preliminary study on the biology of these fishes is considered necessary for conservation and development of fishery. The present preliminary investigation has undertaken on the food and feeding habit of *S. manipurensis* in the Ukhrul district of Manipur.

Materials and Methods

Specimens of *S. manipurensis* were collected from the Challou river during the period from May, 1997 to March, 1998. A total of 65 specimens ranging from 60.0 to 150.0 mm in TL of *S. manipurensis* were examined for the analyses of gut contents. The food of the fish was identified by the "points method" (Hynes, 1950). The gut contents were removed, sorted and identified up to generic level with a compound microscope.

The total length of the gut was measured for determining 'Relative gut length' (RLG) which was calculated as the ratio of intestinal length to total body length. Feeding habits were observed in the field.

For the study of alimentary Canal, the head of the fish were severed from the body, a few inches behind the operculum and cut across the mouth passing through one of the angles of the jaw. The specimen was fixed on tray and structure, and arrangement of the bucco-pharynx was studied. The types of gill rakers were also observed. Intestine was taken out from the body of the fish and studied.

Observations and Results

The alimentary canal of *S. manipurensis* is modified according to its feeding habit. It is morphologically distinguished as a non-tubular part consisting of mouth and bucco- pharynx and tubular region comprising of oesophagus, intestinal bulb, intestine and rectum. A brief description of the alimentary canal is given below:

Mouth

Mouth of the fish is of great wide and located inferiorly. The lower jaw has a exposed cornified cutting edge and free to move anterio-porteriorly.

Buccal Cavity

The mouth leads into a spacious buccal cavity, which is lined by numerous membranes. The anterior part of the roof shows feeble longitudinal markings. Whereas the floor of the cavity is lined by a thick transverse mucous membrane in the middle. The root of the cavity is deeply concave while the floor is nearly flat. Tongue is rudimentary, almost absent in small specimens (<80 mm in TL.). Teeth are completely absent in the buccal cavity of this species. Behind the upper lip, there is a fold of the velum, which is papillated.

Pharynx

The buccal cavity leads into a short pharynx which is provided with minute pharyngeal teeth on the floor. These teeth are borne by the inferior pharyngeal bones and work against a pharyngeal pad on the roof. The plateau of the floor of pharyngeal cavity is provided with numerous taste buds.

Gill Rakers

The ventro-lateral wall of the pharynx is perforated by oblique gill slits, separated by means of four pairs of gills. Gill raker is short, thin, densely packed and arranged in two rows. The rakers of neighbouring arches meet with one another to form a sieve across the gill slits, providing a very efficient filtering apparatus by which the outgoing current of water is filtered.

Oesophagus

The pharynx leads into a short indistinct oesophagus. While entering the body cavity it become slightly expanded.

Intestine

This fish do not possess a true stomach. The intestine is swollen behind the oesophagus for the storage of food and is known as intestinal bulb or intestinal swelling. The intestinal bulb gradually narrow down into the thin walled intestine proper and finally opens outside through the anus. The intestine of the fish is very long and highly coiled.

Gut Length in Relation to Body Length of Fish (RLG)

The relationship of gut length to the total length of the fish of 60.0-150.0 mm was ranges between 3.2865583 and 5.6910344. The value of "r" was found to be 0.9201377, showing a positive correlation of the gut length and total length of the fish.

Food Analysis

The percentage of different groups of food items was calculated from the gut content of about 65 fishes in the size group 60.0-150.0 mm in TL. The food contents of *S. manipurensis* reveal the occurrence of algae, diatoms, zooplanktons, detritus, sand and insects, and its larvae. However, the stomach contents may be divided into the following broad categories.

Plant Matter (Phytoplankton)

The result shows that plant matters constitute the main part of the gut content, which constitute about 70.5-74.0 per cent. It comprises algae and diatoms, which comes under chlorophyceae (green algae), myxophyceae (blue green algae) and bacillariae (diatoms). Desmids have been treated under chlorophycea. The generic constituents are *Zygnema, Gonatozygon, Spirogyra, Cladophora, Mougeotia* of chlorophyceae, *Nostoc, Anabaena, Lyngbya* of Myxophyceae, and *Epithemia, Navicula, Nitzschia, Diatoma* of bacillariae.

Detritus and Sand

Detritus consist of unidentified plant as well as animal matter, occurring regularly in guts throughout the study period, forming about 16.3-20.2 per cent. Sand was found to occur in small amount along with the detritus.

Animal Matter

Zooplankton, Rotifera and crustaceans mainly represented the animal matter. They were recorded in small percentage (4.3-7.5 per cent).

Miscellaneous

Miscellaneous matter was composed mainly of dipteran larvae, insect appendages, oligochaetes. These items were occasionally recorded in insignificant percentage (1.5-3.2 per cent) in the gut content.

Feeding Intensity

The condition of feed was determined from the degree of distension of the stomach and the same has been classified as 'gorged' 'full', '3/4 full' '1/2 full', '1/4 full', 'trace' and 'empty'. It is obvious that the feeding intensity of the fish is appreciably high, a good number of stomachs being either in 'gorged' or full conditions in the month of November. The few empty stomachs observed were from specimens collected during March (Table 9.1).

Table 9.1: Condition of Feed (in percentages) of *S. manipurensis* in Three Different Months During 1997-1998.

Month	*Gorged*	*Full*	*¾ Full*	*½ Full*	*¼ Full*	*Trace*	*Empty*	*N*
July	-	2.4	31.3	49.9	15.2	1.2	-	20
November	15.8	57.7	22.4	4.9	-	-	-	14
March	-	-	-	10.2	49.2	37.8	2.8	31

Discussion

The anatomical features of the alimentary canal in a fish are related to the food and feeding habit. The alimentary canal of *S. manipurensis* is structurally and functionally well adapted to their food and feeding habits. Presence of "Sector mouth" in this carp is of special interest in its feeding habit. According to Howes (1982) "sector mouth" is an inferior mouth which is wide (probably the widest mouth in relation to head length of any cyprinid fish) and lower jaw with an exposed corrified mandibular cutting edge. The lower jaw of this fish has a mobile structure, which is free to move anterio-posteriorly and appears to have the basic function of scrapping or ploughing epilithic material. They scrap off the algae in small instalments making a series of crescentic impressions on the rocks.

It is well known fact that, the stomachless fishes are provided with pharyngeal chewing device, which permit ingested food reach the intestine in a fragmented condition. In *S. manipurensis,* the pharyngeal teeth born by the inferior pharyngeal bones seems to work against a bony callous pad, thereby forming on effective grinding apparatus for crushing the food. Various workers have divided the bucco-pharyngeal region of fishes into an anterior part, the buccal cavity extending upto the first gill arch and a posterior portion, the pharynx. Sinha (1993) while studying the histo-morphology of the alimentary canal in Indian freshwater teleosts remarked that there is a little morpho-histological difference in the two parts and probably there is no hard and fast line to demarcate them. Hence, in the present study the entire region has been described as the bucco-pharyngeal region. The densely packed gill rakers of this fish seem to form an effective sieve to stain plankton from the irrigating water.

The length of the gut in relation to that of the total length varies greatly in different species. In general, the carnivorous species have a shorter intestine, the omnivores a longer one and the herbivores the longest due to obvious reasons emanating from digestion of specific type of food (Sinha, 1993). The fish has a very long, highly coiled and stomachless alimentary canal which confirms that it is a herbivorous fish. However presence of intestinal bulb compensates the absence of stomach in this fish as described by Lalger (1984). According to many workers (Das and Moitra 1963, Biswas, 1985, Biswas and Phukon, 1996), RLG value has direct relationship with the feeding types of fish. A lower value indicates intake of animal matter in the diet whereas higher value suggests dependence on plant matter in large quantities. In the present study RLG value of the fish was high and the value tends to increase as the fish grow. This finding shows herbivorous nature of the fish and increasing of RLG value as the fish grow suggest more dependence on plant matter in large quantities as the fish grow.

The feeding intensity of the species was low during March. It may be attributed to the developments of gonads, which occupy the major space of the abdominal cavity, which is described as 'spawning fast' by many workers (Frost, 1954 and Lawler, 1965). The present observations on the nature of gut content, RLG values, structure of mouth, nature of gill rakers suggests that *S. manipurensis* is an omnivorous bottom feeder, feeding predominantly on planktonic materials.

References

Biswas, S.P. (1985). Studies on the gut length in relation to feeding habit of five commercially important fishes from Assam. *J. Assam Sc. Soc.*, 28(1): 10-13.

Biswas, S.P. and Phukon, J. (1996). On some aspects of an ornamental fish, *Erethistes pussilus* from the Brahmaputra River System. *J. Freshwater Biol.*, 8(1): 27 -32.

Das, S.M. and S.K. Moitra (1963). Studies on food and feeding habit of some freshwater fishes of India, Part-IV. A review on the food and feeding habits with general conclusion. *Ichthyoiogica.*, 2 (1-2): 107-115.

Frost, W.E. (1954). The food of pike *Esox lucius* L., in Windermer. *J. Anim. Ecol.*, 23(2): 339-360.

Howes, G.J. (1982). Anatomy and evolution of the jaws in the semiplotine carps with a review of the genus *Cyprinion* Heckel, 1843 (Teleostei: Cyprinidae). *Bull. Brit. Mus. Nat. Hist. (Zool.)*, 42(4): 299-335.

Hynes, H.N.B. (1950). The food of freshwater strickle backs *(Gasterosteus acculatus* and *Pygosteus pungitius*) with a review of methods used in studies of the food of fishes. *J. Anim. Ecol.*, 39: 36-58.

Lalger, K.F. (1952). Freshwater fishery biology. Win.C. Brown Company (Publishers) Dubuque Lowa.

Lawler, G.H. (1965). The food of the pike, Esox lucius, in Hemming Lake, Manitoba. *J. Fush Res. Board Can.*, 22(6): 1357-1377.

Menon, A.G.K. (1989). Conservation of the ichthyofauna of India. In: Jhingran A.G. and V.V. Sugunam, (Eds). Conservation and management of Inland Capture fisheries resources of India. Inland Fisheries Society of India, Barrackpore: 25-33.

Sinha, G.M. (1993). Histo-morphology of the alimentary canal in Indian freshwater teleost. *Adv. Fish Res.*, 1: 193- 220, pl. 2.

Talwar, P.K. and A.G. Jhingran, (1991). Inland Fishes of India and Adjacent Countries, I. Oxford and IBH Publ. Co. Pvt. Ltd., New Delhi, 541 pp.

Vishwanath, W. and L. Kosygin , (2000). Fishes of the cyprinid genus *Semiplotus* Bleeker, 1859 with description of a new species from Manipur, India. *J. Bombay Nat. Hist. Soc.*, 97(1): 92-102.

10

A Comparative Study on the Problems and Nature of Assistance availed from the Development Department by the Landless Farm Women Labourers in Irrigated and Rainfed Areas of Shimoga District, Karnataka

K.P. Raghu Prasad, T. Geetha Lakshmi* and Dr. B. Sundaraswamy**

Assistant Professor of Agricultural Extension, College of Agriculture, Shimoga
** Lecturer in Home Science, Kamala Nehru College, Shimoga*
*** Professor and Head of Agricultural Extension, UAS, Dharwad*

ABSTRACT

The study revealed that, inadequate food, clothing, wages, medical facilities, drinking water, transportation were the major problems faced by the respondents. Most of the respondents had obtained assistance from the development departments to build a house followed by a ration card from the co-operative society. The major reasons given by the respondents for not making use of the benefits extended by the development departments were that they could not repay the instalments of the loans followed by not aware of the facilities, and officers did not co-operate to help them to get the facilities.

Key Words: Inadequate Food, Clothing, Development Departments, Instalment of Loans.

Introduction

Women play a crucial role in agricultural development. They contribute to about 1/3rd of the labour required for agriculture field. At the same time they have very low socio-economic profile which is largely based on the myth that women are inferior as agents of production. Whereas in reality women work for every minute of their working hours. Their day starts from fetching water to cooking, washing, feeding children and then rush off to a 6-8 hours of back breaking jobs. On their return fires must be lit, food must be cooked and the days chores need to be completed. Inspite of all these, there is no guarantee that food will be adequate for the family. In addition to energy deficiency there is a high incidence of vitamin and mineral deficiencies leading to anemia, glossities and other associated problems. Inadequate food leaves one tired through the day, work output suffers and even the pace of life cannot be quickened. Still pathetic and worst is the condition of those people who possess no land and are purely dependent on agricultural wages.

Though the government has made few attempts through the development programmes to help this section of the people, not many are benefited through these programmes. Utter poverty and illiteracy among the labourers make them think of all probable ways to earn a square meal per day and nothing else. This sector of the people are usually exploited by the upper class people. With this background the study was undertaken to understand the problems, the assistance availed from the development departments and reasons for not making use of the benefits by the respondents.

Methodology

The study was conducted in the northern transitional zone of Shimoga district in the taluks of Bhadravathi and Honnali representing irrigated and rainfed areas respectively. A total of 200 respondents, one hundred each from irrigated and rainfed areas were selected randomly for the study. The data was collected by personal interview method with the help of a specially designed pretested interview schedule. The data has been presented using frequency and percentage.

Results and Discussion

Problems Faced by the Respondents

According to Table 10.1 all the respondents stated that they were facing problems with regard to the basic necessities of life like inadequate food (100 per cent) and inadequate clothing (99.5 per cent) in both the areas. This could be mainly because of the low wages which did not suffice to meet their basic needs like a square meal per day and adequate clothing. Their housing conditions were very poor with leaking roof. Some lived in rented houses, while others lived in thatched shed situated in unhygienic conditions with poor water and sanitary facilities.

Inability to provide education for their children was stated by 76 per cent of the respondents. This might be because of the low economic conditions, and their perception of sending child to school would mean deletion of one earning member.

Inadequate medical facilities was the problem stated by 57 per cent of the respondents. Many of the villages even though had a primary health unit either did not have a doctor residing at the center, or did not have enough medical facilities like medicines and beds to accommodate patients. The doctors usually resided in a nearly town/city and visited the hospital once or twice a week, which would be inconvenient.

Inadequate drinking water facilities was stated by 40 per cent of the respondents. This was a major problem of the rainfed area because of the absence of canals and water distribution system. Even though there were bore wells and mini water supply, water was not sufficient as they failed to work during the peak summer seasons.

Inadequate transportation facilities, no electrification and untouchability were the other problems encountered by the respondents. The above findings were similar to that of Mehta (1988) who revealed that the labourers had problem of inadequate water food and fuel.

Table 10.1: Problems Faced by Respondents

Problems	*Respondents*		
	N = 100 Irrigated	*N = 100 Rainfed*	*N = 200 Total*
Inadequate food	100	100	200 (100)
Inadequate clothing	100	100	200 (100)
Do not own a house	36	56	92 (46)
Low wages	99	98	197 (98)
Inability to provide education to children	70	82	152 (76)
Inadequate medical facilities	39	75	114 (57)
Inadequate drinking water	20	60	80 (40)
No electrification for houses	44	58	102 (51)
Inadequate transport facilities	18	56	74 (37)
Poor housing	31	39	70 (35)
Untouchability	16	33	49 (24.5)

Note: Figures in the Parenthesis Indicate Percentages.

Table 10.2: Nature of Assistance Availed from the Development Departments by the Respondents

Assistance Availed	*Respondents*		
	N = 100 Irrigated	*N = 100 Rainfed*	*N = 200 Total*
Janatha/HUDCO House from the Government	57 (57)	38 (38)	95 (47.5)
Ration Card from the Co-operative Society	53 (53)	36 (36)	89 (44.5)
Loan for Cattle from Co-operatives/Banks	24(24)	11(11)	35(17.5)
Electricity for their house Bhagya Jyothi/ Mandal Jyothi	21 (21)	-	21 (10.5)

Note: Figures in the Parenthesis Indicate Percentages.

Table 10.3: Reasons for not making use of the benefits extended by the Development Departments

Reasons	*Respondents*		
	N = 100 Irrigated	*N = 100 Rainfed*	*N = 200 Total*
Cannot pay the instalment of the loan obtained	40 (40)	35 (35)	75 (35.5)
Not aware of the facilities	33 (33)	14 (14)	47 (23.5)
Officer do not help to get the facilities	10 (10)	8 (8	18 (9)
Did not try to get the facilities	1 (1)	17 (17)	18 (9)
Affluent class and higher caste people dominate over them	1 (1)	16 (16)	17 (8.5)

Note: Figures in the Parenthesis indicate Percentages.

Nature of assistance availed from the Development Departments

It can be observed that 57 and 38 percent of the respondents in irrigated and rainfed areas respectively had obtained assistance to build a house followed by a ration card from the co-operative society. Man usually tends to fulfil his basic needs of food, clothing and shelter and this might be the strong drive behind obtaining assistance for building a house and also to obtain a ration card which would provide them with food grains at a fair price. The slightly greater percentage in irrigated area might be because of the general awareness of the people and better economic conditions. The other assistance like loans for cattle and electricity for their houses was not availed by many of the respondents in both the areas because of the long and cumbersome procedures involved to obtain loans, inability to repay the instalments and maintain the cattle, existence of previous loans which had not been cleared, lack of time to make frequent visits to the banks, and lack of knowledge about the personnel to be approached might be the probable reasons. Their sheer poverty was another important reason.

Reasons for not making use of the benefits extended by the Development Departments

The table shows that 40 per cent in irrigated and 35 per cent in rainfed area gave reason that they could not repay the instalments of the loans, 33 per cent in irrigated and 14 per cent in rainfed area stated that they were not aware of the facilities. The reasons for the above facts could be utter poverty. Less contact with the officers, illiteracy and ignorance might be the reasons for not being aware of the facilities. Officers do not help to get the facilities affluent class and higher caste people dominate over them and did not try to get the facilities were the other reasons stated by few respondents in both areas. The usual tendency of the officers to neglect the weaker sections of the society, the existence of the exploitation of the lower class people by the affluent and upper class thus making use of the facilities available depriving the weaker class the fruits of benefits might be the possible reasons.

Inadequate food, clothing, wages, medical facilities, drinking water, transportation, inability to provide education for their children, no electrification were the major problems faced by the respondents. In this respect measures could be taken to enhance the wages of the labourers to ensure at least the basic amenities like food, clothing and shelter. Drinking water and medical facilities should be given the due attention. Though certain programmes are implemented by the government not many are benefited through these programmes. Many of them are not aware of them and many others were not able to reap the benefits because of the dominating upper class and the corrupt officials. Steps need to be taken to make the people aware of the programmes and see that the benefits reach the beneficiaries.

References

Mehta, A. (1998), The Voice of Rural Women, *Kurukshetra*, 36(11): 6

Shashikala, S. (1990). A Study on the activities and time spent by farm women of rainfed and irrigated areas of Dharwad. M.Sc. Thesis (unpubl), Univ. Agric. Sci., Dharwad.

11

Observations on Mixed Fish Farming of Indian and Exotic Carps in a Seasonal Pond of Jodhpur

Anita Jhajhria

Department of Zoology, Government P.G. College, Jalore - 334 001, Rajasthan

ABSTRACT

Mixed fish farming in Shekhawat pond (0.736 ha) was carried out for a period of 8 months from August 1999 to March 2000 using *Catla catla, Labeo rohita, Cirrhinus mrigala* and *Cyprinus carpio.* No supplementary feed and fertilizers were supplied to the fishes reared. The growth of *Cyprinus carpio* (1173.50 gm.) was best followed by *Catla catla* (578.30 gm.), *Labeo rohita* (263.40 gm.) and *Cirrhinus mrigala* (252.50 gm.). The fish production (4127 kg./ha/8 months) was achieved which is higher than the average fish production (100-400 kg/ha/yr) in similar water bodies of this region.

Key Words: Mixed Fish Farming, Shekhawat Pond, Indian and Exotic Carps, Production.

Introduction

The goal of achieving productivity from pond ecosystem is to sustain high yield of fish rich in protein. In aquatic ecosystems energy culminates at its different trophic levels with considerable loss of energy at each link of food chain. For augmenting high fish productivity of carps, the knowledge of bioecology of fresh water ponds is utmost important as ecology plays vital role in regulating productivity of these fishes.

The basic principle of mixed fish farming or polyculture is that when compatible fishes of different feeding habits are stocked together, they secure for themselves in the most efficient manner, all life requisites available in the pond for fish production without harming each other.

In North-Western Rajasthan, there are 176 seasonal water bodies with water spread area of 5700 ha. in which the fish production is as low as 100-400 kg/ha/yr (Pragati Prativedan, 1995). In Jodhpur township, as well as in its vicinity, there are many ponds of varied dimensions where knowledge on ecology and related fish productivity of fresh water pond ecosystems for utilization of appropriate fish fauna is still lacking. With this consideration, the studies where conducted for assessing the fish production of Shekhawat pond.

Materials and Methods

The fries of three Indian major carps namely, *Catla catla, Labeo rohila* and *Cirrhinus mrigala* alongwith an exotic carp *Cyprinus carpio* were obtained from the induced spawning exercises using the synthetic hormone. These fries were then transported in oxygen packed system of polythene bags in August, 1999 for stocking in the pond. The number of fries to be stocked in the pond was computed by adopting the formula (Jhingran, 1975):

$$\text{Number of fries to be stocked per unit area} = \frac{\text{Total expected increase in weight}}{\text{Expected increase of weight of individual fish}} + \text{Mortality (not more than 10\%)}$$

The Shekhawat pond (26°49′02″ North Latitude and 73°47′03″ East Longitude) is about 10 Km. from the Jodhpur railway station. The area of the pond is 0.7366 ha having a mean depth of 1.89 m. It lies in the vicinity of the army and the army mess is located upstairs to it.

The harvesting was done using cast and drag net in March, 2000. The pond water was collected monthly during the stocking period between 8.30 a.m.-11.30 a.m. and was analysed using the methods of APHA (1985) and Welch (1948).

Results and Discussion

The results of stocking and production are summarised in Table 11.1. The abiotic parameters analysed are shown in the Table 11.2. The production of 4127 kg/ha/8 months was achieved in the pond. The average weight of (1173.50 gm.) *Cyprinus carpio* was highest followed by *Catla catla* (578.30 gm.), *Labeo rohita* (263.40 gill.) and *Cirrhinus mrigala* (252.50 gm.).

Valuable contributions in the field of fish productivity of ecologically different ponds with reference to carps has been done by various workers, Lakshmanan *et al.* (1971) and Singh *et al.* (1972) conducted year long experiments at the pond culture, division of CIFRI combining seven Indian and exotic species of fish and obtained a production varying from 2234-4210 kg/ha/yr Sukumaran (1976) obtained a production of 6191-7332 kg/ha/8 months at Karnal. Sharma *et al.* (1979) reported a production of 4323 kg/ha/yr in fish-cum-duck rearing without adding any fertilizer to the pond and giving any supplementary feed.

Mathew *et al.* (1979) obtained productions with six species combination as 2199-2242 kg/ha/8 months at Poona.

Kumar (2001) recorded maximum growth of catla (9.3 kg.) followed by common carp (3.1 kg.), *rohu* (1.9 kg.) and for *mrigal* (320 gill.) in one year from deep ponds of Patna. Baghel and Saxena (2002) reported a production of 4965.97 kg/ha/9 months in Tarai region of Uttaranchal.

In the present study, the fish production (4127 kg/ha/8 months) relied only on the natural food and the kitchen wastes of the army mess. At least 50 per cent of the total food of carps in the fish ponds comes from the pond itself (Schaeperclaus, 1952). Gaudet (1967) reported that in Israel out of 2000 kg/ha, probably 400 kg. comes from the natural food.

Table 11.1: Species-wise Stocking and Production of Indian and Exotic Carps in Shekhawat Pond (0.076 ha.)

Fish Species	*Fin Formula*	*Habitats*	*Feeding Habits*	*Stocking Density of Fries*	*Total Weight (kg.)*	*Average Weight (gm.)*	*Average Length (cm)*
Catla catla (Catla)	Diii-iv 14-16, Aiii5, Pi 2O, Vi8	Surface Feeder	Planktonic	900	580	578.3	25.23
Labeo rohita (Rohu)	Diii-iv 12-14, Aiii-iii5, Pi 16-18, Vi8	Column Feeder	Herbivorous	2000	450	263.4	19.69
Cirrhinus mrigla (Mrigal)	Diii-iv 12-13, Aiii5, Pi17, Vi8	Bottom Feeder	Ominovorous	3200	410	252.5	20.65
Cyprinus carpio (Common Carp)	Diii-iv 18-20, Aiii5, Pi15, Vi8	Bottom Feeder	Ominovorous	4000	1600	1173.5	40.23

Table 11.2: Physico-chemical Properties of Pond Water During the Mixed Culture Period

Abiotic Parameters	*Stocking Months*							
	August 1999	*September 1999*	*October 1999*	*November 1999*	*December 1999*	*January 2000*	*February 2000*	*March 2000*
Water Temperature (°C)	24.1	24.5	23.6	20.1	16.5	15.5	15.9	20.4
pH	7.6	7.7	7.8	7.8	7.9	8.0	7.8	8.0
Transparency (cm)	14.0	20.0	22.0	24.0	19.0	17.0	26.0	30.0
DO (ml/l)	5.643	4.559	4.453	4.749	4.997	5.088	4.640	4.159
Free CO_2 (ppm)	9.0	8.0	7.0	6.0	4.0	3.0	5.0	4.0
CO_3^{-2} (ppm)	-	-	-	-	-	-	-	-
HCO_3^{-2} (ppm)	84.0	98.0	95.0	102.0	110.0	115.0	120.0	124.0
Total Alkalinity (ppm)	84.0	98.0	95.0	102.0	110.0	115.0	120.0	124.0
Total Hardness (ppm)	70.0	89.0	86.0	110.0	120.0	125.0	135.0	140.0
Chloride (ppt)	0.302	0.526	0.496	0.574	0.608	0.649	0.797	0.803
Salinity (ppt)	0.054	0.950	0.895	0.104	0.109	0.117	0.144	0.145
Phosphate (µg/l)	0.212	0.354	0.410	0.287	0.279	0.250	0.398	0.559
Nitrate (µg/l)	3.336	3.685	4.528	4.104	3.910	3.432	4.728	6.223

Thus, it may be concluded that on the basis of ecobased planning, judicious management and by incorporating the application of supplementary feed and fertilizers a many fold enhancement in fish production of the ponds of this region is possible.

Acknowledgement

The author expresses sincere gratitude to Dr. Devendra Mohan, Associate Professor, Department of Zoology, Jai Narain Vyas University, Jodhpur for evincing keen interest and guidance in the research work.

References

APHA (1985). *Standard methods for the examination of water and waste water*. AWWA, WPCE, New York, 16th edition.

Baghel, D.S and Saxena, V. (2002). Composite fish culture in Tarai region of Uttaranchal. *Geobios.*, 29: 101-104.

Gaudet, M. (1967). The status of warm-water pond fish culture in Europe. FAO, *Fish, Rep.*, 174 p.

Jhingran, V.G. (1975). *Fish and Fisheries of India*. 1st edition, Hindustan Publishing Corporation, New Delhi.

Kumar, D. (2001). Fish productivity of ecologically different ponds with reference to carps and air breathing fishes. *J. Inland. Fish. Soc. India*, 33: 8-16.

Lakshmanan, M.A.V., Sukumaran, K.K., Murty, D.S., Chakraborty, D.P., and Philipose, M.T. (1971). Preliminary observations on intensive fish farming in fresh water ponds by the composite culture of Indian and exotic species. *J. Inland Fish Soc. India*, 2: 1-21.

Mathew, P.M., Singh, B.K. and Chakraborty, D.P. (1979). Stocking of fry in composite fish culture. In: *Symposium on Inland Aquaculture*, CIFRI, Barrackpore. pp. 1-44.

Pragati Prativedan (Raj.) (1995). Inter State Fishery Planning Cell (1994-95). State Fisheries Department, Rajasthan, Jaipur.

Schaperclaus, W. (1952). *Die* karpfenteichswirtschaft in der Deutsche Demokratischen Republik. *Sitz. Ber.* VII (7), Berlin, Deutsche Ahad. Landwirtschaftwiss.

Sharma, B.K., Kumar, D., Das, S.R. and Chakraborty, D.P. (1979). Observations on swine-dung recycling in composite fish culture. In: Symposium on Inland Aquaculture (Abstract) CIFRI, Barrackpore. February 12-14, pp. 99.

Singh, S.B., Sukumaran, K.K., Chakrabarti, P.C. and Bagchi, M.M. (1972). Observations on composite culture of exotic carps. *J. Inland Fish. Soc. India*, 4: 38-50.

Sukumaran, K.K. (1976). Progress report of the Karnal subcentre of Haryana for the period November 1972 to August (1975). In: *Third workshop on the All India Coordinated Research project on composite fish culture and fish seed production*. February 26-27, CIFRI, Puri.

Welch, P.S. (1948). *Limnological Methods*. The Blakiston, Co., Philadelphia.

12

Water Quality of Well and Bore Well of Ten Selected Locations of Chitrakoot Region

S.S. Garg

Deputy Manager (Environment), Sanghi Industries Ltd. (Cement Division), Sanghipuram, P.O. Motiber, District Kutch, Gujarat - 370 655

ABSTRACT

Physico-chemical parameters of ground water from ten sampling locations of Chitrakoot Region were monitored for four seasons during year 2000 and value obtained were compared with standards prescribed by Bureau of Indian Standards (BIS). Analysis of ground water showed that water is good for drinking as well as irrigation purposes.

Key Words: Water Quality, Physico-chemical Parameter, Chitrakoot Region, Correlation and Regression Analysis.

Introduction

The great solvent power of water has been making the creation of absolutely pure water a theoretical rather than a practical goal. Even the highest quality distilled water is having dissolved gases and to a slight degree solids. The problem, therefore, has been one of determining what quality of water has been put – washing, irrigation, flushing away wastes, cooling, making paper etc. something to the water. In fact for centuries rivers and lakes have been used for dumping grounds for human sewage and industrial wastes of every conceivable kind, many of them have been highly toxic. Added to this have been the materials leached and transported from land by water percolating through the soil and running off its surface to aquatic ecosystems.

Materials and Methods

Preliminary survey was carried out to identify ground water sampling location considering its uses for domestic, drinking, agriculture and other purposes. Based on these ten different tube wells, in three litre clean polythene cans and transported to the laboratory for analysis. The parameter Dissolved Oxygen was analysed at the sampling site itself.

pH

pH of water sample was measured by using Systronics make digital pH meter.

Turbidity

Turbidity of water has measured by using Systronics make Nephelometer calibrated with Hydrazene Sulphate and Hexa Methylenetetramine turbidity solution and the values are expressed in NTU.

Total Dissolved Solids

The filtrate in sample (whatman 42) was collected in a clean pre-weighted porcelain dish and dried in hot air oven at in desecrator and weighed. The weight difference was used to calculate the TDS as mg/l.

Electrical Conductivity

EC of water sample was measured by using Systronics make conductivity meter with 1 cm width of cell and results are expressed in µmhos/cm.

Alkalinity

Fifty ml sample was taken and titer with 0.02 N Hcl using phenolphthalein and methyl orange indicator.

Total Hardness

50 ml of sample was buffered at pH 10(NH_4Cl and NH_4OH) and titillated against 0.02 N EDTA using Eriochrome Black T as indicator expressed as $CaCO_3$ mg/l.

Calcium and Magnesium

Calcium and magnesium was calculated from total hardness as mg/l.

Sodium and Potassium

Sodium and potassium was annualised using flamephotometer.

Chloride

Hundred ml of the sample was titillated against 0.0141 M $AgNO_3$ to a pinkish yellow using K_2CrO_4 as indicator.

Sulphate

To 100 ml sample, 20 ml buffer solution (magnesium chloride and sodium acetate) was added and a spoon of one minute the optical density was measured at 520 NM using Systronics make Spectrophotometer. The SO_4^{--} was determined from calibration curve in mg/l.

Phosphate

To 100 ml sample, 4 ml after 10 minutes the optical density was measured at 690 nm. From the calibration curve phosphate concentration was determined as mg/l.

Heavy Metals

Heavy metals were analysed using atomic absorption spectrophotometer.

Bacteria (MPN Fecal Coliform)

The water sample for microbial examination was collected in a clean and sterilized bottle. The bottle was immersed immediately, the sample was transported to laboratory for the analysis.

To test fecal coliform Al medium was used which contained the following ingredients:

Tryptose	:	20 g
Lactose	:	5 g
Bile Salts	:	1.5 g
K_2HPO_4	:	4 g
KH_2PO_4	:	1.5 g
NaCl	:	5 g
Reagent Grade H_2O	:	1 litre

The ingredients were heated to dissolve and pH was adjusted to 6.9 ± 0.1. Before sterilization, a vial medium. The tubes were capped with cotton plugs and sterilized.

Fermentation tubes were arranged in rows of five tubes each in test tube racks. In each of five tubes 10, 1, 0.1 ml sample was carefully inoculated complying all precautions and incubated for 24 hrs. at 37°C ± 0.2°C. After 24 hrs. gas production in vials were observed and positives counted. The MPN of faecal coliforms were determined using the following formula:

$$\text{MPN/100 ml} = \frac{\text{Number of positive tubes}}{\sqrt{\text{ml sample in negative tubes} \times \text{ml sample in all tubes}}} \times 100$$

Results and Discussion

pH

Measurement of pH is one of the most important and frequently used tests in water chemistry. Practically every phase of water supply and water treatment, *e.g.*, acid-base neutralization, water softening, precipitation, coagulation, disaffection and corrosion control, is pH dependent. pH is used in alkalinity and carbon dioxide measurement and many other acid-base equlibria. At a given temperature the intensity of the acidic or basic character of a solution is indicated by pH or hydrogen ion activity. The permissible limit of pH in drinking water is within 6.5 to 8.5 according to Bureau of Indian Standards (BIS) .The value of pH in all categories of water is within the permissible range. The value of pH in ground water samples of the study area ranges between 7.29 to 8.12.The seasonal variation shows the pH values fluctuating maximum during Post-monsoon- and minimum in summer at all the locations. The seasonal fluctuation in ground water quality of the region shows the maximum pH value in monsoon (7.76 ± 0.22) and minimum in winter (7.65 ± 0.18). This may be attributed due to the leaching of surface soil from surrounding areas in rainwater.

Turbidity

Turbidity in water is caused by suspended matter, such as clay, silt, finely divided organic and inorganic matter, soluble coloured organic compounds, and plankton and other microscopic, while in all surface water sampling stations at exceed from the limit. The value of turbidity in ground water samples of the study area ranges from 0.91 NTU to 5.00 NTU.

The seasonal fluctuation was maximum in monsoon and minimum in summer.

Total Dissolved Solids

Solids refer to matter suspended or dissolved in water or wastewater. Solids may affect water or effluent quality adversely in a number of ways. Waters with high dissolved solids generally are at inferior waters. The value of total dissolved solids of the study area ranges from 321 mg/l to 720 mg/l in ground water samples. The seasonal fluctuation shows that total dissolved solids are higher in monsoon and minimum in summer in most of the locations. The higher range during monsoon may be due to leaching of surrounding rainwater.

Conductivity

Conductivity is a numerical expression of the ability of an aqueous solution to carry an electric current. This ability depends on the presence of ions, their total concentration, mobility, valance and μmho/cm in ground water, however the maximum seasonal variation was in monsoon and the minimum in summer and winter. The seasonal average conductivity values show maximum in monsoon (875.2 ± 161.3) and minimum during summer (764.3 ± 141.0).

Alkalinity

Alkalinity of water is its acid-neutralization capacity. It is the sum of all the titrable bases the measured value may vary significantly with the end point pH used. Alkalinity is a measure of the chemical composition of the sample is known. Alkalinity in excess of alkaline value of alkalinity in drinking water is 200 mg/l and 600 mg/l respectively, according to BIS. The value of alkalinity in the study area ranges between 98 mg/l to 219 mg/l. The alkalinity values for all the seasons fluctuate from 156.3 mg/l to 160.5 mg/l. This shows there is very little fluctuation in the alkalinity throughout the year.

Total Hardness

Originally, water hardness was understood to be a measure of the capacity of water to precipitate soap. Soap is precipitated chiefly by the calcium and magnesium ions present. Other polyvalent in conformity with current practice, total hardness is defined as the sum of the calcium and magnesium concentrations, both expressed as calcium carbonate, in milligrams per liter. The maximum permissible and allowable value of total hardness in drinking water is 30 mg/l and 6000 mg/l respectively, according to BIS. The value of total hardness in the study area ranges between 32 mg/l to 31 mg/l. The seasonal fluctuation was maximum in monsoon and the minimum in winter and Post-monsoon. The seasonal fluctuation in total hardness of the region shows the variation from 182.3 ± 38.7 during Post-monsoon to 218.0 ± 57.3 in summer. This higher seasonal value in summer is mainly attributed due to rising temperature thereby increasing the solubility of calcium and magnesium salts.

Calcium

The presence of calcium (fifth among the elements in order of abundance) in water supplies results from passage through or over deposits of limestone, dolomite, gypsum, and gypsiferous shale. The maximum permissible and allowable concentration of calcium in drinking water is study area ranges between 29 mg/l to 81 mg/l and 28 mg/l to 72 mg/l. Calcium is an essential constituent of human being. The low content of calcium in drinking water may cause ricket and defective teeth. It is essential for nervous system, cardiac function and coagulation of blood.

Magnesium

The concentration of magnesium in water is comparatively less than calcium possibly due to lesser occurrence of a laxative effect result, therefore, some caution must be exercise with it. The maximum permissible concentration of magnesium in drinking water is 30 mg/l, according to BIS.

The concentration of magnesium in the study area ranges between 5 mg/l to 25 mg/l.

Sodium

Sodium ranks sixth among the harmed by a high sodium ratio. Person afflicted with certain diseases requires low sodium concentration. The concentration of sodium in the study area ranges between 5 mg/l to 12 mg/l in all the samples.

Potassium

Potassium ranks seventh among the elements in order of abundance yet its concentration in most drinking waters seldom reaches 20 mg/l. However the concentration of potassium were analysed from 2 mg/l to 16 mg/l for all the seasons.

Chloride

Chloride (Cl^-) is one of the major inorganic anions in water and wastewater. In potable water, the salty taste produced by chloride concentration is variable and dependent on the chemical composition of water. Some waters containing 250 mg/l may have a detectable salty taste if the cation is sodium. On the other hand, the typical salty taste may be absent in waters/l, according to BIS. The concentration of chloride in the study area ranges between 46 mg/l to 121 mg/l. The seasonal average value shows maximum in summer (85.2 ± 24.7) and minimum during monsoon (72.1 ± 15.1) in the study area.

Sulphate

Sulphate (SO_4^{-2}) is widely distributed in and Magnesium sulphate exert a cathartic action. The maximum permissible and allowable concentration of sulphate in drinking water is 200 mg/l and 400 mg/l respectively, according to BIS. The concentration of sulphate ranges between 9 mg/l to 51 mg/l in all samples. The maximum seasonal value was in winter (23.0 ± 8.3) and minimum during Post-monsoon (17.9 ± 6.5).

Phosphate (PO_4^{-2})

Phosphorous occurs in natural water almost solely as phosphates. Phosphorous is essential to the growth at organisms and can be the nutrient that limits the primary productivity of a body of water and to that water may stimulate the growth of photosynthetic aquatic micro and macro organisms in ,nuisance quantities. The concentration of phosphate in the study area ranges between Nil to 0.20 mg/l in ground water.

Cadmium

Natural sources of Cd in the drinking water system are industrial discharge. Mining wastes, metal plating and water pipes. It causes destruction of testicular tissue, red blood cells disorder of bone marrow etc. the maximum permissible level of cadmium in drinking water is 0.05 mg/l according to WHO and ICMR. It was found that, the average values of cadmium in all the water samples ranges from ND to 0.02 mg/l.

Chromium

Chromium may exist in water supplies in both valent from rarely occurs in potable water. The maximum permissible concentration of lead in drinking water is 0.05 mg/l according to BIS. The values of Chromium content in all the water samples were not detectable.

Copper

Copper is essential to humans, the daily adult requirement has been estimated at 2.0 mg. Copper salts are used in water supply system to control biological growths in reservoir and distribution pipes and to catalyze the oxidation of manganese. The maximum permissible and allowable to BIS. The concentration of copper in the study area ranges between ND to 0.01 mg/l.

Table 12.1: Physico-chemical Characteristics

Sl. No.	*Location*	*Season*	*pH*	*Turbidity (NTU)*	*Total Dissolved Solids (mg/l)*	*Conductivity (µmho/cm)*	*Alkalinity (as CaCO³ mg/l)*
1.	Kamta	Winter	7.51	1.00	598	956	212
		Summer	7.62	0.91	556	888	210
		Monsoon	7.49	2.62	610	976	198
		Post-monsoon	7.89	2.12	608	970	165
2.	Pokharwar	Winter	8.00	1.61	481	762	151
		Summer	7.81	1.11	488	780	146
		Monsoon	7.62	3.00	492	782	139
		Post-monsoon	7.29	2.00	460	736	141
3.	Nayagaon	Winter	7.62	1.51	478	762	126
		Summer	7.58	1.50	461	738	212
		Monsoon	8.00	2.56	520	830	156
		Post-monsoon	8.11	2.50	511	817	200
4.	Mohakamgarh	Winter	7.82	2.00	407	651	219
		Summer	7.61	1.61	382	590	146
		Monsoon	7.91	2.56	446	669	147
		Post-monsoon	7.80	2.11	432	681	159
5.	Rajola	Winter	7.66	2.00	321	513	166
		Summer	7.65	3.61	425	688	126
		Monsoon	7.77	5.00	492	781	98
		Post-monsoon	7.75	3.26	519	830	190
6.	Pathra	Winter	7.68	1.61	695	1010	167
		Summer	7.80	1.00	613	972	175
		Monsoon	7.79	2.65	514	825	200
		Post-monsoon	7.85	2.11	597	950	168
7.	Paldeo	Winter	7.70	1.21	621	994	159
		Summer	7.81	1.10	592	946	200
		Monsoon	7.83	2.61	713	1140	169
		Post-monsoon	7.82	2.12	648	1040	151
8.	Pausarha	Winter	7.62	1.41	52	832	141
		Summer	7.56	1.50	490	813	132
		Monsoon	7.69	2.50	531	828	149
		Post-monsoon	7.72	1.41	50	813	140
9.	Choubepur	Winter	7.50	1.40	396	834	122
		Summer	7.60	1.00	360	576	129
		Monsoon	8.12	1.43	480	770	161
		Post-monsoon	7.35	1.40	396	631	149
10.	Guptagodavari	Winter	7.35	1.00	389	622	131
		Summer	7.47	0.91	596	652	129
		Monsoon	7.40	1.21	720	1151	146
		Post-monsoon	7.62	1.00	613	980	128

Table 12.1-Contd...

Sl. No.	Location	Season	Total Hardness (mg/l)	Ca (mg/l)	Mg (mg/l)	Na (mg/l)	K (mg/l)	Cl (mg/l)	SO_4 (mg/l)	PO_4 (mg/l)
1.	Kamta	Winter	169	52	10	14	8	110	26	NIL
		Summer	200	58	12	10	6	98	51	NIL
		Monsoon	152	46	9	10	5	68	21	NIL
		Post-monsoon	146	39	7	10'	2	100	11	NIL
2.	Pokharwar	Winter	132	32	10	9	10	75	17	0.06
		Summer	216	41	8	9	6	96	18	NIL
		Monsoon	202	56	6	9	5	71	16	NIL
		Post-monsoon	200	50	10	8	4	66	21	0.02
3.	Nayagaon	Winter	169	41	11	7	7	67	41	NIL
		Summer	165	32	10	11	6	110	31.	NIL
		Monsoon	141	39	9	12	5	71	26	0.05
		Post-monsoon	146	30	8	9	9	60	21	0.04
4.	Mohakamgarh	Winter	226	40	6	12	10	68	17	NIL
		Summer	231	42	5	11	6	61	20	NIL
		Monsoon	210	38	9	9	5	65	25	0.02
		Post-monsoon	196	32	15	8	7	60	31	NIL
5.	Rajola	Winter	269	69	21	7	10	58	30	0.05
		Summer	315	60	25	5	12	61	26	NIL
		Monsoon	251	42	10	6	10	76	29	NIL
		Post-monsoon	249	38	9	4	8	100	21	NIL
6.	Pathra	Winter	246	30	8	10	11	106	16	0.02
		Summer	240	29	7	9	10	75	22	NIL
		Monsoon	221	36	11	10	8	77	18	NIL
		Post-monsoon	196	29	10	9	16	65	20	0.04
7.	Paldeo	Winter	229	38	9	8	10	67	26	NIL
		Summer	300	46	12	8	8	121	22	NIL
		Monsoon	291	69	15	7	6	100	11	0.02
		Post-monsoon	226	61	14	7	5	110	19	NIL
8.	Pausarha	Winter	256	29	7	9	6	105	25	NIL
		Summer	216	41	12	8.	8	69	18	NIL
		Monsoon	210	76	16	10	9	89	16	0.05
		Post-monsoon	186	62	11	7	10	95	10	0.02
9.	Choubepur	Winter	210	61	10	12	14	96	16	0.01
		Summer	151	45	15	11	10	110	11	0.20
		Monsoon	139	41	9	9	8	58	9	0.11
		Post-monsoon	136	40	16	7	6	61	12	NIL
10.	Guptagodavari	Winter	252	81	25	11	16	69	16	NIL
		Summer	146	61	10	10	12	51	15	NIL
		Monsoon	132	56	16	12	10	46	10	0.01
		Post-monsoon	142	42	11	11	8	50	13	NIL

Iron

Iron in drinking water may be present as geological sources, industrial wastes and domestic discharges and also from mining products. Excess amount of Fe (more than 10 mg/kg) causes rapid increase in respiration, pulse rate and coagulation of blood vessels, hypertension and 1.0 mg/l and 0.3 mg/l respectively. The concentration of iron in all water samples of the study area ranges from 0.11 mg/l to 1.00 mg/l. The maximum seasonal value of iron in the region was in monsoon (0.52 ± 0.21) and the minimum in Post-monsoon (0.41 ± 0.16).

Nickle

The values of Nickle were analysed not detectable in ground water samples.

Lead

Lead enters in the drinking water from industrial and mining wastes, smelter discharges, dissolution of old lead plumbing and than 800 mg in the body, produces coma and death. The maximum permissible concentration of lead in drinking water is 0.1 mg/l, according to WHO and ICMR. The average values of lead content in all ground water samples were not detectable.

Table 12.2: Heavy Metals Characteristics

Sl. No.	Location	Season	Cd (mg/l)	Cr (mg/l)	Cu (mg/l)	Fe (mg/l)	Ni (mg/l)	Pb (mg/l)	Zn (mg/l)
1.	Kamta	Winter	ND	ND	0.01	0.25	ND	ND	ND
		Summer	ND	ND	ND	0.61	ND	ND	ND
		Monsoon	ND	ND	ND	0.21	ND	ND	ND
		Post-monsoon	ND	ND	0.01	0.11	ND	ND	ND
2.	Pokharwar	Winter	ND	0.61	ND	0.61	ND	ND	ND
		Summer	ND	0.52	ND	0.52	ND	ND	ND
		Monsoon	ND	0.68	ND	0.68	ND	ND	ND
		Post-monsoon	ND	0	ND	0.55	ND	ND	ND
3.	Nayagaon	Winter	0.01	ND	ND	0.41	ND	ND	ND
		Summer	ND	ND	ND	0.26	ND	ND	ND
		Monsoon	0.02	ND	ND	0.56	ND	ND	ND
		Post-monsoon	ND	ND	ND	0.41	ND	ND	ND
4.	Mohakarngarh	Winter	0.02	ND	ND	0.26	ND	ND	ND
		Summer	ND	ND	ND	0.42	ND	ND	ND
		Monsoon	ND	ND	ND	0.62	ND	ND	ND
		Post-monsoon	ND	ND	ND	0.40	ND	ND	ND
5.	Rajola	Winter	ND	ND	ND	0.36	ND	ND	ND
		Summer	ND	ND	ND	0.29	ND	ND	ND
		Monsoon	ND	ND	ND	0.39	ND	ND	ND
		Post-monsoon	ND	ND	ND	0.30	ND	ND	ND
6.	Pathra	Winter	ND	ND	ND	0.26	ND	ND	ND
		Summer	ND	ND	ND	0.21	ND	ND	ND
		Monsoon	ND	ND	ND	0.29	ND	ND	ND
		Post-monsoon	ND	ND	ND	0.25	ND	ND	ND
7.	Paldeo	Winter	ND	ND	ND	0.20	ND	ND	ND
		Summer	ND	ND	ND	0.25	ND	ND	ND
		Monsoon	ND	ND	ND	0.31	ND	ND	ND
		Post-monsoon	ND	ND	ND	0.41	ND	ND	ND
8.	Pausarha	Winter	ND	ND	ND	0.56	ND	ND	ND
		Summer	ND	ND	ND	0.32	ND	ND	ND
		Monsoon	ND	ND	ND	0.56	ND	ND	ND
		Post-monsoon	ND	ND	ND	0.39	ND	ND	ND
9.	Choubepur	Winter	ND	ND	ND	0.29	ND	ND	ND
		Summer	ND	ND	ND	1.00	ND	ND	ND
		Monsoon	ND	ND	ND	0.86	ND	ND	ND
		Post-monsoon	ND	ND	ND	0.68	ND	ND	ND
10.	Guptagodavari	Winter	ND	ND	ND	1.00	ND	ND	ND
		Summer	ND	ND	ND	0.98	ND	ND	ND
		Monsoon	ND	ND	ND	0.68	ND	ND	ND

		Post-monsoon	ND	ND	ND	0.56	ND	ND	ND

Zinc

The maximum allowable concentration and permissible concentration of zinc in drinking water is 15.0 mg/l and 5.0 mg/l respectively, according to ICMR. But according to BIS, the value is 10.0 mg/l and 5.0 mg/l. known adverse physiological effects upon men. In fact, it is an essential and beneficial element in human nutrition. In all the locations zinc were analysed in ND.

Table 12.3: Bacteriological Characteristics

Sl. No.	*Location*	*Season*	*Feacal Coliform MPN/100 ml*
1.	Kamta	Winter	2
		Summer	8
		Monsoon	10
		Post-monsoon	5
2.	Pokharwar	Winter	5
		Summer	NIL
		Monsoon	8
		Post-monsoon	7
3.	Nayagaon	Winter	NIL
		Summer	NIL
		Monsoon	10
		Post-monsoon	8
4.	Mohakamgarh	Winter	3
		Summer	8
		Monsoon	10
		Post-monsoon	6
5.	Rajola	Winter	NIL
		Summer	5
		Monsoon	6
		Post-monsoon	NIL
6.	Pathra	Winter	NIL
		Summer	10
		Monsoon	11
		Post-monsoon	6
7.	Paldeo	Winter	NIL
		Summer	NIL
		Monsoon	5
		Post-monsoon	NIL
8.	Pausarha	Winter	7
		Summer	5
		Monsoon	8
		Post-monsoon	4
9.	Choubepur	Winter	NIL
		Summer	11
		Monsoon	12
		Post-monsoon	8
10.	Guptagodava	Winter	NIL
		Summer	5
		Monsoon	7

	Post-monsoon	NIL

Table 12.4: Seasonal Fluctuation in Water Quality

Sl.No	Season	Parameter							
		pH	Conductivity (µmho/cm)	Alkalinity (mg/l)	Total Hardness (mg/l)	Cl mg/l	SO_4 mg/l	Fe (mg/l)	MPN
1.	Winter	7.65±0.18	793.6±165.6	159.4±33.6	215.8±45.2	82.1±22.3	23.0±8.3	0.42±0.25	1.7±2.5
2.	Summer	7.66±0.1	764.3±141	160.5±35.4	218.0±57.3	85.2±24.7	22.3±11.2	0.49±0.29	5.2±4.1
3.	Monsoon	7.76±0.22	875.2±161.3	156.3±29.4	194.9±53.1	72.1±15.1	18.1±7.0	0.52±0.21	8.7±2.3
4.	Post-monsoon	7.72±0.25	844.8±137.2	159.1±22.5	182.3±38.7	76.7±21.9	17.9±6.5	0.41±0.16	4.4±3.3

Table 12.5: Drinking Water Quality Standard - IS: 10500, 1991

Sl. No.	Substance or Characteristics	Requirement (Desirable Limit)	Permissible Limit in the Absence of Alternate Source
1.	Turbidity NTU, Max.	5	10
2.	pH Value	6.5 -8.5	No Relaxation
3.	Total Hardness (as $CaCo_3$) mg/l, Max.	300	600
4.	Iron (Fe) mg/l, Max.	0.3	1.0
5.	Chloride (as Cl) mg/l, Max.	250	1000
6.	Dissolved Solids, mg/l, Max.	507	2000
7.	Calcium (as Ca), mg/l, Max.	75	200
8.	Magnesium (as mg), mg/l, Max.	305	100
9.	Copper (as Cu), mg/l, Max.	0.05	1.5
10.	Sulphate (as SO_4), mg/l, Max.	2800	400
11.	Cadmium (as Cd), mg/l, Max.	0.013	No Relaxation
12.	Lead (as Pb), mg/l, Max.	0.015	No Relaxation
13.	Chromium (as cr^{6+}), mg/l, Max.	0.05	No Relaxation
14.	Alkalinity, mg/l, Max.	200	600
15.	Total Coliform, MPN	1/100 ml	No Relaxation

MPN

Form public health standpoint the bacteriological quality of water is as important as the chemical quality. The principal indicator in polluted water where pathogen are or might be present, unable to multiply under conditions when pathogens do not multiply. The value of MPN ranges from Nil to 12/100 ml in ground water. The maximum seasonal value in the region was (8.7 ± 2.3) in monsoon and the minimum (1.7 ± 2.5) in winter.

References

APHA (1985). *Standard Methods For the examination of water and wastewater*, , Washington U.S. A. 16th

Edition.

CPCB (1990). *Water Quality Statistic of India 1990*, CPCB, New Delhi, India.

CPCB (1994-95). *Laboratory Analytical Techniques*, Series: Lats/9/1994-95

G. Kamalak Kannan (1995). Limnological Studies on River Mandakani (Paisuni) with Intention to Evolve an Approach for Conservation and Management, Ph.D. Thesis, MGCGV, Chitrakoot.

Hynes, H.B.N. (1960). *The Biology of Polluted Water*, Liverpool Univ. Press, 202 p.

Khadilkar, C.H. (1962). *Text Book on Water Supply.*

Mohapatra, S.P., Saxena, S.K., and Arifall (1992). Occurrence of Coliform bacteria in channels receiving municipal sewage. *Ind. Jr. of Env. Prot.*, 12(7): 509-511.

Saxena, M.M. (1990). *Environmental Analysis of Water, Soil and Air*. Agro-Botanical Publishers, India.

Todd, D.K. (1980). *Ground Water Hydrology*, 2nd Edition, John Wiley Sons, New York.

Trivedy, R.K. and Goel, P.K. (1984). *Chemical and Biological Methods for Water Pollution Studies,* Env. Pub. Karad.

13

Physico-chemical and Biological Quality of Ground Water in Mysore City

H.R. Meenakumari and S.P. Hosmani*

DOS in Botany, University of Mysore Manasagangotri, Mysore - 570 006, Karnataka
**DOS in Environmental Science, University of Mysore Manasagangotri, Mysore - 570 006, Karnataka*

ABSTRACT

Water is one of the most essential components for survival of life on the earth. One of the major problems of ill health in the underdeveloped countries is largely due to lack of safe drinking water. The physico-chemical and Biological characteristics of ground water in various parts of Mysore city were studied during summer (February to May, 2002) seasons. The correlation between nine chemical parameter's has been calculated and regression analysis carried out. Cluster's of linked and non-linked variables were separated out following average linkage method. Most of the physico-chemical constituents are not within maximum permissible limits of BIS and WHO standards for drinking water. However, some of the samples, show pH value of 6.5. The analysis also reveals less concentration of fluoride than the desirable limit (< 0.6 mg/l). Hardness, Total Dissolved Solid and Calcium are very high, excess Iron is also observed. The results revealed that ground water at all the sampling zones have a vide variation and are above the upper permissible limit. The suitability of water for domestic or drinking purpose, indicates that 50 per cent of samples are far from the standards prescribed for portable waters.

Key Words: Physico-chemical and Biological Parameters, Portable Water, Groundwater, Cluster Analysis.

Introduction

Ground water accounts for nearly 100 per cent of drinking water supply. It has become an important water resource due to increasing trend of pollution in surface water. This problem is more acute in the areas which are densely populated and inadequately supplied with potable drinking water. Ground water study has been conducted by Pathak and Badve (1999), Mariappan *et al.* (1999), Naik and

Porohit (2001), Das *et al.* (2000), Narayana and Lokes (1999) and Pather *et al.* (2001). These studies indicate that the quality of ground water in many places have detoriated and there is a need for monitoring such waters by examining various chemical and biological parameters.

Mysore city has been witnessing inflow of population and there is an increasing demand for potable drinking water. This is harnessed mainly through a network of open wells and bore wells which are being alarmingly contaminated. The objective of the present investigation was to assess the water quality in Mysore city and to determine the suitability for these drinking purposes.

Materials and Methods

The ground water samples which include open wells and bore wells were collected in clean polythene bottles from each ward which include 100 samples. The chemical and biological parameters essential to determine the quality of drinking water were analysed as per the methods described in Standard Methods for the Examination of waters and wastewater, (1995) and Trivedy and Goel (1984). The data obtained was subjected to cluster analysis to understand the interrelationship between the various parameters. The values of physico-chemical and biological parameters are represented in Table 13.1. About 100 representative samples were collected from different localities of the study area. All the samples were divided into four zones, and based on the percentage of utilization of water, the areas have been classified. These have been compared to standard values of BIS and WHO.

Results and Discussion

The values of temperature varied from sample to sample. pH is the most important parameter that serves as an index for pollution. The range of pH is between 6.5 and 8.0. The Electrical conductivity values range between 1516 and 2803.58 µS/cm. The values of chlorides were found to be with in the permissible limit (80.94 to 310.98 mg/l) in all the samples, except in places like J.P. Nagar (310.98 mg/l) and Nyachanahally palya (298.20 mg/l) in southern zone. A high chloride content imparts a salty taste of water. High chloride in drinking water are subjected to laxative effects. The observed values of Alkalinity were found to be very high and ranged between 344 to 510 mg/l. Higher values of alkalinity were observed in all the samples. Still waters containing less than 100 mg/l of alkalinity are desirable for domestic use. Alkalinity is an important parameter involved in corrosion control. Alkalinity itself is not harmful to human beings, although alkalinity values exceeded permissible limits. The alkalinity values provide an idea of the nature of salt present.

Total hardness values were recorded within 238.80 to 868 mg/l. The values of total hardness in all the samples are high except in places like, Mandimohalla in Northern zone, Agrahar in Southern zone, T.K. Layout in Eastern zone, and Jyothi Nagar in Western zone. According to Kannan (1991) water with hardness values more than 180 mg/l is vary hard. The standard unit for the total hardness specified is 300 mg/l and is considered portable, but beyond this limit it produces gastrointestinal irritation (ICMR, 1975). In ground water of the area under study, calcium ranged between 84.08 and 262 mg/l respectively. The standard unit for the calcium specified is 75 mg/l. The results indicate that ground water is very hard and is not suitable for domestic purpose. Although calcium is an essential constituent for various functioning of the human body its low content may cause defective teeth, nevertheless it was too high in these waters. The values of Iron were quite near to the portable values, only in Eastern part, while they were sufficiently high in southern and western zone. Although iron is little concerned for health it is still considered a nuiscence in excessive quantities and most of the ground water sample of Mysore City have excess iron. Nitrate values were quite less as compared to

Table 13.1: Physico-chemical and Biological Analysis of Ground Water of Mysore City

Area	*pH*	*EC*	Cl_2^-	*Alk*	*TH*	*Ca*	*Fe*	NO_3	*F*	SO_4	*TDS*	*(%) Usage of Water*	*MPN per 100 ml*
North													
Mandi Mohalla	6.94	1929.70	110.76	344.00	266.40	1t4.034	0.360	3.304	0.244	203.60	776.00	75	≥ 2400 - ≥ 2400
N.R. Mohalla	7.34	2406.31	201.64	370.00	576.00	163.758	0.830	0. 776	0.040	464.00	859.36	90	≥ 2400 - ≥ 2400
Metagahalli	7.18	2078.08	110.76	376.00	331.20	150.840	0.770	0.070	0.038	233.20	900.00	30	210 - ≥ 2400
Kesare	7.72	2594.79	192.55	420.00	824.00	211.112	1.390	0.554	0.068	711.80	926.69	90	28 - ≥ 400
Banimantap	7.18	2586.82	107.93	370.00	640.00	180.96	0.450	0.546	Nil	452.00	943.84	30	3 - ≥ 2400
South													
Agrahar	7.26	1516.80	117.86	355.00	232.80	87.320	0.428	0.636	0.378	168.40	621.52	30	21 - 260
Laxmipuram	7.46	1586.56	160.46	400.00	387.00	113.600	0.438	1.030	0.202	245.00	646.621	30	3 - 340
Vidyaranyapuram	7.52	1699.80	126.38	405.00	432.00	127.390	0.430	0.914	0.050	382.60	728.96	75	3 - 20
J.P. Nagar	8.10	2552.60	310.98	475.00	393.00	183.820	1.202	0.566	0.058	702.80	1011.60	90	28 - 1100
Nachanally Palya	7.80	2692.40	298.20	445.00	868.00	189.520	1.012	0.292	0.008	776.00	1063.12	90	3 - 28
East													
Saraswathipuram	7.50	2103.15	80.94	415.00	456.00	197.280	0.360	0.584	0.100	349.80	1004.00	30	21 - ≥ 2400
Kuvempu Nagar	8.16	2380.28	100.82	510.00	464.00	156.620	0.360	1.094	0.016	368.00	1120.00	50	28 - ≥ 2400
Ramakrish Nagar	7.48	2410.34	124.06	420.00	448.00	155.500	0.938	0.806	0.050	368.00	1178.00	75	≥ 2400 - ≥ 2400
Jayanagar	7.66	2312.20	132.06	410.00	344.00	173.780	0.240	1.096	0.112	301.40	905.64	30	3 - 210
T.K. Layout	7.28	1804.20	115.02	395.00	295.00	132.320	0.874	0.904	0.128	254.60	744.48	30	20 - ≥ 2400
West													
Siddaratha Layout	8.12	2444.57	203.06	465.00	404.00	199.700	1.196	1.360	0.022	318.20	976.05	90	3 - 20
Jyothi Nagar	7.20	1582.31	89.45	380.00	289.00	84.080	0.382	1.084	Nil	201.80	665.11	90	3 - 7
Nazarbad	7.06	2207.92	133.48	350.00	417.00	171.780	1.098	1.090	0.042	368.80	889.94	50	3 - 7
Gayathripuram	7.60	2043.66	112.18	410.00	426.00	202.660	1.154	1.070	0.028	384.20	1038.62	90	3 - 4
Kaythama Ranahalli	7.18	2803.58	200.22	360.00	740.00	262.000	0.080	0.362	0.028	682.00	1045.08	90	4.15
SPL BIS (ppm)	6.5 - 8.5	-	250	-	300	75	0.3	45	0.6	150	500	-	1/100 ml
SPL WHO	7.85	-	200	-	-	75	0.3	50	0.5	200	500	-	1/100 ml

SPL: Standard Permissible Limits, BIS: Bureau of Indian Standard, WHO: World Health Organization.
Note: All Parameters expressed in (mg/l), Except pH, Conductivity (µS/cm)
MPN = Most Probable Number of Coliform (No./100 ml).

standard values prescribed for drinking water.

The standard values prescribed for fluoride in drinking water is between 0.5 and 0.6 mg/l. The water samples from all zones concentration of fluorides (0 to 0.378 mg/l) is less than 0.6 mg/l. Less concentration of fluoride causes staining of teeth. Although the values of sulphate were found to be very high (168.40 and 776 mg/l) in all the samples, except in places like Agrahar in Southern zone. The standard unit for the sulphate specified is 200 mg/l. High concentration of sulphate may induce diarrhea. The values of total dissolved solid are very high in all samples and are beyond permissible limit of WHO and BIS. Except in places like Agrahar and Laxmipuram in the southern zone, Jyothinagar in the Western zone all other places have high concentration of total dissolved solid. High concentration of total dissolved solid which may produce distress in cattle and live stock.

The determination of *E. coli* by MPN count indicates that the open wells and bore wells from the southern and western zone of Mysore city were less contaminated. While water samples from northern and eastern zone were more contaminated. The highest contamination occurred in places like Mandi Mohalla, N.R. Mohalla, Metagahalli and Kesare in Northern zone, Saraswathipuram, Ramakrishna Nagar and Kuvempunagar in eastern zone, were the usage of bore wells was upto 75 per cent, the standard values prescribed for MPN or *E. coli* 1/100 ml.

To trace the interrelationship between the various chemical parameters the data was subjected to Pearsion's Correlation's Matrix Table 13.2 and cluster analysis to obtain a dendrogram. This facilitates to arrive at the closet parameter which are linked and dependent with other parameters. In the entire city it is the concentration of Chlorides, alkalinity, Total Hardness and Nitrate that are playing an important role in determining the quality of water. From the dendrogram it is clear that iron and Chloride are the most important chemical parameters that are operating in this water, while these in turn are related to the Total Dissolved Solids. Most of the water samples are contaminated to a greater extent and proper monitoring is very essential. It is very necessary to maintain propehygenic conditions among the bore-wells.

Table 13.2: Pearson's Correlation Matrix

	Cl_2	*Alk*	*TH*	*Ca*	*Fe*	*NO_3*	*F*	*SO_4*	*TDS*
Cl_2	1000								
Alk	- 061	1000							
TH	295**	278**	1000						
Ca	- 185	- 058	0.119	1000					
Fe	0.017	019	- 004	- 006	1000				
NO_3	- 171	- 164	- 264**	- 110	0.096	1000			
F	- 044	- 067	- 020	- 012	- 087	- 049	1000		
SO_4	249*	283**	- 980**	0.115	0.016	- 237*	- 004	1000	
TDS	344**	- 008	0.092	0.147	0.115	0.180	- 023	090	1000

*Correlation is significant at the 0.01 level (2-tailed)
*Correlation is significant at the 0.05 level (2- tailed)
Cl_2: Chloride, Alk: Total Alkalinity, TH: Total Hardness,
Ca: Calcium, Fe, Iron, NO_3: Nitrate, F: Fluoride, SO_4: Sulphate, TDS: Total Dissolved Solids.

Acknowledgement

The authors wish to acknowledge the Chairman DOS in Botany and to Coordinator, DOS in

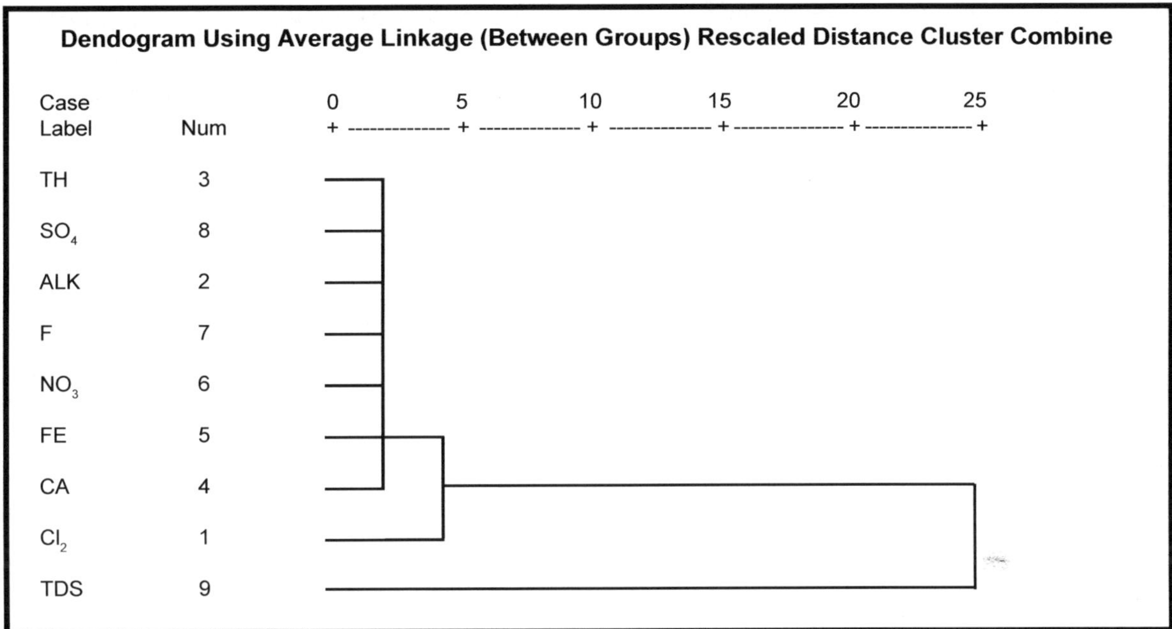

Environmental Science for providing laboratory facilities.

References

APHA, AWWA, WPCF (1995). *Standard Methods of Examination of Water and Waste Water* (19th edition). American Public Health Association, Washington D.C.

Beena, T. Abraham and T.M. Sajeela Beegum (2001). Comparison of Water Quality of well and Bore well of Moothakunnam-Maliankara area with USPHS standards. *Indian J. Environ. and Ecoplan.,* 5(3): 593-596.

Dasgupta, M. Adak and K.M. Purohit (2001). Status of Surface and Ground Water Quality of Mandiakudar-part I: Physico-chemical parameters. *Poll. Res.,* 20(1): 103-110.

Deepali Sohani, Sapna Pande and V.S. Srivastava (2001). Geo Quality at Tribal Town: Nandurbar (Maharashtra). *Indian. J. Environ. and Ecoplan.,* 5(2): 475-479.

Dilip, B. Patil, Rajendra V. Tijare and S.B. Rewatar (2001). Physico-chemical Characteristics, of Ground water of Armori Town of Gadchiroli District, Maharashtra, *Poll. Res.,* 20(2): 207-209.

ICMR (1975). *Manual of Standards of Quality for Drinking Water Supplies*. ICMR, New Delhi.

Kannan, K. (1991). *Fundamentals of Environmental Pollution.* S. Chand and Co. Ltd. New Delhi.

Mariappan, P., V. Yeganaraman and T. Vasudevan (2000). Occurrence and Removal Possibilities of Fluoride in Ground Waters of India. *Poll. Res.,* 19(2) 165-177.

Mariappan, P., T. Vasudevan and V. Yegnaraman (1999). Fluoride Distribution in Ground Waters of Salem District. 1 *JEP* 19(9): 681-688.

Mohapatra, D., B. Das and V. Chakravorty (2001). A correlation Study on Physico-chemical characteristics of Ground Water in Paradip Areas. *Poll. Res.,* 20(3): 401-406.

Narayana, K. and K.N. Lokesh (1999). Quality of Ground Water of Bore wells in M.I.T., Campus, Manipal, Karnataka. *J. Environ. Hlth.*, Vol. 41, No.2, p.144- 148.

Naik, S.K.and K.M. Purohit (2001). Studies of Water Quality at Bandamui Roukela Industrial Complex part -II. Metallic Parameters. *Indian J. Environ. and Ecoplan.*, 5(1): 115-118.

Noor, Md. Alam (2001). Studies on Variations in the Physico-Chemical parameters of a Pond at Hathwa (Bihar). *J. of Environ. and Pollu.*, 8(2): 179-181.

Pathak, M.D. and Badve (1999). Occurrence of Fluoride in Ground Water of Maharashtra. *J. IAEM.* 26: 168-171.

Preeti Mukherjee and M.L. Naik (1997). Bacteriological Water Quality of the Ponds of Raipur city. *Proc. Acad. Environ. Bioil.*, 6(1): 77-79.

Richariya, L.K. and Mishra, Ravi (1998). MPN and *E. coli* in Ground Water Samples by Rewa Area (M.P.) *Indian Environ. and Poll.*, 5(1): 73-77.

Somashekar, R.K., V. Rameshaiah and A. Chethana Suvarna (2000). Ground Water Chemistry of Channapatna Taluk (Bangalore Rural District): Regression and Cluster Analysis, *Journal of Environ. and Poll.*, 7(2): 1010-1119.

Trivedy, R.K. and Geol, P.K. (1983). *Chemical and Biological Methods for Water Pollution Studies.* Environmental Publication, Karad.

14

Physico-chemical Characteristics of Dal Lake Water

Shoukat Ara, M.A. Khan and M.Y. Zargar

Division of Environmental Sciences
S.K. University of Agricultural Sciences and Technology (Kashmir), Shalimar, Srinagar - 191 121

ABSTRACT

Multiple anthropogenic pressures have deteriorated water quality of Kashmir valley Dal lake. To investigate present position of the lake water, study on its physico-chemical characteristics was undertaken. Analysis exhibited richness in chloride, nitrogen (nitrate, nitrite and ammonical) and orthophosphate, in addition to high value of total dissolved solids, electrical conductivity, total alkalinity, biochemical oxygen demand and low transparency value. The cation composition of lake is Ca > Na > Mg.

Key Words: Pollution, Physico-chemical Characteristics, Dal Lake.

Introduction

Though the defilement of water and deterioration of aquatic systems is as old as civilization, however escalating industrialization, urbanization, developmental and agricultural activities have brought irreversible change in such systems. Unplanned and excessive exploitation and mounting anthropogenic influences in and around aquatic ecosystems have resulted in pollution problems. Lakes being fragile ecosystems are vulnerable to such problems. Pollution caused by plethora of human activities primarily affects physico-chemical characteristics of water, leading to destruction of community disrupting delicate foodwebs, deteriorating lake environment.

Several of the lakes including famous Dal Lake in the Kashmir Himalayan Valley have environmentally deteriorated. Located at an elevation of 1,587 m covering an area of 21 km^2 and situated in the lacustrine basin of intermontane Kashmir Valley, Dal (Lat. 34° 9′ 34° 10′ N and

Long. 74° 8′-74° 9′ E) is a post glaciale lake of shallow depth. This multi-basined, open drainage type lake (Zutshi and Khan, 1978) is under intense pressure of tourism, food supply and waste disposal. Flotilla of house boats, sewage from peripherial habitations, agro-chemicals from floating gardens, land mass and catchment enrich nutrient status of lake thereby deteriorating its water quality. Though lot of work has been done on physico-chemical characteristics of Dal lake water (Zutshi and Vass, 1973, 1978, Vass and Zutshi, 1983, Ishaq and Kaul, 1989, Khan, 1993, 1996). However, to investigate the present water quality status of lake, present study was undertaken.

Materials and Methods

Physico-chemical characteristics of lake water were studied from March, 2000-February, 2001. Water samples from surface (max. depth. 10 cm) were collected monthly with necessary precautions in plastic bottles from different sites of lake. Samples for biochemical oxygen demand were collected in BOD bottles. Samples were analysed as per standard methods recommended by APHA (1995), Trivedy and Goel (1984) and Mackereth (1963). Analysis was done within 48 hours after sampling, however dissolved oxygen was analysed immediately and pH recorded soon after collection.

Results and Discussion

Physico-chemical characteristics of water of an aquatic system reflect not only the quality of system but also the type and density of its biota. Analysis of such characters generate information regarding pollution pattern and magnitude of pollutant loading of aquatic system.

One of the physical parameters which is directly related with chemical and biochemical reactions is temperature. Temperature of water in the present study, ranged between 6-28°C, highest being recorded in August (Table 14.1).

Tabl 14.1: Physico-chemical Characteristics of Dal Water

Month	*Parameters*									
	Temp. (°C)	*Transparency (cm)*	*pH*	*Total dissolved solids (mg l^{-1})*	*Electrical conductivity (μScm^{-1})*	*Total alkalinity (mg l^{-1})*	*Free carbon dioxide (mg l^{-1})*	*Dissolved organic matter (mg l^{-1})*	*Dissolved oxygen (mg l^{-1})*	*Bio-chemical oxygen demand (mg l^{-1})*
March	15.0	0.75	7.3	275	200	71.5	46.0	21.0	11.4	3.9
April	22.0	0.68	7.8	251	161	71.0	47.0	22.25	11.1	4.2
May	24.0	0.55	7.3	235	150	77.0	60.5	26.0	9.75	4.7
June	23.0	0.23	7.2	326	250	85.0	70.5	33.25	9.65	5.8
July	24.0	0.19	7.3	345	300	90.0	80.0	38.0	9.5	6.2
August	28.0	0.13	7.4	475	400	117.0	85.0	45.5	9.45	7.4
Sep	23.0	0.32	7.3	305	245	98.0	68.5	42.0	10.45	6.65
Oct	18.0	0.63	7.6	206	200	95.0	56.0	36.0	10.9	5.5
Nov	12.0	0.66	6.7	295	240	94.0	Absent	28.5	11.85	4.5
Dec	7.0	0.54	7.3	310	230	86.0	Absent	21.0	11.9	4.15
Jan	6.0	1.0	7.3	261	165	53.0	Absent	17.0	12.4	3.8
Feb	7.0	0.21	6.9	260	185	64.0	Absent	19.5	11.85	4.1

contd...

Table 14.1-Contd...

Month	*Parameters*										
	Chloride ($mg\ l^{-1}$)	*Silicate ($mg\ l^{-1}$)*	*Nitrate -N ($\mu g\ l^{-1}$)*	*Nitrite -N ($\mu g\ l^{-1}$)*	*Ammonical -N ($\mu g\ l^{-1}$)*	*Total Phos-phorous ($\mu g\ l^{-1}$)*	*Orthopho-sphate ($\mu g\ l^{-1}$)*	*Ca ($mg\ l^{-1}$)*	*Mg ($mg\ l^{-1}$)*	*Na ($mg\ l^{-1}$)*	*K ($mg\ l^{-1}$)*
March	19.0	1.35	202.5	14.0	319.0	370.0	23.0	18.1	5.55	8.5	3.25
April	22.25	1.3	190.0	8.0	320.0	465.0	25.5	18.5	6.9	9.0	3.75
May	23.25	1.1	164.0	4.5	383.0	562.5	21.5	21.75	8.5	9.75	4.75
June	28.5	2.7	188.5	1.5	416.5	645.0	27.5	28.2	9.75	11.1	5.25
July	32.5	2.7	175.0	2.25	492.5	775.0	28.0	28.5	10.75	13.0	5.75
Aug	36.5	2.65	88.5	1.75	502.5	815.0	32.5	30.0	11.0	13.65	6.5
Sep	33.25	2.5	245.0	2.5	460.0	680.0	28.5	29.25	9.7	11.6	6.25
Oct	28.25	2.35	287.5	6.0	405.0	640.0	24.5	26.0	7.5	10.75	5.15
Nov	27.25	1.8	297.5	8.0	370.0	510.0	22.0	21.0	5.75	9.25	3.5
Dec	23.65	1.55	322.5	13.5	370.0	370.0	21.0	19.0	5.5	8.0	3.0
Jan	17.5	1.15	367.5	22.5	268.5	248.5	20.5	16.5	4.7	6.0	1.0
Feb	18.5	1.2	290.0	20.5	250.0	325.0	22.0	16.5	5.7	7.0	3.0

Transparency value varied from 0.13-1.0 m. Low value was recorded during June, July and August. Nutrient influx from catchment agrifields and floating gardens triggers growth of submerged macrophytes especially during high temperatures reducing transparency as the latter is negatively proportional to primary productivity. Comparison of the values with earlier records of Khan (1996) reflect decrease in transparency indicating mounting nutrient influx in Lake.

Total dissolved solids composed mainly of carbonates, bicarbonates, Cl^-, SO_4^-S, Po_4^-P, No_3^-N, Ca^+, Na^+ and K^+ (Trivedy and Goel, 1984) governing physico-chemical properties of water ranged between 235-475 mg l^{-1}. Nutrient enrichment due to fertilizers and wastes from inhabitations of the lake enhance total dissolved solids increasing electrical conductivity as the later is directly related to total dissolved solids (Mishra and Saksena, 1993).

Electrical conductivity ranged between 150-400 μs cm^{-1}. The values when compared with earlier work (Vass and Zutshi, 1983, Khan, 1996) showed a drastic increase.

Total alkalinity varied between 53-117 mg l^{-1}. High productivity during June, July and August enhances concentration of bicarbonates (Water, 1975). Prasad *et al.* (1985) attributed higher values of total alkalinity to polluted condition of waters. Higher values of alkalinity raise pH which during present study ranged between 6.7-7.8.

Though absent during November, December, January and February, free carbondioxide the only source of carbon assimilated and incorporated into the skeleton of living matter of aquatic autotrophs, ranged between 46-85 mg l^{-1}.

Dissolved organic matter showed a value of 17-45.5 mg l^{-1}. Increased value supports high bacterial activity, decreasing dissolved oxygen (Badge and Verma, 1985, Sangu and Sharma, 1987) and increasing biochemical oxygen demand.

Dissolved oxygen varied from 9.45 to 12.4 mg l^{-1}. Regulated primarily by free diffusion of oxygen air to water, production through photosynthesis, consumption by biota, dissolved oxygen affects the solubility and availability of many nutrients and therefore productivity of aquatic ecosystem (Wetzel, 1983). Highest value (12.4 mg l^{-1}) recorded during January may be due to circulation by cooling and draw down of dissolved oxygen in water (Hunnan, 1979). Reduction in the value was observed with increase in temperature which is attributed to high microbial activity. Low content of dissolved oxygen though a sign of organic pollution, is also due to inorganic reductants like hydrogen sulphide, ammonia, nitrates and ferrous iron. Other oxidisable substances also tend to decrease this value.

Biochemical oxygen demand ranged from 3.8-7.4 mg l^{-1}. Minimum value (3.8 mg l^{-1}) recorded during January may be due to retarded microbial activity and high value (7.4 mg l^{-1}) during August is attributed to high organic decomposition and enhanced pollution load.

Chloride and silicate values ranged between 17.5-36.5 and 1.1-2.7 mg l^{-1}, respectively. High chloride value may be due to organic wastes of animal origin and domestic wastes. About four fold increase in chloride value from earlier values of Vass and Zutshi (1983) reflect magnitude of organic waste deposition in Lake.

Nitrate (N) in the present study varied from 88.5-367.5 µg l^{-1}. Though nitrate, the major form of nitrogen in oxidizing water is the product of aerobic decomposition of organic nitrogenous matter, however main source of elevated nitrate concentration may be inorganic fertilizers used indiscriminately in and around the lake. Direct relation exists between the degree of pollution and concentration of nitrates (Agarwal *et al.*, 1976).

Nitrite (N) an intermediate and unstable state of nitrogen in the present study varied from 1.5-22.5 µg l^{-1} indicating presence of organic matter entering the lake. Low concentration during June, July, August and September may be due to utilization by cyanophycean members as they consume nitrites highly. Microbial activity can also be responsible for low value (Sharma *et al.*, 1981).

Ammonical-N rapidly taken up by phytoplanktons and other hydrophytes (Toetz, 1971) showed a value of 250-502.5 µg l^{-1} and is present naturally in surface water. However, high value of ammonia is a sign of pollution (Bruce, 1958) as in addition to aquatic animals in the form of excretory product it is also generated by degradation of organic nitrogenous matter (Wetzel, 1983) and hydrolysis of urea.

Inorganic phosphorus in the form of orthophosphate (PO_4^{-3}) plays a dynamic role in lake ecosystem as it is readily taken up by phytoplanktons or lost to the sediment. The values varied from 20.5 to 32.5 µg l^{-1}, highest being , recorded during August and lowest during January. Highest value is due to runoff from surrounding crop fields and floating gardens and land mass within lake fertilized with phosphate. Increase may also be due to decayed phytoplanktons and concentration of zooplankton excreta (Heron, 1961). Use of detergents with long chain phosphate compounds and use of lake as receptacle for waste disposal have also resulted in excessive phosphorus loading.

Calcium the dominant cation ranged between 16.5-30 mg l^{-1} followed by sodium. Richness of calcium is due to lacustrine deposits in the valley.

Sodium and magnesium fluctuated between 6-13.65 and 4.7-11 mg l^{-1}, respectively and potassium which acts as enzyme activator and plays a vital role in metabolism of fresh water environment varied from 1-6.5 mg l^{-1}.

On the basis of present investigation the cation composition could be arranged as:

Ca > Na > Mg- contradictory to standard composition given by Rodhe (1949). Further, the present work clearly depicted significant changes in physico-chemical characteristics which the lake water has undergone during last two decades.

References

Agarwal, D.K., Guar, S.R., Tiwari, I.C., Narayanswami, N. and Marwah, S.M. (1976). Physico-chemical characteristics of Ganges water at Varanasi. *Indian J. Environ. Hlth.*, 18: 201-206.

APHA, AWWA, WPCW (1995). *Standard methods for examination of water and waste water.* 19th edition. American Public Health Association. Washington, D.C.

Badge, U.S. and Verma, A.K. (1985). Limnological studies in J.N.U. lake, New Delhi, India. *Bull. Bot. Soc. Sagar*, 32: 16-23.

Bruce, A. (1958). Report on a biological and chemical investigation of the waters in the Aven and Hethcata Rivers: Reprinted by the pollution advisory council Morimo Department Wellington, New Zealand.

Heron, J. (1961). Phosphorus adsorption by lake sediments. *Limnology and Oceanogr.*, 6: 338.

Hunna, H. (1979). Chemical modification in reservior regulated streams. In: *The ecology of regulated streams.* Ed. J.W. Wart and J.A. Stanford. Plenum Corporation Publication, 75-94.

Ishaq, M. and Kaul, V. (1989). Phosphorus load-concentration relationship in lake Dal, high altitude marl lake in the Kashmir Himalayas. *Int. Revue Ges. Hydrobiol.*, 74(3): 321-328.

Khan, M.A. (1993). Euglenoid red-bloom contributing to the environmental pollution of Dal Lake, Kashmir Himalaya. *Environ. Conserv.*, (Switzerland), 20: 352-356.

Khan, M.A. (1996). Recent bio-limnological pollution trends in the Kashmir Himalayan Dal Lake ecosystem: Development of Red-bloom. *Ecol. Environ. Energy*, pp. 41-77.

Mackereth, F.J.H. (1963). Water analysis for limnologists. Freshwater Biological Association. 21: 70.

Mishra, S.P. and Saksena, D.N. (1993). Planktonic fauna in relation to physico- chemical characteristics of Gauri Tank at Bhind, M.P. India. *Advances in Limnology*, Narendra Publishing House, New Delhi, pp. 57-61.

Prasad, B.N., Jaily, Y.S. and Singh, Y. (1985). Periodicity and inter-relationships of physico-chemical factors in ponds. Proc. Nat. Symp. Pure and Appl. Limnol. (ed.) A.D. Adoni. *Bull. Bot. Soc. Sagar*, 32: 1-11.

Rodhe, W. (1949). The ionic composition of Lake waters. *Verh. Internat. Veretn. Limnol.*, 10: 377-386.

Sangu, R.P.S. and Sharma, S.K. (1987). An assessment of water quality of river Ganga at Garmukteshwar (Ghaziabad). *Ind. J. Ecol.*, 14(2): 278-287.

Sharma, R.D., Neeru Lal and Pathak, P.D. (1981). Water quality of sewage drains entering Yamuna at Agra. *Indian J. Environ. Hlth.*, 23(2): 118-122.

Toetz, D.W. (1971). Effects of pH, phosphate and ammonia on the rate of uptake of nitrogen and ammonia by fresh water phytoplankton. *Ecology*, 52(4): 903-908.

Trivedy, R.K. and Goel, P.K. (1984). *Chemical and biological methods for water pollution studies.* Environmental Publications, Karad, India.

Vass, K.K. and Zutshi, D.P. (1983). Energy flow, trophic evolution and ecosystem management of

Kashmir Himalayan Lake. *Arch. Hydrobiol.*, 97(1): 39-59.

Wetzel, R.G. (1983). *Limnology.* Second edition Saunders College Publishing USA: 767 p.

Zutshi, D.P. and Khan, M.A. (1978). On lake typology of Kashmir. *Environ. Physiol. Ecol. Plants,* pp. 465-472.

Zutshi, D.P. and Vass, K.K. (1973). Variation in the water quality of some Kashmir lakes. *Trop. Ecol.*, 14: 182-196.

Zutshi, D.P. and Vass, K.K. (1978). Limological studies on Dal Lake: Chemical features. *Indian J. Ecol.*, 5(1): 90-97.

15

Ecological Study of Algae of Aner Dam (Maharashtra)

S.N. Nandan and M.R. Kumavat*

S.S.V.P.S.L.K. Dr. P.R. Ghogrey Science College, Dhule (M.S.)
**S.S.V.P.S's Late S.D. Patil Arts, Commerce and L.B.M.D.S. Science College, Shindkheda, Dist. Dhule - 425 406 (M.S.)*

ABSTRACT

In India earlier workers studied limnological aspects of algae from different rivers and lakes. But it is observed that very few workers paid attention on ecological aspects of algae from dam habitats. The present investigation was carried out by selecting the three stations of Aner dam of Maharashtra as a study area. An attempt was made to find out the relationship of algal communities and physicochemical parameters of Aner dam of Maharashtra.

Key Words: Ecology, Physico-chemical Parameters, Algae.

Introduction

In the past considerable contributions on algal flora and ecology have been made in India (Ganapati, 1940, Gonzalves and Joshi, 1946, Singh, 1960, Prasad and Singh, 1980). In Maharashtra state very few workers had paid attention on limnological study of dams and rivers. Therefore, the present work was undertaken for limnological study of algae of Aner dam of Maharashtra.

Materials and Methods

Aner dam is situated in border of Dhule and Jalgaon districts (21°118 N latitude, 75°E longitude with 193 MSL). The length of dam is about 2130 meters and height is about 45 meters. The four groups of algae *viz.* Bacillariophyceae, Cyanophyceae, Chlorophyceae and Euglenineae were estimated according to Whitton (1969). The chemical analysis of water samples of 3 stations of dam was carried out by standard methods of APHA (1975).

Results and Discussions

The range values of physico-chemical parameters of 3 stations of Aner dam are shown in Table 15.1. The range, average and total population of 4 groups of algae are shown in Table 15.2. The total population of 4 groups of algae from 3 stations of Aner Dam is shown in percentage (Fig. 15.1).

Table 15.1: Physico-chemical Parameters at 3 Stations of Aner Dam During January 2000 to December 2000

Sl. No.	*Physico-chemical Parameters*	*Stations*		
		DS-I	*DS-II*	*DS-III*
1.	Water Temperature	17.8-30.2°C	18.1-29.5°C	20.5-30.7°C
2.	pH	6.71-8.41	7.01-8.14	3.37-8.22
3.	Dissolved Oxygen	2.38-11.75	4.57-11.74	2.43-12.15
4.	Free CO_2	13.2-30.8	13.2-30.8	8.8-44
5.	Bicarbonate	160-240	150-230	160-240
6.	Total Alkalinity	160-240	150-230	160-240
7.	Calcium	15.23-63.32	12.82-64.92	13.17-72.14
8.	Magnesium	3.88-22.95	0.96-26.34	10.23-32.69
9.	Hardness	126-198	126-188	128-192
10.	Chloride	19.88-38.34	21.3-34.08	21.3-39.76

All parameters are expressed in mg/l except pH and water temperature.
DS-I: Aner Dam Station-I, DS-II: Aner Dam Station-II, DS-III: Aner Dam Station-III.

Table 15.2: The Range and Average Population of 4 Groups of Algae from 3 Stations of Aner Dam During January 2000 to December 2000

Sl.No.	*Algal Groups*		*Stations*		
			DS-I	*DS-II*	*DS-III*
1.	Cyanophyceae	Range	12.5-42	12.5-51.5	17.5-48.5
		Average	25.8	31.7	29.1
		Total Population	310	381	349
2.	Chlorophyceae	Range	20-52	24.5-43.5	17.5-48.5
		Average	32.2	31.9	30.5
		Total Population	386	383	367
3.	Bacillariophyceae	Range	17.5-45	17.5-48.5	18-51.5
		Average	33.3	31.66	36.2
		Total Population	400	380	435
4.	Euglenineae	Range	1-5	1-7	1-5
		Average	1.3	1.8	1.8
		Total Population	16	22	22

All figures are in cell no/ml × 10^4.
DS-I: Aner Dam Station-I, DS-II: Aner Dam Station-II, DS-III: Aner Dam Station-III.

The population of algae was maximum in summer season indicating water temperature played important role to increase the population of algae as agreed with earlier observations of Nazneen

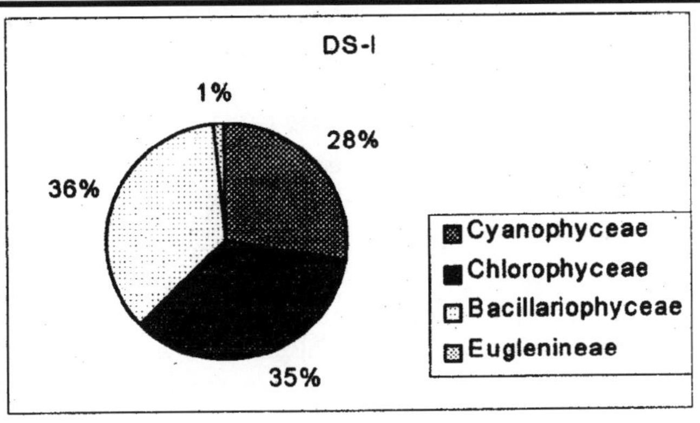

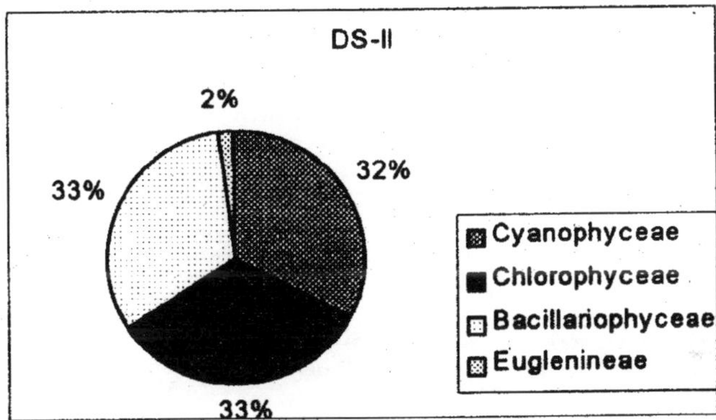

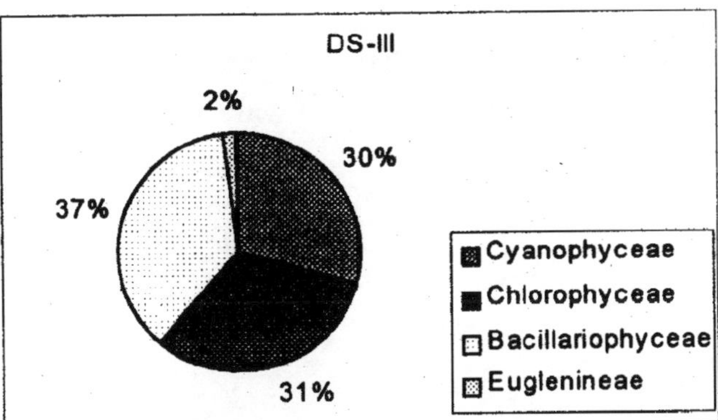

Fig. 15.1: Population of 4 Groups of Algae from 3 Stations of Aner Dam During the Period from January 2000 to December 2000.

(1980). The population of diatoms was greater at stations DSI and DSII as compared to other groups of algae in present study. Similarly total population of Chlorophyceae was greater at all 3 stations as compared to those of other groups. Euglenoids were more or less uniform in population. Its percentage was very less as compared to other groups.

Seasonal variations of water temperature were well marked with respect to different seasons, water temperature played an important role in controlling the abundance of algae as agreed with views of earlier workers (Singh, 1960, Nazneen, 1980, Nandan and Patel, 1984).

In present study the percentage of diatoms and green algae was greater as compared to other groups. This might be due to the higher concentration of dissolved oxygen, high temperature, total alkalinity and high concentration of pH as agreed with the views of Hosmani and Bharati (1980).

Acknowledgement

Authors are grateful to Principal B.M. Patil, S.S.V.P.S's L.K. Dr. P.R. Ghogrey Science College, Dhule and Principal Dr. B.V. Kamble, S.S.V.P.S's Arts, Commerce and Science College, Shindkheda for providing the facility of laboratory.

References

APHA (1975). *Standard method for the examination of water and waste water*, 14th Ed. American Public Hlth. Asso., New York.

Ganapati, S.V. (1940). The ecology of temple tank containing a permanent bloom of *Mycrocystis aeruginosa* (Kuetz) Henfr. *J. Bombay Nat. Hist. Soc.*, 42: 64-77.

Gonzalves, E.A. and D.B. Joshi (1946). Freshwater algae near Bombay. *J. Bombay Nat. Hist. Soc.*, 46: 144-176.

Hosmani, S.P. and S.G. Bharati (1980). Limnological studies in Ponds and lakes of Dharwar Comparative phytoplankton ecology of four water bodies. *Phykos*, 19(1): 27-43.

Nandan, S.N. and R.J. Patel (1984). Ecological studies on algal flora of Vishwamitri river, Baroda, Gujarat. *Indian J. Plant Nature*, 1: 17-32.

Nazneen, S. (1980). Influence of hydrobiological factors of seasonal abundance of phytoplankton in Kinjhaor lake, Pakistan. *Int. Revu. Ges. Hydrobiol.*, 65(2): 269-280.

Prasad, B.N. and Y. Singh (1980). Algal hydrobiology in India. *Proc. Nat. Acad. Sci.* India. Golden Jubilee Commemoration Volume, pp. 271-300.

Singh, V.P. (1960). Phytoplankton ecology of inland waters of Uttar Pradesh. *Proc. Symp. Algol.* ICAR: 243-271.

Whiton, B.A. (1969). Seasonal Changes in the phytoplankton of St. Jame's Park Lake, London. *The London Naturalist*, 48: 14-39.

16

Ground Water Quality of Block-V Srisailam Right Branch Canal Command Area, Kurnool District, Andhra Pradesh

B. Kotaiah and S. Sreedhar Reddy*

HOD, Dept. of Civil Engineering, S.V.U.C.E., Tirupathi - 517 504
**Center for Environmental Engineering, VIT, Vellore - 632 014*

ABSTRACT

The present study deals, with hydro geochemistry for ground water of Block-V. SRBC region, where major ion distribution and the process controlling the seasonal variation of these ions were studied. Rock water interaction was found to be the mechanism controlling the major ion chemistry.

Key Words: Ground Water, Hydro Geochemistry.

Introduction

The block-V of Srisailam right branch canal command region is one among the four isolated domains of the SRBC along the East Coast and has an Areal Extent of 2259.4 Ha. The climate of the region is humid and tropical and receives an average annual rainfall of 656 mm which is mostly contributed by the North-East monsoon. Ground water occurs in this region both under water table as well as confined conditions and is being developed by dug wells, dug-cum-bore wells and tube wells. Ground water is the sole source of water supply for domestic and agricultural activities. Agricultural being the major economic activity or rural population, it also has its effects on the ground water quality of this region.

Scant attention has been paid so far on ground water quality of Block-V, SRBC region. The aquifer for this region is over stressed in order to meet the increase in demand for fresh water due to the non-availability of surface water. This may lead to the depletion of water resources and increase in major and trace Element concentrations and salinity in ground water of this region. Though a few activities have been conducted to assess the water resources in Block-V, SRBC region, no Extensive investigations have been made on the quality aspects.

Thus, it has become necessary to assess the quantity and monitor the quality variances in ground water, caused by natural and anthropogenic activities in the urban as well as in the rural Environment of Block-V, SRBC region. This study is a consorted effort towards the understanding of the various natural and anthropogenic processes influencing the ground water of Block-V, SRBC region and to develop effective management strategies for the future.

Location of Study Area

The Block-V of Srisailam right branch canal command comprising the villages of (1) Part of Ramathirtham (2) Part of Gonavaram (3) Part of Nandivargam and (4) Part of Anupur has a total extent of 2259.4 Ha. The area lies between coordinates of N 15° 23′ 50″ to N 15° 26′ 55″ and E 78° 18′ to E 78° 23′ 10″ in survey of India topo sheet No. 57 1/7 (A1 , A2, B1 , B2).

Materials and Methods

Water has the ability to dissolve a greater range of substances than any other liquid precipitation reaching the earth contains only small amounts of dissolved mineral matter. Once it reaches on earth, during percolation, it reacts with minerals of the soil and rock in contact with it. The quality and type of mineral matter dissolved depends on the chemical composition and physical structure of the rocks as well as the hydrogen ion concentration (PH) and the redox potential (Eh) of the water, The ability of water to dissolve minerals determines the chemical nature of the ground water.

Therefore it needs a constant monitoring of chemical parameters throughout the year in all seasons. For any regional hydro chemical studies, a set of observation wells is to be selected and sampling has to be done periodically. Accordingly, in the study area, 24 bore wells, have been selected for sampling (Fig. 16.1), covers various sections such as domestic and irrigation wells.

Water samples were collected in acid washed new one liter polyethylene bottles from bore wells that were flowing continuously for a period of one year in pre monsoon and post monsoon seasons. All sample bottles were flushed with several volumes of water before the samples were collected. Separate 500 ml samples were collected for analysis of cations and anions and filtrated immediately through 0.45 mm Millipore filters. Temperature, PH and Electrical conductivity were measured at the site using portable meter (CHECKMATE 90).

In the laboratory water samples were anlaysed for major ion concentrations such as Ca^{2+}, Mg^{2+}, Na^{+}, K^{+}, Hco^{-}_{3}, Cl^{-} and F^{+} etc., as per the procedures laid down in standard methods (APHA, 1993).

Results and Discussion

Geo Chemical Characteristics of Ground Water

A complete chemical analysis may determine the suitability of ground water for domestic, agriculture and industrial use. The analysis of a ground water sample includes the determination of the concentration of the inorganic constituents present, in addition to the measurement of pH, electrical

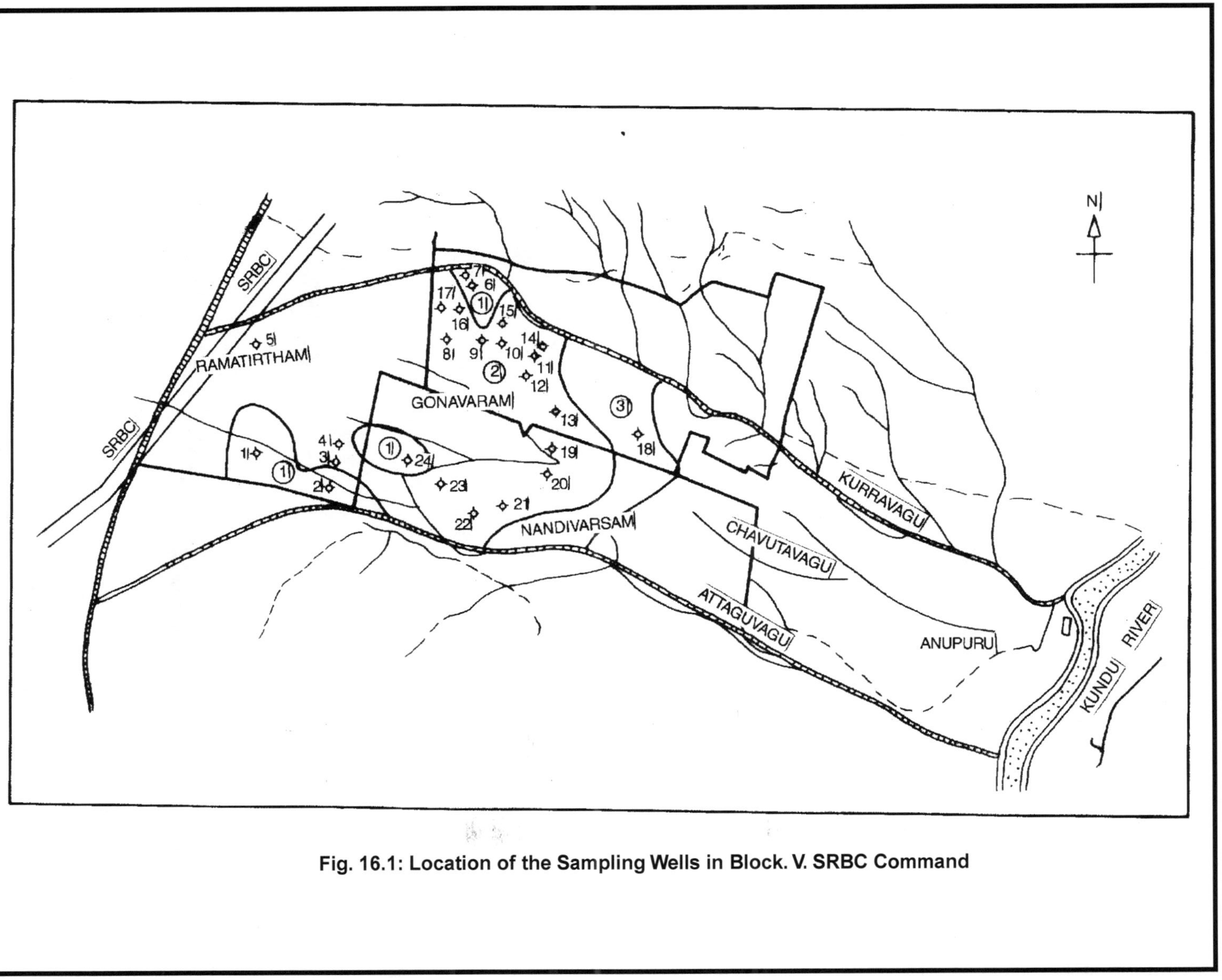

Fig. 16.1: Location of the Sampling Wells in Block. V. SRBC Command

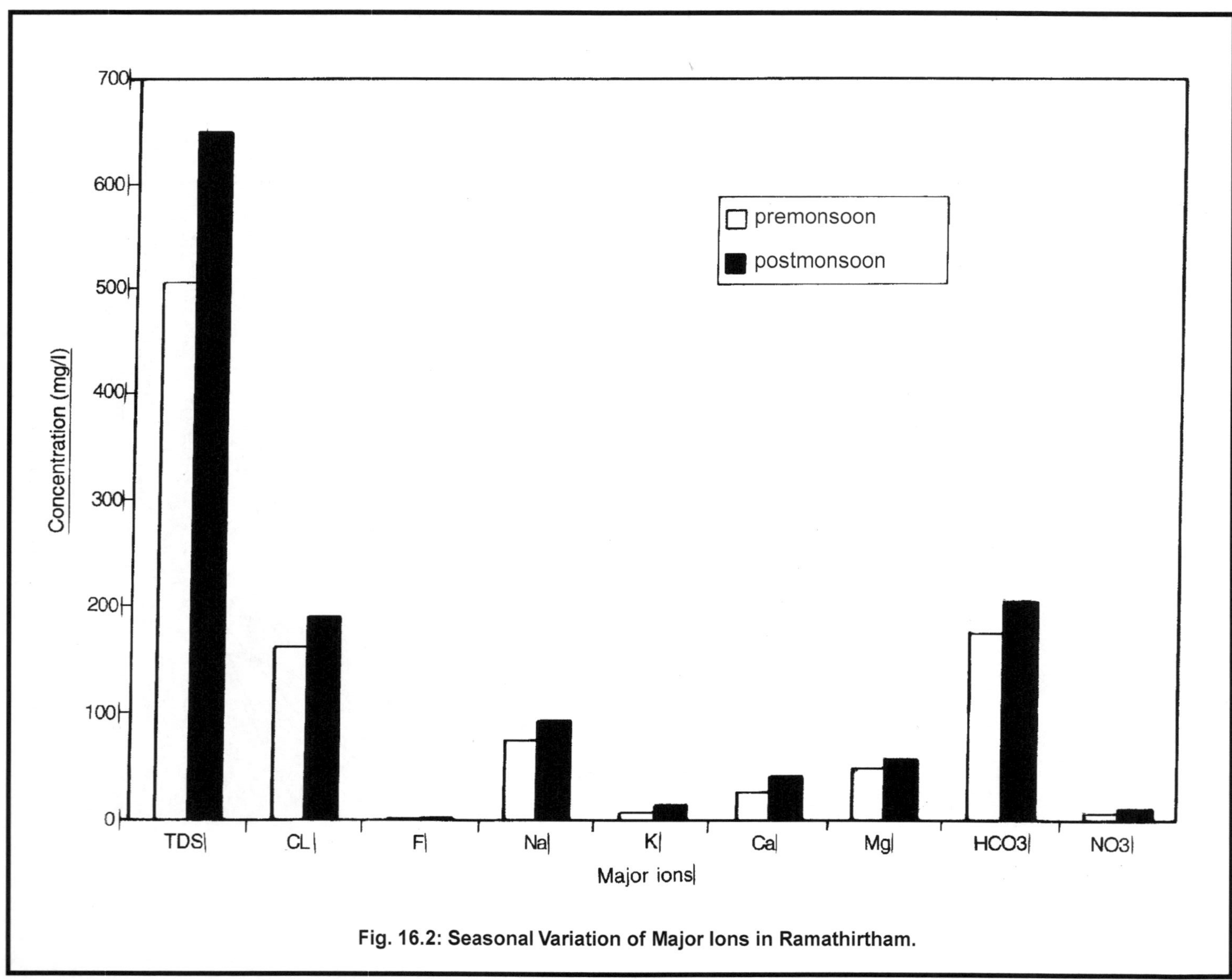

Fig. 16.2: Seasonal Variation of Major Ions in Ramathirtham.

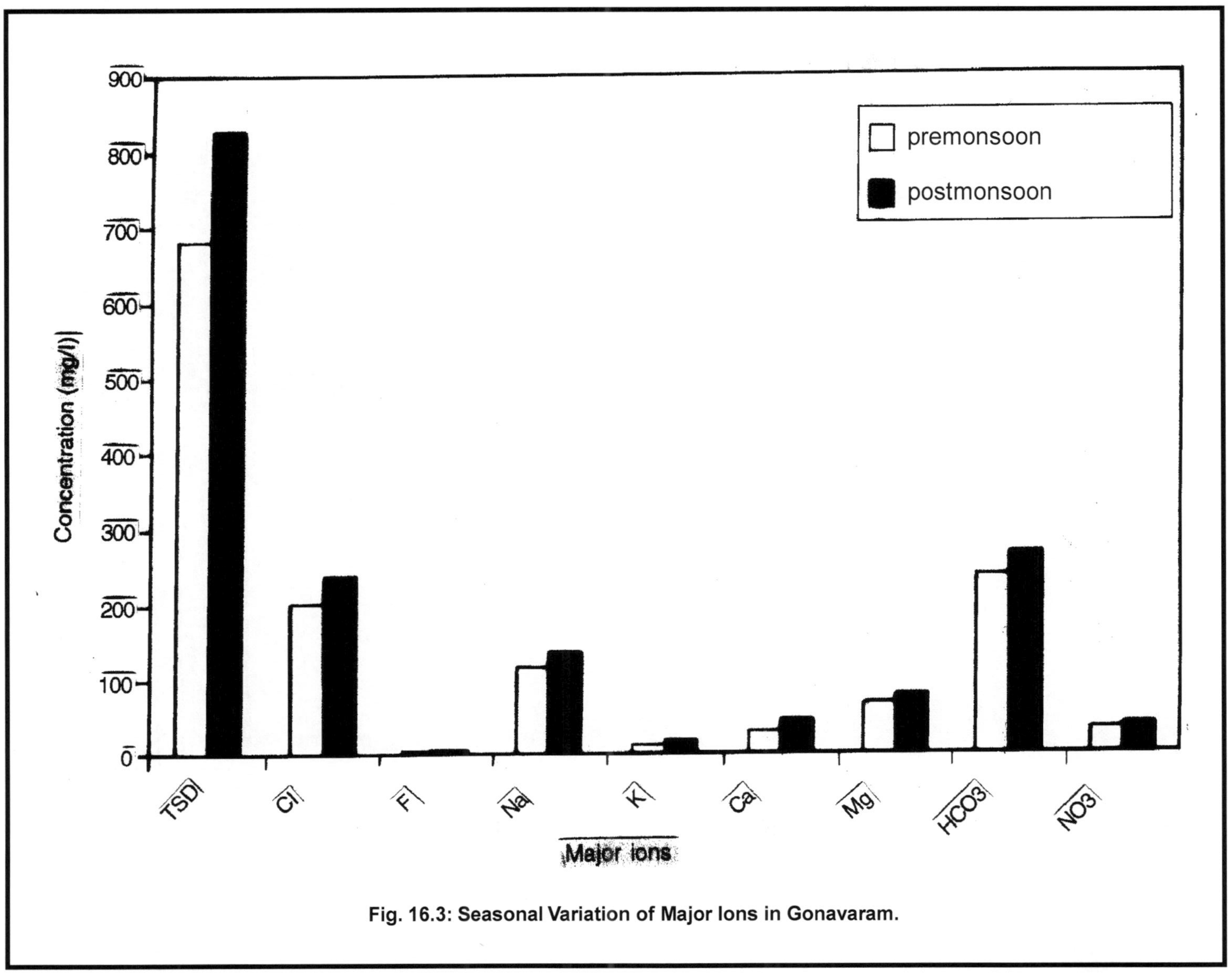

Fig. 16.3: Seasonal Variation of Major Ions in Gonavaram.

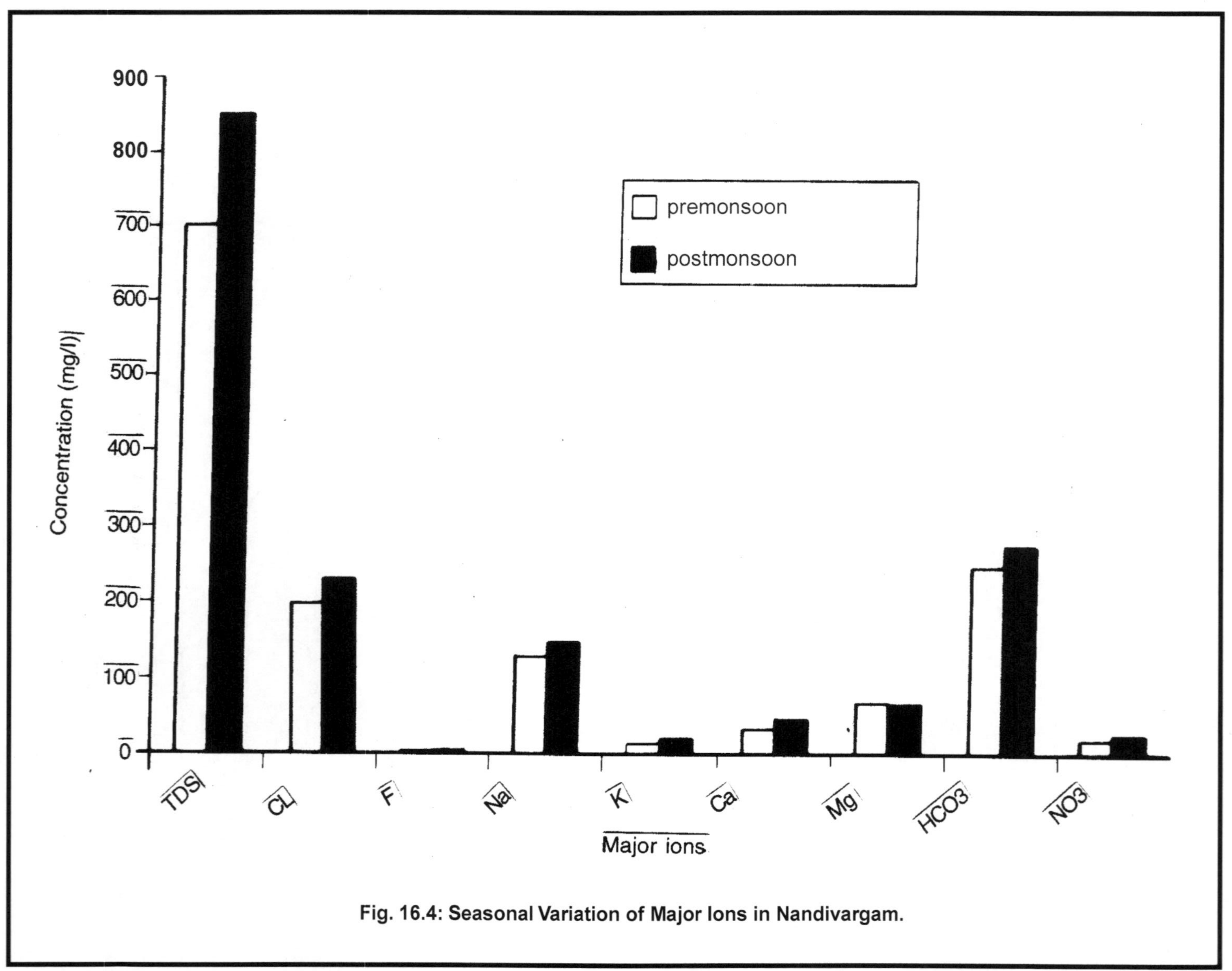

Fig. 16.4: Seasonal Variation of Major Ions in Nandivargam.

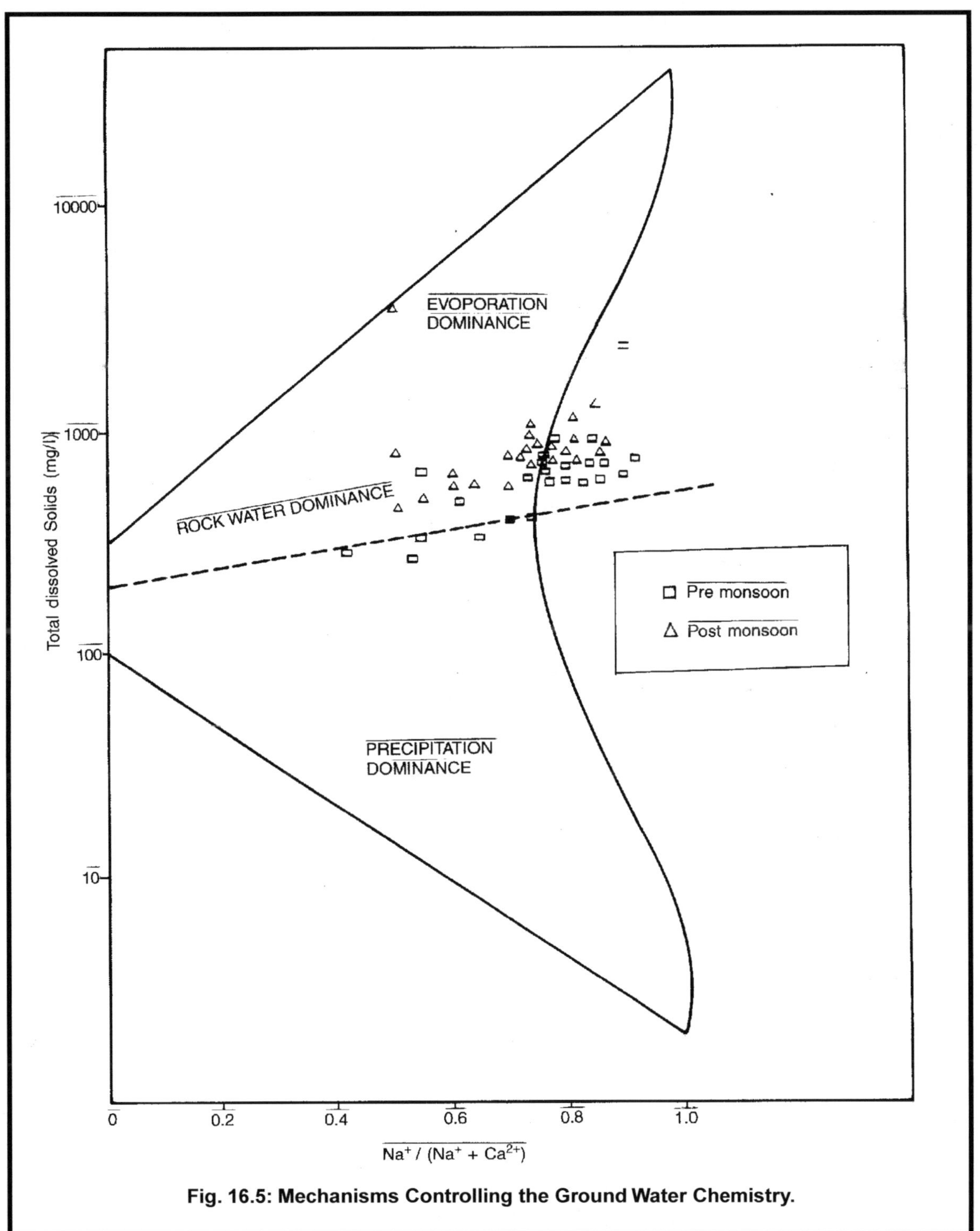

Fig. 16.5: Mechanisms Controlling the Ground Water Chemistry.

conductance, total dissolved solids and other minor constituents. Each of these properties is useful in evaluating the chemical character of ground water. Ground water, quality is also influenced by meteorological factor such as rainfall, evaporation, etc. Therefore it needs a constant monitoring of chemical parameters through out the year. In the present study, ground water from 24 bore wells tapping various aquifer formations in Block-V, SRBC command area, have been sampled and analysed for a period of one year during both pre and post monsoon seasons.

The temporal variation in the concentrations of major ions is shown in Fig. 16.2-16.4. From the figures, it is evident that the concentrations of all the ions in pre monsoon season were low and exhibiting increasing trend to post monsoon season. The reason for these changes could be the dissolution of salts and minerals which are present in soil due to the rise in water table during monsoon period. Ramesam (1973) and Kripanidhi (1984) have reported similar trend in ground water of a typical hard rock terrain and pollution in village wells in Karnataka state, India, respectively. The examination of the seasonal data of major ion concentration also reveals that the Hco_3 is the dominant ion irrespective of the season. Because, it is mainly derived from rock weathering and only minor variation is observed in F, Ca and Mg ion concentrations due to their low mobility in ground water system.

Mechanisms Controlling the Chemistry of Ground Water

Conway (1984), Garham (1961), Christ and Garrels (1965, 1966), Gibbs (1970) and Ramesam and Barua (1973) have discussed in detail the mechanisms, controlling the chemistry of fresh water. The hydro chemical studies are being used to establish the relationship of water composition to acquifer lithology. This helps not only to explain the origin and distribution of the dissolved constituents, but also to elucidate the factors controlling the ground water chemistry.

As per the classification of Gibbs (1970), the major natural mechanisms controlling world surface-ground water chemistry are atmospheric precipitation, rock weathering and evaporation and factional crystallization. A boomerang shaped diagram resulted when Gibbs plotted the ratio of two major cations as $Na^+/(Na^+ + Ca^{2+})$, versus TDS. For Block-V, SRBC command region the ratio of $Na^+/(Na^+ + Ca^{2+})$ of the ground water samples have been plotted against TdS and is shown in Fig. 16.5. The density of distribution of points is maximum above the center like of the enveloping curve and indicates that mechanism controlling the chemistry of ground water is predominantly due to rock water interaction and also partly due to evaporation in the subsurface environment.

Conclusion

In this study an attempt has been made to identify the path ways and concentration of major ions in the ground water of Block-V, SRBC command region, Kurnool district, Andhra Pradesh, over exploitation of ground water and improper management of natural resources led to the unequal distribution of major ions in nature. Also it can be seen that the major ion concentrations are predominantly influenced by natural agencies than anthropogenic activities.

References

APHA (1993). *Standard Methods for the examination of water and waste water*, APHA, AWWA, WPCF, New York.

Conway, E.J (1984). Mean geo-chemical data in relation to oceanic evolution. *Proc. Irish Acad. Sect. B*, Vol. 48: 119-159.

Gibbs, R.J. (1970). *Mechanisms controlling water chemistry*, 170: 1088-1090.

Garham, E. (1961). Factors influencing supply of major ions to inland water with special reference to the atmospheric, *Bull. Geo. Soc. Amer.*, 72: 795-840.

Garrels, R.M. and Christ, C.L. (1965). Solutions, Minerals and Equilibria, Harper row, New York.

Kripanidhi, K.V.J.R. (1984). Mechanism of ground water pollution in Village wells, *Bull. of Geological Society of India*, 25: 301-302.

Ramesam, V. and Barua, S.K. (1973). Preliminary studies on the mechanism controlling salinity in the north western regions of India, *Ind. Geohydrology*, 9: 10-18.

17

Ecological Status of Sitapur Pond at Haridwar (Uttaranchal)

D.R. Khanna and Rakesh Bhutiani

Department of Zoology and Environmental Sciences,
Gurukula Kangri Vishwavidyalaya, Haridwar - 249 404

ABSTRACT

The physico-chemical characteristics of the aqueous phase influence on the types and distribution of aquatic biota. Conversely, they are also influenced by the activity of the aquatic biota. Normally water is never pure in chemical sense. It contains gases (Dissolved oxygen, free CO_2, N_2 etc.). Dissolved minerals (Ca, Mg, Na salts etc.), suspended matter (Clay, Silt, Sand etc.) and even microbes. These are natural impurities derived from atmosphere catchment areas and the soil. Thus the importance of the study of water quality is obvious.

In the present study an attempt has been made to quantify the present ecological status of Sitapur pond also efforts has been made to find out the indicators of water quality in terms of planktonic studies.

Key Words: Ecological Status, Pond, Phytoplankton.

Introduction

Water is a very essential and vitally important substance. It is the medium which gave birth to the first primitive molecule and without it no life can exist.

Ponds are small bodies of water in which the littoral zone is relatively large and the limnetic and profundal regions are small or absent.

The village Sitapur is situated near Jwalapur in the district Haridwar of Uttaranchal and situated about 4 km. far from Haridwar. The pond is completely filled with water for the whole year. The pond receives waste water from the nearby houses and also the tube well water mixes in to it.

Sitapur pond contains a number of phytoplankton including Diatoms, blue green algae, green algae etc.

In the present study an attempt has been made to find out the Ecological status of Sitapur pond.

Materials and Methods

The samples were collected monthly during morning hours (7.00 A.M to 11.00 A.M) the study was conducted for one year (1999-2000). The physico-chemical and biological parameters were analyzed according to the standard methods of APHA (1998), Trivedy and Goel (1986) and Khanna (1993).

Table 17.1: Seasonal Variation in Physico-chemical Parameters of Sitapur Pond

Parameters	*Winter*	*Summer*	*Monsoon*	*Average*
Temperature (°C)	12.6 ± 0.80	22.30 ± 0.63	17.34 ± 0.91	17.41 ± 3.96
Transparency (cm)	57.70 ± 0.13	39.80 ± 0.41	29.95 ± 0.02	42.48 ± 11.48
Turbidity (JTU)	27.65 ± 0.26	19.80:10.42	42.35 ± 0.33	29.93 ± 9.34
Velocity (m/s)	1.12 ± 0.31	1.31 ± 0.29	1.97 ± 0.39	1.46 ± 0.36
Total Solids (mg/l)	402.73 ± 10.76	473.36 ± 7.86	800.60 ± 13.20	558.89 ± 173.32
pH	7.10 ± 0.08	8.10 ± 0.05	7.90 ± 0.12	7.70 ± 0.43
Alkalinity (mg/l)	367.30 ± 5.32	231.61 ± 1.79	215.70 ± 1.69	271.53 ± 68.02
Free CO_2 (mg/1)	3.01 ± 0.38	4.80 ± 0.20	5.88 ± 0.35	4.56 ± 1.18
BOD (mg/l)	4.97 ± 0.14	5.30 ± 0.11	5.89 ± 0.19	5.38 ± 0.37
Hardness (mg/l)	293.67 ± 1.66	247.00:1:3.61	239.61 ± 0.51	260.09 ± 23.93

± : Standard Deviation.

Table 17.2: Seasonal Variation of Chlorophyceae at Sitapur Pond

Parameters	*Winter*	*Summer*	*Monsoon*	*Average*
Clostrium	118 ± 2.50	88.0 ± 5.20	0.00 ± 0.00	68.66 ± 50.07
Ankistrodesmus	140 ± 1.36	92.0 ± 2.70	0.00 ± 0.00	77.33 ± 58.08
Ulothrix	189.0 ± 18.25	85.0 ± 9.05	87.0 ± 7.24	120.33 ± 48.56
Cosmarium	130.0 ± 5.25	81.0 ± 4.12	88.0 ± 1.25	99.66 ± 21.66
Cladophora	123.0 ± 0.70	82.0 ± 0.96	0.00 ± 0.00	68.33 ± 51.13
Chaetophora	128.0 ± 0.86	90.0 ± 3.25	0.00 ± 0.00	72.66 ± 53.67
Euglena	0.00 ± 0.00	60 ± 0.98	71 ± 0.66	43.66 ± 18.40
Microcystis	0.00 ± 0.00	00 ± 0.00	88 ± 8.45	29.33 ± 41.48
Oocystis	0.00 ± 0.00	48 ± 8.76	84.0 ± 8.75	44.00 ± 34.40
Volvox	0.00 ± 0.00	00 ± 0.00	72 ± 0.85	24.00 ± 33.94
Stigeocolonium	0.00 ± 0.00	00 ± 0.00	60.0 ± 1.84	20.00 ± 28.28

± : Standard Deviation.

Table 17.3: Seasonal Variation and Bacillariophyceae at Sitapur Pond

Parameters	*Winter*	*Summer*	*Monsoon*	*Average*
Desmidium	87.0 ± 4.27	87.0.0 ± 1.27	118.0 ± 3.13	97.33 ± 14.61
Nitzschia	82.0 ± 8.24	88.0 ± 3.34	96.0 ± 0.56	88.66 ± 5.73
Navicula	63.0 ± 0.96	95.0 ± 0.95	0.00 ± 0.00	52.66 ± 39.46
Cymbella	49.0 ± 3.36	0.00 ± 0.00	115.0 ± 8.25	54.66 ± 47.11

± : Standard Deviation.

Table 17.4: Seasonal Variation in Cyanophyceae at Sitapur Pond

Parameters	*Winter*	*Summer*	*Monsoon*	*Average*
Oscillatoria	94.0 ± 4.50	89.0 ± 2.24	96.0 ± 1.35	93.00 ± 2.94
Nostoc	57.0 ± 3.40	55.0 ± 0.96	0.00 ± 0.00	37.33 ± 26.41
Anabena	89.0 ± 3.70	64.0 ± 1.75	92.0 ± 3.90	81.66 ± 12.55

± : Standard Deviation.

Results

The results of various physico-chemical and biological parameters are show in Table 17.1-17.4. The average value of water temperature recorded for Sitapur pond was 17.41°C. The average value of total solids was observed as 558.89 mg/l. It shows an increase after winter due to increase in turbidity. Turbidity ranged between 27.65 JTU to 42.35 JTU, turbidity increases during monsoon season. The annual average of transparency was recorded as 42.48 cm. The value of pH ranged between 7.10 to 8.10, a very little variation in pH was recorded. The values of alkalinity, Hardness and DO fluctuated between 215.70 mg/l to 367.30 mg/l, 239.61 mg/l to 293.67 mg/l and 5.97 mg/l to 9.67 mg/l respectively. The other chemical parameters *i.e.* free CO_2 and BOD were continuously increasing from winter to monsoon.

Chlorophyceae and Cyanophyceae were recorded highest in winter while Bacillaropllyceae were recorded highest in monsoon season.

Discussion

The unity of organism and its environment is a basic principle of ecology. The organism cannot exist without the environment, any organism, population or species lives at the expense of its environment without this interaction it cease to exist (Nikolski, 1963). The aquatic life of Sitapur pond have direct effect on ecological condition of water.

In the present study temperature showed all inverse relationship with dissolved oxygen in all the seasons similar kind of relationship was also reported by (Khanna *et al.*, 1997).

At high temperature during summer and monsoon seasons due to high rate of decomposition of organic matter, resulted in decrease of pH, release of CO_2 as also reported by Husmani and Bharti (1980), Yadav *et al.* (1987).

Monawar, (1970), Jindal and Vashist (1985) have also reported the free CO_2 influence the alkalinity and pH of water.

Maximum DO was recorded in winter but it reduces onwards due to turbidity. Maximum DO was due to phytoplankton activity as also reported by Gautam (1990). Total solids, turbidity and dissolved matters are found closely interrelated with one another and cause common effect upon the pond and its aquatic life as also reported by Khanna (1993). The similar decreasing trend in transparency rate was also reported by Swaroop and Singh (1979). Turbidity was found minimum in summer and maximum in monsoon same type of results were given by Khanna (1993). The fluctuations of pH lies in alkaline range same results were observed by Bagde and Verma (1985). Alkalinity showed a decreasing tendency during the course of study Khanna (1993). Hardness ranged between (234.8 mg/l to 170.16 mg/l) same results were observed by Paka and Narsingh Rao (1997).

In the present study Chlorophyceae was found to be dominating group as also reported by Verma and Mohanty (1995) and Khanna and Singh (2000). Chlorophyceae was recorded highest in winter and lowest in monsoon season, among Chlorophyceae total 11 genera were recorded as *Ankistrodesmus, Chaetophora, Cladophora, Clostrium, Euglena, Cosmarium, Microcystis, Oocystic, Volvox, Stigeocolonium and Ulothrix.*

Among Bacillariophyceae the highest count was recorded in monsoon while lowest in summer, a total of 4 genera were recorded as *Cymbella, Desmidium, Navicula* and *Nitzschia.*

The Cyanophyceae were represented by only 3 genera which are *Oscillatoria, Anabena* and *Nostoc,* these were recorded highest in winter and lowest in monsoon season.

The general observed in study probably show high organic pollution according to Algal Genus index, Palmer (1980). The number of plankton were also affected by the turbidity.

The increased turbidity reduces the phytoplankton production as also reported by Das and Upadhayaya (1979) and Khanna *et al.* (1997).

Conclusion

At last we must conclude that some physico-chemical and biological parameters of Sitapur pond are within the range and some parameters crosses the limits and this pond was more high osmotic pollution probable due to high organic pollution so at last we must conclude that water of Sitapur pond is not usable for drinking purposes.

References

APHA, AWWA and WPCF (1998). *Standard methods for the examination of water and waste water*, American Public Health Association, Washington D.C. 20th edition, New York.

Bagde, V.S. and Verma, A.K. (1985). Physico-chemical characteristics of water of JNU lake at New Delhi. *Ind. J. Ecol.,* 12(1): 151-156.

Das, S.M. and Upadhayaya, J.C. (1979). Studies on qualitative and quantitative fluctuations of plankton in two Kumaon lakes, Nainital and Bhimtal (India). *Acta Hydrobiol.,* 21(1): 9-17.

Gautam, A. (1990). Ecology and pollution of mountain waters, Ashish Publishing House, Delhi, 1-209.

Husmani, S.P. and Bharti, S.G. (1980). Limnologic studies in ponds and lake of Dharwar comparative phytoplakton ecology of four water bodies. *Phyco.,* 19(1): 27-43.

Jindal, R. and Vashist, H.S. (1985). Ecology of fresh water pond at Nabha Punjab, India. *Zoological Orientalis,* 18: 74-85.

Khanna, D.R. (1993). Ecology and pollution of river Ganga. Ashish Publishing House, Delhi, 1-241.

Khanna, D.R., Malik, D.S. and Badola, S.P. (1997). Population of green algae in relation to physico-chemical factors of River Ganga at Lalji Wala, Haridwar, Uttar Pradesh. *J. Zool.,* Vol. 17(3): 237-240.

Khanna, D.R., Singh, R.K. (2000). Seasonal fluctuations in the plankton of Suswa River at Raiwala (Dehradun), *Env. Cons. J.,* 1(2&3): 89-92.

Monawar, M. (1970). Limnological studies on fresh water ponds of Hyderabad India, the biotape. *Hydrobiologia,* 35(1): 127-162.

Nikolski, G.V. (1963). The ecology of fishes. Academic Press, London, 1-352.

Paka, S. and Rao, A.N. (1997). Interrelationship of physico-chemical factors of a pond. *J. of Env. Biol.* 18(1): 67-72.

Palmer, C. Mervin (1980). Algae and Water pollution. Palmer Corn (1969). *J. Phycol.,* 5: 78-82.

Swarup, K. and Singh, S.R. (1979). Limnologic studies of Suraha lake (Ballia). *J. of Inland Fish Soc.,* India, 2(1): 22-23.

Trivedy, R.K. and Goel, P.K. (1986). Chemical and Biological methods for water pollution studies, Enviromnental publications, Karad.

Verma, J.P. and Mohanty, R.C. (1995). Phytoplankton of Malyanta pond of Laxmisagar and its correlation with physico-chemical parameters. *Enviromedia pollution research,* 14(2): 243-253.

Yadav, Y.S., Singh, R.K., Choudhary, M. and Kalekar, V. (1987). Limnology and productivity of Dilghali Bhel (Assam). *India Trop. Ecol.,* 28(2): 137-147.

18

Seasonal Variation in Biochemical Composition in the Muscle Tissue of *Mugil cephalus*, Linnaeus and *Liza macrolepis* (Smith) in Relation to their Maturity Stages

G. Sivani and L.M. Rao

Department of Zoology, Andhra University, Visakhapatnam - 530 003

ABSTRACT

A detailed comparative study on the variations in the biochemical parameters such as protein, lipid and carbohydrate besides moisture content in muscle of *Mugil cephalus*, Linnaeus and *Liza macrolepis* (Smith), in relation to the maturity stages, at Gosthani Estuary near Visakhapatnam, has been carried out during November 1993 to October (1994). The percentage composition of protein, carbohydrate and lipid were maximum in post-spawning period and minimum in spawning period. These parameters are inversely related to moisture and directly related to each other. The results are judiciously interpreted and discussed in detail.

Key Words: *Mugil cephalus, Liza macrolepis,* Proteins, Lipids, Carbohydrates, Maturity.

Introduction

Many scientists all over the world worked out on the proximate composition of various groups of fishes in detail. These studies are of considerable use in understanding the ecology and overall food value at the different trophic levels of the environment. The frequency of changes in the composition of biochemical constituents of biota vary not only with the environmental changes but also with seasons

(Green, 1915, Milroy, 1908, Saha and Guha, 1940, Damberge, 1964, Somvanshi, 1983, Jafri *et al.*, 1964, Gras *et al.*, 1967, Jafri *et al.*, 1964, Jafri, 1968, Bligh *et al.*, 1973, Masurekar and Pai, 1977 and Saleem Mustafa and Jafri, 1978). Moreover the changes of biochemical constituents have also been attributed to various physiological and other factors such as maturation, spawning, feeding etc. (Love, 1957 and Jacquot, 1961). However, the biochemical composition in individuals of a single species have been reported in relation to their age, sex, habitat and seasons. Hence, in the present study the variations in the biochemical parameters in relation with maturity stages in *Mugil cephalus* and *Liza macrolepis* have been evaluated and discussed in detail.

Materials and Methods

The fish were collected at monthly intervals throughout the year (November 1993 to October, 1994) from Gosthani estuary, Bheemunipatnam. Freshly caught specimens of the two species were separated into males and females. The muscle samples of each sex were analyzed separately so as to find out the difference, if any, in between the sexes and the species in relation to seasons.

To analyze the biochemical constituents, the muscle samples in wet condition were first ground thoroughly and the individual samples were weighed for the quantitative estimation of proteins, carbohydrates, lipids and moisture. The total protein content was estimated by using the method of Lowry *et al.*, (1951), the carbohydrates by Caroll *et al.*, Method (1956) and total lipid content was estimated by Folch *et al.*, (1957) Method. All these values were expressed as percentage.

Results

The results of all the biochemical parameters in the tissue of the two species are given in Tables 18.1 and 18.2. Moisture forms the main bulk in the composition of muscles. Lowest moisture values (70.2 per cent, 72.37 per cent) in both the sexes corresponds to the post-spawning period, whereas higher values (73.9), 75.15 per cent) coincides with the pre-spawning period. The values gradually decreased as gonads attain maturity. Similar observations were made in *Sillago sihama* (Sivani, 1995), *Amblypharyngodon mola* (Ravisankar Piska and Waghrey, 1998) and *Salmostoma bacaila* (Piska and Kumar, 1989). In *Mugil cephalus* the muscle protein content ranged from a minimum of 7.68 per cent to a maximum of 18.49 per cent in males, while it varied from a minimum of 9.98 per cent to a maximum of 24.17 per cent in females. In both males and females the protein content was low in March and high in November. The protein content remained higher in females than in males through out the study period. The seasonal fluctuations observed in biochemical constituents are presented in Table 18.1. The low values (11.45 per cent, 13.48 per cent) were observed during pre-monsoon period and high values (16.96 per cent, 21.31 per cent) were recorded during winter months in both males and females. In *Liza macrolepis* the percentage protein varied from 13.16 per cent (May) to 22.18 per cent (January, 94) in males and in females it varied from 16.51 per cent (in May) to 23.19 per cent (in January). Seasonally there was a difference in the protein levels. The low levels were observed during the pre-monsoon period and monsoon periods (19.52 per cent females, 16.81 per cent males) and high values during post-monsoon period (19.90 per cent males, 22.30 per cent females).

The total carbohydrate content in *M. cephalus* varied between 1.99 per cent (January, 94) to 3.64 per cent (October, 94) in males whereas in females it varied from 2.28 per cent (January, 94) to 3.98 per cent (April, 94). The carbohydrate content in males remained lower than females almost throughout the year, except in the month of June when it was slightly higher in males than in females. In the post-monsoon season the carbohydrate level was high (3.48 per cent in males and 3.53 per cent in females) whereas in the winter season the level was low (2.07 per cent in males and 2.45 per cent in females).

Table 18.1: Seasonal Variations Observed in Biochemical Constituents of Muscle of *M. Cephalus* During 93-94 (per cent).

Season	*Moisture*		*Protein*		*Carbohydrate*		*Lipid*	
	Male	*Female*	*Male*	*Female*	*Male*	*Female*	*Male*	*Female*
Pre-monsoon (Feb-May)	71.23	72.3	11.45	13.48	2.90	3.30	1.51	1.80
Monsoon (June-Sep)	70.2	72.37	11.48	18.40	2.60	2.68	1.95	2.67
Post-monsson (Oct-Nov)	72.7	74.4	15.00	19.61	3.48	3.53	2.50	2.76
Winter (Dec-Jan)	73.9	75.15	16.96	21.31	2.07	2.45	2.66	3.65

Table 18.2: Seasonal Variations Observed in Bio-chemical Constituents of Muscle of *L. macrolepis* during 93-94 (per cent).

Season	*Moisture*		*Protein*		*Carbohydrate*		*Lipid*	
	Male	*Female*	*Male*	*Female*	*Male*	*Female*	*Male*	*Female*
Pre-monsson (Feb-May)	69.95	73.15	17.55	19.52	3.04	2.82	1.52	1.86
Monsoon (June-Sep)	68.52	71.13	16.81	20.71	3.87	3.12	1.95	2.20
Post-monsson (Oct-Nov)	70.10	72.34	17.35	22.30	3.28	3.07	1.99	2.56
Winter (Dec-Jan)	69.8	72.10	19.90	21.74	2.96	2.86	3.85	4.09

In *L. macrolepis* the average value of carbohydrate in male was slightly higher than those of females. In males the range varied between 1.73 per cent (January, 94) and 4.18 per cent (December, 94). Seasonally high values were observed during monsoon months (3.87 per cent males, 3.12 per cent females) and low values during winter months (2.96 per cent males, 2.86 per cent females).

The total lipid content in *M. cephalus* varied between 0.78 per cent (September, 94) and 3.98 per cent (June, 94) in males, whereas in females it varied from 1.12 per cent (April, 94) to 4.01 per cent (June, 94). Seasonally there was a marked difference in the total lipid content, high values were recorded during winter season (2.66 per cent females, 3.65 per cent females) and low values (1.51 per cent females, 1.80 per cent females) during the pre-monsoon period. In *L. macrolepis* the lipid levels showed conspicuous variations in different months. The values ranged between 0.98 per cent (April, 94) and 3.96 per cent (January, 94) in males and 1.18 per cent (April, 94) and 4.15 per cent (January, 94) in females. High lipid values (3.85 per cent in males, 4.09 per cent females) were observed during winter months and low lipid values (1.52 per cent in males, 1.86 per cent in females) were observed during pre-monsoon season.

Discussion

The protein content, in the two species was low in the pre-monsoon months and high in the late post-monsoon and winter months, coinciding with the spawning and recovery (maturation) periods.

The range of protein content in the two fishes is up to the mark and agrees with the findings of Love (1970), in different fishes from different environs. The quantity of protein, the body building food, showed an increase along with the growth and maturity but decreased during the spawning periods. In *M. cephalus* the feeding intensity was more during the period of maturation. As this species is a coastal breeder, only a few specimens in spawning stage could be collected from this estuary and hence no details of chemical composition could be estimated. As such the data indicate that the juveniles of the fish enter the estuary sometime in the pre-monsoon months, continue to grow in size up to October-November to attain maturity by December and then migrate to the sea for spawning. Our earlier observations, elsewhere on the maturity stages of *M. cephalus* from Gosthani estuary, strengthen this observation. In *L. Macrolepis* the maximum values of gastro somatic indices were observed in maturity stage-II condition. From this observation it is clear that while attaining maturity the feeding intensity increases and the food that is consumed by the fish is used in building up of the muscle.

Variations in carbohydrate content also showed correlation with the feeding intensity and spawning seasons as in the case of proteins. The high carbohydrate content can be correlated with the high feeding intensity (observed when the fish is attaining maturity) during post-monsoon period and the low carbohydrate values during winter season can be correlated with the low metabolic rate. During winter months, as the metabolic rate becomes low and the utilization of energy decreases, the surplus energy gets accumulated in the form of lipids. From the earlier observation (Sivani, 1994), it is clear that *L. macrolepis* has two spawning seasons one in January-February and the other in July-August. It is further observed that while attaining maturity the feeding intensity increases (Sivani and Rao, 1996). This could be the reason for the high values of carbohydrates recorded in November-December and January-July. The sudden fall in the carbohydrate content correlates the spawning period, when the animal would be subjected to heavy stress and strain with simultaneous cessation of feeding.

According to George and Berger (1966), the lipids are non-caloric metabolic proteins and their content is directly related to the intensity of feeding as observed by Channon (1939) and Hickling (1947).

Acknowledgements

The authors are thankful to the Head, Department of Zoology, Andhra University, for the facilities provided. The first author is grateful to the Council of Scientific and Industrial Research for awarding the SRF.

References

Bligh, E.G., Dyer, W.J., Regier, L.W., Dingle, J.R., and Legendre, R. (1973). Functional and Nutritive food proteins from fish. In: Fishery Products. Kreuzer, R (ed) Pub. By Fishing News. *Proc. of F AO Technical Conference on fishery products* held at Tokyo.

Caroll, W.V., R.W., Longley, and J.M. Roe (1956). The determination of glycogen in the liver and muscle by the use of anthrone reagent. *J. Biol. Chem.* 220: 586-593.

Channon, H.J. (1939). Fat metabolism of the herring. I. Preliminary Survey. *Biochem. J.,* 26: 2021-2034.

Damberg, N. (1964). Extractives of fish muscle. IV. Seasonal variations of fat, water soluble proteins and water in cool (*Gadus morhua*) fillets. *J. Fish. Res. Bd. Canada*. 21: 703.

Folch, J., Lees and G.H. Stone Stanley (1957). A simple method for the isolation an purification of total lipid from animal tissues. *J. Biol. Chem.,* 266: 497-509.

Gras, J., R. Reynaud, I. Ganoity, J. Frey and J.C. Henry (1967). Biochemical studies on fishes. I. Six month variations in the content of water proteins in the muscle tissue of Rainbow trout (*Salmo gairdinerii* Rich). *Experintia*, 23: 430-431.

Green, C.W. (1919). Biochemical changes in the muscle tissue of king salmon during the fast spawning of migration. *J. Biol. Chemi.* 39: 435-456.

Hahn, (1967). *Bioch. Physiol.*, 23: 83-93.

Hickling, C.F. (1947). The seasonal cycle in the corn fish *Pilchard sardina* pilchardus wolf. *J. Mar. Bioi. Ass.*, U.K., 26: 115-138.

Jachquot, R. (1961). Organic Constituents of fish and other aquatic animal foods. In: *Fish as Food.* (Ed. George Borgstorm) Academic Press Inc., N.Y. 1, pp. 145-209.

Jafri, A.K. (1968). Seasonal changes in the biochemical composition of the freshwater murrel "*Ophiocephalus punctatus*" (Bloch). *Hydrobiologia*, 32: 206-213.

Jafri, A.K., D.K. Khawaja and S.Z. Qasim (1964). *Studies on the biochemical composition of some freshwater fishes.*

Love, R.M. (1957). The bio-chemical composition of fish. In: *The Physiology of Fishes* (Ed. Brown, M.E), Academic Press, London and N.Y., 1, pp. 401-418.

Love, R.M. (1970). The Chemical biology of fishes. pp. 1-547.

Lowry, O.H., H. Rosenborough and R. Pandau (1951). Protein measurement with the Folin-Phenol Reagent. *J. Biol. Chem.*, 193: 1-265.

Masurekar, V.B. and S.R. Pai (1977).Observations on the fluctuations in protein, fat and water content in *Cyprinus carpio* (Linnaeus) in relation to the stages of maturity. *Ind. J. Fish*, 30: 217-224.

Milroy, T.H. (1908). Changes in the chemical composition in the herring during the reproductive period. *Biochemc. J.*, 3: 366-390.

Rao, L.M. and Sivani, G. (1996). The food and feeding habits of five commercially important fishes of Gosthani estuary. *Indian J. Fishes*, 44.

Ravi shankar Piska and Salara Waghrey (1989). Biochemical variations of reproductive tissues of *Amblypharyngodon mola* (Ham) with reference to spawning cycle. *Ind. J. Fish.*, 36(4).

Saleem mustafa and A.K. Jafri (1978). Some biochemical constituents and the caloric values of different regions of the body musculature of pond murrel *Channa punctatus* Bloch. *Fish. Technol.*, 15: 57-59.

Sivani, G. (1994). Studies on Biology, Nutritive value and Fishery of some commercially important fishes of Gosthani estuary near Visakhapatnam. Ph. D Thesis.

Somavanshi, V.S. (1989). Seasonal changes ill the biochemical composition of a hill-stream fish: *Garra mullya* (Sykes). *Ind. J. Fish*, 30(1).

19

Water Quality and Phytoplankton Abundance in the South Indian River Tamiraparani

Martin P. and M.A. Haniffa

Centre for Aquaculture and Research Extension,
St. Xavier's College (Autonomous), Palayamkottai - 627 002

Introduction

Rapid industrialization and urbanisation destroy natural waters by adding their effluents and wastes (Sabata and Nayar, 1987). Urban population is increasing day-by-day and discharging domestic sewage into aquatic systems. Indians are emotionally attached to rivers and they take bath during festivals to worship God (Sankaran, 1988). In the Tamiraparani river bank areas there are 5 temples and the river is considered as sacred in South Indian history.

Since water pollution is a biological phenomenon and algae occur in a wide variety of polluted and unpolluted water, it is proper that being the primary producer of aquatic environment, the knowledge of the algal indicator species provides a useful solution to this problem. Distinct pulses in the population of phytoplankton have been attributed to the periodicity of various physicochemical factors (Strom, 1928, Pearsall, 1932). Algae have been found to be one of the best parameters for the assessment of the trophical status of the water body (Kaushik *et al.*, 1991b). Kolkwitz and Marsson (1909) for the first time described the pollution oriented changes in composition and algal communities in river. Biological methods to study the ecology of waste water algae of river systems throw light on indicator species as emphasized by Hosmani and Bharati (1981) and Palmer (1969, 1980). Wielgolaski (1975) studied the pollution aspects of river water and assessed the degree of pollution on the basis of phytoplankton abundance. Phytoplankton are the basic members in the aquatic ecosystem and hence changes in phytoplankton population have a direct link with the alterations in water quality in any aquatic environment. This account summarises the phytoplankton abundance and distribution, indicator species and dominant species of unpolluted and polluted (by industry and sewage) stations in the Tamiraparani river basin. Sorensen's index of similarity (Sorensen, 1948) was used to assess the effects of urban and industrial wastes on the distribution of phytoplanton communities in the basin.

Materials and Methods

The Tamiraparani is a medium river basin in Southern Tamil Nadu with a catchment of 5482 km^2. The river takes its origin from Podihai hills of Pababasam mountain ranges of Western Ghats (8° 42′ N and 7° 24′ E) at a height of 2074 m. The river meanders through a distance of 120 km (24 km in hilly terrain and 96 km in plains) in Tirunelveli and Tuticorin districts. Eleven sampling stations were selected on the basis of point sources (PS) and non-point sources (NPS) of pollution (Fig. 19.1).

Four stations (2, 6, 8 and 11) were selected for the phytoplankton analysis. Station 2 was highly polluted by the textile-mill effluent and station 6 was a non-polluted one. Stations 8 and 11 were moderately polluted by the domestic sewage. The sampling was done in four months in the year 1991, March, June, October and January representing the pre-monsoon, southwest monsoon, north-east monsoon and post monsoon. Phytoplankton samples were collected and preserved immediately in Lugol's iodine (10 ml/100 ml) solution and after sedimentation and centrifugation sub-samples of 10 ml were collected and few drops of formalin and glycerine were also added for their long-term preservation. The identification of phytoplankton was done with the help of standard texts and monographs (Smith, 1950, Desikachary, 1959 and Philipose, 1967). Statistical analysis of co-efficient of correlation was carried out between the physicochemical parameters and the phytoplankton families.

Inter-station Relationship

The phytoplankton samples collected from the 4 stations of the river were compared by Sorensen's index for similarity (Sorensen, 1948)

$$S = \frac{2j}{a+b}$$

Where S = Coefficient of association between sites A and B.
j = Number of species common to both sites A and B.
a = Number of species present at site A.
b = Number of species present at site B.

Results

Pollution in the River

River Thamiraparani is polluted by industrial, domestic and agricultural wastes which are directly or indirectly discharged into the river. The Madura Fabrics is the major textile-industry situated at Vikramasingapuram, on it banks near its origin. This industry discharges about 7,700 kl of effluent and 1,380 kl of sewage daily into the river (BOD-572 mg/l, COD - 1628 mg/l, Mn - 0.26 mg/l, Cu - 7.6 mg/l, Cr 4.8 mg/l and Zn - 6.9 mg/l). The service station of Tamil Nadu State Transport Corporation at Tirunelveli is discharging oily drainage into the river. In addition fertilizers, pesticides and sewage are also being drained into their river from the cultivated land village along its course.

Species Number

Phytoplankton samples collected from the 4 selected sampling stations of the river belonged to six taxonomic divisions, namely Chlorophyta, Bacillariophyta, Cyanophyta, Euglenophyta, Chrysophyta, and Xanthophyta. A total of 39 species has been identified of which 15 species belonged to Chlorophyta, 11 to Bacillariophyta, 10 to Cyanophyta and the remaining 3 to each of Euglenophyta,

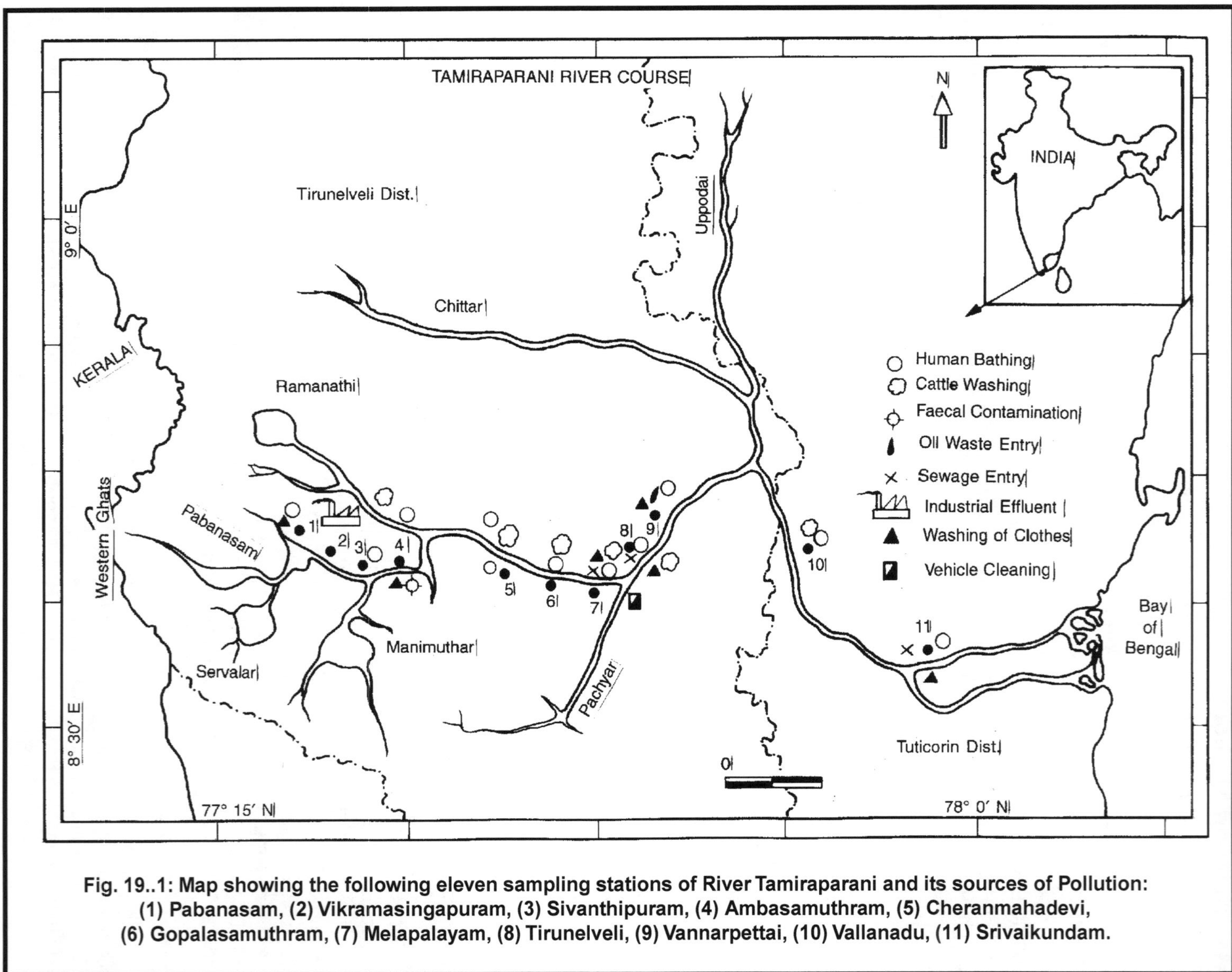

Fig. 19..1: Map showing the following eleven sampling stations of River Tamiraparani and its sources of Pollution: (1) Pabanasam, (2) Vikramasingapuram, (3) Sivanthipuram, (4) Ambasamuthram, (5) Cheranmahadevi, (6) Gopalasamuthram, (7) Melapalayam, (8) Tirunelveli, (9) Vannarpettai, (10) Vallanadu, (11) Srivaikundam.

Chrysophyta and Xanthophyta.

Species Composition and Abundance of Phytoplankton

Among the Chlorophycean members, *Scenedesmus diamorphus* was noticed in all stations in all seasons. All the species of Chlorophyceae were dominant in the pre-monsoon month (March) in all the stations except in station 11. Volvox species and Scenedsmus dimorphus were noticed dominant in the post monsoon period (January). Generally during the south-west and north-east monsoon the Chlorophycean species showed a decline. In station 2, only two Chlorophycean species (*Scenedesmus dimorphus* and *Closterium*) were identified. Closterium species was noticed in the premonsoon month (18 cells/ltr).

In station 6, five Chlorophycean species were identified, *Scenedesmus dimorphus* was the predominant species and *Coelastrum* sp. was the least abundant one. In the pre-monsoon month, each species number was noticed viz.: *Volvox* sp. (86 cells/l), *S. dimorphus* (120 cells/l), *S. quadricauda* (52 cells/l). In station 8, twelve species of Chlorophyceae family were identified. The *Volvox* sp. was the predominant one (60 to 180 cells/l) and the *Hydrodictyon* sp. was the least abundant species (10 to 18 cells/l). The perennial species found during all the seasons at station 8 were *Volvox* sp., *S. dimorphus, Ulothrix* sp., *Cladophora* sp., *Oedogonium* sp., *Spirogyra* sp., *Ankistrodesmus* sp., *Coelastrum* sp., *Closterium* sp., *Hydrodictyon* sp., *Tetraspora* sp. and *Chlorella* sp. In station 11 also twelve species were noticed. Volvox sp. was the predominant species with more number (48 to 120 cells/l) whereas the *Coelastrum* sp. (6 to 18 cells/l) and *Pediastrum* sp. (8 to 18 cells/l) were less in number. The perennial species found in all the seasons were *Volvox* sp., *S. dimorphus, S. quadricauda, Cladophora* sp., *Oedogonium* sp., *Ankistrodesmus* sp., *Coelastrum* sp., *Hydrodictyon* sp., *Pediastrum* sp, *Chlorococcum* sp., *Tetraspora* sp. and *Chlorella* sp.

The sole Bacillariophycean member present in station 2 was *Navicula cryptocephala* noticed in all the seasons (12 to 18 cells/l) except during north-east monsoon month (October). In station 6, four species were identified as perennial namely *Synedra* sp. (16 to 42 cells/l) *Fragilaria* sp. (10 to 38 cells/l) *Stauroneis* sp. (30 to 58 cells/l) and *Concinodiscus* sp. (26 to 38 cells/l). Here the Bacillariophyceae was noticed as dominant in pre-monsoon month (March) and post monsoon month (January). In station 8 seven Bacillariophycean species were noticed. Among them six species were recognised as perennial species viz. *Tabellaria* sp. (20 to 28 cells/l) *N. cryptocephala* (10 to 26 cells/l), *Diploneis* sp. (26 to 48 cells/l) *Stauroneis* sp. (44 to 68 cells/l) *Gomphonema clevoi* (26 to 48 cells/l) and *Concinodiscus* sp. (12 to 26 cells/l). In station 11, *Tabellaria* sp. was the predominant one. Five species were identified in this station. Out of these 4 were found as perennial species namely *Synedra* sp., (12 to 36 cells/l), *Tabellaria* sp. (26 to 56 cells/l), *Fragilaria* sp. (30 to 56 cells/l) and *G. gracile* (16 to 38 cells/l).

Two species of Cyanophyceae family were noticed in station 2. They were *Oscillatoria princeps* (6 to 12 cells/l) and *Anabaena* sp. (22 to 26 cells/l). The former was dominant in the post monsoon month whereas the latter was in the pre-monsoon month. In station 6, three perennial species were identified namely *Spirulina* sp. (40 to 124 cells/l), *O. curviceps* (32 to 84 cells/l) and *Microcystis aeruginosa* (26 to 86 cells/l). In station 8 the Cyanophycean member *Spirulina* sp. was the predominant species. Eight perennial species were identified in this stations viz.: *Spirulina* sp. (64 to 148 cells/l), *O. princeps* (42 to 124 cells/l), *O. curviceps* (32 to 98 cells/l), *Lyngbya* sp. (10 to 34 cells/l), *Nostoc* sp. (40 to 120 cells l), *Anabaena* sp. (40 to 82 cells/l), *M. aeruginosa* (36 to 64 cells/l) and *Pseudoanabaena* (12 to 36 cells/l). *Spirulina* sp. was the predominant one in station 11. In this station eight perennial species were noticed namely *Spirulina* sp. (92 to 210 cells/l), *O. princeps* (40 to 150 cells/l), *O. curviceps* (28 to 62 cells/l), *Nostoc* sp. (50 to 116 cells/l), *Anabaena* sp. (32 to 92 cells/l), *Merimopedia* (14 to 28 cells/l), *Phoramidium* sp. (22 to 48 cells/l) and *Pseudoanabaena* sp. (14 to 28 cells/l).

In Euglenophyceae, only one species viz., *Phacus* sp. was found in all the four stations. But in station 8, it was found as a perennial species throughout the year. The number of species ranged from 12 to 16 cells/l in stations 2, 14 to 36 cells/l in station 6, 20 to 38 cells/l in station 8 and 12 to 26 cells/l in station 11. *Phacus* sp. was dominant in the pre-monsoon month (March) in all the stations.

The sole Xanthophycean member *Botryococcus* sp., was found only in station 8. In this station too it was not available during the north-east monsoon and post monsoon. The species count in the other seasons were 22 cells/l in pre-monsoon, and 8 cells/l in south-west monsoon.

Chromulina sp. was a single member of Chrysophyceae found in stations 8 and 11. The predominancy was observed in the pre-monsoon month (March) as 18 cells/l (Station 8) and 22 cells/l (Station 11).

In the pre-monsoon month (March) the total population of Phytoplankton was high i.e., 110 cells in station 2; 852 cells in station 6; 1762 cells in the station 8 and 1528 cells in station 11. In the north-east monsoon season phytoplankton was recorded less in number in all the sampling stations viz.: 16 cells in station 2, 472 cells in station 6, 804 cells in station 8 and 630 cells in station 11. In the south-west monsoon the total phytoplankton population was high in station 8 (987 cells) followed by station 11 (760 cells). Stations 6 and 2 showed the population of 522 and 44 cells respectively. In the post monsoon i.e., at the month of January the highest number of phytoplankton was recorded in station 8 (1132 cells) and the lowest in station 2 (76 cells). In stations 6 and 11, the total phytoplankton population were 547 cells and 982 cells respectively (Table 19.1, Fig. 19.2).

Table 19.1: Seasonal Variation in Phytoplankton Density at chosen stations in River Tamiraparani (values given in parenthesis indicate percentage).

	Station 2				*Station 6*				*Station 8*				*Station 11*			
Genera	*Mar.*	*Jun.*	*Oct.*	*Jan.*	*Mar.*	*Jun.*	*Oct.*	*Jan.*	*Mar.*	*Jun.*	*Oct.*	*Jan.*	*Mar.*	*Jun.*	*Oct.*	*Jan.*
Chlorophyceae	46 (41.82)	20 (45.46)	16 (100)	14 (18.42)	364 (42.72)	264 (50.58)	276 (58.48)	295 (53.92)	706 (44.07)	444 (44.99)	342 (42.54)	538 (47.53)	528 (34.56)	286 (37.63)	250 (39.68)	454 (46.23)
Bacillariophyceae	12 (10.91)	18 (40.90)	0	16 (21.05)	176 (20.66)	92 (17.63)	84 (17.79)	126 (23.04)	272 (15.43)	184 (18.64)	146 (18.16)	180 (15.90)	218 (14.27)	116 (15.26)	84 (13.33)	136 (13.85)
Cyanophyceae	36 (32.73)	6 (13.64)	0	34 (44.74)	276 (32.39)	150 (28.73)	98 (20.76)	126 (23.04)	706 (40.07)	318 (32.22)	284 (35.32)	392 (34.63)	734 (48.04)	334 (43.95)	296 (46.99)	378 (38.49)
Euglenophyceae	16 (14.54)	0	0	12 (15.79)	36 (04.23)	16 (03.06)	14 (02.16)	0	38 (02.16)	24 (02.43)	20 (02.49)	22 (01.94)	26 (01.70)	12 (01.58)	0	14 (01.43)
Chrysophyceae	0	0	0	0	0	0	0	0	18 (01.02)	9 (00.91)	12 (01.49)	0	22 (01.43)	12 (01.58)	0	0
Xanthophyceae	0	0	0	0	0	0	0	0	22	8 (01.25)	0 (00.81)	0	0	0	0	0

The quality of water and its biota not only differ from river to river but also along the course of the same river. The quality of the river water is subjected to widest variation according to geochemistry of land over which it flows, the discharge of sewage and effluents into it and the influx of water from various tributaries. These variations in water chemistry bring about the changes in the distributional

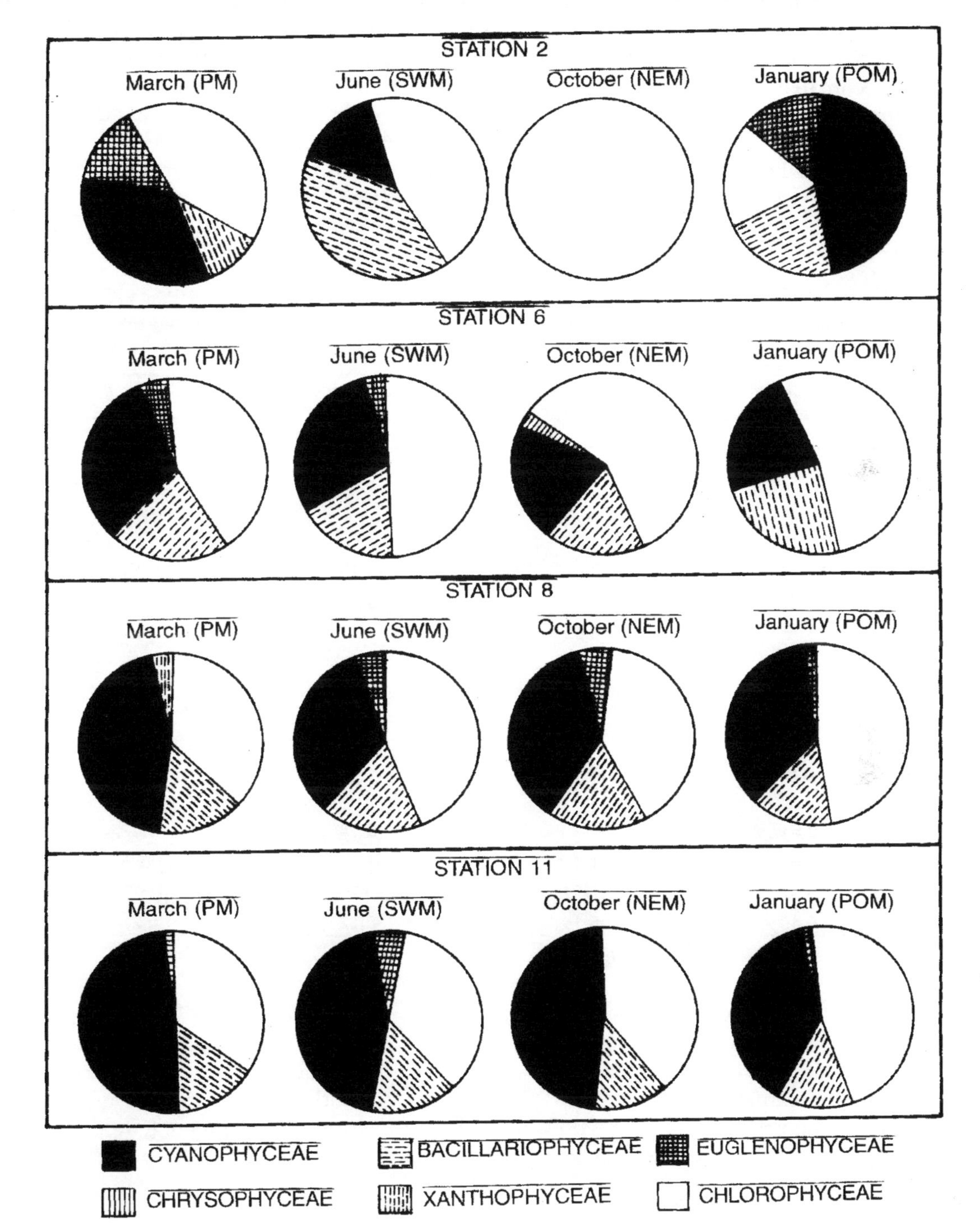

Fig. 19.2: Seasonal variations of Phytoplankton families in chosen stations of river Tamiraparani

pattern of phytoplankton in the river. In station 2, total solids, total alkalinity and dissolved Oxygen showed the negative correlation with Chlorophyceae (r = - 0.96, - 0.80 and - 0.81) whereas the total suspended solids, free CO_2 and BOD showed a positive relationship (r = 0.76, 0.61 and 0.95). Bacillariophycean members showed a high negative correlation with water temperature (r = - 0.89), sediment temperature (r = - 0.80), total hardness (r = - 0.77) and chemical oxygen demand (r = 0.78). Total suspended solids showed a positive relationship (r = 0.63). Cyanophycean members showed a negative correlation with physico-chemical parameters such as sediment temperature (r = - 0.58), total dissolved solids (r = - 0.94) total solids (r = - 0.97) and a positive correlation with total suspended solids (r = 0.86). In the case of Euglenophycean members, the physico-chemical parameters such as total dissolved solids and total solids showed negative relationship (r = - 0.96) and total suspended solids and BOD showed positive relationship (r = 0.85 and 0.62) (Table 19.2).

Table 19.2: Test of Significance Attempted for Correlation Coefficient between Physico-Chemical Parameters and Phytoplankton Density in Station 2 in River Tamiraparani

Physico-chemical Parameter	*Chlorophyceae*		*Bacillariophyceae*		*Cyanophyceae*		*Euglenophyceae*	
	r	*t*	*r*	*t*	*r*	*t*	*r*	*t*
Water Temperature	0.31	1.04	-0.89	6.41***	-0.38	1.30	-0.18	0.61
Sediment temperature	0.25	0.84	-0.80	4.27***	0.58	2.29*	0.41	1.43
pH	0.57	2.17	0.37	1.27	-0.13	0.44	-0.11	0.37
TDS	-0.52	1.91	-.052	1.93	-0.94	9.04****	-0.96	9.42****
TSS	0.76	3.70***	0.63	2.60*	0.86	5.40****	0.85	5.15****
TS	-0.96	11.60***	-0.40	0.13	-0.97	14.21****	-0.96	10.59****
Total hardness	0.47	1.69	-0.77	3.84***	-0.34	1.17	-0.15	0.49
Total Alkalinity	-0.80	4.16***	-0.23	0.78	-0.05	0.14	0.11	0.38
Dissolved Oxygen	-0.81	4.36***	0.24	0.81	0.09	0.27	-0.07	0.22
FCO_2	0.61	2.45*	-0.44	1.53	-0.04	0.13	-0.02	0.07
BOD	0.95	9.94***	-0.19	0.61	0.47	1.68	0.62	2.52*
COD	0.51	1.85	-0.78	3.98**	-0.26	0.87	0.07	0.21
Conductivity	0.45	1.58	0.14	0.44	-0.32	1.06	-0.29	0.96

*P<0.05 **P<0.01 ***P<0.005 ****P<0.001.

In station 6, Chlorophycean member showed a positive relationship with water temperature (r = 0.93), sediment temperature (r = 0.91) and total hardness (r = 0.63) and a negative correlation with conductivity. Bacillariophycean members showed a positive correlation with water temperature (r = 0.91), sediment temperature (r = 0.88), total hardness (r = 0.81) and conductivity (r = 0.89). Cyanophycean members showed a positive correlation with water temperature (r = 0.99), sediment temperature (r = 0.99), pH (r = 0.74) and dissolved oxygen (r = 0.58) and a negative correlation with conductivity (r= - 0.59) (Table 19.3).

Table 19.3: Test of Significance Attempted for Correlation Coefficient between Physico-Chemical Parameters and Phytoplankton Density in Station 6 in River Tamiraparani

Physico-chemical Parameter	*Chlorophyceae*		*Bacillariophyceae*		*Cyanophyceae*		*Euglenophyceae*	
	r	*t*	*r*	*t*	*r*	*t*	*r*	*t*
Water Temperature	0.93	7.86****	0.91	7.11****	0.99	40.92***	0.85	5.16****
Sediment temperature	0.91	6.93****	0.88	5.95***	0.99	31.44	0.89	2.92****
pH	0.37	1.25	0.40	1.38	0.74	3.50**	0.74	3.48**
TDS	0.10	0.34	0.11	0.37	-0.33	1.10	-0.54	2.00
TSS	-0.07	0.05	-0.13	0.41	-0.51	1.85	-0.54	2.02
TS	0.01	0.04	-0.01	0.04	-0.42	1.47	-0.55	2.06
Total hardness	0.63	3.35**	0.81	4.29***	0.63	1.98	0.13	0.42
Total Alkalinity	0.01	0.03	0.18	0.54	-0.24	0.76	-0.69	2.98*
Dissolved Oxygen	-0.58 -0.23	2.21 0.73	0.36 -0.24	1.23 0.79	0.58 0.21	2.25* 0.68	-0.89 0.45	6.04**** 1.61
FCO_2	0.36	1.02	0.13	0.42	0.07	0.23	0.34	0.89
BOD	-0.35	1.17	0.54	2.06	-0.57	2.22	-0.20	0.65
COD Conductivity	-0.85****	5.03	0.89****	6.11	-0.59	2.31*	-0.23	0.73

*P<0.05 **P<0.01 ***P<0.005 ****P<0.001.

In station 8, Chlorophycean members showed a positive relationship with pH (r = 0.63), total dissolved solids (r = 0.86), total solids, COD (r = 0.75) and a negative correlation with dissolved oxygen (r = – 0.95) and BOD (r = – 0.77). Bacillariophycean members showed positive correlation with pH (r = 0.83), total dissolved solids (r = 0.84), total solids (r = 0.58) total hardness (r = 0.66) and negative correlation with dissolved oxygen (r = – 0.97) and BOD (r = – 0.58). Cyanophycean members showed a positive relationship with pH (r = 0.82), total dissolved solids (r = 0.95) and total hardness (r = 0.58) and a negative correlation with dissolved oxygen (r = – 0.99). Euglenophycean members showed a negative correlation with dissolved oxygen (r = – 0.96) and conductivity (r = – 0.66) and a positive correlation with pH (r = 0.90), total dissolved solids (r = 0.85) and total hardness (r = 0.75). Chrysophycean members showed a positive correlation with pH (r = 0.93), total hardness (r = 0.95) and conductivity (r = 0.90). Xanthophycean members showed a positive correlation with pH (r = 0.93) total dissolved solids (r = 0.72) and total hardness (r = 0.86) and a negative correlation with dissolved Oxygen (r = – 0.88) and conductivity (r = – 0.80) (Table 19.4).

In station 11 Chlorophycean members showed a positive correlation with the water temperature (r = 0.59), sediment temperature (r = 0.60), total dissolved solids (r = 0.98), total suspended solids (r = 0.93) and total solids (r = 0.96). Bacilariophycean members showed a positive relationship with pH (r = 0.58), total dissolved solids (r = 0.83), total suspended solids (r = 0.70), total solids (r = 0.77) and total harness (r = 0.82). Cyanophycean members showed a positive relationship with total dissolved solids (r = 0.75), total suspended solids (r = 0.61), total solids (r = 0.69) and total hardness (r = 0.88). Euglenophycean member showed a positive relationship with pH (r = 0.92), total hardness (r = 0.99) and dissolved oxygen (r = 0.84) and a negative correlation with total alkalinity (r = – 0.57) (Table 19.5).

Table 19.4 : Test of Significance Attempted for Correlation Coefficient between Physico-Chemical Parameters and Phytoplankton Density in Station 8 in River Tamiraparani

Physico-chemical Parameter	*Chlorophyceae*		*Bacillariophyceae*		*Cyanophyceae*		*Euglenophyceae*		*Chrysophyceae*		*Xanthophyceae*	
	r	*t*	*r*	*t*	*r*	*t*	*r*	*t*	*r*	*t*	*r*	*t*
Water Temperature	0.26	0.84	-0.01	0.03	0.19	0.60	-0.07	0.22	0.43	1.51	-0.29	0.94
Sediment temperature	0.20	0.63	-0.04	0.13	0.18	0.56	-0.08	4.21***	0.33	1.11	0.29	0.95
pH	0.63	2.55**	0.83	4.70****	0.82	4.50**	0.90	6.83***	0.93	7.81****	0.93	8.14***
TDS	0.86	6.44****	0.84	4.91****	0.95	9.13**	0.85	5.15***	0.51	1.89	0.72	3.32**
TSS	0.08	0.25	0.04	0.12	-0.21	0.67	-0.17	0.54	0.62	2.49	-0.11	0.36
TS	0.72	3.23**	0.58	2.24*	0.47	1.70	0.44	1.57	-0.29	0.95	0.41	1.42
Total hardness	0.38	1.30	0.66	2.75*	0.58	2.24*	0.75	3.58**	0.95	9.13***	0.86	5.20***
Total Alkalinity	0.06	0.20	0.10	0.32	-0.13	0.43	0.09	0.27	0.12	0.39	0.28	0.92
Dissolved Oxygen	-0.95	9.43***	-0.97	11.98***	-0.99	70.15*	-0.96	11.27**	-0.55	2.09	-0.88	5.85**
FCO_2	-0.20	0.63	0.02	0.08	-0.28	0.94	0.05	0.15	0.27	0.88	0.26	0.88
BOD	-0.77	3.83***	-0.58	2.22*	-0.33	1.09	-0.44	1.54	0.35	1.18	-0.35	1.17
COD	0.75	5.32***	-0.53	1.98	-0.50	1.84	-0.39	1.33	0.40	1.39	-0.28	0.94
Conductivity	-0.28	0.91	-0.57	2.19	-0.46	1.63	-0.66	2.80	0.90	3.16**	-0.80	4.16**

*P<0.05 **P<0.01 ***P<0.005 ****P<0.001.

Table 19.5 : Test of Significance Attempted for Correlation Coefficient between Physico-Chemical Parameters and Phytoplankton Density in Station 11 in River Tamiraparani

Physico-chemical Parameter	*Chlorophyceae*		*Bacillariophyceae*		*Cyanophyceae*		*Euglenophyceae*		*Chrysophyceae*	
	r	*t*	*r*	*t*	*r*	*t*	*r*	*t*	*r*	*t*
Water Temperature	0.59*	2.30*	0.21	0.66	0.10	0.33	-0.33	1.17	-0.40	1.50
Sediment Temperature	0.60*	2.34*	0.22	0.70	0.24	0.79	-0.34	1.13	0.39	1.34
pH	0.24	0.95	0.58	2.23*	0.56	2.16	0.88	5.85***	0.92	7.62***
TDS	0.98	17.37***	0.83	4.61***	0.75	3.60***	0.40	2.36	0.34	1.15
TSS	0.93	7.93***	0.70	3.07*	0.61	2.42*	0.21	0.68	0.15	0.48
TS	0.96	11.27***	0.77	3.83***	0.69	3.01*	0.32	1.05	0.26	0.84
Total hardness	0.52	1.90	0.82	4.59****	0.88	5.71***	0.99	28.56***	0.99	17.99***
Total Alkalinity	0.21	0.67	0.18	0.56	-0.36	1.20	0.33	1.12	-0.57	2.22
Dissolved Oxygen	0.05	0.05	-0.36	1.23	0.44	1.55	0.80	4.25***	0.84	4.80***
FCO_2	-0.45	1.58	-0.13	0.42	0.14	0.46	0.36	1.12	0.43	1.51
BOD	-0.01	0.04	-0.05	0.16	0.13	0.40	-0.16	0.53	-0.23	0.75
COD	-0.30	0.98	-0.17	0.53	0.02	0.08	0.20	0.65	-0.26	0.85
Conductivity	0.23	0.76	-0.14	0.45	-0.33	1.09	-0.54	2.04	-0.54	2.03

*P<0.05 **P<0.01 ***P<0.005 ****P<0.001.

Inter-station Relationship

Among the Chlorophycean members the highest index value (0.75) was found between stations 8 and 11, which are polluted by sewage water and this was followed by stations 6 and 11 (0.59), stations 6 and 8 (0.47), stations 2 and 6 and stations 2 and 8 (0.29) and the lowest index value was estimated between stations 2 and 11 (0.14) (Table 19.6).

The Cyanophycean members also showed a highest similarity index (0.75) between stations 8 and 11 whereas the lowest index was noticed between staions 6 and 11 (0.36). Stations 6 and 8 showed a similarity of 0.55 and stations 2 and 8 and 2 and 11 showed 0.4 similarity (Table 19.6).

In the Bacillariophyceae family, the highest index value was found between stations 6 and 11 (0.44) followed by stations 6 and 8 (0.36) and stations 2 and 11 (0.33). The index value was 'nil' between stations 2 and 6 and 2 and 8. In the Xanthophyceae family only one species was recorded. The highest value (1) was found between stations 8 and 11. The value estimated between other stations were nil. For Euglenophyceae family only one species (*Phacus* sp.) was found in all the stations. So the similarity index was one. In the family of Chrysophyceae, *Chromulina* sp. was found in stations 8 and 11. The inter site relationship between stations 8 and 11 was one. Relationship between remaining stations (2 and 6, 2 and 8, 6 and 8 and 6 and 11) was nil (Table 19.6).

Table 19.6: Inter-station Relationship of Phytoplankton Families by Sorensen's Index of Similarity

Chlorophyceae					*Cyanophyceae*					*Bacillariophyceae*				
Station	*2*	*6*	*8*	*11*	*Station*	*2*	*6*	*8*	*11*	*Station*	*2*	*6*	*8*	*11*
2		0.29	0.29	0.14	2		0	0.4	0.4	2		0	0	0.33
6	0.29		0.47	0.59	6	0		0.55	0.36	6	0		0.36	0.44
8	0.29	0.47		0.75	8	0.4	0.55		0.75	8	0	0.36		0.17
11	0.14	0.59	0.75		11	0.4	0.36	0.75		11	0.33	0.44	0.17	
Duglenophyceae					*Chrysophyceae*					*Xanthophyceae*				
Station	*2*	*6*	*8*	*11*	*Station*	*2*	*6*	*8*	*11*	*Station*	*2*	*6*	*8*	*11*
2		1	1	1	2		0	0	0	2		0	0	0
6	1		1	1	6	0		0	0	6	0		0	0
8	1	1		1	8	0	0		1	8	0	0		1
11	1	1	1		11	0	0	1		11	0	0	1	

Generally, high index values were obtained between the stations polluted by sewage water (Stations 8 and 11). Only one species of Euglenophyceae was found in all stations and so the index value observed is one. The idex value was moderate when the relationship between clear water station and the polluted stations was correlated. The lower values were obtained between stations 2 and other stations since station 2 was severely polluted by the textile-mill effluent.

Discussion

In the present study 39 species of phytoplankton were recorded from the Tamiraparani river basin belonging to Chlorophyceae, Bacillariophyceae, Cyanophyceae, Euglenophyceae, Chrysophyceae and Xanthophyceae. A minimum of 6 species (station 2) and a maximum of 30 species (station 8) were noticed. In stations 6 and 11 the species number was 13 and 27 respectively. In the Adayar river, luxuriant phytoplankton population of 167 taxa was recorded with a minimum of 104 species in station 1 and a maximum of 131 in station 5 (Govindan *et al.,* 1987). Kulshrestha *et al.,* (1987) reported that 41 taxa were recorded at prepollution point and 14 taxa at post pollution point of the river Chambal polluted by the effluents from a Rayon and chemical factory. Kaushik *et al.* (1989) reported 68 planktonic algae in the Narmada river, out of which 40 species were Chlorophyceae 12 species were Bacillariophyceae, 10 species were Euglenophyceae and 6 species were Cyanophyceae.

Pollution, especially organic and industrial pollution, normally eliminates rare and intolerant species and causes certain species (tolerant species) to become excessively common. In the industrially polluted area, 6 commonly available tolerant species wee noticed, namely *Oscillatoria parinceps, Anabaena* sp., *Phacus* sp., *Navicula cryptocephala, Scenedesmus dimorphus* and *Closterium* sp. In the Chambal river, Rayon and chemical effluent discharge point was completely devoid of phytoplankton community. This was due to high toxicity and the quantity of discharge (Kulshrestha *et al.,* 1987). In Ambikapur water bodies of Madhya Pradesh, 58 species of planktonic algae were identified out of which 10 belonged to Cyanophyceae, 18 each to Chlorophyceae and Bacillariophyceae and 12 to Euglenophyceae (Kaushik *et al.,* 1991a). Eighty seven species of planktonic algae (Chlorophyceae, Cyanophyceae, Euglenopyceae, Bacillariophyceae and Dinophyceae) were identified in the Chambal river (Kaushik *et al.,* 1991b). Sixty three species of planktonic algal forms were identified in Vivek Nagar pond at Gwalior. Out which 26 species belonged to Chlorophyceae, 19 species to Euglenophyceae 9 species each to Cyanophyceae and Bacillariophyceae (Kaushik *et al.,* 1991c). A total of 21 species of phytoplankton was collected from Matsya Saropver, Gwalior (Kaushik *et al.,* 1991d). Kaushik *et al.* (1992) recorded 66 species of phytoplankton in Saank, Asaun, Kurai and Chambal, in which 30 species belonged to Chlorophyceae, 12 species to Bacillariophyceae and 9 species to Cyanophyceae.

The presence and abundance of species at each station in the river/stream is one of the methods for biological assessment of water quality. Wilhm (1975) suggested that the application of a more sophisticated biological assessment of water quality is extremely wasteful if readily available information can indicate river water quality. In the present study mere presence and absence of species were not used to indicate the water quality, but a comparative study of the total number of species with that of the other species was carried out in relation to a particular area of pollution. The dominant species at one station need not be present in the remaining stations (Nather Khan, 1991). In the present study species commonly available in polluted waters also occurred in clean water areas. The relative abundance of each species within each station and between different stations is considered as a criterion for the assessment of water quality.

In the Tamiraparani river basin at textile-mill effluent discharge point (Station 2) 6 pollution tolerant species *viz. S. dimorphus, Closterium, N. cryptocephala, O. princeps, Anabaena* and *Phacus* sp. were identified. The species abundance was noticed in the following order: Chlorophyceae>Cyanophyceae>Bacillariophyceae>Euglenophyceae. *O. princepes* was found throughout the year except in the north-east monsoon period whereas the *Anabaena* sp. and *Phacus* sp. were found in the pre-monsoon and post monsoon periods. *N. cryptocephala* was found in all the seasons except in the north-east monsoon. The Chlorophyceaen member *S. dimorphus* appeared throughout the year while the *Closterium* sp. was noticed only in the pre-monsoon period.

Kumar *et al.* (1974) while studying the algae of industrial effluents found that diversity in physico-chemical characteristics leads to the variety in biological characteristics. Rama Rao *et al.* (1978) have found *Euglena acus, E. viridis, E. intermedia, Phacus quinquimarginatus, P. sueicea, P. oryx, Lepocincis ocm, Trachelomonas volvocina, Oscillatoria terebriformis, O. rubescenis, O. chlorina, O. tenuis, O. willei, O. quadripunctulata, O. chalybea, O. laetevirens, Melosira varians, Navicula radiosa, N. cryptocephala, Ankistrodesmus falcatus* and *Chlorella vulgaris* as dominant algal species in the out fall industrial wastes and sewage in Khan river near Indore. Govindan and Sundaresan (1979) were able to grow *Microcystis aeruginosa, Chlorella phyrenoidosa, C. vulgaris, Merismopedia tenuissima, Ankistrodesmus falcatus, S. quadricauda* and *E. acus* in textile-mill waste and sewage waste mixture during effluent treatment. Vashist and Sra (1979) found *Oscillatoria* as dominant algae in the mixed factory effluent at Chandigarh.

Algal forms such as *Microcystis, Closterium, Lyngbya* and *Oscillatoria* were observed in rubber factory effuents (Sharma and Bendre, 1982). Somashekhar (1984b) identified 30 species of algae belonging to 17 genera in the waste water from a textile-mill. Trivedy *et al.* (1986) described phytoplanktonic algae in textile mill waste water and their seasonal variations. They have studied algae from the textile-mill effluent treatment plant and found *Phacus* sp., *Phormidium* sp., *Oscillatoria* sp., *Cyclotella* sp., *Microcystis* sp., *Euglena* sp., *Spirulina* sp., and two species of diatoms in mixed effluent tanks.

In station 6 (non-polluted station) 13 species were identified namely *Volvox* sp., *S. dimorphus, S. quadricauda, Coelastrum* and *Hydrodictyon* under Chlorophyceae, *Synedra* sp., *Fragilaria* sp. and *Stauroneis* sp., *Concinodiscus* sp. under Bacillariophyceae, *Spirulina* sp., *O. curviceps, M. aeroginosa,* under Cyanophyceae and *Phacus* sp. under Euglenophyceae. Among the Chlorophycean members *S. dimorophus* was the dominant species while in Bacillariophyceae *Stauroneis* sp. was the dominant species, *Spirulina* sp. was the dominant species in Cyanophyceae. *Phacus* sp. was the sole member in Euglenophyceae.

Stations 8 and 11 were polluted by sewage water which contain high organic matter. Both the stations showed the presence of 12 species of Chlorophyceae of which *Volvox* sp. was the predominant one. *Volvox* sp. is an indicator of septic conditions in water. The dominant species of Bacillaripophyceae was *Stauroneis* sp. in station 8 and *Tabellaria* sp., in station 11. In both stations 8 and 11, Cyanophyceae member *Spirulina* sp. was the predominant one followed by *Oscillatoria* sp. Among the Eglenophyceae, *Phacus* sp. was the sole species found in both stations.

Algae are common and normal inhabitants of surface layers in water bodies exposed to sunlight. The pollution indicators, species distribution and abundance are reported by different authors in different rivers. Patrick (1943), Schlichting and Gearheart (1966) were of the opinion that Spirogyra and diatom species are indicators of eutrophication. Jaag (1956) and Plamer (1969) noticed that growth of *Oscillatoria* sp. indicated the high level of organic pollution. On the basis of the study of algal communities Hutchinson (1967) reported that *Synedra, Stephanodiscus, Melosira, Aphanizomerion, Anabaena, Oscillatoria, Staurastrum, Consmarium, Pediastrum, Scenedesmus* and *Ankistrodesmus* are present in nutrient rich waters.

In the present study, *Sprulina* sp., *Oscillatoria* spp., *Nostoc* sp., *Anabaena* sp., *Microcystis aeruginosa, Phacus* sp., *Stauroneis* sp., *Tabellaria* sp., *Volvox* sp., *Scenedesmus dimorphus, Spirogyra* sp., *Ankistrodesmus* sp., *Closterium* sp., *Scenedesmus* sp., *and Chlorella* sp. were noticed in the sewage and industrially polluted stations. *Chlorella, Ankistrodesmus, Scenedesmus, Euglena, Chlamydomonas. Oscillatoria, Micrecthinium, Golenkinic* and *Oocystis* are representative of polluted waters (Palmer, 1980). Hosmani and Bharti (1981) studied phytoplankton of Karnataka state and reported that the *Euglena, Phacus, Scenedesmus, Closterium, Pediastrum, Navicula, Spirulina, Oscillatoria, Phormidium, Lepocinclis* and *Trachelomonas* occurred in organically polluted waters. Philipose (1960) commented that *Scenedesmus quadricauda, S. abundance, S. longus, Merismopedia* spp. along with *Navicula* spp. and *Oscillatoria* spp. Volvocales and Chlorococcales remain dominant in sewage fed water tanks. Patil *et al.* (1983), Sengar and Singh (1986) and Desai (1987) found dominance of the members of Cyanophyceae and Bacillariophyceae over chlorophyceae in sewage polluted waters. Cyanophyceae and Bacillariophyceae were shown to be correlated with the intensity of pollution by Palharya and Malviya (1988). They reported that the *Scenedesmus, Spriogyra, Closterium, Oscillatoria, Microcystis, Anabaena, Synedra, Navicula, Melosira, Cymbella* can be considered pollution tolerant forms. The pollution tolerant groups and species and their share in a total phytoplankton community have been used as biological indices for assessing the level of water pollution. Rama Rao *et al.* (1978) have designated green algae as the indicators of highly polluted waters. Prasad and Singh (1980) are of the opinion that sewage contamination favours growth of chlorophyceae, they are indices for assessing the level of water pollution. Ramanibai and Ravichandran (1987) have also shown increase in nutrients accelerating the growth of phytoplankton due to sewage discharge.

In the present study, *Scenedesmus quadricauda* (Chlorophyceae) *Phacus* (Euglenophyceae) and *Oscillatoria curviceps* (Cyanophyceae) were noticed as the faculative forms. Similar type of observation was noticed by Kaushik *et al.* (1991c) in the Vivek Nagar Pond at Gwalior.

During the late pre-monsoon month (March), both the water and atmospheric temperature increased and consequently caused accumulation of nutrients due to high rate of evaporation of water. This condition was favourable for the growth of the members of Chlorophyceae, Bacillariophyceae, Cyanophycee, Euglenophyceae, Xanthophyceae and Chrysophyceae when compared to other seasons. Similar condition was documented in the River Chambal Tal during summer (Kaushik *et al.*, 1991b).

During the post monsoon the phytoplankton population was lower than that of pre-monsoon month. During the south-west monsoon month and north-east monsoon month the phytoplankton population fluctuated and the values are lower than that of the pre-monsoon and post monsoon month. *Scenedesmus* spp. occurred throughout the year in fair amount. Cooke and Hirsh (1958) observed that *Spirogyra* and *Oscillatoria princeps* show abundant growth in the summer season.

In the present study, 39 species were identified in the Tamiraparani river basin. Only two species viz., *Scenedesmus dimorphus* and *Phacus* sp. were observed in all the stations. Rana and Bhati (1984) also reported that *Anabaena* sp. was the important constituent during the winter season. Kaushik *et al.* (1992) reported the distribution of algal species in different rivers Chambal (42 species), Saank (39 species), Kuari (27 species) and Asaun (15 species).

In the present investigation, in station 2 phytoplankton families showed the negative correlation with water temperature (Bacillariophyceae: $P < 0.001$) and sediment temperature (Bacillariophyceae $P < 0.005$: Cyanophyceae $P < 0.05$). Similarly the total dissolved solids, total solids, total hardness, total alkalinity and COD showed the negative correlation. While the total suspended solids (Chlorophyceae: $P < 0.005$ Bacillariophyceae: $P < 0.05$ Cyanophyceae and Euglenophyceae $P < 0.001$), free carbon dioxide (Chlorophyceae: $P < 0.05$) and bio chemical oxygen demand (Chlorophyceae: $P < 0.001$) showed the statistically significant value with positive relationship (Table 19.2). Here the textile-mill effluent influenced the water quality parameters resulting in phytoplankton abundance.

In station 6, all the phytoplankton members showed a statistically significant ($P < 0.001$) relationship with water and sediment temperature. There was no influence of effluent or sewage on water quality parameters. Rao (1975) and Mohan (1987) have also reported that temperature ranging from 28°C to 34°C was favourable for the growth of desmids. Cyanophycean and Euglenophycean members only showed statistically significant relationship when the values are compared to pH ($P < 0.01$). Total hardness values showed a positive relationship with phytoplankton families (Chlorophyceae: $P < 0.01$, Bacillariophyceae: $P < 0.005$). Cyanophyceae members showed a positive relationship with dissolved oxygen ($P < 0.05$). Similarly the Bacillariphyceae showed a positive relationship with conductivity ($P < 0.001$) (Table 19.3).

In station 8, pH showed the statistically significant ($P < 0.05$) relationship with all the phytoplankton families. Similarly the total dissolved solids showed statistically significant relationship with phytoplankton families except Cyanophyceae. Total solids values compared to phytoplankton families showed the statistically significant values (Chlorophyceae: $P < 0.01$, Bacillariophyceae: $P < 0.05$). Total hardness verses phytoplankton families showed statistically significant t values except chlorophycea. Chemical Oxygen Demand showed a positive relationship with Chlorophycean members ($P < 0.001$) (Table 19.4). The sewage entries into the receiving water body of station 8 influenced the abundance of phytoplankton.

In station 11, water and sediment temperature and total solids showed a positive relationship with Chlorophycean members ($P < 0.05$). pH showed the statistically significant relationship with Bacillariophyceae ($P = 0.05$) Euglenophyceae and Chlorophyceae ($P < 0.001$). Total dissolved solids when compared with phytoplankton family members showed the statistically significant correlation ($P < 0.05$) except Chrysophyceae. Similarly the total suspended solids showed the positive relationship with Chlorophycean members ($P < 0.05$) Bacillariophycean and Cyanophycean members ($P < 0.001$). Total solids showed the statistically significant correlation with Bacillariophyceae ($P < 0.001$) and Cyanophyceae ($P < 0.05$). Test of significance for the correlation coefficient for total hardness verses phytoplankton families (Bacillariophyceae, Cyanophyceae, Euglenophyceae and Chrysophyceae) showed the statistically significant value ($P < 0.001$). Dissolved oxygen showed a positive relationship with the test of significance at 0.005 level (Euglenophyceae) 0.001 level (Chrysophyceae) (Table 19.5).

Most of the Indian rivers are reported to be slightly to moderately alkaline in nature (Mitra, 1982 and Venkateswarlu, 1986). In the present study pH of river water was noticed circumneutral ($pH < 5$). Mitra (1982) and Somashekhar (1986) concluded that alkalinity water is dependent upon the concentration of anions. Neutral and slightly alkaline waters are said to harbour a wide range of algal species (Bhargava, 1985 and Somashekhar, 1986). The nutrients, organic matter, dissolved oxygen, pH and other physical factors play a role in the ecological distribution of Bacillriophycean algae (Jindal and Vasisht, 1981). Hickman and Penn (1977) have found that oxygen deficit environment is favourable for the growth of desmids and euglenoids. Munawar (1970) concluded that higher pH, more oxygen and nitrogenous organic matter are favourable for the growth of chlorococcales. Difference in chemical nature of water affects the distribution of different groups of algae and species difference can be used as water quality indicator (Hutchinson, 1967). In general water temperature (except station 8), sediment temperature, pH (except station 2), total dissolved solids and total solids (except station 6), total hardness, dissolved Oxygen, and biochemical oxygen demand (except stations 6 and 11) are positively correlated to the phytoplankton abundance.

The Sorensen index measures the relative similarity of two stations or habitats in terms of species composition (Mountford, 1962). The index value is based on the presence or absence data. However, it is also affected by sample size, since there is an increase in chance of encounter of rare species in correspondence to the sample size. The index gives equal weight to each species irrespective of abundance and 'over values' the rare species relative to the dominant species (Whittaker and Fairbanks,

1958). Sorensen's index is still used because of its operational simplicity (Davis, 1963). The index value between stations 8 and 11 were generally high for all the families of phytoplankton. These two stations received organic wastes from the domestic sewage water. This nutrient automatically enhanced the algal production. The index value is generally high except for the families Xanthophyceae and Chrysophyceae due to smaller number of species involved and their wide distribution.

This explanation finds the support from the findings of Bishop (1973), Ho (1973), Nather Khan (1990) has evaluated Sorensen's index in the Linggi river basin. In the present study, the low index values (Sorensen's index of similarity) were obtained between industrially polluted station (Station 2) and the remaining stations. The high index values were obtained between the sewage polluted stations (Stations 8 and 11) and the medium index values were obtained between non-polluted station (Station 6) when compared with other stations.

References

Bharati, A., Saxena., R.P. and Pandey. G.N. (1981). Role of microflora in the assessment of pollution level of river Ganga at Kanpur. *J. Institution of Engineers* (India) 75, 112-122.

Bhargava, D.S. (1985). Water quality variation and control technology of Yamuna river. *Envtl. Pollut.* 37 (Section B), 355-376.

*Bishop, J.E. (1973). *Limnology of a small Malayan River - Sungai Gombak*. (June, The Hagve).

Cooke, W.B. and Hirsh, A. 1958. Continuous sampling of trickling filter populations. *Sewage and Indust. Wastes*. 30, 138.

Davis, B.N.K. (1963). A study of micro-arthropod communities in mineral soils near corby, Northants. *Journal of Animal Ecology*. 32, 49-71.

Desai, P.V. (1987). The effect of mining on the lotic and lentic environments of Goa. *Poll. Res.* 6(2), 83-86.

Desikachary, T.V. (1959). *Cyanophyta,* I.C.A.R., New Delhi.

Govindan, V.S., and Sundarensan, B.B. (1979). Seasonal succession of algal flora in polluted region of Adyar river. *Indian J. Environ. Hlth*. 21, 131-142.

Govindan, V.S. Ahana Lakshmi, N. and Sampathkumar, M. (1987). Urban water quality impact assessment a case study. *Proc. Nat. Con. Env. Impact on Biosystem,* Loyola College, Madras, pp. 31-33.

Hickmen, M. and Penn, I.D. (1977). The relationship between planktonic algae and bacteria in a small lake. *Hydrobiologia.* 52 (2-3), 213-219.

*Ho, S.C. (1973). The ecology of the lowland stream, Sungei Renggam with special reference to pollution. M.Sc. thesis, University of Malaya, Kuala Lumpur.

Hosmani, S.P. and Bharati, S.G. (1981). Algae as indicators of organic pollution. *Phykos*. 19 (1), 23-26.

*Hutchinson, G.E. (1967). *A Treatise on Limnology*, Vol. 2. Introduction to lake biology and limnoplankton. John Wiley and Sons, New York. USA.

*Jaag, O. (1956). *The pollution of surface and ground water in Switzerland*. 4th European seminar for Sanitary Engineers, Geneva.

Jindal, R. and Vasisht, H.S. (1981). Hydrobiological studies of a tributary of Sirhind canal at Sangrur

(Punjab, India). In, *Proc. Symp. Ecol. Anim. Popul. 2001. Surv. India* (Ed.) 2, 1-17.

Kaushik, S., Agarker, M.S. and Saksena, D.N. (1989). Phytoplanktonic flora of Narmada river at Amarkantak. *Bionature*. 2, 81-83.

Kaushik, S., Agarkar, M.S. and Saksena, D.N. (1991a). A report on the algae of water bodies at Ambikapur, Madhya Pradesh. *Phykos*. 30 (1 & 2), 115-122.

Kaushik, S., Agarkar, M.S. and Saksena, D.N. (1991b). Water quality and Periodicity of Phytoplanktonic algae in Chambal Tal, Gwalior, Madhya Pradesh. *Bionature*. 11(2), 87-94

Kaushik, S., Agarkar, M.S. and Saksena, D.N. (1991c). On the planktonic algae of sewage fed Vivek Nagar Pond at Gwalior. *Poll. Res*. 10 (1), 25-31.

Kaushik, S., Saxena, M.N. and Saksena, D.N. (1991d). Phytoplankton population dynamics in relation to environmental parameters in Matsya Sarovar at Gwalior (M.P.) Acta. *Botanica Indica*. 19, 13-116.

Kaushik, S., Agarkar, M.S. and Saksena, D.N. (1992). Distribution of phytoplankton in riverine waters in Chambal command area, Madhya Pradesh. *Bionature*. 12 (1 & 2), 1-7.

*Kolkwitz, R. and Marsson, M. (1909). Okologie de Planzichem saprobein, Deustch. *Bot. Ges*. 26, 505-519.

Kulshrestha, S.K., Adholia, U.N. Khan, A.A. Bhatnagar, A. and Baghail, M. (1987). Pollution studies on river Chambal near Nagda with reference to phytoplankton community. *Proc. Trends in Environmental Pollution and Pesticide Toxicology*, pp. 119-123.

Kulshrestha, S.K., Adholia, U.N. Bhatnagar, A. Vhan, A. Saxsena, M. and Baghail, M. (1989). Studies on pollution in river Kshipra zooplankton in relation to water quality. *Int. J. Ecol. Envt. Sci*. 15, 27-36.

Kumar, H.D., Bisaria, G.P. Bhandari, L.M., Rana, B.C. and Sharma Vasundhara. (1974). Ecological studies on algae isolated from the effluents of an oil refinery a fertilizer factory and a Brewery. *Ind. J. Environ. Hlth*. 16, 247-255.

Mitra, A.K. (1982). Chemical characteristics of surface water at a selected gauging station in the river Godavari, Krishna and Tungabhadra. *Indian J. Environ. Hlth*. 24, 165-179.

Mountford, M.D. (1962). An index of similarity and its application to classificatory problems. In, *Progress in soil zoology*. (Ed. P.W. Murphy), pp. 43-50.

Munawar, M. (1970). Limnological studies of freshwater ponds of Hyderabad, India. II. The Biocenosa. *Hydrobiologia* 36 (1), 105-128.

Nather Khan, I.S.A. (1990). Diatom distribution and inter-site relationship in Linggi river basin, Peninsular Malaysia. *Malayan Nature Journal*. 44, 85-95.

Nather Khan, I.S.A. (1991). Effect of urban and industrial wastes on species diversity of the diatom community in a tropical river Malaysia. *Hydrobiologia* 224, 175-184.

Palmer, C.M. (1969). A composite rating of algae tolerating organic pollution. *J. Physiol*. 6, 78-82.

Palmer, C.M. (1980). *Algae and water pollution*. Castle House Publications Ltd., p. 123.

Palharya, K.J.P. and Malviya, S., (1988). Pollution of the Narmada river at Hoshangabad in Madhya Pradesh and Suggeted measures for control. In, *Ecology and Pollution of Indian rivers* (Ed. Trivedi) Ashish Publishing House, New Delhi.

Patil, S.G., Singh, D.B. and Harshey, D.K. (1983). Ranital (Jabalpur) a sewage polluted water body as evidenced by chemical and biological indicators of pollution. *J. Environ. Biol.* 4(2), 43-49.

Patrick, R. (1943). The diatoms of Linsley Pond. *Proc. Acad. Natural Sci*. Philadelphia 95, 53-110.

*Pearsall, W.H. (1932). *Rev. Algol.* 20, 241.

Philipose, M.T. (1960). Freshwater phytoplankton of inland fisheries. *Proc. Symp. Algol.* ICAR, New Delhi, 272-291.

Philipose, M.T. (1967). *Chlorococcales*. I.C.A.R. New Delhi, p. 365.

Prasad, B.N. and Singh, T. (1980). Algal hydrobiology in India. In, *Proc. Nat. Acad. Sci. India*, Golden Jubilee, Commemoration Volume, 271-300.

Rama Rao, S.V. Singh, V.P. and Mall, L.P. (1978). Pollution studies of river Khan (Indore) India. I. Biological Assessment of Pollution. *Water Research*, 12, 555 - 559.

Ramanibai, P.S. and Ravichandran, S. (1987). Limnology of an urban pond at Madras, India. *Poll. Res.* 6(2), 77-81.

Rana, K.S. and Bhati, D.P.S. (1984). Occurrence of permanent algae bloom in Bharatpur Fort Moat with reference to pollution. *J. Environ. Res.* 5 (1 & 2), 29-34.

Sabata, B.C. and Nayar, M.P. (1987). Water pollution studies in River Hooghly with relation to phytoplankton. *Proc. Nat. Con. Env. Impact of Biosystem*, Loyola College, Madras, pp. 1-5.

Sankaran, Y. (1988). Pollution studies in Cauvery and Adayar rivers in Tamil Nadu. In, *Ecology and Pollution of Indian Rivers* (Ed. Trivedy), Ashish Publishing House, New Delhi.

Schlichting, H.E. and Gearheart, R.A. (1966). Some effects of sewage effluents upon phyco-periphyton in lake Murray Okalhoma. *Proc. Okla. Acad. Sci*. 46, 19-24.

Sengar, R.M.S. and Singh, S. (1986). Pollution status of Keelham lake at Agra. *Geobios*. 13, 56-61

Sharma, A. and Bendre, A.M. (1982). Algal characteristics of the effluents from a rubber factory. *IAWPC Tech. Annual*. 9, 167-171.

Smith, G.M. (1950). *Fresh Water Algae of United States*. 2nd Edn. McGraw-Hill Co., New York, pp. 1-719.

Somashekhar, R.K. (1984). Algae of textile-mill waste waters. *Comp. Physiol. Ecol.* 9, 267-271.

Somashekhar, R.K. (1986). *Ecological Studies on the Two Major Rivers of Karnataka*. (Ed. Trivdy), Ashish Publishing House. New Delhi, India, pp. 271-286.

*Sorensen, T. (1948). A method of establishing groups of equal amplitude in plant sociology on similarity of species content, and its application to analyses of the vegetation of Danish commons. K. danske vidensk, *Selsk*. 5, 1-34.

Strom, K.M. (1927-28). Recent advances in limnology, *Proc. Limn. Soc. Lund*. 140, 96.

Trivedy, R.K., Khatavkar, S.D. and Goel, P.K. (1986). Characterization, treatment and disposal of waste water in a textile industry. *J. Indian Poll. Cont*., 2, 1-12.

Vashist, H.S. and Sra, G.S. (1979). The biological characteristics of Chandigarh waste water in relation to the physico-chemical factors. *Proc. Symp. Environ. Biol.*, pp. 429-440.

Venkateswarlu, V. (1986). Ecological studies on the rivers of Andhra Pradesh with special reference to water quality and pollution. *Proc. Indian. Sci. Acad.* 96(6), 495-508.

*Whittaker, R.H. and Fairbanks, C.W. (1958). Study of plankton copepod communities in the Columbia basin, South-eastern, Washington, *Ecology* 29, 46-64.

Wielgolaski, F.E. (1975). Biological indicators on pollution. *Urban Ecol.*, 163-79.

Wilhm, J.L. (1975). Biological indicators of pollution. In, *River Ecology*, (B.A. Whitton ed) Blackwell Scientific Publications, Oxford, p. 725.

* Originals not referred.

20

Macrobenthic Molluscan Spectrum in the Coastal West Bengal

Abhijit Mitra, Amitava Aich, Amalesh Choudhury* and D.P. Bhattacharyya**

Department of Marine Science, University of Calcutta, 35, B.C. Road, Calcutta - 700 019
**S.D. Marine Biological Research Institute, Sagar Island (Sundarbans), 24 Paraganas (S), W.B.*
***Department of Theoretical Physics, Indian Association for the Cultivation of Science, Jadavpur, Calcutta - 700 032*

ABSTRACT

The present study highlights the distribution of macrobenthic molluscs in relation to physico-chemical variables (surface water salintiy, pH, temperature and dissolved oxygen) and dissolved heavy metals (Zn, Cu and Pb) in the ambient aquatic phase of ten sampling stations selected in the North eastern coast of the Bay of Bengal during April, 1999. The Shannon Weaner species diversity index was found to be more in the high saline mangrove dominated stations indicating the intimate association of the molluscan community with mangrove vegetation in terms of shelter, food and favourable habitat for reproduction. Simple correlation analysis performed between the $\overline{H}$ and the selected environmental variables showed significant negative stress imparted by dissolved Cu & Pb on the value of $\overline{H}$. The role of aquatic salinity and pH was also found to be considerable in determining the community structure of macrobenthic molluscs.

Key words: Macrobenthic molluscs, Shannon Weaner species diversity index, dissolved heavy metal.

Introduction

West Bengal is a maritime state in the north-eastern part of the country, adjacent to Bangladesh whose coastal zone spreads over about 5,777.7 sq km and is restricted within the latitude, 21°30' N to 22°30'N and longitude 87°25'E to 89°10' E (Anonymous, 1992). The river Harinbari or Heronbhanga of the Indian Sundarbans (India-Bangladesh Border) is the easternmost border, while the new Digha Coast of Orissa-West Bengal border constitutes the western boundary of coastal West Bengal. With a considerable degree of marine characteristics in major protion of the ecosystem, the important morphotypes of coastal West Bengal are sand flats, coastal dunes, beaches, mudflats, estuaries, creeks, inlets and mangrove flats.

Benthic molluscan communities thrive in this dynamic ecosystem which exhibit pronounced seasonal variation. The lowering of aquatic salinity due to increase of dilution factor in monsoon months of the state (July to October) results in the death of oyster population, which cannot withstand very low salinity. As a result of this, the value of ecological indices also changes with the oscillation of physico-chemical variables.

Study on the community structure of the molluscan assemblage is very important to understand the tolerance of each species or the species group with respect to changing physico-chemical variables of the ambient environment. Such study is important to develop a sound management policy with respect to molluscan conservation in the north-east coast of the Bay of Bengal as they are not only the source of lime, but certain edible species (*e.g., Saccostrea cucullata* and *Crassostrea cuttackensis*) have excellent export prospect and can open the gateway of an alternative source of income for the coastal population. The present study aims to highlight the spatial variation of community structure of macrobenthic molluscs inhabiting hard substrata in relation to selective environmental variables like surface water salinity, pH, temperature, dissovled oxygen and concentrations of dissolved Zn, Cu and Pb in the ambient aquatic phase during April, 1999.

Materials and Methods

The sampling station Haldia (station 1), Kakdwip (station 2), Kachuberia (station 3), Chemaguri (station 4), Sagar South (station 5), Frazergaunge (station 6), Jambu Island (station 7), Lothian Island (station 8), Lower Long Sand (station 9) and Shankarpur (station 10) were selected in the coastal and estuarine zone of north-eastern Bay of Bengal. Existence of sluice gates, ruins of old defunct light house (at station 5), jetties, cross spurs, guide wall and mangrove trees made the colonization of benthic molluscs possible at the sampling sites.

Population estimation of macrobenthic molluscs inhabiting hard substrata was done at the selected sampling stations using a 1 m^2 metal frame. The number of species, the total number of individuals per species and all the species were used to compute the Shannon Weaner species diversity index (H), an expression to understand the health of the ambient environment.

The mean value of 50 quardrants were taken randomly in each of the selected stations to estimate the mean $\overline{H}$ value in each station.

During the high tide period, sample water was collected in clean TARSON bottles for analysis of dissolved Zn, Cu and Pb. Analyses of these heavy metals in the water samples were done as per the procedure outlined by Chakraborty *et al.* (1987). All analyses were done in duplicate by direct aspiration into Perkin-Elmer Atomic Absorption Spectrophotometer (Model 3030) equipped with a HGA 500 graphite furnace atomizer and a deuterium background corrector. The results obtained were expressed in μgl^{-1}. Surface water salinity was analysed by argentometric method. The surface water pH and temperature were estimated by a portable pH meter (sensitivity = ± 0.02) and a Celsius thermometer respectively. Dissolved oxygen of the surface water was analysed by Winker's method.

A simple correlation was performed to find the inter-relationship between $\overline{H}$ of macrobenthic molluscan community inhabiting the sampling stations of the north eastern Bay of Bengal and each of the selected environmental variables by using MICROSTAT package.

Results and Discussion

The selected stations showed considerable variation with respect to Shannon Weaner species diversity index computed from the molluscan population density occupying the hard substrata of the selected stations (Table 20.1). Lowest value of molluscan diversity (H) in the Haldia zone may be attributed to low salinity and intensive industrial activities, which includes refining of crude oil, manufacture of automotive, industrial and submarine storage batteries etc. The molluscan divesity index was also low at Kakdwip (Station 2) and Kachuberia (Station 3), where there is almost no mangrove vegetation except few individuals of *Sonneratia apetala.* It was reported earlier that there is an increase in marine invertebrate species diversity with the increase in number of plant species and their structural complexity and extent (Hutchings and Recher, 1983). Infact the detritus provided by the mangrove vegetation charges the ambient water with nutrients, which triggers the growth of plankton in the ambient media and supplies food for the molluscan species. This may explain the relatively higher diversity value in the mangrove dominated stations at the north eastern Bay of Bengal coast (stations 6, 7, 8 and 9).

Table 20.1: Variation of Shannon Weaner species diversity index at the selected sampling stations of the north-eastern Bay of Bengal

Species (No/m^2)	*Stn. 1*	*Stn. 2*	*Stn. 3*	*Stn. 4*	*Stn. 5*	*Stn. 6*	*Stn. 7*	*Stn. 8*	*Stn. 9*	*Stn. 10*
Nerita articulata	1	2	4	7	11	15	14	21	20	23
Nertiina smithi	1	0	0	0	0	0	0	0	0	0
Nertitina violacea	2	1	1	2	3	0	6	4	7	0
Telescopium telescopium	0	2	0	9	11	13	17	19	21	9
Cerithedia cingulata	0	2	2	16	8	29	33	47	49	51
Cerithedia obtusa	0	0	0	7	9	11	13	9	17	0
Enigmonea aenigmatica	0	0	0	0	1	3	8	11	13	4
Saccostrea cucullata	0	0	8	29	46	59	61	60	42	67
Cymia lacera	0	0	0	0	7	11	13	17	11	14
N	4	7	15	70	96	141	165	188	180	168
H	1.038	1.344	1.130	1.521	1.440	1.620	1.789	1.786	1.887	1.442

A strong positive relation of Shannon Weaner species diversity index with surface water salinity and pH (Table 20.3) reveals the preference of molluscan species towards high saline habitat in the coastal and estuarine zone. This typical condition is observed at stations 5 to 10 (Table 20.2).

The high concentrations of dissolved Zn, Cu and Pb in the surface waters around the sampling stations 1 (Haldia), 2 (Kakdwip), 6 (Frazergaunge) and 10 (Shankarpur) may be due to run-off from the adjacent landmasses which include an port-cum-industrial complex near station 1 and fish landing stations near 2, 6 and 10 (Table 20.2). Considerable concentrations of Zn and Cu at station 3 (Kachuberia) may be attributed to the presence of a passenger vessel jetty at this sampling zone. The existence of highly urbanized and industrialised city of Calcutta, Howrah and Haldia complex is another major source of Pb in the coastal waters (Mitra and Choudhury, 1993). Zn, Cu and Pb in the coastal stations find their origin from the industrial effluents of painting units, galvanizing units, cosmetics manufacturing units and antifouling paints used from conditioning fishing vessels, trawlers, ships etc. The significant negative relationship between disolved Cu and Pb with macrobenthic molluscan diversity (Table 20.3) indicates considerable stress exerted by these metals on the community and emphasizes frequent monitoring for these parameters in the coastal region if the ecosystem is to be kept ecologically sound.

Table 20.2: Variations of Physico-chemical variables at the selected sampling stations of the North eastern Bay of Bengal

Parameters	*Stn. 1*	*Stn. 2*	*Stn. 3*	*Stn. 4*	*Stn. 5*	*Stn. 6*	*Stn. 7*	*Stn. 8*	*Stn. 9*	*Stn. 10*
Surface Water Temperature (°C)	33.8	33.2	32.0	34.0	34.1	33.9	33.6	34.2	32.9	34.1
pH	8.1	8.17	8.23	8.3	8.34	8.36	8.4	8.39	8.4	8.4
Salinity (‰)	6.29	12.23	14.96	21.08	27.63	28.02	28.31	28.68	29.92	30.37
D.O. (mgl^{-1})	4.8	5.25	6.63	6.08	6.27	5.85	6.75	5.66	7.31	6.89
Dissolved Zn (μgl^{-1})	414.3	361.20	348.55	313.39	298.20	366.29	201.85	196.34	272.29	401.28
Dissolved Cu (μgl^{-1})	360.29	312.65	300.47	266.83	169.20	288.3	176.45	102.51	261.13	301.99
Dissolved Pb (μgl^{-1})	32.85	16.62	9.90	5.67	4.30	28.67	2.06	2.11	7.65	26.29

Table 20.3: Inter-relationship between the selected physico-chemical variables and computed from the macrobenthic molluscan community occupying the hard substrata at the north eastern Bay of Bengal

Combination	r-value	p-value
$\bar{H}$ × Salinity	0.8261	<0.01
$\bar{H}$ × Surface water temperature	-0.0685	IS
$\bar{H}$ × pH	0.8477	<0.01
$\bar{H}$ × Dissolved oxygen	0.4974	<0.05
$\bar{H}$ × Dissolved zinc	0.4470	<0.05
$\bar{H}$ × Dissolved Cu	-0.6557	<0.01
$\bar{H}$ × Dissolved Pb	-0.5221	<0.05

"IS" means insignificant.

References

Anonymous. (1992). *Coastal Environment*, Space Application Center (ISRO), Ahmedabad.

Chakraborty, D., Adams, F., Van Mol, W. and Irgolic, J.K. (1987). Determination of trace metals in natural waters at Nanogram per litre levels by electrothermal atomic absorption spectrometry after extraction with sodium diethyldithio carbamate. *Analytica Chem Acta*. 196: 23-31.

Hutchings, P.A. and Recher, H.F. (1983). The faunal communities of Australian mangroves. In: *Tasks for Vegetation Science*. Vol. 8. (H.J. Teas, ed) W. Junk Publishers. The Hague.

Mitra, A., Choudhury, A, and Jamaddar, Y.A. (1992). Seasonal Variation in metal content in the gastropod *Cerithedia* (*Cerithidiopsis*) *cingulata*. *Proceedings of Zoological Society*, Calcutta 45: 497-500.

Mitra, A. and Choudhury, A. (1993). Trace metals macrobenthic molluscs of the Hooghly estuary, India. *Mar. Pollut. Bull.* 26: 521-522.

21

Status of Andaman Sea Ecology: Past, Present and Future

I.K. Pai

Department of Zoology, Goa University, Goa

ABSTRACT

In recent years, marine resources have become increasingly interesting, as they are rightly predicted to be a major source of food, reservoir of minerals, major suppliers of oxygen, regulator of climate and also ultimate dumping ground for the mounting burden of human waste material.

Due to unchecked human activity in recent years, physico-chemical as well biological status of these marine environments have changed significantly, which in turn resulted in the changing environmental scenario of the world.

The Andaman Sea is known to be a part of Bay of Bengal. This Bay is one of the largest Bay in the world and is known to receive large flow of sediments from several rivers and other water bodies from India, Bhutan, Bangladesh, Myanmar, Indonesia etc. Many of these rivers bring along with them, a large quantity of efluents from cities/town located on either side of these rivers, thus making the Bay nutrient rich. This Bay also plays a major role in determining the climatic conditions of India and other South East Asian countries. Thus, its ecology is of paramount interest.

Apart from the above, the Bay is also known for its oligotrophic nature as well low productivity, thus resulting in high diversity of flora and fauna. As the ecological status of the Bay has a direct bearing on lifescape of the Bay, a study was undertaken to understand the ecological status of the Bay.

In addition to the above, it is also known that there are certain member of zooplankton species, which are known to act as bioindicators, indicating the quality of water in environment. Thus, an attempt has also been made in the present studies to understand such bioindicators, which will throw light on the status of abiotic factors of the Bay.

Based on the observations, the study recommends certain appropriate measures to be taken to conserve the ecology of one of the largest Bay in the world.

Key words: Ecology, Andaman sea, zooplankton.

Introduction

Though, 71 per cent of the surface of our planet is covered with marine waters, whose average depth is 3.8 km and volume is about 1370 x 10^6 Km^3 and innumerable number of species of organism live in it since millions of year, only during recent years, it has attracted the attention of the world as a promising major source of food, reservoir of minerals, major supplier of oxygen, regulator of climate and ultimate dumping ground, for the mounting burden of human waste material, for years to come. It is also known that, 32 of 33 animal phyla exist, in sea. It is also said 173 animal classes live in sea, 35 in fresh water and 33 on land (Nicoll, 1971) for which, possible reasons are listed by May (1994). While, Grassie *et al.*, (1991) have reported that 13 out of 28 phyla found in marine environment are endemic to marine environment, and only one out of 11 phyla is endemic to terrestrial ecosystem. Thus, highlighting the importance of marine environment with regard to life scape. Thus, making it a single largest repository of organisms, where the water is the substance that surrounds all marine organisms and it composes the bulk of the bodies of the marine plant and animals, and is also the medium in which various chemical reactions takes place, both inside and outside living organism (Nybakken, 1997).

In general, sea water consist of an average of 35 g/1000 ml of dissolved compounds collectively called as salts or practival salinity units (psu) which include Cl^-(55.04%), Na^+(30.61%), SO_4^{2-} (7.68%), Mg^{2+}(3.69%), Ca^{2+}(1.16), K^+(1.10%) as major constituents and HCO_3^-(0.41%), Br (0.19%), H_3BO_3 (0.07%) and St^{2+} (0.04%) apart from 0.01 per cent of dissolved substances of several inorganic salts needed for living of the organisms in sea.

It is also known that some of the organisms like diatoms and radiolarians show their existence in a place where there is availability of silicon dioxide, which is needed for them to construct their skeleton. Thus, acting as bioindicator. Similarly in contrast to most ions, NO^{3-} (nitrates), and PO^{4-} (phosphates) do not exist in constant ratio with other elements or ions, and tend to be in short supply in surface waters. Thus, such a kind of varying in abundance will result in biological activity as many a times become a limiting factor mainly for plant production. While, the other elements such as Fe, Mn, Ca, Cu though may exist in trace amounts, but can very well act as limiting factor for sustainance of life (Martin, 1994). Among gases, O_2 and CO_2 dissolved in sea water has metabolic importance. Their solubility also varies and depends on temperature of the water, because, cold waters will have high solubility, hence O_2 is more in cold waters.

Unlike in the places like Arctic sea, temperate northern Pacific, which show summer bloom and temperate north Atlantic sea which exhibit spring bloom (Parsons *et al.*, 1984), as tropical seas show thermal stratification, their productivity and biomass in general and zooplankton in particular is some what constant.

Further, there is a new wide spread recognition that chemical monitoring is not enough and that pollution is essentially biological phenomenon because of its impact is on the living organisms (Wright *et al.*, 1994), and the need for biological methods has been accepted (Newman *et. al.*, 1992., Rosenberg and Resh, 1993). At community level too, use of biological approach is already well established and accepted (Cairns and Pratt, 1993). Further, the advantage of using these bioindicators has been listed and discussed by Rosenberg and Resh (1993).

It is known that three can be three major categories of environmental stress viz., natural, imposed and environmental manipulation (mainly by anthropogenic activities) and will be reflected in the biotic system. Pollutants effect at various levels like at cellular/molecular level, they may interfere with hormonal system and may responsible for production of more of male steroids in females. They may also have their effect on development of reproductive system leading to induction of sterility or even death of females (consequent upon which, the population size diminishes). While, at the population level, they reduce the number of a particular species drastically and at a later stage may show the impact on community too. To look into all these aspects even software packages have been developed to predict and classify the system (Wright *et al.*, 1994).

McAllister *et al.* (1994), while analysing global distribution of coral reef fishes have reported that in Indian subcontinent, Laccadive-Maldives-Chagos and Sri Lankan region have high diversity and have also reported that sampling is weak in western Sumartra i.e., Eastern Andaman sea in particular and Andaman sea in general.

Keeping in view of the above, to fill the lacunae of knowledge ecology of Andaman Sea, an attempt has been made to contrast the past ecological status of Andaman Sea with that of present and fore see the future of the same in the light of available literature.

Material and Methods

During the Department of Ocean Development (DOD) (Government of India), and National Institute of Oceanography (NIO) organised multidisciplinary cruise SK-118, on ORV Sagar Kanya, which had a predetermined area of operation as Andaman Sea, 18 stations, covering entire Andaman groups of Islands, both near shore and off shore were selected for the present studies (Fig. 21.1)

At every station, conductivity-temperature-depth (CTD) profile system with rosette samplers was lowered to 30m depth and apart from recording the relevant data the water samples were collected from that depth by triggering the sample bottles from control panel on deck, Bethythermography, thermosalinograph were also run at all the stations to record the parameters. The water thus collected was analysed for various physico-chemical parameters such as temperature, pH, oxygen contents, salinity, chlorides, sodium, sulphates, magnesium, calcium, potassium, bicarbonates, bromide, boric acid, strontium etc., by following standard analyses methods. Results obtained for three samples each, at every station was pooled.

Simultaneously, at every station, sampling for zooplankton was also done by both vertical (30 m to surface) and horizontal hauls (on water surface) by using bongo net (dia, 0.6 m, length 2.5 m, mesh width 300 μm). A pre-caliberated flow meter (T.S. Flow meter no. 4512), was also attached to the net mouth, to calculate the actual quantity of water filtered during the operation. Thus, collected samples were brought to the deck and later isolated and separated in the laboratory on board of the vessel. Later, the samples were preserved in 4 per cent formalin and were brought to land laboratory for taxonomic identification and classification by following available literature (Kasturirangan, 1963, Mori 1964, Daniel, 1985, Zheng Zong, 1989, Santanam and Srinivasan, 1994).

Results

Figure 21.1 provides information on the stations, where the sampling was undertaken. It can be seen from the figure, that both near shore and off-shore samplings was done to have a more reliable picture about the Andaman sea.

Table 21.1 provides information on physico-chemical parameters of Andaman sea. The data obtained do not differ much with results obtained elsewhere by working at other marine environment (Nybakken, 1997).

Table 21.2 exhibits the data on number of species observed/collected in each from the present survey at the Andaman sea.

The various data obtained on biotic and abiotic factors thereby were compared with that of available earlier reports (Anonymous, 1981, Madhupratap, 1981, Vijayalaxmi, 1981) and the possibility of making the use of modern tools such as ANPP (Anal Net Primary Productivity), AVRIS (Air borne visible Infra-Red Imaging Spectrometer), BIOCLIM (Biological Climate analyses and prediction system), ERIN (Environmental Resources Information Net work, GEMS (Global Environmental Monitoring System), GRID (Global Resource Information Database), HRV/MLA (High Resolution Visible Multispectral Linear Array), MSCP (Multiple Species Conservation Plan), MSS (Multiple Spectral Scanner), RAP (Rapid Assessment Procedures) etc., to have a constant monitoring of the ecology of this sea.

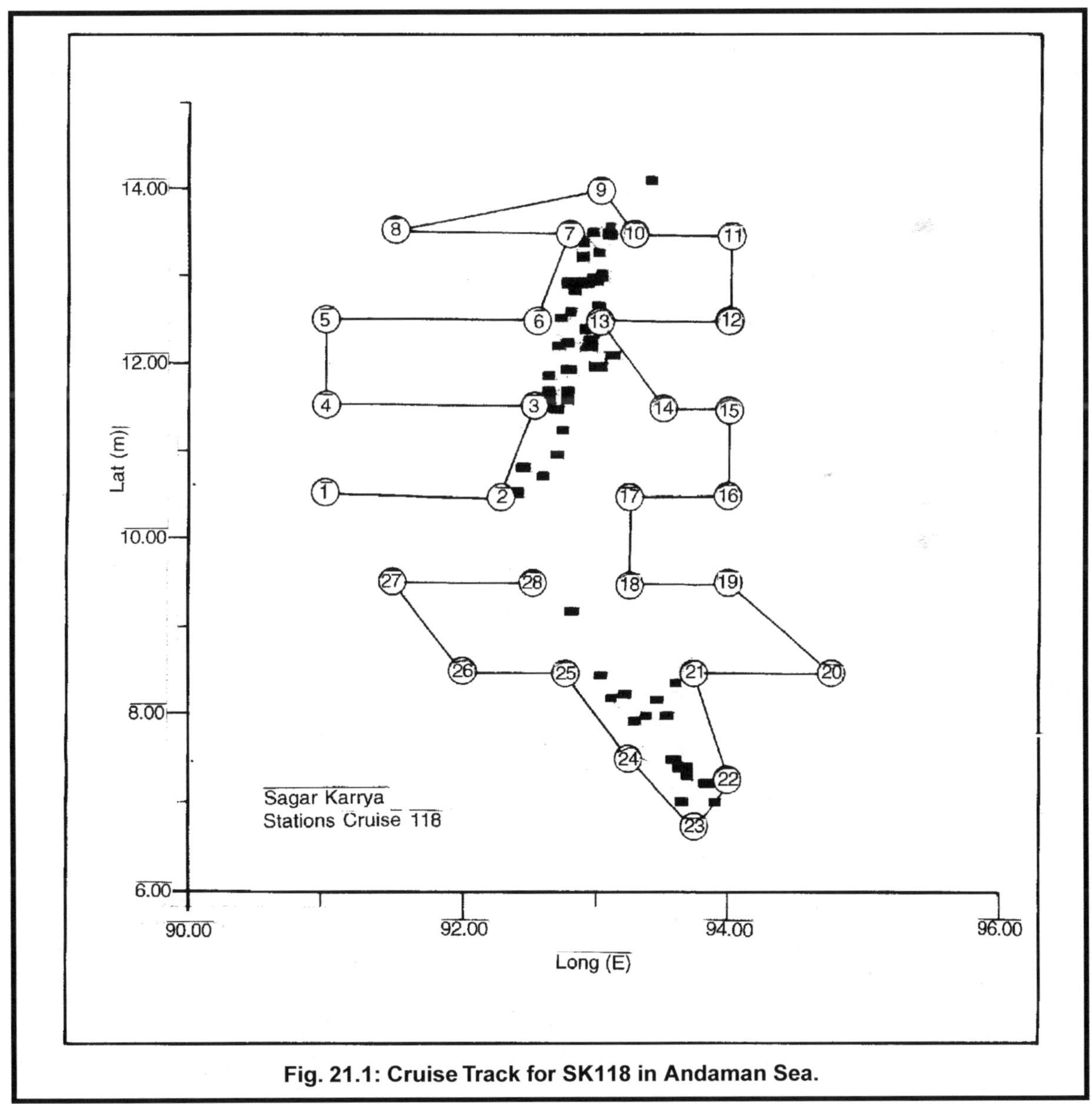

Fig. 21.1: Cruise Track for SK118 in Andaman Sea.

Table 21.1: Physico-chemical Parameters of Andaman Sea

Temperature	2.0 ± 50C	Magnecium (Mg^{2+})	3.66 ± 0.1%
pH	7.9 ± 0.4	Calcium (Ca^{2+})	1.15 ± 0.03%
Oxygen	5.1 10.6ml/lit.	Potassium (K^{+})	1.16 ± 0.02%
Salinity	34.3 ± 0.7 psu	Bicarbonate	0.46 ± 0.002%
Chlorides (Cl)	55.09 ± 1.3%	Bromide (Br)	0.21 ± 0.001%
Sodium (Na+)	30.11 ± 0.6 %	Boric acid (H_3BO_3)	0.08 ± 0.0003%
Sulphate (8042)	7.70 ± 0.2%	Strontium	0.03 ± 0.0002%

Table 21.2: Number of Zooplankton Species Observed in Andaman Sea

Taxonomic Group		*Number of species observed*
Protozoa	Foraminifera	16
	Radiolaria	06
	Tintinnida	
Coelenterata	Hydrozoa	15
	Schyphozoa	21
	Ctenophora	14
Nemertinia	Enopla	06
Annelida		26
Chaetognatha	Errentia	18
Arthropoda	Cladocera	06
	Ostracoda	19
	Calanoida	120
	Cyclopoida	27
	Herpecticoida	09
	Monstrilloida	13
	Mysidae	19
	Hyperildea	12
	Gemmaridea	19
	Euphausiacea	22
	Decapoda	17
Mollusca	Heteropoda	03
	Pteropoda	12
Chordata	Prochordata	11
	Appendicularia	13
	Thallacea	05

Discussion

It is well known that, quality of an ecosystem can be assessed by analysing its components namely abiotic and biotic ones. In a marine environment it is not only oxygen, salinity and chlorides are important yard sticks as major components, but even minor components such as calcium, strontium, pottasium, bicarbonate, bromide also would hold a key to the success of flora and fauna of the area as limiting factors (Nybakken, 1997). In the present studies, the results (Table 21.1) indicate that physico-chemical parameters analysed are all on per with other ideal, unpolluted marine ecosystems. Further, the comparison of the present data with the earlier results obtained by earlier workers does not show significant variation. Thus, indicate the Andaman sea has not changed significantly as well not polluted so far.

Bioindicators at lower levels of organisation correlates more directly with environmental levels of known stress than those at the higher level. Many organisms have been used as bioindicators, apart from being used as sources of bioindicator molecules such as metallothioneins (Langston & Zhou, 1986) providers of cellular indices (Moore *et al.*, 1982, Moore, 1991) or at individual level (Widdows *et al.*, 1980).

Using of gastropods, branacles in general, *Mytilus edulis* in particular as bio-indicator is in vogue since 1939 (Moore and Kitching, 1939, Southward and Crisp, 1954, 1956), Dogwelks *Nucelia lapillus* and *Nassarius obselata* are helpful in analysing Tributyline indused pollution (Hawkins *et al.*, 1994), mussel egg has been identified as an indicator of mutagen (Dixon and Pascoe, 1994). *Patella vulgata, P. dispersa, Monodonta lineata, Littorina* spp. etc., are some of the well known bio-indicators of oil spill and red tides (Southward and Southward, 1978). Further, Southward (1984) indicated the role of *Sagitta setosa* and *S. elegans* as bio-indicators particularly for phosphates.

As can be seen from Table 21.2, a rich fauna of zooplankton in general and bioindicator species such as chaetognaths and other molluscs are present in abundance in Andaman sea. It can be judged that biotic factors also functioning perfectly well in this sea. It is quite understandable that, when zooplankton are present in abundance, there must be sufficient phytoplanktons to feed on and inturn there must also be sufficient secondary consumers, like fish and other higher organisms, which feeds on these zooplankton, thus completing a marine food chain systematically. This shows that at Andaman sea the ecosystem is a mature, complete, self regulating and self sustaining one.

Although the fact remains that a large number of rivers from adjoining countries bring in large run-offs along with pollutants to Bay of Bengal, the reasons for not recording noticeable pollution in this area may be due to the distance from shore to this Andaman Sea area. It may also be due to degradation of most of the pollutant before they reach this area from coasts of Myanmar, India, Bhutan, Bangladesh, Indonesia etc. It can also be suspected that due to sinking of most of the heavy pollutants to the bottom of the sea, from where they cannot disperse further due to almost stagnant conditions of water. Apart from the above, one more plausible reason is that the countries surrounding the Andaman sea have become industrialised at relatively recent period, and the quantity of effluents released is not so much so that it could pollute Andaman sea, which is quite far from the coasts of these countries.

However, as the Andaman sea environment is also prone to pollution at the rate at which the coastal areas are becoming industrialised, one has to have a close, regular and careful monitoring of the ecology of Andaman sea. It could be done by using modern techniques such as ANPP, AVIRIS, BIOCLIM, ERIN, GEMS, GRID, HRV/MLA, MSCP, MSS, RAP etc., apart from regular survey, sampling for biological organism as well for physico, chemical parameters to see that this prestine environment remains unpolluted for years to come.

Acknowledgement

It is pleasure to thank Department of Ocean Development (Government of India) and National Institute of Oceanography, Donapaula, Goa for the cruise facilities. Support extended by Dr. P.V. Narvekar, Chief Scientist, NIO is appreciated.

References

Anonymous. (1981). Andaman Sea. *Ind. J. Mar. Sci.* 10(7): 209-210.

Cairns, J. and J.R. Pratt. (1993). A history of biological monitoring using benthic macroinvertebrates. In: *Freshwater Biomonitoring and Benthic Macroinvertebrates*. (Eds. Rosenberg, D.M. and V.H. Resh), Chapman and Hall, New York: 10-27.

Daniel, R. (1985). *Fauna of India: Coelenterata, hydrozoa, siphonophora*, Z.S.I. Calcutta.

Dixon, D.R. and P.L. Pascos. (1994). Mussel eggs as indicators of mutagon exposure in coastal and estuarine marine environment. In*: Water quality and stress indicators in marine and fresh water Systems: Linking levels of organisation* (Ed: D.W. Sulchife), Freshwater Biological Association, UK: 124-137.

Grassie, J.F., Lessrre, P., Mcintyre, A.D. and G.C. Ray. (1991). Marine biodiversity and ecosystem function. *Biol. Internet*. Special issue 23: 1-19.

Hawkins, S.J. Proud, S.V., Spence S.K. and A.J. Southward. (1994). From the individuals to the community and beyond, water analysis, stress indicators and key species in coastal ecosystem. In: *Water quality and stress indicators in marine and fresh water systems: Linking levels of organisation* (Ed: D.W. Sutchife), Freshwater Biological Association, (UK): 35-62.

Kasturirangan, L.R. (1963). A key for the identification of the more common planktonic copepods of Indian coastal waters. CSIR, New Delhi, India

Langaston, W. J. and M. Zhou. (1986). Evaluation of significance of metal binding proteins in gastropods *Littorina littorea*. *Mar. Biol*., 9: 505-515.

Madhupratap, M. (1981). Thermocline and zooplankton distribution. *Ind. J. Mar. Sci.* 10(7): 262-265.

Martin, J. (1994). Testing the iron hypothesis in ecosystem of the equatorial Pacific. *Nature*. 371:123-129.

May, R.M. (1994). Biological diversity, difference between land and Sea. Phil. Trans. Roy. Soc. Lond. B 343: 105-111.

McAllister, D.E., Schueler, F.W., Roberts, C.M. and J.P. Hawkins. (1994). Mapping and GIS analysis of the global distribution of coral reef fishes on an equal area grid. In: *Mapping the diversity of Nature* (Ed: Miller, R.I.), Chapman and Hall Publ. London., pp. 155-175.

Moore, M.N. (1991). Lysosomal changes in the response of molluscan hepatopancreatic cells to extracellular signals. *Histochemical Journal*. 23: 495-500.

Moore, M.N. and J.A. Kitching. (1939). The biology of *Chthamalus stella* (Poll.) *J. Mar. Biol. Assn*. UK, 23: 521-541.

Moore, MN., Pipe, R.K. and S.V. Farrar. (1982). Lysosomal and microsomal responses to environmental factors in *Littorina littorea* from Sullom vol. *Mar. Poll. Bull.*, 13: 340-345.

Mon, T. (1964). *The pelagic copepods from neighbouring waters of Japan*, Tokyo, The Soyo Co.,

Newman, P.J., Piavaux, M. A. and R.A. Sweeting (Ed). (1992). River water quality - *Ecological assessment and control CEC*, Brussels, p. 751.

Nicol, D. (1971). Species class and phylum diversity of animals. *Q. JI. Fla. Acad. Sci.* 34: 191-194.

Nybakken, J.W. (1997). *Marine Biology: An Ecological Approach*. 4th ed. Addison Wesley Longman Inc., California.

Parsons, T.R., Takahashi, M. and B. Hargrove. (1984). *Biological Oceanographic Processes*. 3rd ed. Pergmon Press, New York.

Rosenberg, D.M. and V.H. Resh (Ed). (1993). *Freshwater Biomonitoring and Benthic Macroinvertebrates*. Chapman & Hall, New York, p. 488.

Santhanam, R. and A. Srinivasan, (1994). *A Manual of Marine Zooplankton*, Oxford and IBH Publ. Bombay.

Southward, A.J. (1984). Fluctuations in the indicator chaetognath *Sagitta elegans* and *Sagitta setosa* in the westen channel. *Oceanaoloagia acta*. 7: 229-239.

Southward A.J. and D.J. Crisp. (1954). Recent change in distribution of the inter-tidal barnacles *Chthamalus stelatus* Poli, And *Balananus halanoides*. L. in the British Isle. *J. Ani. Ecol.* 23: 163-177.

Southward, A.J. and D.J. Crisp. (1956). Fluctuations in the distribution and abundance of intertidal barnacles. *J. Mar. Biol. Assn.* UK. 35: 211-299.

Southward A.J. and E.C. Southward. (1978). Recolonisation of rockyshores in Cornwell after use of toxic dispersant to clean up the Torrey canyon spill. *J. Fish Res. Board. Canada,* 35: 682-706.

Vijayalaxmi. (1981). Chaetognatha of Andaman Sea. *Ind. J. Mar. Sci.* 10(3): 270-273.

Widdows, J., Phelps, D.K. and W. Galloway. (1980). Measurement of physiological conditions of mussle transplanted along a pollution gradient in Narragensett Bay. *Mar. Env. Res.* 4: 181-194.

Wright, J.F., Furse, M.T. and P.D. Armitage. (1994). Use of macroinvertebrate communities to detect environmental stress in running water. In: *Water quality and stress indicators in madne and fresh water systems: Linking levels of organisation* (Ed. D.W. Sutchife), Freshwater Biological Association, UK, pp. 15-34

Zheng Zhong. (1989). *Marine Planktology*. China Ocean Press, Beijing.

22

Variations of Some Abiotic and Biotic Factors of Fish Culture Ponds Treated with Neem Cake

S.K. Sarkar

Department of Zoology, Netaji Nagar College (Day), Regent Estate, Calcutta - 700 092

ABSTRACT

Four freshwater fish culture ponds were treated with neem cake at the rates of 500 and 1,000 kg/h/6-month. Variations of some physico-chemical parameters and their relationships between these factors and biotic factors were studied. Some abiotic factors of water were found conducive for plankton and bottom fauna productivity. For good quality ecosystem, use of neem cake at high rate should of avoided.

Introduction

A number of abiotic and biotic factors have an important role in fish culture ponds. The relationship between abiotic factors and living organisms in fish culture ponds is far from the unidirectional because fish population dramaticaly affects the trophic status of other organisms and alters the water conditions in various ways (Boyd, 1982). Biologically, organic manure has some attributes. It provides a wide variety of nutrients along with organic matter that improves the physical characteristics of soils. Its beneficial effects on fish growth area sometimes difficult to duplicate with other materials. In spite of its high handling and labour costs and low analysis, organic manure remains a most valuable soil organic resource.

In most Asian countries, fish culturists commonly use animal manures at rates far in excess of those employed in commercial fish culture. Their aim is to create a productive ecosystem that does not result in water pollution, and the cost is a secondary matter. Manure applications can be made with little concern about adding toxic quantities of the nutrients. However, use of animal manure could result in different production rates because manures collected from different animals vary in quality. Therefore, relationship between oil cakes and abiotic-biotic factors of water must be evaluated. Neem cake (a product of seeds of the plant *Azadiracta indica*) is a great demand as a fertilizer in agricultural soil (Amorose, 1995). The present study was carried out to measure the influence of neem cake on the water quality variables and bottom fauna concentrations in ponds.

Materials and Methods

Experiments were conducted over a period of six months in six ponds (ponds varied between 0.02 and 0.04 h with depth varying between 1.0 and 1.6 m). These ponds were used for three species composite fish culture in irregular way under the management of fish farmers. These ponds are generally stocked with *Labeo rohita, Catla catla,* and *Cirrhinus mrigala.* Experiments followed a randomized block design with two replicates each of six treatments with accompanying controls. Beginning in the first week of June 2000, neem cake was uniformly broadcast over pond water (except controls) at the rates of 83 and 167 kg/h. The same application rates and procedures were repeated 20 days after the initial application, and this was continued upto six months (first week of December, 2000) for a total of six applications. Proximate analysis of the neem case indicated that nitrogen, phosphorus and potash constituted 5.2, 1.1 and 1.5 per cent respectively (Jhingran, 1991). Water samples were collected at 30 days inteval at 1300 hrs and physico-chemical parameters such as plankton concentration, pH, dissolved oxygen, transparency, total ammonia nitrogen, total alkalinity, filtrable orthophosphate, and bottom fauna concentration were analyzed following standard methods (APHA, 1989). In every month, bottom fauna was collected with a bottom sampler from four different areas (2 m^2 each) in ponds and analyzed by direct count and wet weight methods.

Results and Discussion

Plankton Concentration: At different rates of neem cake, mean phytoplankton and zooplankton concentrations were increased by 38-90 and 210-525 per cent of control, respectively (Table 22.1, $P < 0.05$-0.001). Concentrations of zooplankton were always higher than that of phytoplankton. Dominant species of phytoplankton were *Chlorella vulgaris* (20.7%), *Coelestrum* spp. (19.8%), *Pandorina morum* (17.5%), *Scenedesmus* spp. (12.3%), *Selenestrum* spp. (13.5%), *Anabaena* spp. (10.0%), *Microsystis aeruginosa* (6.2%), *Navicula* spp. (5.7%), *Euglena* spp. (3.5%), and *Phacus* spp. (0.8%) and that of zooplankton were *Brachionus* spp. (21.6%), *Keratella* spp. (16.0%), *Filinia longiseta* (8.5%), *Trichocera* spp. (11.7%), *Daphnia magna* (20.0%), *Bosmina longirostris* (10.0%), *Mesocyclops hyalinus* (5.5%) and *Heliodiaptomus viduus* (4.5%). During the period of study, the percentage of total plankton (both phyto and zooplankton) in different treatments is variable (Fig. 22.1).

Bottom Fauna: Mean weight of different species of bottom fauna (*Viviparus bengalensis, Chaoborus* sp., *Planorbis exustus, Branchiura sowerbii, Chironomus* sp., and unidentified Odonata) in control pond was 11.0 g/m^2. At 500 kg neem cake/h, bottom fauna concentration increased by 145 per cent of control value ($P<0.01$). At 1,000 kg neem cake/h, it was decreased by 45.5 per cent of control value (Fig. 22.2, $P<0.01$).

Water Parameters: At 500 and 1,000 kg neem cake/h, mean dissolved oxygen concentrations increased by 16.2 and 58.8 per cent of control, respectively (Table 22.2, $P < 0.05$-0.01). At different rates of neem cake, pH of water was comparable to that of control value. At 500 kg neem cake/h, transparency of water was comparable to that of control value. At 1,000 kg neem cake/h, transparency decreased by 19.6 per cent of control ($P < 0.05$). At different treatment rates of neem cake, total ammonia nitrogen, total alkalinities and filtrable orthophosphate concentrations of water were increased by 40-190, 20.1-59.2 and 71.4-328.6 per cent of control value, respectively ($P < 0.01$-0.001). During the period of study, temperature of water varied between 20.2 and 27.5°C.

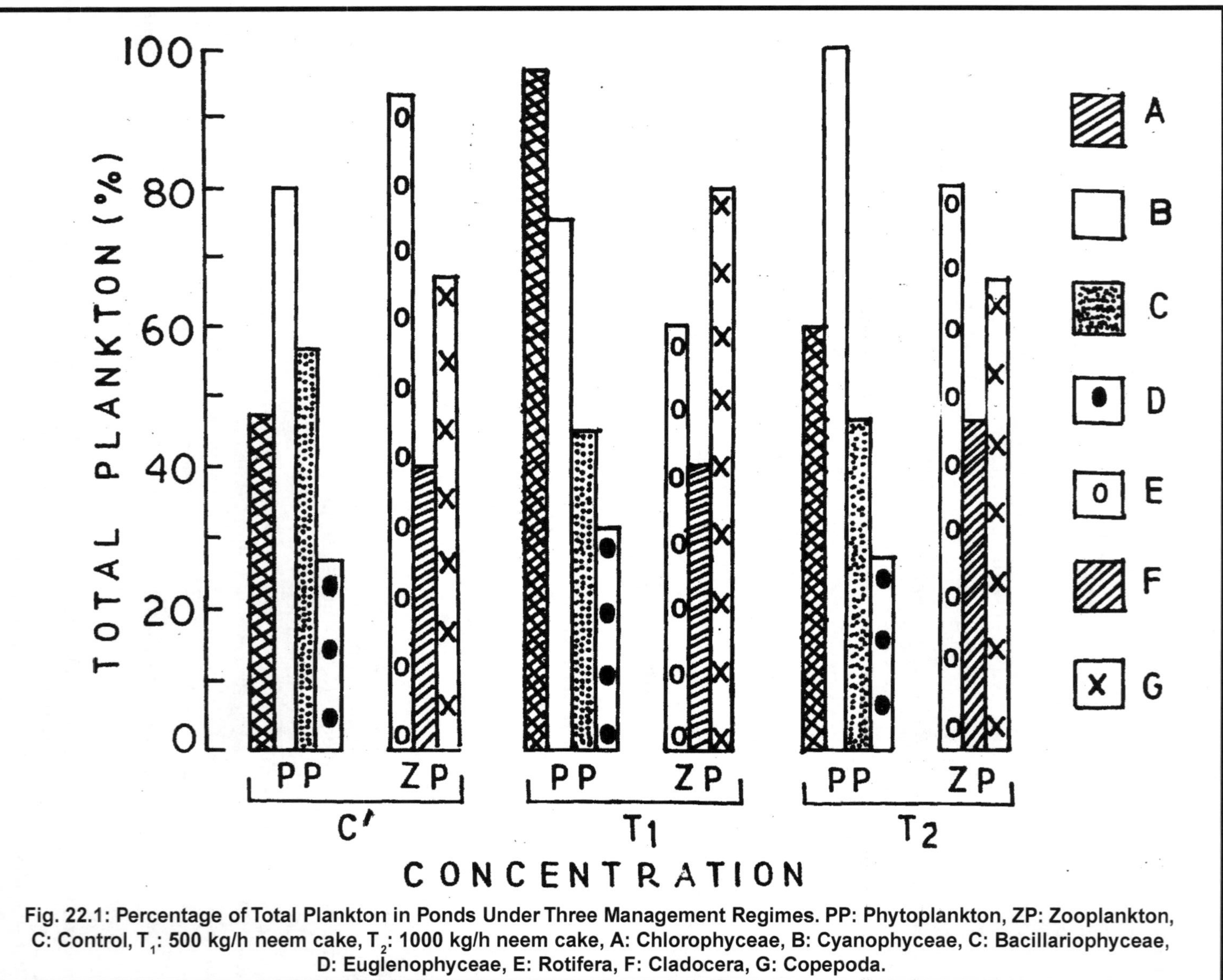

Fig. 22.1: Percentage of Total Plankton in Ponds Under Three Management Regimes. PP: Phytoplankton, ZP: Zooplankton, C: Control, T_1: 500 kg/h neem cake, T_2: 1000 kg/h neem cake, A: Chlorophyceae, B: Cyanophyceae, C: Bacillariophyceae, D: Euglenophyceae, E: Rotifera, F: Cladocera, G: Copepoda.

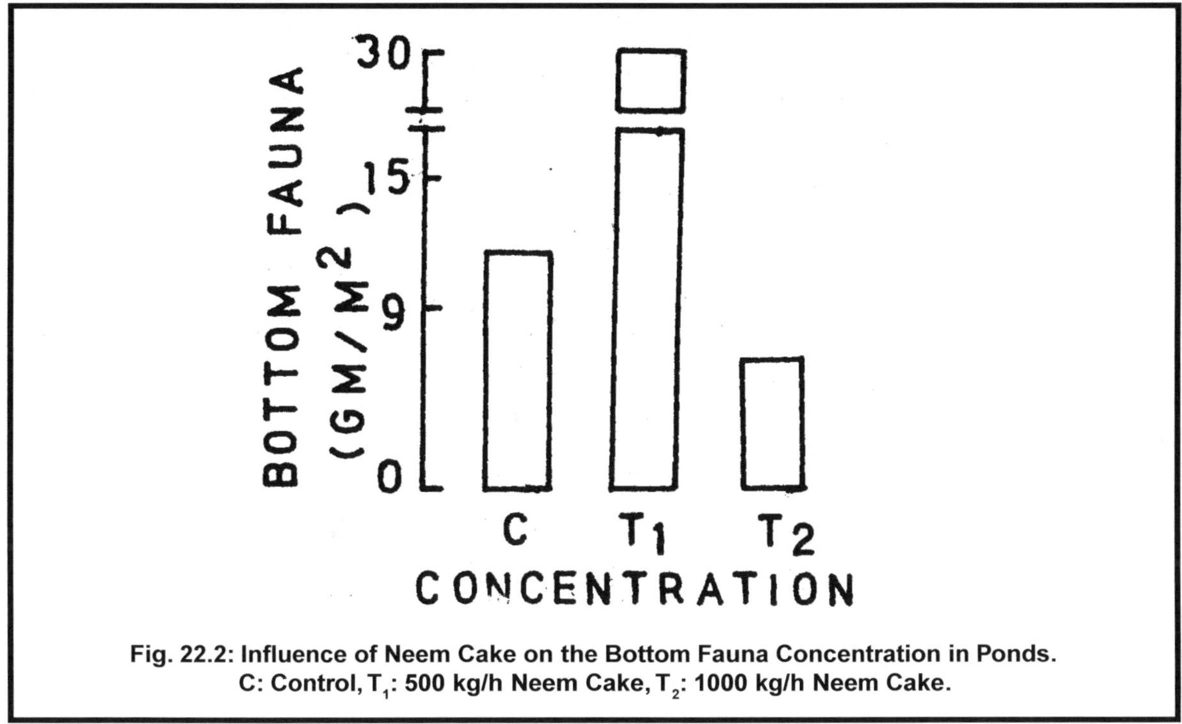

Fig. 22.2: Influence of Neem Cake on the Bottom Fauna Concentration in Ponds. C: Control, T_1: 500 kg/h Neem Cake, T_2: 1000 kg/h Neem Cake.

Application of neem cake at different rates steadily increased various abiotic factors of water such as dissolved oxygen, total alkalinity, total ammonia nitrogen and filtrable orthophosphate. Neem cake at the rate of 500 and 1,000 kg/h did not interfere with dissolved oxygen during initial day of its application until there was heavy growth of phytoplankton, when dissolved oxygen increased. Similar dissolved oxygen fluctuations were observed after organic fertilization in ponds by Sarkar (1983). On the other hand, dissolved oxygen decreased gradually after initial application of neem cake at higher rate (1,000 kg/h) but concentration of dissolved oxygen was restored to normal level soon after the septic condition of water was over. Therefore, application of neem cake to ponds at heavy instalments should be avoided if ponds are already stocked with fish.

Although fluctuations of dissolved oxygen may not be directly responsible for any adverse impact on fish production (Sarkar, 1988), the effect of total ammonia nitrogen concentration is important. The total ammonia nitrogn concentration exhibited adverse effects on fish growth even at threshold concentrations (Colt and Tchobanoglous, 1976). Because chronic ammonia poisoning is a serious concern in fish culture ponds it is realistic to expect adverse impact on fish in ponds treated with oil cakes at high rates. Use of glycocomponents from extracts of Yucca plants are useful to control ammonia (Sarkar, 1999).

Application of neem cake streadily increased the concentrations of phytoplakton and zooplankton, the latter being more abundant than the former. It was further observed that at different rates of neem cake, total alkalinity significantly increased. Sarkar (1992) established that there is potential gain of alkalinity from the use of organic manure. Boyd (1982) observed that water with higher alkalinity favoured plankton growth than waters of lower alkalinity. Enhancement of various abiotic factors of water triggered to increase primary productivity (Sarkar, 1983). Consequently, enough plankton organisms grew in ponds in the present study thus reducing the transparency of water.

Table 22.1: Influence of Neem Cake (NC) on the Plankton Concentration of Water. Values are Means and Ranges of 13 Observations

kg NC/h	Plankton concentration (Number per ltr)	
	Phytoplankton	Zooplankton
Control	110 70 - 150	80 35 - 125
500	209 135 - 283	248 180 - 316
1,000	152 80 - 222	282 180 - 384

Role of bottom fauna in determining productivity of ponds is well documented. Production of various fish food organisms depends largely on the availability of different nutrients. Dynamics of availability of nutrients is, in turn, determined by the conditions prevailing in the pond soil. However, concentrations of different species of bottom fauna at neem cake level of 1,000 kg/hectare significantly decreased. Generally bottom fauna is increased by the application of mustard and mahua cakes (Sarkar, 1983, 1992) but great development of phytoplankton blooms caused by application of at high rates lowered the bottom fauna by shedding (Moss, 1976). Variations of dissolved oxygen, total alkalinity and orthophosphate of water have significant influence upon the concentrations of phytoplankton, zooplankton and bottom fauna (Table 22.3).

Table 22.2: Influence of Neem Cake (NC) on Water Qualities in Ponds. Values are Means of 13 Observations ± SD

kg NC/h	Water parameters					
	pH	Dissolved oxygen (mg/l)	Total alkalinity (mg/l)	Total ammonia nitrogen (mg/l)	Transparency (cm)	Filtrable orthophosphate (mg/l)
Control	7.3	6.8	179	0.10	51	0.07
500	7.4	9.5	241	0.18	52	0.18
1,000	7.4	10.8	285	0.29	37	0.30
SD (all treatments)	0.03	0.05	3.53	0.08	1.64	0.04

Table 22.3: Correlation Coefficient Between Various Abiotic Factors of Water and Biotic Community
A: Significant at 5 per cent Level

	Abiotic factor					
Biotic Community (Group)	Temperature	pH	Dissolved oxygen	Total alkalinity	Total ammonia nitrogen	Filtrable orthophosphate
Phytoplankton	-0.105	0.207	0.747[a]	0.731[a]	0.266	0.777[a]
Zooplankton	0.171	0.240	0.699[a]	0.762[a]	0.234	0.762[a]
Bottom fauna	- 0.120	0.310	0.788[a]	- 0.155	0.177	0.816[a]

Acknowledgement

Thanks are due to the Principal of Netaji Nagar College (Day) for providing laboratory facilities and to Shri Kashna Shau, Laboratory Assistant, for helping the author during the period of study. The author is also thankful to the University Grants Commission for granting financial assistance (Research Project No. F.PSW-007/00-01/ERO).

References

Amorose, T. (1995). Efficiency of neem oil and defatted cake *Culex quinquafasciatus. Geobios.* 22: 160-163.

APHA. (1989). *Standard Methods for the Examination of Water and Wastewater.* American Public Health Association, Washington, DC.

Boyd, C.E. (1982). *Water Quality Management for Pond Fish Culture.* Development in Aquaculture and Fisheries Science, Vol. 9, New York.

Colt, J. and Tchobanoglous. (1976). G. Evaluation of toxicity of nitrogen compounds to *Ictalurus punctatus. Aquaculture,* 1976, 8: 209-224.

Jhingran, V.G. (1991). *Fish and Fisheries of India.* Hindustan Publishing Corporation, New Delhi.

Moss, B. (1976). Fertilization on community structure and biomass of aquatic macrophytes and algal populations. *J. Ecol.,* 64: 313-342.

Sarkar, S.K. (1983). Role of cowdung and mustard oil cake on the effectiveness of fertilizers in fish production. *Environ. Ecol.,* 1: 31-40.

Sarkar, S.K. (1988). Effects of organic manures and fertilizers on fish. Fert. News, 33: 15-19.

Sarkar, S.K. (1992). Phosphorus availability in fish pond soil treated with mahua cake, poultry manure and superphosphate. *Proc. Zool. Soc.* (Calcutta), 45: 95-100.

Sarkar, S.K. (1999). Role of glycocomponents (De-Odorase) on water parameters in fish ponds. *J. Environ. Biol.,* 20: 131-134.

23

Enzymatic Evaluation of a Heavily Polluted Lake in Mysore

T.B. Mruthunjaya and S.P. Hosmani

DOS in Botany, University of Mysore, Manasagangotri, Mysore - 570 006

ABSTRACT

A one year study of a heavily polluted lake (Dalvoi lake) was undertaken to study the behaviour of five enzymes along with other physico-chemical parameters and their relation to plankton population. These enzyme include Amylase, Phosphatase, Protease, Catalase and Urease activity. Urease activity was almost nil. Total bacteria averaged between 52 and 100 x 10^7 organisms per litre. Euglenaceae and Myxophyceae dominated most of the plankton population. BOD was quite low and reduced to nil during winter months. Phosphatase activity seemed to be high. Correlation matrix and cluster analysis appear to be handy tools in determining the important factors that control activity in polluted waters.

Key words: Enzyme activity, Dalvoi lake, Organic pollution, Sewage Matrix, Cluster analysis.

Introduction

Information on the enzymatic activity of sewage oxidation ponds are available as evident from the publications of Patil (1975), Patil *et al.* (1972), Gaddad (1983), Hosetti *et al.* (1985), Hosetti (1988), Stahl and May (1967). Fitzgerald and Rohlich (1958), Hosetti and Frost (1995), but literature pertaining to enzymatic activity in polluted lakes is lacking. The present work was undertaken to study the overall performance in the polluted Dalvoi lake for a period of one year (March 2000 to February 2001), with special reference to five enzymes, apart from other important physico-chemical parameters and bacteria, phytoplankton and zooplankton.

Materials and Methods

Mysore district is in the southern most portion of Karnataka State and is situated between 13°36′ and 13° 35′ north latitude, 75°55′ and 77° 20′ east longitude. Mysore city is situated 12.5° north of equator in the northern subtropic region and 76.39° east. It is a tourist center and the river Cauvery is on the northern side and its tributary Kabani is on the south west side. Mysore city and its surrounding have a large number of small and large reservoirs which are prone to domestic pollution receiving the city's sewage.

Dalvoi lake is a large water body in the near vicinity of the city measuring about 16 sq m cubic ft, with a maximum depth when full being 8 m, and a normal depth of six m. The city sewage is continuously left into this lake, the water is almost green in colour and emits a foul smell. Macrophytic vegetation is not much but the lake experiences occasional blooms of Chlorococcales and regular blooms of Myxophyceae and is organically polluted. Phosphates, nitrate, albuminoid ammonia, TDS, dissolved organic matter, ammonia nitrogen, dissolved oxgyen and BOD were analysed as per the methods described in standard methods for the examination of water and waste water (1995), water temperature was measured as °C using mercury thermometer and pH was determined with the help of ITL pH, meter. Alkaline phosphatase activity and Amylase activity was measured as per the method of Verstrate *et al.* (1976), Protease activity and Urease activity by the method of Lenhard, (1967), Catalase activity by the method of Sridhar and Pillai, (1969). Plankton was estimated as per the methods described by Hosmani and Vasanth Kumar (1996), Bacterial analysis closely follows the method of De (1989).

The one year data on the physico-chemical parameters, enzymatic activity and plankton population was subjected to statistical analysis and a Pearsons correlation matrix was obtained. From this data, contained and non-contained clusters were derived by the ramifying linkage method (Miller and Khan, 1962) and the data obtained is used for the study.

Results and Discussion

The monthly variations in physico-chemical parameters, enzymatic activity and plankton population is presented in Table 23.1, Pearson's correlation matrix in Table 23.2 and the list of non-contained clusters of variables in Table 23.4. Table 23.3 depicts "cards for extracting clusters by the ramifying linkage method".

Although Dalvoi lake is not really an oxidation pond, its features and the type of water entering into it brings it very close to a sewage oxidation pond. The floating and settling organic matter is broken down which is aided by algae and microbial activities in the lake. Algae when in blooms consume large quantities of nitrogen and phosphorous two common nutrients found in waste water and therefore these two nutrients tend to vary seasonally in this lake. BOD determines how much wastes still remains to be degraded. Referring to Table 23.1 during the months of November, January and February, BOD is nil although dissolved oxygen is in appreciable amounts, so also is the quantity of dissolved organic matter.

The enzyme activity also shows considerable variations throughout the period of collection and hence pose greater difficulty in relating definite activities in the lake. Under such circumstances the correlation matrix and the cluster analysis seems to be an appropriate method to quantify the performance of enzymatic activity and to trace the relationship of various parameters. From the correlation matrix, cards for extracting clusters by ramifying linkage method (Miller and Khan, 1962) are derived, from these cards, first, contained clusters are obtained. It is obvious to explain the contained clusters, since one or the other parameters is already contained in the small or large clusters. From the cards, non-contained clusters are obtained. Non-contained clusters explain the relations between

Table 23.1: Monthly Variations in Physico-chemical Complexes, Enzymatic Activity and Plankton Population in Dalvoi Lake (March 2000 to February 2001)

	Parameters	2000										2001	
		Mar	April	May	June	July	Aug	Sept	Oct	Nov	Dec	Jan	Feb
1.	Water Temperature °C	29.0	30.0	31.0	28.0	25.0	28.0	27.0	29.0	26.0	28.0	26.0	23.0
2.	pH	7.9	8.3	7.4	7.9	7.6	8.2	8.0	8.0	6.9	7.0	7.0	8.0
3.	Dissolved Oxygen mg/l	5.6	6.8	6.0	7.2	2.8	8.4	7.2	6.0	4.4	3.6	3.2	4.4
4.	Biochemical Oxygen Demand mg/l	3.2	3.6	3.2	4.0	1.6	7.6	3.6	4.4	0.0	2.8	0.0	0.0
5.	Nitrate mg/l	0.31	0.06	0.6	0.05	0.33	0.12	0.15	0.08	0.04	0.06	0.003	0.0
6.	Phosphate mg/l	0.59	1.2	1.8	2.2	2.8	0.28	1.4	1.5	0.26	0.39	0.02	0.57
7.	Ammonia nitrogen mg/l	16.7	14.5	16.7	14.8	17.8	14.8	11.7	11.6	17.5	8.3	27.3	23.0
8.	Dissolved organic matter mg/l	4.3	7.3	5.4	2.2	5.0	4.9	3.0	5.7	7.9	3.7	7.7	8
9.	Albuminoid ammonia mg/l	22.0	60.8	21.60	31.2	28.0	0.00	22.40	10.8	30.4	7.6	10.0	30.4
10.	Total solids mg/l	1.4	1.4	0.9	0.2	0.6	.5	0.7	0.6	0.2	0.5	0.5	0.6
11.	Amylase (1)	14.2	13.8	16.8	18.8	8.5	9.6	15.9	17.2	16.8	10.4	11.8	10.3
12.	Phosphatase (2)	16.2	86.5	27.3	18.8	70.0	26.9	72.5	54.4	87.0	6.35	34.35	18.8
13.	Protease (3)	2.34	2.83	3.14	2.69	2.70	6.18	3.78	4.14	4.74	0.22	0.27	0.72
14.	Catalase (4)	19.6	9.6	20.8	28.0	24.0	32.0	20.0	15.6	13.2	21.0	28.0	25.2
15.	Urease (5)	Nil	Nil	Nil	Nil	Nil	Nil	Nil	Nil	0.029	Nil	Nil	Nil
16.	Total bacteria x 10^7 (o/l)	80	50	93	63	58	66	52	58	54	96	90	100
17.	Desmids (o/l)	0	0	0	0	10	10	0	0	0	0	0	0
18.	Diatoms (o/l)	20	25	20	12	15	15	0	210	0	0	0	0
19.	Euglenaceae (o/l)	10	12	840	4200	210	0	252	0	1470	0	0	0
20.	Colorococcales (o/l)	1596	1596	126	60	40	84	420	1470	2352	1638	126	0
21.	Myxophyceae (o/l)	4674	4424	76034	26668	0	1008	2016	2352	10	1071	1480	9261
22.	Zooplankton	10	1016	42	294	84	21	42	0	0	126	126	26

(1) mg of maltose liberated per minute per 100 ml of sample.

(2) μg of paranitrophenol liberated per minute per 100 ml sample.

(3) μg of thyptophan liberated per minute for 100 ml of sample.

(4) μg moles of H_2O_2 decomposed per minute per 100 ml of sample.

(5) Activity measured by determining ammonia produced in 2 hrs when sample was allowed to react with 7.5 per cent urea at pH 6.5, 30°C. All parameters of physico-chemical complexes are average of three readings.

Table 23.2: Pearsons Correlation Matrix: Physico-chemical Complexes, Enzymatic Activity and Plankton Population (Dalvoi Lake)

	1	*2*	*3*	*4*	*5*	*6*	*7*	*8*	*9*	*10*	*11*	*12*	*13*	*14*	*15*	*16*	*17*	*18*	*19*	*20*	*21*	*22*
1	1.00	.365	.552	.510	.405	-.199	-.387	-.147	.157	.581	.417	-.235	.132	-.319	-.330	.128	-.356	.309	.598	.164	-.004	.345
2		1.00	.708	.608	-.021	-.120	-.224	-.206	.292	.438	.030	.072	.369	.009	-.500	-.373	.205	.282	-.075	-.236	-.061	.345
3			1.00	.814	.042	-.085	-.421	-.390	.058	.178	.426	.016	.658	.033	-.188	-.409	.035	.160	.200	-.100	.252	.223
4				1.00	.168	-.176	-.615	-.563	-.268	.166	.012	-.156	.608	.182	-.404	-.318	.374	.302	.082	-.104	.033	.106
5					1.00	.296	-.087	-.250	-.038	.396	.095	-.086	.166	-.017	-.194	-.140	.197	-.032	.700	-.214	-.069	-.214
6						1.00	-.138	-.111	.409	-.271	.421	.464	.350	-.312	.450	-.561	.126	.094	.225	.017	.526	.008
7							1.00	.637	.068	-.065	-.248	-.058	-.345	.339	.079	.362	.007	-.308	.034	-.402	-.066	-.104
8								1.00	.259	.077	-.222	.331	-.120	-.291	.397	.145	-.113	.023	-.136	.194	-.400	.151
9									1.00	.415	.231	.528	-.075	-.587	.150	.375	-.265	-.179	.070	.204	.217	.758
10										1.00	-.047	.077	-.093	-.443	-.385	.007	-.150	.048	.103	.209	-.476	.458
11											1.00	.184	.294	-.403	.284	-.332	-.622	.331	.418	.275	.603	.056
12												1.00	.433	-.631	.472	-.797	.083	.129	-.287	.333	-.098	.355
13													1.00	-.105	.337	-.681	.422	.279	.005	.112	.116	-.108
14														1.00	-.317	.404	.472	-.307	.068	-.791	.172	-.433
15															1.00	-.295	-.135	-.142	-.155	.570	.228	-.164
16																1.00	-.240	-.265	.347	-.268	-.206	-.331
17																	1.00	-.091	-.219	-.396	-.182	-.158
18																		1.00	-.061	.242	-.144	-.043
19																			1.00	-.339	.344	-.043
20																				1.00	-.136	.181
21																					1.00	.07
22																						1.00

Significance at 5% level: Arbitrary value 0.2 and above for obtaining clusters various parameters referring to Table 23.3 certain important characters of lake activity can be observed.

Table 23.3: Dalvoi Lake: Cards for Extractig Clusters by the Ramifying Linkage Method (Miller and Khan, 1962)

Card	Parameters
1	2, 3, 4, 5, 6, 7, 8, 9, 10, 11, 12, 14, 15, 17, 18, 19, 20, 22
2	3, 4, 7, 8, 9, 10, 13, 15, 16, 17, 18, 20, 22
3	4, 7, 8, 10, 11, 13, 15, 16, 18, 19, 21, 22
4	5, 6, 7, 8, 9, 10, 12, 13, 14, 15, 16, 17, 18
5	6, 8, 10, 13, 15, 17, 19, 20, 22
6	9, 10, 11, 12, 13, 14, 15, 16, 19, 21
7	8, 11, 13, 14, 16, 18, 20
8	9, 11, 12, 14, 15, 16, 20, 21, 22
9	10, 11, 12, 14, 15, 16, 17, 18, 20, 21, 22
10	14, 15, 17, 20, 21, 22
11	12, 13, 14, 15, 16, 17, 18, 19, 20, 21
12	13, 14, 15, 16, 19, 20, 22
13	15, 16, 17, 18
14	15, 16, 17, 18, 20, 21, 22
15	16, 19, 20, 21, 22
16	17, 18, 19, 20, 21, 22
17	19, 20, 21, 22
18	20, 21
19	20, 21
20	22

No. 1-22 indicates various parameters.

Table 23.4: List of non-contained clusters of variables (Dalvoi lake)

Sl. No.	Variable No.	Abbreviation
1.	1-17-11	WT-NH_3-Alase
2.	2-7	pH-NH_3
3.	3-7-11	Do-NH_3-Alase
4.	4-7	BOD-NH_3
5.	5-8	NO_2-DOM
6.	6-11	PO_4-Alase
7.	7-11-14-18	NH_3-Alase-Case-Dia
8.	8-11-14	DOM-Alase-Case
9.	9-11	ANH_3-Alase
10.	10-17-21	TS-Des-Myxo
11.	11-14	Alase-Case
12.	12-14	Pase-Case
13.	13-18	Prase-Dia
14.	14-18-21	Case-Dia-Myxo
15.	15-19-21	Case-Eng-Myxo
16.	16-18-21	TB-Dia-Myxo
17.	17-21	Des-Myxo
18.	18-21	Dia-Myxo
19.	19-21	Eng-Myxo

There are eleven small clusters of variables containing 2 parameters each indicating a direct variation e.g. ammonia is controlled by change in pH and BOD, Amylase controls phosphate and Alb-NH_3 which in turn is activated by enzyme catalase. Similarly, there are seven variables of 3 parameters and only one variable of four cluster. Catalase and amylase activity are the most common enzymes that are controlling the lake performance. It is also controlling the growth of diatoms to a little extent and Euglenaceae and Myxophyceae to a greater extent. Wherever Myxophyceae grow in abundance, they retard the growth of diatoms in general and desmids in particular. Hosetti and Patil (1988) have reported that variations in the activity of enzymes seemed to be directly related to bacterial number and further catalase and phosphatase also seemed to be related to microbial density in a sewage oxidation pond. However, in the present study the three most prominent features (Table 23.3) are the enzymes amylase, catalase and the regular bloom forming Myxophyceae which is represented throughout the year by the species of *Microcystis aeruginosa*. Therefore, the two enzymes and the algal species go hand in hand and maintain a steady growth, throughout the year. The members of the Euglenaceae and Chlorococcales appearing as sporadic blooms during monsoon and summer months, catalase amylase activity are high during monsoon and winter months. Gaddad *et al.* (1982) and Hosetti and Patil (1986b) have observed that the reduction in the activity of enzymes and BOD values are indications of the improvement of the quality of the lake water. But the observation of Sridhar and Pillai (1969) that catalase activity in waters is fairly a measure of organic pollution caused by microbial activity holds good to this study also. The microbial activity here may be that of Myxophyceae members. Similar importance of plankton are reported by Dodakundi and Rodgi (1977) and Hosetti *et al.* (1984).

Heavily polluted lakes due to sewage act like oxidation ponds. Heterotrophic bacteria degrade organic matter in the sewage which results in the production of cellular material and minerals. The production of these supports the growth of algae, which allows further decomposition of organic matter by producing oxygen. There waters like oxidation ponds tend to fill, due to the settling of the bacterial and algal cells formed during decomposition of sewage. Enzymatic activity in this lake is high. Catalse and amylase activity regulate the growth of Myxophyceae as a bloom which in turn controls diatoms. Catalase activity in this lake is a direct measure of organic pollution caused by Myxophyceae (*Microcystis aeruginosa*). These results are alarming and indicate that the fresh water lake is rapidy acquiring characters of a sewage oxidation pond.

Acknowledgement

The authors acknowledge Dr. B.B. Hosetti, Kuvempu University, Shimoga for giving access to his Personal Library.

References

APHA, AWWA, WPCF (1995). *Standard methods for the examination of water and waste water*, 19th Edition, New York.

Broota, K.D. (1989). *Experimental design in behavioural research*, New Delhi, Eastern Willey.

Dodakundi, G.B. and S.S. Rodgi (1977). Catalase and Protease activities in sewage stabilization pond. *J. Kar, Uni.* 22: 152-158.

* Fitzgerald, G.B. and G.A. Rohlich (1958). An evaluation of pond performance. *Sew. Ind. Wastes,* 30: 1213-1224.

Gaddad, S.M, Y.M. Jayaraj and S.S. Rodgi (1982). Catalase and Protease activities in relation to BOD removal and Bacterial growth in Sewage. *Ind. J. Environ. Health,* 24: 321-325.

* Gaddad, S.M. (1983). Studies on enzymes in relation to microbial growth and activity in sewage and stabilization ponds, Ph.D. thesis, Gulbarga University, Gulbarga.

Hosetti, B.B., H.S. Patil and S.S. Rodgi (1984). Vertical distribution of microbes in oxidation pond in relation to changes in physico-chemical parameters. *Environ and Ecol.,* 2: 159-161.

Hosetti, B.B., H.S. Patil, S.S. Rodgi and S.M. Gaddad (1985). Effect of detention period on the biochemical activities of sewage stabilization ponds - A laboratory study. *J. Environ. Biol.,* 6: 1-6.

Hosetti, B.B., and H.S. Patil (1986). Impact of *Lemna minor* on the effluent quality of sewage stabilization ponds. *Geobios,* 13: 244-247.

Hosetti B.B.and H.S. Patil (1988). Enzymatic evaluation of oxidation pond performance. *Int. Revue. Ger. Hydrobiol.* 73: 6: 641-647.

Hosmani S.P. and L. Vasanth Kumar (1996). Calcium carbonate saturaton index and its influence on phytoplankton. Pollution Research 15(3): 285-288.

Lenhard G. (1967). Determination of protease activity in bottom deposits of sewage stabilization ponds *Hydrobiologia,* 29: 67-74.

Miller, R.L. and J.S. Khan (1962). Statistical analysis in the geological sciences, John Willey and Sons, New York.

Patil, H.S. (1985). Studies on the physico-chemical and biological characteristics of sewage stabilization ponds. *J. Env. Biol.* 6: 93-102.

Patil, H.S., G.B. Doddakundi and S.S. Rodgi (1972). Succession and stratification of organisms in sewage oxidation ponds. *Curr. Sci.,* 41: 615.

Sridhar, M.K.C. and S.C. Pillai (1969). Catalase activity in polluted waters. *Effl. Water Treat.* J., 9: 81-88.

Stahl, J.B. and D.S. May (1967). Microstratification in waste stabilization ponds. *J. Wat. Pollut. Cont. Fed.,* 39: 72-88.

* Not seen in Original.

24

Immunoresponse of Aquatic Molluscs in Biounsafe Environment

Sajal Ray

Department of Zoology, University of Calcutta, 35, Bellygunge Circular Road, Calcutta - 700 019

Introduction

Immunology – the science of defence had almost dealt with the mechanism and functioning of immunity of vertebrates especially in human. Animal kingdom is comprised of approximately two million species of which only about seventy thousand are vertebrates. A bulk of animal life consists of invertebrates. Invertebrates like molluscs and insects serve as hosts for parasites of medical and economic importance. Mollusca exhibits diverse degrees of organisation and shows characteristic features in formation of foot, mantle and strongly centralised nervous system. Mollusca imparts a profound impact on human society for its key role on agriculture, veterinary science and human health problems. The group is important for its contribution to nutrition. Molluscs transmit pathogens to man and livestock. One of the examples is the intermediate host snail of the genus *Biomphalaria* where numerous cercariae of *Schistosoma mansoni* reproduce and get matured. As such, molluscs received bulk of attention in the studies of internal defence for its importance as a source of diet and as a pest or vectors of pathogen where the internal immunological response must have to be overcome under the challenge of pathogen and environmental xenobiotics.

External Physicochemical Barriers

Invertebrates apparently lack immunoglobulins and well interactive lymphocyte subpopulations and lymphoid organs as observed in vertebrates. The effective external barriers in molluscs includes tough and rigid exoskeletion in the form of shell and mucus. Effective physicochemical barriers of the body surface offer an effective line of defence against the invaders. Apart from such invasion the external physical threat which may jeopardise the survival of these animals of lower groups include extreme weather condition, continuous loss of water to result desication, detrimental mechanical forces that may damage the body structure and continuous exposure to ultraviolet radiation which may effect the genetic makeup of the organism. Calcareous hard covering in the form of shell covers the most vulnerable part of the body of molluscs and it can be withdrawn into the shell as and when necessary. Instance of periwinkle in *Littorina* capable of withstanding external adversity is another example of molluscan first line of external defence. Array of immuno competent cell types evolved in molluscs are capable of functioning in a cooperative manner. Cell cooperation is well documented among different types of cell including progenitor cell, hemostatic granular cell, phagocytes, nutritive cell and pigmented cell.

Extracellular Cytotoxicity

Hemocytes exhibit two mechanical ways of eliminating the foreign invaders. The first possible mechanism of killing is an external one. Encapsulation around schistosome larvae by hemocytes is documented under electron microscope represents classical phenomena of extracellular cytotoxicity. Successful encapsulaion reaction towards the invading pathogen leads to a asphyxiation and physical starvation. Hemocytic encapsulation around parasite creates a compartmentalisation that compells the organisms to be inactive and leads to a possible state of death. Such a killing mechanism in only effective where the invader expresses higher demands for nutrition, oxygen and waste product than do the hemocytes.

Lysosomal Enzymes

Limiting membrances of the invading target is interacted with the lysosomal hydrolytic enzymes of hemocytes. Integrity of membrance and the ionic balance of target cell is affected by the action of hydrolytic enzymes. Once the membrane integrity is lost, the enzymes make entry into the interior of the cell. Various sugar residues like mannan, glycan, mucopolysaccharides and glyeopopetide murien form a sheath around bacterial cell envelop. Muraminolytic lysosyme and glycosidases are secreted by hemocytes are demonstrated in the phagolysosme. Optimal physicochemical milieu is maintained within the phagolysosome to bring out an effective and instant intracellular killing. When a bacterial population is injected in molluscs, an increase in the activity of lysosomal enzymes indicated the efficacy of the strategy of lysosomal enzyme mediated killing.

Respiratory Burst

On stimulation by foreign particles or humoral substances phagocytes generate intracellular superoxide anion radical (O_2) a reactive defence molecule against intruding micro-organisms. Hemocytes of pacific oyster exhibit enhanced activity of phagocytosis against various foreign particles and microbes and show an effective antimicrobial activity.

Strategy of oxygen radical mediated killing is based on the premises of toxicity evolved in high concentration of molecular oxygen. Many of the oxygen derived molecules are extremely toxic. These derived molecules are reactive oxygen intermediates or oxygen radicals. Toxicity of reactive oxygen intermediate is due to their high order of chemical reactivity. Oxygen derivatives are generated at the cell surface when the consumption of oxygen in the cell is increased due to respiratiory burst following the contact of the phagocyte to micro-organism. Enzyme NADPH oxidase of plasma membrane oxidises NADPH to NADP and transfers electrons to outer orbital of molecular oxygen (O_2) to yield superoxide anion (O_2). Dismutation reaction converts superoxide anions to hydrogen peroxide at the exterior of the cell.

$$O_2^- + O_2^- + 2H^\bullet \rightarrow H_2O_2 + O_2$$

$$O_2^- + H_2O_2 \rightarrow OH^\bullet + OH^- + {}^1O_2$$

Singlet oxygen generated in the reaction is the most significant reactive species formed. Singlet oxygen formed from the reaction between superoxide anion and hydrogen peroxide.

$$O_2^- + OH^\bullet \rightarrow {}^1O_2 + OH^-$$

Ions may react with hydrogen peroxide to produce oxygen singlet. Production of superoxide in hemocyte can be demonstrated by reduction of nitroblue tetrazolium (NBT) to a blue coloured insoluble complex formazan. Nitroblue tetrazolium is reduced by O^{2-} in presence of superoxide dismutase. Generation of superoxide anion by hemocytes of pacific oyster *Crassostrea gigas* was reported using electron spin resonance spin trapping with the spin trap 5,5 dimethyl - 1 - pyrroline N-oxide (DMPO) and by chemiluminescence (Takahashasi *et al.*, 1993). In both the methods employed hemocytes released O_2^- on stimulation with phorbol myristate. Data is suggestive of existence of superoxide generating system in *Crassostrea gigas* which becomes functional when receives optimum stimulation. Microscopical and morphological confirmation of formazan deposition in the hemocytes of *Crassostrea virginica* is in record (Anderson *et al.*, 1992). Proper quantification of the percentage of phagocytes was also recorded. A comparative study of formation of superoxide anion by oyster hemocyte and human leukocyte was carried out by Gifford and Malawista (1970, 1972). Upon exposure to NBT, about 40 per cent adherent human and oyster granulocyte showed precipitation of blue stain with a strong reaction at the periphery of the cell. A typical appearance of formazan cell with enlarged appearance and extensive spreading is recorded.

Gifford and Malawista (1972) reported that the formation of formazan cell did not originate due to loss of intergrity of membrane and cell degeneration followed the production of intracellular formazan. NBT is capable to gain access to the cytoplasm durign engulfment but in the absence of phagocytosis, NBT may enter hemocyte during its attempt to interact the nonself glass surface to which it get adhered. Dikkeboom *et al.* (1987, 1988) reported reduction of NBT by hemocytes of diverse snail species. Faint staining of unstimulated hemocyte as evident from background staining is suggestive to requirement of stimulation for the process of generation of superoxide anion. Microscopic observation of NBT reduction is reported in several species of molluscs. Experimental evidence of different workers support the idea that the tendency to generate cytotoxic reactive oxygen intermediates is a general property of leukocytes of invertebrates and is probably involved in internal defence reaction. Modulation of hemocytic NBT reduction may be used as a tool for biomarkering of cellular stress induced by sublethal concentration of environmental xenobiotics.

There is a positive correlation of increased production of reactive oxygen intermediate and progression of disease in *Crassostrea virginica* is reported by Anderson *et al.* (1992). *Perkinsus marinus* is a protozoan parasite responsible for an infectious disease of *Crassostrea virginica.* Report described the extent of hemocyte to generate reactive oxygen intermediate during the progression of infection. Hemocytes contribute to a great extent as a mediator of major internal defence effeuctor in molluscs. Production of the oxygen radicals in both resting and stimulated hemocytes was measured using luminol augmented chemiluminescence. Hemocytes harvested from infected oysters showed significantly higher levels of chemiluninescence than that from the oysters with low level of infection. Advanced cases of the disease is characterised by hemocyte activation or stimulation with hyperproduction of cell dervied reactive oxygen intermediates. In this study, almost all the animals collected from natural habitat exhibited low or moderate degree of infection. A low level of background signal indicated the generation of reactive oxygen intermediate is supportive to an innate immunoresponse contributed by the hemocytes.

Antioxidant Enzymes

Internal defence of molluscs is dependent on circulating hemocytes capable of phagocyting pahtogenic organisms. In performing the intracellular killing of ingested micro-organism, hemocytes are capable of generating reactive oxygen intermediates. In order to protect the tissue of the self from reactive oxygen metabolite produced during respiratory burst, various potentially active antioxidant enzymes are reported in the mussel *Mytilus edulis* (Pipe *et al.*, 1993). Enzymes like catalase, superoxide dismutase and glutathion peroxidase are localised in the hemocytes employing immunocytochemical methods. Catalase is localised in minute electron dense granules known as peroxisomes. Superoxide dismutase were found to be assoicated with plasma membrane of hyalinocytes. Biochemical assay of these enzymes is supportive to immunocytochemical localisation of antioxidant enzymes. Glutathion peroxidase activity was predominantly present in hemolymph whereas catalase and superoxide dismutase activity were detectable only in hemocytes.

Prophenoloxidase Induction

Importance of phenoloxidase in different invertebrate phyla as a defence enzyme has increased considerably since its localisation was achieved in hemocytes and hemolymph. The enzyme has been reported in various animal tissues and cell types including circulating blood cells, neurones and melanocytes. The enzymes is involved in the defence mechanism of many invertebrates by catalysing the oxidation involved in sclerotisation and wound healing and participate in the process of encapsulation and melanisation of foreign bodies. A further possible role for phenoloxidase in invertebrate defence has recently been reported. The enzyme usually exists in the cell as prophenoloxidae - an inactive precursor. It has been demonstrated that β 1,3 glucan (Leonard *et al.*, 1985) and lipopolysacchaside of bacterial cell wall can activate prophenoloxidase in invertebrates (Brookman *et al.*, 1989). The reaction process is very specific and sentitive, involves the activation of enzyme and initiates a cascade of serine protease activation. Serine protease in turn, proteolytically cleaves prophenolxidase and thus activates the precursor to an active phenoloxidase. Activation of prophenoloxidase by serine protease minics alternative pathway of complement activation in mammals (Soderhall, 1982). Proteins involved in the aggregation reaction are also induced during activation of cascade. Physicochemical factors like pH and Ca^{+2} concentration activating the prophenoloxidase cascade initiate and augment other classical defence reactions namely phagocytosis and exocytosis by circulating hemocytes. Induction of phenoloxidase is established in invertebrates but little research is carried out in non-arthorpod phyla. In Mollusca, hemolymph expressed a low level of activity (Smith and Soderhall, 1991) of phenoloxidase. Cytochemical and biochemical detection of the enzyme was reported using a substrate dihydroxyphenylalanine (L-dopa) (Coles and Pipe, 1994) in the hemocytes and hemolymph of marine mussel *Mytilus edulis.* Hemocytes of *Mytilus edulis* were also tested for perodxidase activity using the substrate diaminobenzidine. Activiation of the phenoloxidase enzyme was demonstrated biochemically under preincubation of cells with 0.025 per cent zymosan supernatant in the same animal. Cytochemical localisation indicated the presence of enzyme in the hemocytes of *Mytilus edulis* which is capable of transforming L-dopa to quinones and melanin *in vitro.* Histological investigation confirmed the presence of enzyme in the granular eosinophilic hemocytes of the mussel. Ultrastructural localisation of the reaction product was reported in the larger cytoplasmic granules of such type of cells of molluscs. However, the percentage of occurrence of positive cell vary among individual animals and further study is demanded to elucidate the definite role of phenoloxidase in molluscan immunity.

Xenobiotics

Molluscs occupy diverse habitats and their strategy of survival is correlated to multidimensional immune responses. Molluscs inhabiting in biounsafe environment are under constant threat of physiological stress caused by different species of xenobiotics. Animals exhibit immunological response both against xenobiotics and parasitic infection. Many of the disease conditions of marine molluscs and augmented by pollution (Pipe and Coles, 1995). Studies on modulation of immune response in bivalves by xenobiotics have been concentrated on the effect of toxic metals and polycyclic aromatic hydrocarbons. There is a world wide concern about the pollution of marine biota with toxic chemicals. When exposed to sublethal concentrations of chemical pollutants the marine organisms are subjected to various forms of disease conditions and neoplasms. Habitat of American oyster *Crassostrea virginica* was contaminated by polycyclic aromatic hydrocarbon near Virginia, USA in 1991. Cytometric characteristics of hemocytes of the affected animals were determined using multichannel Coulter counter (Sami *et al.*, 1992). Sharp modulation of cytometric characteristics were indicative to the possibility of development of sensitive biomarker against the exposure to polycyclic aromatic hydrocarbon.

Immunotoxicity of cadmium and copper was investigated in marine bivalves pre-exposed to *Vibrio tubisashi* (Pipe and Coles, 1995). Mussels exposed to cadmium for 7 days followed by exposure to *V. tubiashi* exhibited higher number of circulating hemocytes compared to non-vibrio exposed batch. Decrease in percentage of circulating hemocytes was recorded following an exposure to *V. tubiashi* of mussel pre-exposed to copper. Intracellular generation of superoxide by hemocytes was augmented in mussels challenged with *V. tubiashi* with no copper pre-exposure for a short span of time. However, a significant decrease of production of superoxide was demonstrated following a pre-exposure of copper for 7 weeks. Studies on phagocytic capability of hemocytes of molluscs with respect to xenobiotics was carried out by measuring the bacterial clearance or other non-self particulates by hemocytes. Laboratory experimentation has supported to the fact the phgocytic index of molluscan hemocytes is induced after a short-term low level exposure to contaminants (Anderson, 1993, 1981, Cheng and Sullivan, 1984). A correlation between high level of metal toxicity and depressed phagoxcytosis was demonstrated (Pipe *et al.*, 1995). Quantification of the reactive oxygen intermediate meant for killing is another common measure to assess the degree of physiological damage caused by pollutants. Enzymes associated with the release of oxygen radicals including the antioxidant enzymes can be affected by environmental pollutants (Pipe *et al.*, 1993). Bivalves are also equipped with lysosomal and other enzymes capable of deactivating the pathogens (Pipe, 1990). Activity and release of these adaptive enzymes can also be modulated by xenobiotics.

Factors involved in recognition of self and non-self are also modulated by environmental contaminants. Exposure to polluted sediments, copper on lectin binding sites of hemocytes was demonstrated (Sami *et al.*, 1993). *Crassostrea virginica* when exposed to polycyclic aromatic hydrocarbon for 11 weeks exhibited a significant suppression of concanavatin binding sites of hemocytes. Phenoloxidase activity of hemocytes was increased following an exposure of mussels to fluoranthene for 7 days (Coles *et al.*, 1994). Uptake of neutral red by mussel hemocyte is determined following an exposure to cadmium *in vivo* (Coles *et. al.*, 1995). Neutral red is a cationic dye usually deposited in the lysosomal compartment of cell. Its uptake by hemocyte is thought to occur by the process of pinocytosis or passive diffusion. Effect of immune parameters of bacterial challenge following exposure of *Mytilus edulis* to copper and cadmium is investigated. However, the susceptibility of disease like vibriosis is offen modulated by the pre-exposure of copper and cadmium.

Environmental contaminants like hexachlorobenzene and atrazine expressed a strong level of immunomodulation in the fresh water parasitised snail *Lymnea palustris* (Russo & Logadic, 2000). A sharp increase in total number of hemocytes from 2.2 to 8 fold was reported following an exposure of contaminants. No increase in phagaocytosis and production of reactive oxygen intermediate was reported in case of parasitised animal. Parasitism and atrazine treatment resulted a significant increase of lectin strained hemocytes. Percentage of stained and unstained cells did not show any alteration when the animals were exposed to hexachlorobenzene. Effect of fungicide chlorothalonil on oyster hemocyte activation was investigated in dose dependent manner (Anderson and Anderson, 2000). Production of reactive oxygen species by NADPH oxidase like enzyme was considered to contribute to antimicrobial activity in the hemocytes of oyster. Production of reactive oxygen species was suppressed in a dose dependent manner following an exposure to fungicide chlorothalonil. The report suggested that the fungicide might act as inhibitor of NADPH oxydase like enzyme in oyster hemocytes. Gastropods when exposed to sublethal concentrations of methylparathion, an organophosphate pesticide, circulating hemocytes undergo an alternation in surface characteristics (Ray and Chattopadhyay, 1998). Dose dependent increase in aggregation of both round and spreading hemocytes indicated a physiological threat of the animal which propagate in contaminated habitat.

Molluscs when exposed to sublethal concentrations of environmental contaminants are subjected to possible alteration of immune status. Such a modulation of immune response may affect the strategy of defence against invading micro-organisms directly or indirectly. However propagation of the animal in a contaminated and biounsafe environment might lead to a situation of opportunistic growth of parasite in the host animal. On the other hand, the parameters of immunomodulation in molluscs might be exploited as a tool for biomarkering of cell function in altered situation.

Cytokines

Cytokines are peptides synthesised and secreted by both immune and non-immune cells and considered to be major mediators of immune functions in vertebrates. Cytokines expresed hormone like property and affect numerous cell types and organ system involved in host defence and cell communication. Identification of interleukin (IL), and tumour necrosis factor (TNF) in invertebrates indicated the possible presence of other forms of cytokines of phagocytic and T cell origin in these animals (Beck and Habicht, 1991). Host defence in invertebrates is under regulation of a possible network of cytokine functions analogous to vertebrate system. Cytokine like molecules in vertebrates are secreted from phagocytic cell and act upon various aspects of inflammatory and cell communicative responses. In molluscs, isolation of cytokine like molecules does not establish the existence of a cytokine network. Activation of immunoacytes of molluscs by ILI and TNF is suggestive to the possible existence of intricate cell communicative and proliferative response in invertebrate.

Apart from classical immune responses, cytokines are involved in eliciting various other fundamental responses of biological importance. These responses include inflammation, interaction of immune system and neuroendocrine system. A variety of cytokine like molecules of biological importance are reported in the hemocytes of molluscs i.e. *Planorbarius corneus* and *Vivaparus ater* (Ottaviani *et al.*, 1993). There are two types of hemocytes identified in *Planorbarias corneus* namely spreading and round types. Spreading type of hemocyte the resembles the mammalian macrophage whereas round type of cell exhibits few characteristics T lymphocytes. Cell types isolated from the hemolymph of *Planorbarius corneus* and *Viviparous ater* are positive for cytolines like IL 1α, IL 1β, IL 2, IL 6 and TNF α. Moreover, the presence of IL 2 like molecules in the hemocytes expressed phagocytic activity is of evolutionary significance. In the hemolympth of *Mytilus edulis* ILI and TNF like molecules are in report. Experimental data indicates that these molecules express biological and immunological effect on circulating hemocytes and the mode of reaction is similar to that exists in vertebrate system. The presence of IL I ∝ in the glial cells of ganglia of *Mytilus edulis* is confirmed (Ottaviani *et al.*, 1993). Addition of opoids *in vitro* caused an increase of immunoreactive IL I ∝ molecules in the ganglia and hemocytes from *Mytilus edulis* are on record showed parallel finding in vertebrate immune response. Recently immunomolulatory effect of human IL-8 has been reported in the hemocytes of molluscs *Mytilus galloprovincialis* (Ottaviani *et al.*, 2000). Chemotactic cytokines are small peptides associated with mediation of acute inflammation. On the basis of the arrangement in amino acid residues, structurally the IL 8 molecules are of 2 types. IL 8 is a secretary product of various cell types namely monocytes, macrophages and leukocytes. Immune functions like changes in cell shape, chemotaxis, bacterial clearance and mode of cell signalling were studied in hemocytes challenged with recombinant IL 8. Finding suggests that IL 8 is a well conserved molecule and deeply involved in the immunological function in molluscs. Wide ranging activity of the cytokine molecules in initiating the immunoresponse in the critical phage of physiological state suggests that the molecules are highly conserved through the ages of evolution. Structural and biochemical similarity of these molecules in various vertebrate classes strengthen this view of conservation. In the process of evolution of immune system, cytokines played a key role as principal coordinator of immune response. Invertebrates are polyphyletic in origin and parallely their defence mechanisms are diverse and less understood. Recent reports of discovery of cytokine like molecules in molluscs and other invertebrates suggestes that cytokine orchestrated immune response is not solely restricted invertebrate series only.

In the event of cellular cross-talk in invertebrates cytokines play a significant role in signal tranduction as evident from the work done by Ottaviani *et al.* (2000). Recombinant human IL 8 induces the change of shape of hemocytes through protein kinase A and C pathways. The morphological change is under molecular control of reorganisation of acting microfilanents. Changes in conformation, induction of chemotaxis, and bacterial clearance are the immune consequences under the control of IL 8.

Invertebrates probably lack immunoglobulin like molucules. Body fluid of various invertebrates contain a variety of defence molecules with humoral characteristics. Humoral factors reported in the hemolymph includes agglutinins, lysozyme and several types of bactericidines of unknown nature. External physicochemical barriers, which provides the first line of defence in invertebrates may be breached under firm challenge from invader organism. Under this condition, the micro-organisms are subjected to various forms of cellular defence and humoral defence reactions. Cellular immune response involved the recruitment of immunocompetent hemocytes capable of performing phagocytosis, generation of reactive oxygen intermediates, encapsulation reaction etc. Humoral defence reaction is attributed by 2 forms of humoral factors *i.e.*, naturally occurring and inducible form of defence factors. In molluscs, several naturally occurring humoral factors are in record. Phenoloxidase pathway of immune response is under guidance of factors like microbial agents, agglutinin of hemolymph, divalent cations and pH which leads to activation of effective mediators of immunity.

References

Anderson, R.S. (1981). Effects of carcinogenic and non-carcinogenic environmental pollutants on immunological function in marine invertebrates. In: *Phylogenetic approach to cancer* (C. J. Dawe, J.C. Harshbarger, S. Konda, T. Sagimura and S. Takayama eds), pp. 319-331. Tokyo: Japan Sci. Soc.

Anderson, R.S., Oliver, L.M. and Brubacher, L.L. (1992). Superoxide anion generation by *Crassostrea virginica* hemocytes as measured by nitroblue tetrazolium reduction. *J. Invertebr. Pathol*. 591: 303-

307.

Anderson, R.S., Paynter, K.T., Burreson E.M. (1992). Increased reactive oxygen intermediate production by hemocytes withdrawn from *Crassostrea virginica* infected with *Perkinsus marinus*. *Biol. Bull.* 183: 476-481.

Anderson, R.S. (1993). Modulation of non-specific immunity by environmental stressors. In: *Advances in fisheries science, pathobiology of marine and estuarine organisms* (J.A. Couch and J.W. Fournie, eds, pp. 483-510. London: CRC Press.

Beck, G., and Habicht, G.S. (1991). Primitive cytokines: Harbingers of vertebrate defense. *Immunol. Today*, 12(6): 180-183.

Brookman, J.L., Ratcliffe, N.A., and Rowley, A.F. (1989). Studies on the activation of prophenoloxidase system of insect by bacterial cell wall components. *Insect Biochem*. 19: 47-57.

Coles, J.A., Farley, S.R., and Pipe, R.K. (1994). Effects of fluoranthene on the immunocompetence of the common marine mussel *Mytilus edulis. Aquat. Toxicol*. 30: 367-379.

Coles, J.A, and Pipe, R.E. (1994). Phenoloxidase activity in the hemoclymph and hemocytes of marine mussels *Mytilus edulis. Fish and Shellfish Immunol*. 4: 337-352.

Coles, J.A., Farley, S.R. and Pipe, R.K. (1995). Alternation of immune response of the common marine mussel *Mytilus edulis* resulting form exposure to cadmium. *Dis. Aquat. Organisms* 22: 59-65.

Dikkeboom, R. Tijnagel, J.M.G.H., Mulder, E.C., and van der Knaap, W.P.W. (1987). Hemocytes of the pond snail *Lymnea stagnalis* generate reactive forms of oxygen. J. *Invertebr. Pathol.* 49: 321-331.

Dikkeboom, R., van der Knaap, W.P.M. van der Bovenkamp, W., Tijnagel, J.M.G.H., and Bayne, C.J. (1988). The production of toxic oxygen metabolites by hemocytes of different snail species. *Dev. Comp. Immunol.* 12: 509-520.

Gifford. R.H., and Malawista, S.E. (1970). A simple rapid micro-method for detecting chronic granulomatous disease of childhood. *J. Lab. Clin. Med.* 75: 511-519.

Gifford, R.H., and Malawista, S.E. (1972). The nitroblue tetrazolium reaction in human granulocytes adherent to a surface. *Yale J. Biol. Med.* 45: 119-132.

Leonard, C., Soderhall, K and Ratcliffe, N.A. (1985). Studies on prophenoloxidase and protease activity of *Blaberus cranifer* hemocytes. *Insect Biochem*. 15: 803-810.

Ottaviani, E., Franchini, A., and Franceschi., C. (1993). Presence of several cytokine-like molecules in molluscan hemocytes. *Biochem. Biophy. Res. Com.* 195(2): 984-988.

Ottaviani, E., Franchini, A., Malagoli, D., and Genedini, S. (2000). Immunomodulation by recombinant human IL-8 and its signal transduction pathways in invertebrate hemocytes. *Cell. Mol. Life. Sci.* 57: 506-513

Pipe, R.K. (1990). Hydrolytic enzymes associated with the granular hemocytes of the marine musel *Mytilus edulis. Histochem.J.* 22: 595-603.

Pipe, R.K. Porte, C., and Livingstone, D.R. (1993). Antioxidant enzymes associated with the blood cells and hemolymph of the mussel *Mytilus edulis. Fish and Shellfish. Immunol* 3: 221-223.

Pipe, R.K., and Coles, J.A. (1995). Environmental contaminants influencing immune functioning in marine bivalve molluscs. *Fish and Shellfish Immunol*. 5: 581-595.

Pipe, R.K., Coles, J.A., Thomas, M.E., Fassato, V.V. and Pulsford, A.L. (1995). Evidence for environmentally derived immunomodulation in mussels form the Venice Lagoon. *Aquat. Toxical*. 32: 59-73.

Ray, S., and Chattopadhyay, S. (1998). Alteration of aggregation behaviours of hemocytes of *Bellamya bengalensis* exposed to methylparathion. *Proc. World. Cong. Malacol.* Smithsonian Institution. Washington D.C. USA: 271 pp.

Russo, J and Logadic, L. (2000). Effects of parasitism and pesticide exposure on characteristics and functions of hemocytes population in the fresh water snail *Lymnea palusris Cell. Biol. Toxicol*. 16(1): 15-30.

Sami, S., Faisal, M. and Huggett, R.J. (1992). Alteration in cytometric characteristics of hemocytes form the American oyster *Crassostrea virginica* exposed to polycyclic aromatic hydrocarbon (PAH) contaminated environment. *Mar. Biol*. 113: 247-252.

Sami, S., Faisal, M and Huggett, R.J. (1993). Effects of laboratory exposure to sediments contaminated with polycyclic aromatic hydrocarbons on the hemocytes of the American oyster *Crassostrea virginica. Mar. Environ. Res*. 35 131-135.

Smith, V.J., and Soderhall, K. (1991). A comparison of phenoloxidase activity in the blood of marine invertebrates. *Dev. Comp. Immunol*. 15: 251-261.

Soderhall, K. (1982). The porphenoloxidase activating system and melanisation: A recognition phenomena in arthropods? A review. *Dev. Comp. Immunol*. 6: 601-611.

Takahashasi, K., Akaike, T., Sato, K., Mori, K. and Maeda, H. (1993). Superoxide anion generation by pacific oyster (*Crassostrea gigas*) hemocytes: Identification by electron spin resonance spin trapping and chemiluminescense analysis. *Comp. Biochem. Physiol*. 105B (1): 35-41.

25

Range of Extension of Cat Fish *Batasio travancoria* Hora and Law to River Thunga in Karnataka Part of Western Ghats, India

Arunachalam, M., A. Manimekalan, A. Johnson, R. Soranam, A. Sankaranarayanan and P. Sivakumar

S.P.K. Centre for Environmental Sciences, Manonmaniam Sundaranar University, Alwarkurichi -627 412, Tamil Nadu

Introduction

During an intensive study in Kudremukh Wildlife Sanctuary as a part of National Agricultural Technology Programme on "Germplasm inventory, evaluation and gene banking of freshwater fishes of stream/rivers of peninsular India" two specimens of cat fish *Batasio travancoria* Hora and Law was collected from Khuremuk Wildlife Sanctuary, Karnataka. This species is commonly called as Travancore batasio. Hora and Law (1941) originally described this species from Perunteraruvi, a tributary of Pamba river at Edakadathy, Kerala. Later Silas (1951) reported this species from Anamalai Hills, Jayaram *et al.* (1976) reported from Cardamom and Agastya hills of the Western Ghats and Raghunathan (1989) reported from Coorg district, Karnataka. Talwar and Jhingran (1991) reported the distribution of this species upto Kerala part of Ghats Western. Presently this species was collected from Vimalanathi at Munsar in Thunga river basin, Kudremukh National Park.

Many workers were studied the fish fauna of river Thunga and Bhadra (Chacko and Kuriyan, 1948, Chacko and Venkataraman, 1945, Chacko, 1948, Day, 1868, Hora and Misra, 1940, Hora, 1937, Rahimullah, 1943, Rao, 1920, Anuradha, 2000, Chandrashekhariah *et al.*, 2000). But this species was not reported by any of these workers from river Thunga and Bhadra. So the present collection shows the range of extension to the river Thunga river basin.

Diagnosis

D I/7, A iii-iv 9-11, P I 7-9, Vi 5

Table 25.1: Morphometric Characteristic of *Batasio travancoria* Hora and Law

Sl.no.	Morphometric Character	Measurements (in mm)		Proportion to SL	Mean	SD	Proportion to HL	Mean	SD
		1	2	Range			Range	n	
1.	Total Length	118.00	106.00						
2.	Standard Length	98.50	90.00						
3.	Body Depth	16.09	14.89	6.12 - 6.04	6.08	0.05	1.62 - 1.53	1.58	0.07
4.	Head Length	26.11	22.78	3.77 - 3.95	3.86	0.13	1.00 - 1.00	1.00	0.00
5.	Snout Length	12.65	11.44	7.79 - 7.87	7.83	0.06	2.06 - 1.99	2.03	0.05
6.	Mouth Width	5.67	5.13	17.37 - 17.54	17.46	0.12	4.60 - 4.44	4.52	0.12
7.	Mouth Height	9.21	8.85	10.69 - 10.17	10.43	0.37	2.83 - 2.57	2.70	0.18
8.	Eye Diameter	5.10	4.10	19.31 - 21.95	20.63	1.86	5.12 - 5.56	5.34	0.31
9.	Inter Orbital Width	5.15	4.26	19.13 - 21.13	20.13	1.41	5.07 - 5.35	5.21	0.20
10.	Pre Dorsal Length	34.77	34.55	2.83 - 2.60	2.72	0.16	0.75 - 0.66	0.71	0.06
11.	Post Dorsal Length	56.23	53.62	1.75 - 1.68	1.72	0.05	0.46 - 0.42	0.44	0.03
12.	Pre Pelvic Length	46.54	42.89	2.12 - 2.10	2.11	0.01	0.56 - 0.53	0.55	0.02
13.	Distance Between Pectoral and Pelvic	24.58	21.85	4.01 - 4.12	4.06	0.08	1.06 - 1.04	1.05	0.01
14.	Distance Between Pectoral and Vent	31.67	28.78	3.11 - 3.13	3.12	0.01	0.82 - 0.79	0.81	0.02
15.	Distance Between Pelvic and Anal Fin	19.10	17.91	5.16 - 5.03	5.09	0.09	1.37 - 1.27	1.32	0.07
16	Distance Between Pelvic Fin and Vent	15.10	7.09	6.52 - 12.69	9.61	4.36	1.73 - 3.21	2.47	1.05
17.	Distance Between Anal Fin and Vent	15.15	13.26	6.50 - 6.79	6.64	0.20	1.72 - 1.72	1.72	0.00
18.	Dorsal Fin Height	13.58	9.66	7.25 - 9.32	8.29	1.46	1.92 - 2.36	2.14	0.31
19.	Pectoral Fin Length	16.23	13.32	6.07 - 6.76	6.41	0.49	1.61 - 1.71	1.66	0.07
20.	Pelvic Fin Length	15.90	11.92	6.19 - 7.55	6.87	0.96	1.64 - 1.91	1.78	0.19
21.	Anal Fin Height	11.84	6.30	8.32 - 14.29	11.30	4.22	2.21 - 3.62	2.91	1.00
22.	Anal Fin Length	22.48	17.82	4.38 - 5.05	4.72	0.47	1.16 - 1.28	1.22	0.08
23.	Caudal Peduncle Length	16.08	14.28	6.13 - 6.30	6.21	0.13	1.62 - 1.60	1.61	0.02
24.	Caudal Peduncle Depth	6.33	6.12	15.56 - 14.71	15.13	0.60	4.12 - 3.72	3.92	0.28
25.	Caudal Fin Length	22.00	16.00	4.48 - 5.63	5.05	0.81	1.19 - 1.42	1.31	0.17

Body elongate, head globular, occipital process not reaching the basal bone of dorsal fin, 8 barbels, dorsal spine weak, caudal fin deeply forked, colour-greyish with a narrow lateral streak along the lateral sides. Morphometric characters are given in Table 25.1.

Ecology

This species was collected from bank undercut with more tree roots, instream cover, with good canopy cover. Substrate type was muddy with detritus, with low flow. Habitat factors are given in Table 25.2.

Table 25.2: Habitat Characteristics of Vimalanathi (Munsar)

Habitat Factors	
Altitude (m)	628
Position	13°19'21" N, 75°7'32.1"E
Stream Order	2
Gradient (per cent)	2
Mean Width (m)	14.2
Mean Depth (cm)	62.75
Flow (m/sec)	0.434
Riparian Cover (per cent)	60
Substrates (per cent)	
Bedrock	15
Boulder	30
Cobble	28.75
Gravel	13.75
Sand	7.5
Leaf Litter	5
Fine Sand	0

Acknowledgements

Senior author (M. Arunachalam) is grateful for the financial assistance from NATP under the mission mode programme of Germplasam Inventory and Gene banking of Freshwater Fishes (Sanction No. 27(281/98/NATP/MM-III 18 Dated 23-12-1999). We also thank the Mission Leader and Director Dr. D. Kapoor and Dr. S.P. Singh, Principal Investigator of the Lead Centre, National Bureau of Fish Genetic Resources, Lucknow for their leadership in this programme. We thank Sri Chakrabarti, Principal Chief Conservator of Forests, Karnataka for the permission and moral support.

References

Bhat, Anuradha (2000). Fish germplasm inventory of Sharavathi, Aghanashni, Bedti and Kali rivers, Uttara Kannada, pp. 148-151. In: Ponniah, A.G. and Gopalakrishnan, A. (Eds). *Endemic Fish Diversity of Western Ghats.* NBFGR-NATP Publication-1 347. p. National Bureau of Fish Genetic Resources, Lucknow, U.P., India.

Chacko, P.I. (1948). A survey of the fisheries of the Tungabhadra river. *Proc. Indian Acad. Sci.*, 28: 166-175.

Chacko, P.I., Kuriyan, G.K. (1948). A survey of the fisheries of the Tungabhadra river. *Proc. Indian Acad. Sci.*, 28(5): 165-176.

Chandrashekhariah, H.N., Rahman, M.F. and Lakshmi Raghavan, S. (2000). Status of fish fauna in Karnataka. pp. 98-135. In: Ponniah, A.G. and Gopalakrishnan, A. (Eds). *Endemic Fish Diversity of Western Ghats*. NBFGR-NATP Publication-1, 347. p. National Bureau of Fish Genetic Resources, Lucknow, U.P., India

Day, F (1868). Absernations on some Indian fishes. *Proc. Zool. Soc.*, 580-585.

Hora, S.L. (1937). Notes on fishes in the Indian Museum. XXVIII. On three collections of fish from Mysore and Coorg, South India. *Rec. Indian Mus.*, 9(1): 5-28.

Hora, S.L. (1937). The game fishes of India II. The Cachhwa or Butchwa, *Eutropiichthys vacha* (Hamilton). *Bambay Nat. Hist. Soc.*, 39(3): 431-446.

Hora, S.L. and Law, N.C. (1941). The freshwater fish of Travancore. *Rec. Indian Mus.*, 43(2): 233-256.

Hora, S.L. and Misra, K.S. (1940). On fishes of the genus Rohtee Sykes. *Rec. Indian Mus.*, 42(1): 155-172, 1 pl.

Jayaram, K.C., Indra, T.J. and Sunder Singh, M. (1976). On a collection of fish from the Cardamon Hills, South India. *Madras J. Fish.*, 7: 1-7.

Raghunathan, M.B. (1989). A study on the fish fauna of Coorg district, Karnataka. Fishery Tech., 26(1): 19-21.

Rahimullah, M. (1943). Fish survey of Hyderabad state Part 1. A preliminary report of fishes found in the Godavari, Purna, Kistna, Tungabhadra and Siddha rivers. *J. Bombay Nat. Hist. Soc.*, 43: 648-653.

Rao, C.R.N. (1920). Some new species of Cyprinoid fish from Mysore. *Ann. Mag. Hat. Hist.*, (9)6: 45-64.

Silas, E.G. (1951). On a collection of fish from the Anamalai and Nelliampathi hill ranges (Western Ghats) with notes on its zoogeographical significances. *J. Bombay Hat. Hist. Soc.*, 49(4): 670-681.

26

Relative Condition Factor (Kn) of the Bighead Carp, *Aristichthys nobilis* (Richardson)

B.K. Mahapatra and N.C. Datta*

Division of Fisheries, ICAR Research Complex for NEH Region,
Umroi Road, Barapani -793 103, Meghalaya
**Fishery and Ecology Research Unit, Department of Zoology,*
University of Calcutta, 35 Ballygunge Circular Road, Kolkata - 700 019

ABSTRACT

The analysis of length-weight relationship of *Aristichthys nobilis* reveals that cube law is not applicable to evaluate its well-being and hence, the relative condition factor (Kn) has been calculated from the data collected. The Kn value has been found to vary between 0.8492 -1.3300.

Key Words: Relative Condition Factor, Bighead Carp, *Aristichthys nobilis.*

Introduction

Assessment of mathematical relationship between length and weight of fish has wide practical applications in fishery biology. The relationship is also useful in evaluating the general condition or well-being of the fish species through the study of condition factor (K) or relative condition factor (Kn).

The condition factor or ponderal index and relative condition or relative condition factor or condition index or condition co-efficient of fish have been studied by a number of fishery biologists (Le Cren, 1951, Qasim, 1957, Bal and Jones, 1960, Qayyam and Qasim, 1964, Bhatt, 1968, Babu and Nair, 1983, De, 1986, Bandopadhyay and Datta, 1987). However, probably no such attempt has yet been made on bighead carp in a comprehensive manner. It is revealed from the study of length-weight relationship that this fish does not strictly follow the conventional Cube law (Mahapatra and Datta, 1998). Therefore, the present study of the relative condition factor of this carp in different size groups, sexes, seasons and total sample has been made since this species has immense cultural possibility.

Materials and Methods

The samples of *Aristichthys nobilis* were collected from the farm ponds of R.K. Ashram, K.V.K., Sundarbans (21°30′N to 23°N latitude and 88°E to 89°E longitude) during 1988 to 1993. A total of 358 fish were used for computation. By using the log length-weight relationship, the relative condition factor (Kn) was calculated as w/w^1 where w is the observed weight and w^1 is the calculated weight for the observed length.

The relative condition factor (Kn) was calculated after categorising the total sample into following groups:

1. Different length groups
2. Fingerling, male, female and combined group
3. Different length groups of females
4. Females of different seasons
5. Females of different months

Results

Different Length Groups

The relative condition factor in different length groups is presented in Fig. 26.1. The Kn values in different length groups are found to vary between 0.8492-1.3300. The lowest and highest Kn values have been recorded 0.8492 in 50 mm and 1.3300 in 150 mm.

Fingerling, Male, Female and Combined Group

The average Kn values of fingerlings, males, females, and combined group are shown in Fig. 26.2. The Kn values in sequence from higher to lower, are 1.0194, 1.0075, 1.0065 and 1.0020 in fingerlings, males, combined groups and females respectively.

Females of Different Length Groups

The Kn values of different length groups of female are shown in Fig. 26.3. The relative condition factor in different length groups is found to vary between 0.8874 and 1.1430. The highest value is in 510 mm and the lowest in 350 mm length groups.

Females of Different Seasons

The variation of Kn values in different seasons is shown in Fig. 26.4. It is found that the highest value (1.0137) was recorded in monsoon whereas the lowest (0.9770) in winter.

Females of Different Months

Monthly variation of Kn values of female is shown in Fig. 26.5. The values of relative condition factor varies between 0.8958 and 1.1037. The highest value (1.1037) is recorded in May and lowest (0.8958) in February.

Discussion

The gonad weight and feeding rate as influencing factor for the condition of fish have been emphasised by some previous workers (Le Cren, 1951, Bhatt, 1968, Babu and Nair, 1983, De, 1986). In *A. nobilis*, fluctuation of Kn values in between different length groups has been observed. In the Kn

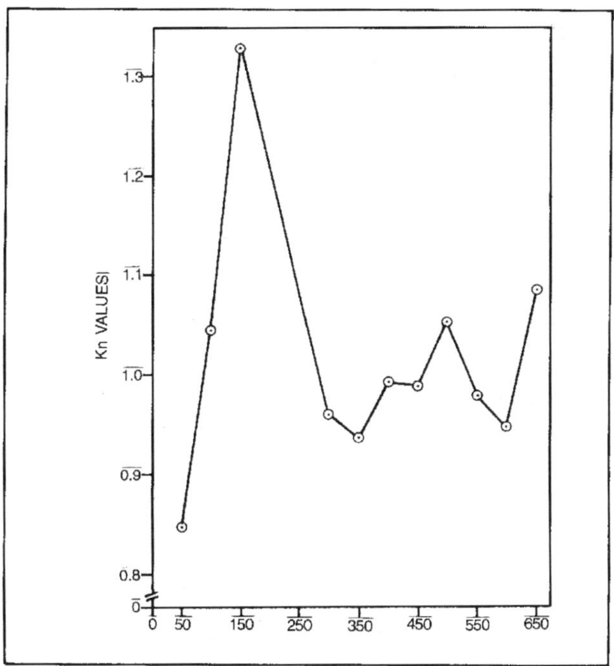

Fig. 26.1: Mean Kn Values in Different Length of *A. nobilis*.

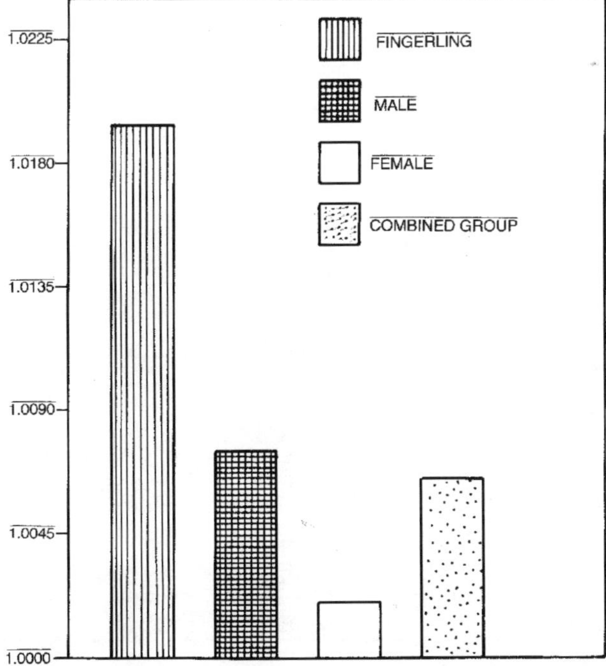

Fig. 26.2: Mean Kn Values in Fingerling, Male, Female and Combined Group of *A. nobilis*.

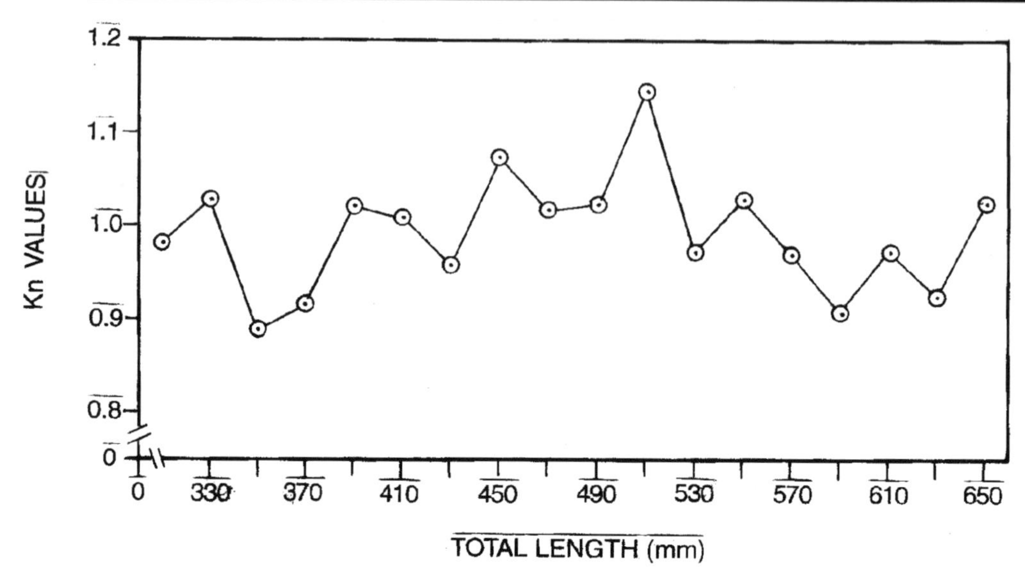

Fig. 26.3: Mean Kn Values in Different Length of *A. nobilis* (Female).

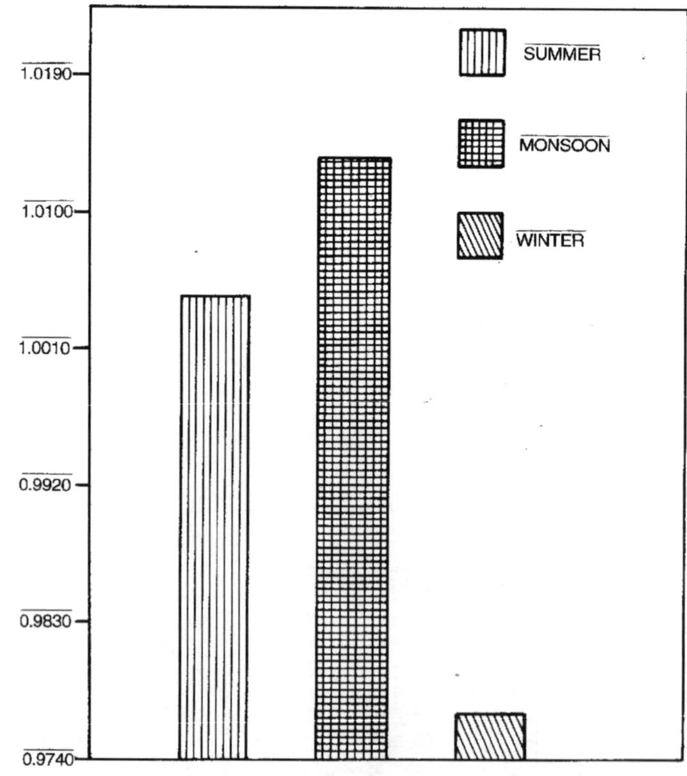

Fig. 26.4: Mean Kn Values in Three Different Seasons of *A. nobilis* (Female).

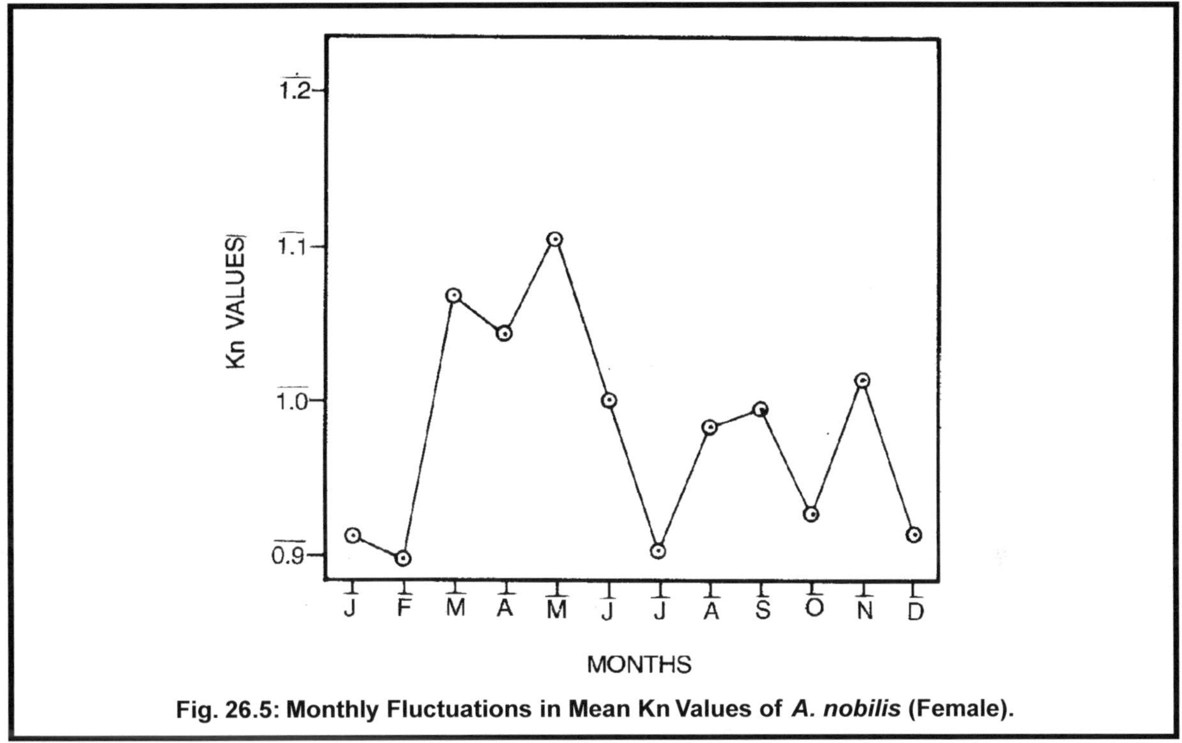

Fig. 26.5: Monthly Fluctuations in Mean Kn Values of *A. nobilis* (Female).

curve (Fig. 26.1) several peaks were observed including one at 150 mm and the second at 500 mm length groups. The first peak of the increase of the Kn value may be attributed to the breadth-wise growth. The second peak is due to the attainment of first maturity.

The Kn values among the fingerlings, males, females and combined groups (Fig. 26.2) clearly show that higher Kn values in case of fingerlings is probably due to breadth-wise growth and voracious feeding habit in that stage. Low mean Kn value was noticed in females which may be due to spawning strain and spent condition.

The Kn curve (Fig. 26.3) in different length of females shows several peaks but the highest peak was noticed in 500 mm length group which is probably due to attainment of first maturity.

Higher Kn values in different seasons (Fig. 26.4) were noticed in monsoon whereas low Kn values in winter. Higher Kn value is probably due to maturity of gonads and low Kn value due to spawning strain, spent condition and low feeding rate during winter.

The Kn curve (Fig. 26.5) in different months of female shows that highest Kn value in May is due to development of gonads as well as less strain and lowest in July possibly due to spent condition. Very likely metabolic strain due to spawning activities has not yet been regained.

Thus, it may be stated that feeding rate, physiological factors like maturity and spawning play an important role on the condition of *A. nobilis* which may be true in other fishes too.

Acknowledgement

The authors are greatly indebted to the authorities of Ramkrishna Ashram Krishi Vigyan Kendra, Nimpith for providing facilities to carry out the research work. The first author is also grateful to the Director, ICAR Research Complex for NEH Region for providing necessary facilities.

References

Babu, N. and Nair, N.B. (1983). Condition factor of *Amblypharyngodon chackaiensis* Babu and Nair in Chackai boat channel (Trivandrurn, Kerala, India). *Proc. Indian Acad. Sci., (Anim. Sci.),* 92(1): 1-10.

Bat, J.W. and Jones, J.W. (1960). On the growth of brown trout of Llyn Tegid. *Proc. Zool. Soc. Lond.,* 134: 1-41.

Bandopadhyay, B.K. and Datta, N.C. (1987). A study on the length-weight relationship and condition factor of *Scatophagus argus* (Linnaeus) from the brackish water impoundments of Hooghly-Matlah Estuary, West Bengal. Abstract, VIIth All India Seminar on Ichthyology, Visva-Bharati: 3.

Bhatt, V.S. (1968). Studies on the biology of some freshwater fishes. Part III: *Heteropneustes fossilis* (Bloch). *Indian J. Fish.,* 15: 99-115.

De, D.K. (1986). Studies on the food and feeding habit of *Hilsa ilisha* (Hamilton) of the Hooghly estuarine system and some aspects of its biology. Ph.D. Thesis, p. 285, University of Calcutta.

Le Cren, E.D. (1951). The length-weight relationship and seasonal cycle in gonad weight and condition factor in the perch *Perca fluviatilis. J. Anim. Ecol.,* 20(2): 201-219.

Mahapatra, B.K. and Datta, N.C. (1998). Length-weight relationship of bighead carp, *Aristichthys nobilis* (Richardson). *J. Inland Fish. Soc., India,* 30(1): 101-104.

Qasim, S.Z. (1957). The biology of *Blennius pholis* L. (Teleostei). *Prac. Zool. Soc. Lond.,* 128(2): 161-208.

Qayyam, A. and Qasim, S.Z. (1964). Studies on the biology of some freshwater fishes. Part I: *Ophiocephalus puntatus* (Bloch). *J. Bombay Nat. Hist. Soc.,* 61: 330- 347.

27

Ground Concentration of Conventional Pollutants from Major Industrial Sources of Visakhapatnam

A.S.N. Murty, N.S. Prasad and Krushna Chandra Gouda

Department of Marine Science, Berhampur University, Berhampur - 760 007

ABSTRACT

In the present article the application of a diffusion model for prediction of ground concentrations from industrial sources has been described. The industrial sources can be considered as point and stationary conditions and using power law profiles for wind speed and coefficient of turbulence, the diffusion equation has been solved which yields ground concentrations as a combination of series of Bessel functions, the solution has been simplified and working formulae for different stability conditions *viz.*, near neutral, stable and unstable conditions have been derived. This has been applied for Visakhapatnam which was divided into one square kilometer grid and the concentration from each individual major industrial sources has been estimated. The overall concentration was derived by graphical integration. The industrial concentrations for different seasons namely, winter, pre-monsoon, monsoon and post-monsoon has been presented.

Introduction

The concentration of pollutants that are thrown into the atmosphere depend upon the source strength, meteorological conditions and topography. Visakhapatnam (17°42′N and 82°18′ E) (Fig. 27.1) is a rapidly developing urban city situated on the east coast of India. It is situated between two hill ranges, the Kailasa in the north and the Yarada in the south and other small hills in west and north east. Thus the city is like a basin surrounded by hills on three sides and the Bay of Bengal in the east. This peculiar topography with its valleys and uplands further complicates the disposal of gaseous wastes released by major industries like Hindustan Polymers Ltd. (HPL), Hindustan Petrochemicals

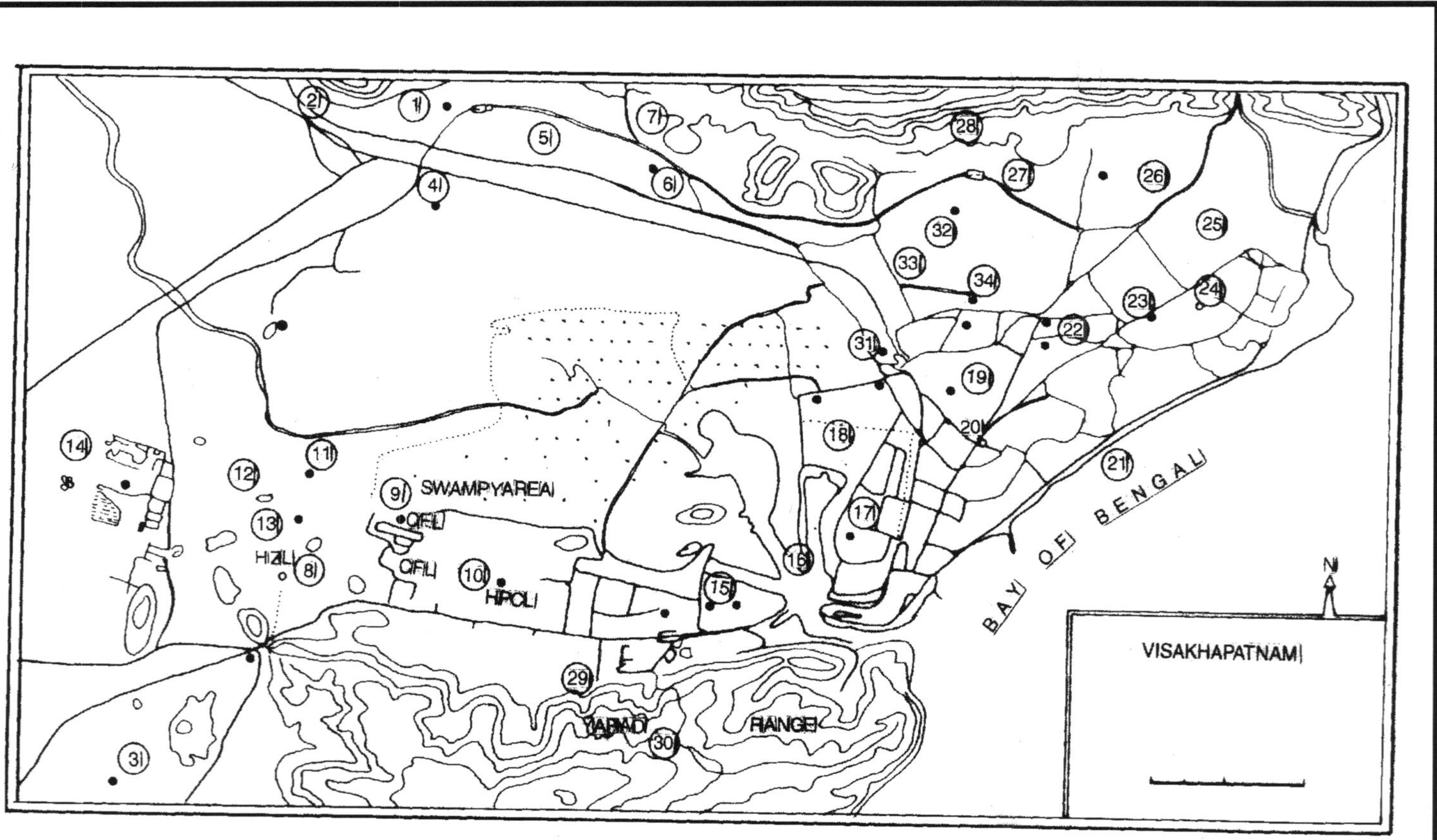

Fig. 27.1: Visakhapatnam City Map

(1) N.S.T.L., (2) Gopalapuram, (3) Steel Plant, (4) N.A.D., (5) Marripalem, (6) Industrial Estate, (7) Madhava Dhara, (8) Hindustan Zinc Ltd., (9) Coramandal Fertilizers, (10) Hindustan Petroleum Corporation Ltd., (11) Chukkavani Palem, (12) Mind, (13) Mulagada, (14) B.H.P.V., (15) Shipyard, (16) Harbour, (17) Port Hospital, (18) Conveyor Belt, (19) Dabagardens, (20) Yellamma Thota/Jagadamba, (21) R.K. Beach, (22) Ramnagar/Judge Court, (23) Andhra Univ./Eng. College, (24) China Waltair, (25) Peda Waltair, (26) Maddila Palem, (27) Sitamma Dhara, (28) Kailasa Range, (29) Nausena Baag, (30) Yarada Range, (31) Waltair (R.S.), (32) Akkayya Palem, (33) New Colony, (34) Asil Metta/RTC Complex.

Ltd. (HPCL), Hindustan Zinc Ltd. (HZL) and Coramandal Fertilizers Ltd. (CFL) and other many small industries.

The main gaseous pollutant coming out of the several industrial pollutants is sulphur dioxide whose concentration in the atmosphere is taken as an index of pollution. The wide spread respiratory diseases like bronchitis and asthma have been reported widely in Visakhapatnam in the recent years which are directly related to SO_2 concentration (Agarwal *et al.*, 1976).

So the present paper deals with the estimation of seasonal concentration of SO_2 pollution which is estimated with the help of a theoretical model developed by Prasad (1995).

Material and Methods

Using the diffusion equations for stationary conditions and using power law profiles for wind speed and coefficient or turbulence, the diffusion equation has been solved, which yields ground concentrations as a combination of series of Bessel functions (Laikhtman 1970 and Vitthal Murty 1972). The solution has been simplified and working formulae, as mentioned below for different stability conditions were developed.

$$Q = \frac{103M}{X^{2.1}U} \exp \frac{-23H^{0.89}}{X} \exp, ^{-y2} / 2y^{-2} \quad \text{(Unstable)}$$

$$Q = \frac{57M}{X^{1.9}U} \exp \frac{-15H^{1.18}}{X} \exp, ^{-y2} / 2y^{-2} \quad \text{(Stable)}$$

$$Q = \frac{68M}{X^{2}U} \exp \frac{-17H^{1.04}}{X} \exp, ^{-y2} / 2y^{-2} \quad \text{(Neutral)}$$

Where Q is the concentration of gaseous pollutant (g/m^3) averaged for ten minutes,
U is the velocity at normal height (m/s),
Y is the distance perpendicular to wind (m),
X is the down wind distance from the physical stack height (m), and
M is the emission rate (g/s).

As the accurate estimation of pollution concentration depends on effective stack height H, ($H = h + \Delta h$) which is dependent on plume rise (Δh), it is essential to estimate the plume rise accurately. This is estimated using the Holland's plume rise formula here.

The entire city was divided into number of grids of one square kilometer and the concentration at the grid points from each industrial source was calculated using the stack data and meteorological data in winter (January), Pre-monsoon (April), Monsoon (August), and post-monsoon (October). Finally the total concentration of all the three industries (HZL, HPCL and CFL) were summed up and plotted in the city map and presented in the Figs. 27.2 through 27.13.

Results and Discussion

The total ground concentrations in four different seasons and for different stability conditions were presented in Figs. 27.2 through 27.13. While the maximum concentration is at the centre

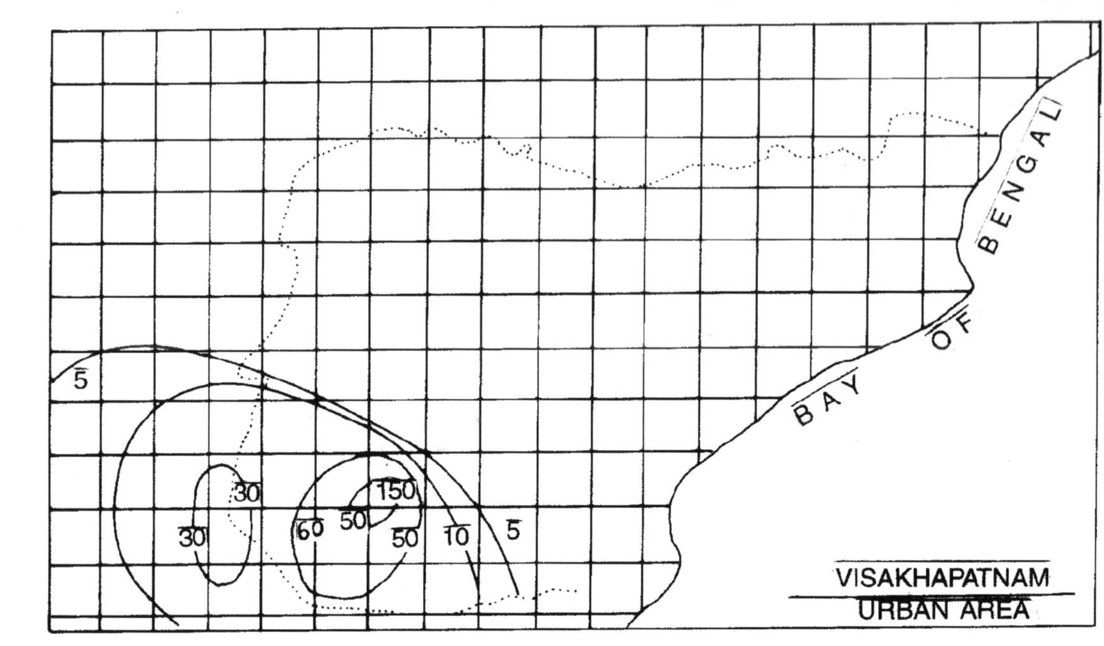

Fig. 27.2: Jan.-Total-Unstable.

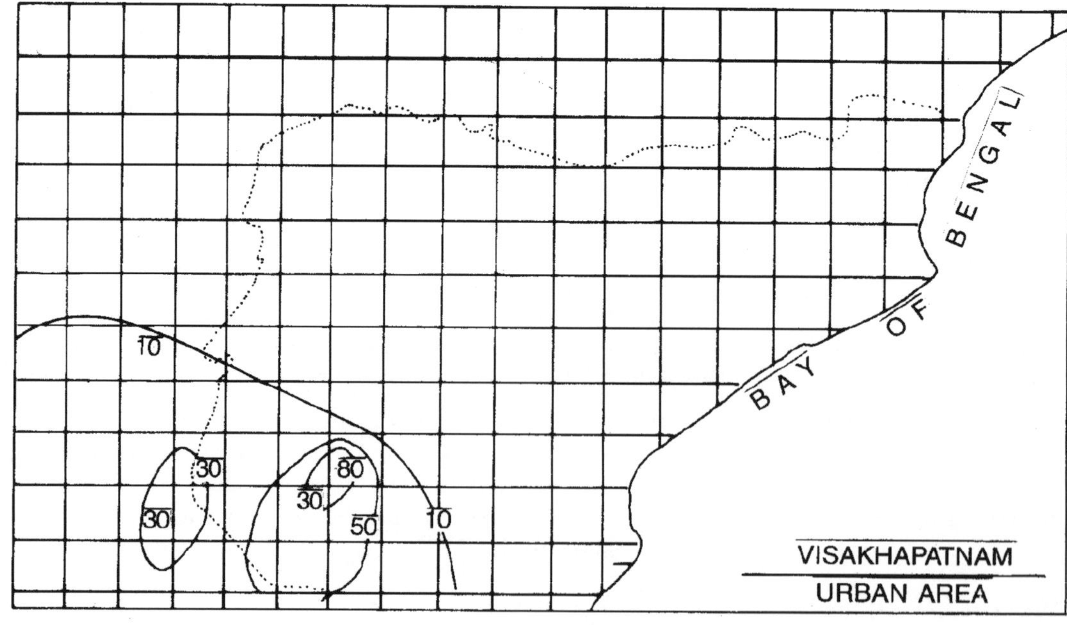

Fig. 27.3: Jan.-Total-Neutral.

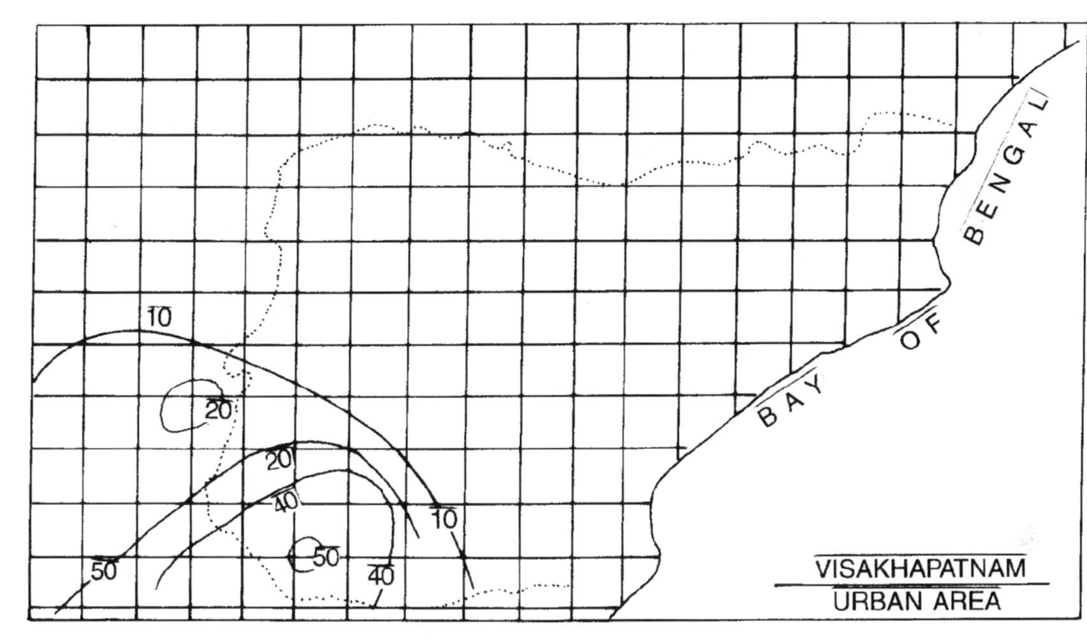

Fig. 27.4: Jan.-Total-Stable.

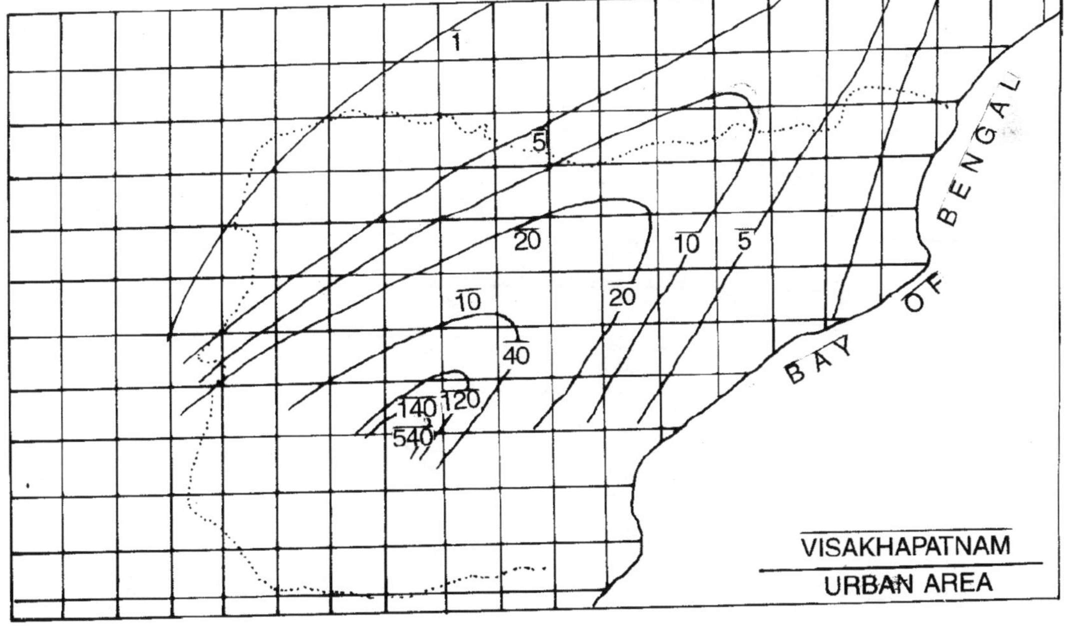

Fig. 27.5: Apr.-Total-Unstable.

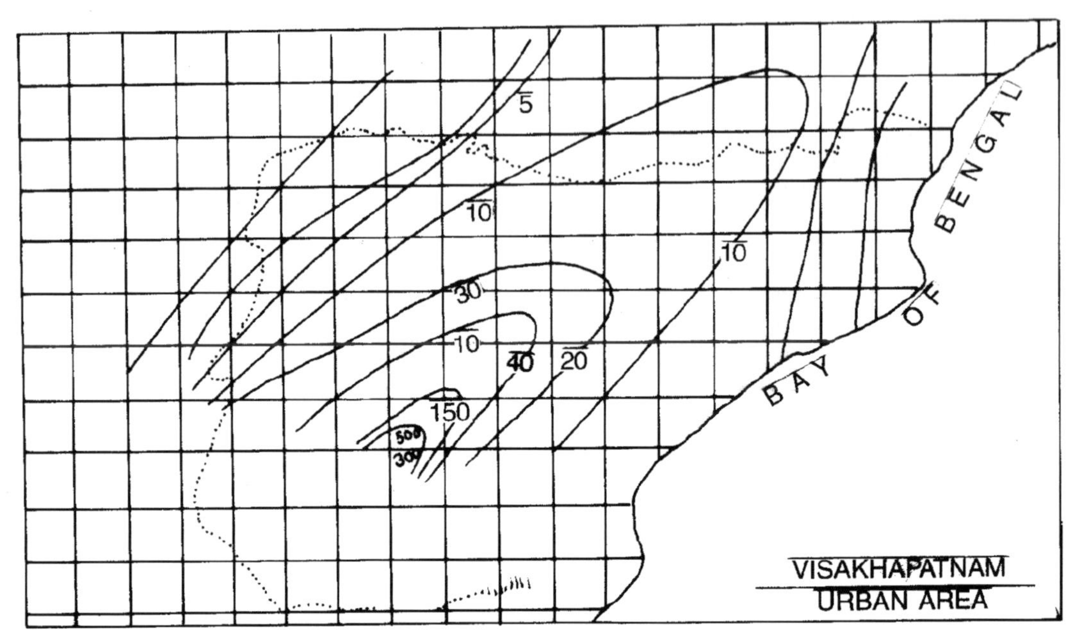

Fig. 27.6: Apr.-Total-Neutral.

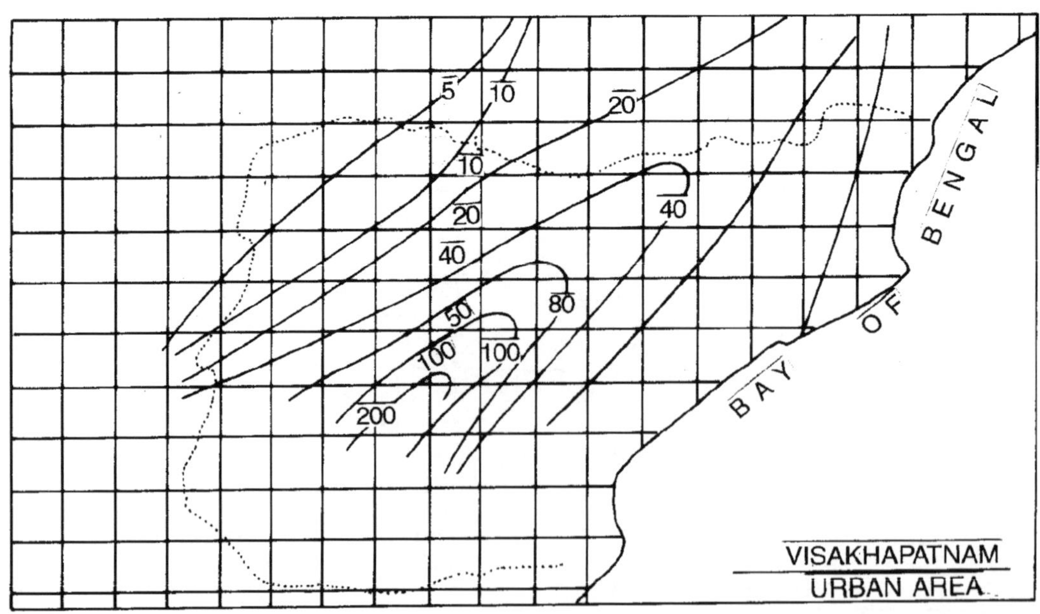

Fig. 27.7: Apr.-Total-Stable.

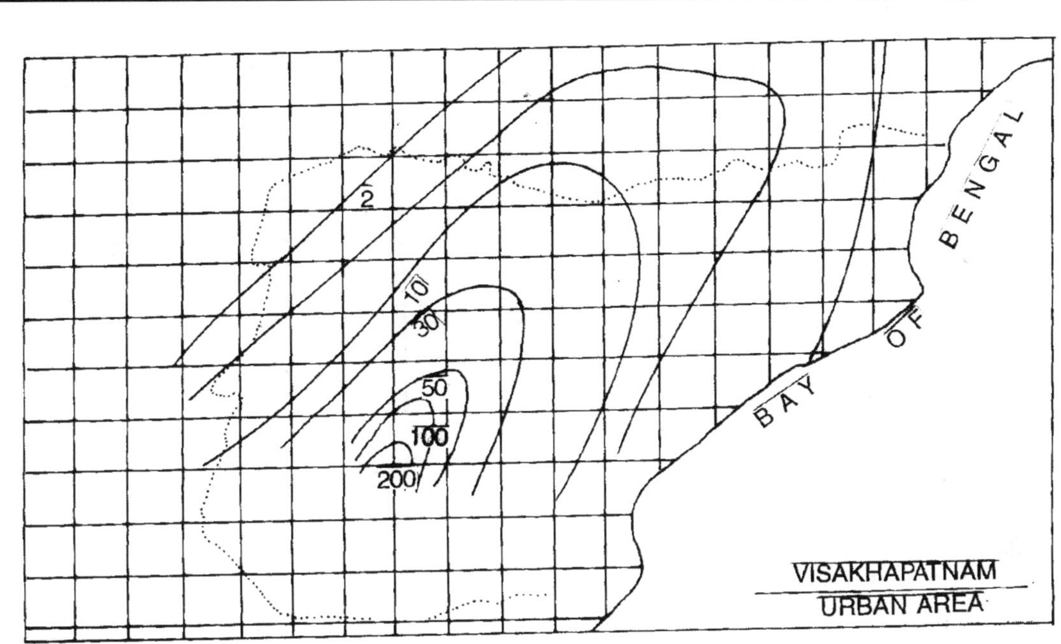

Fig. 27.8: Aug.-Total-Unstable.

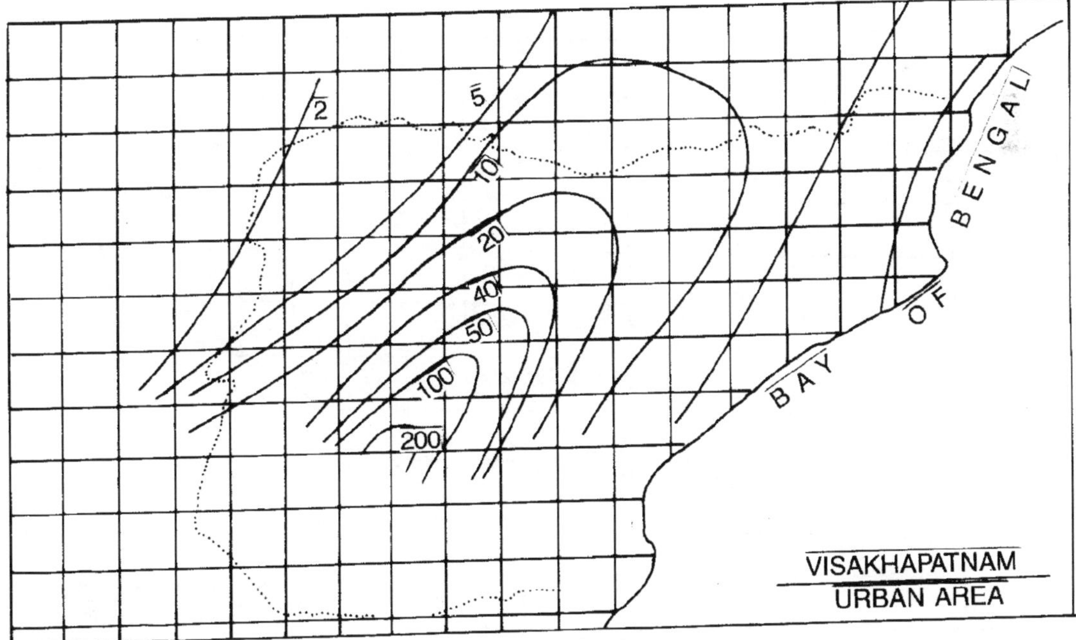

Fig. 27.9: Aug.-Total-Neutral.

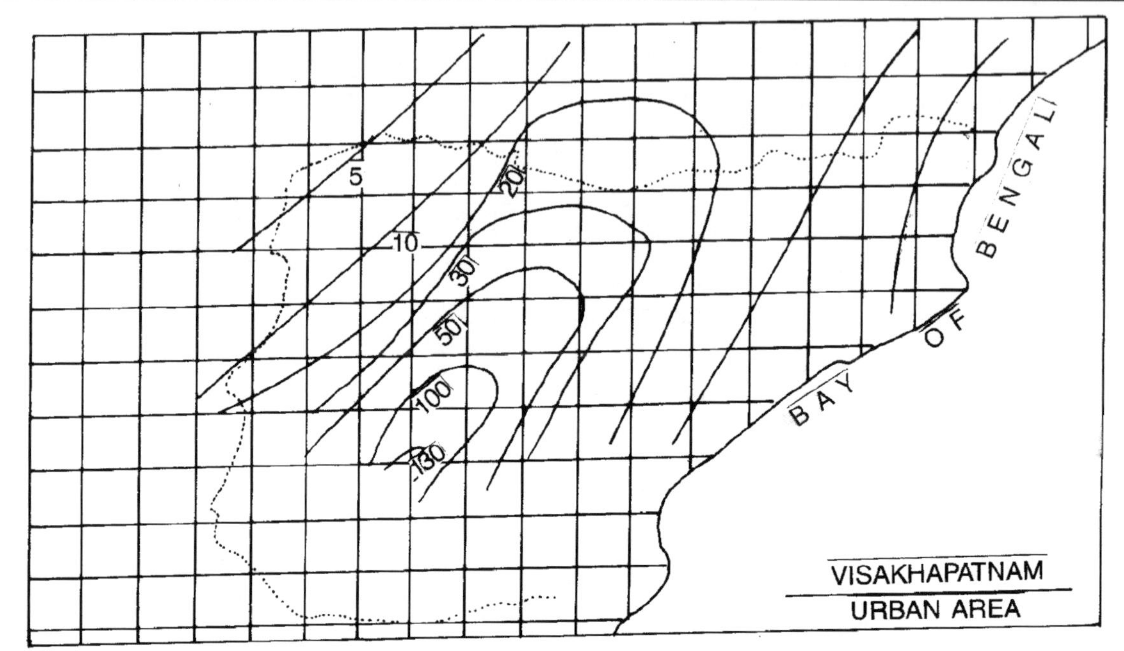

Fig. 27.10: Aug.-Total-Stable.

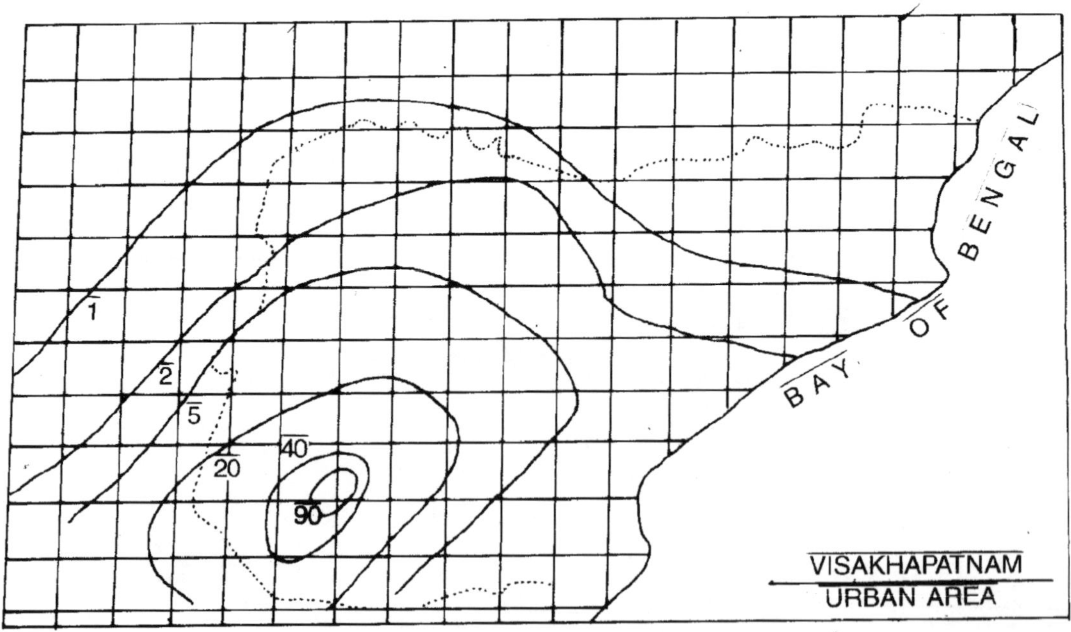

Fig. 27.11: Oct.-Total-Unstable.

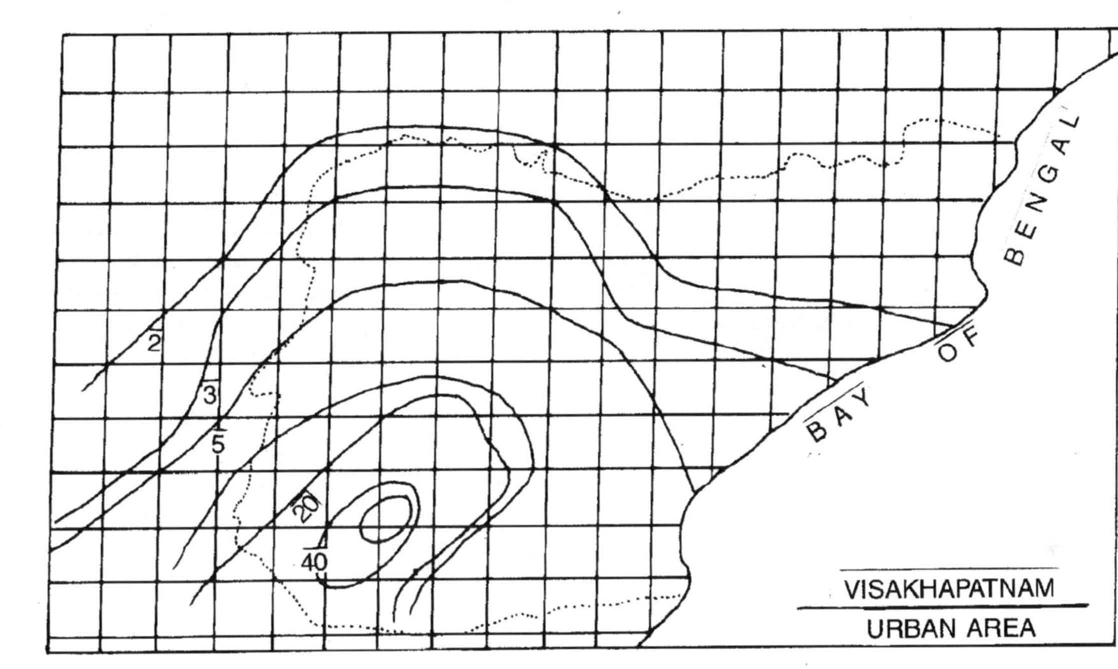

Fig. 27.12: Oct.-Total-Neutral.

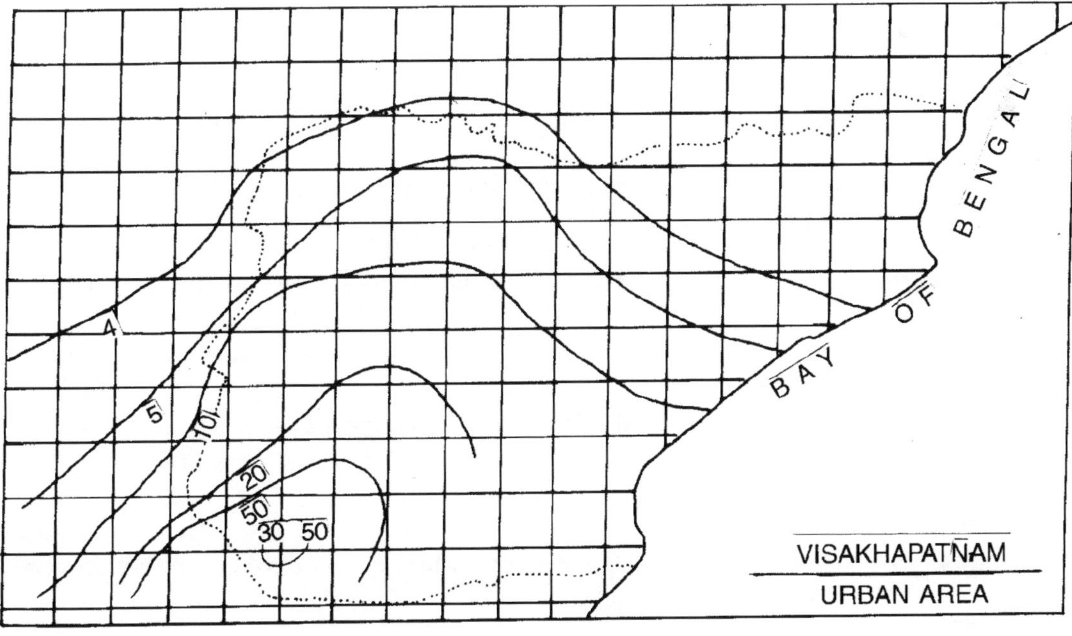

Fig. 27.13: Oct.-Total-Stable.

surrounding the industries, it decreases away from it. The gradient of concentration shows a marked decrease in identifying the maximum and minimum affected areas.

January

In this season (Winter) a maximum concentration of 150 μg/m^3, 80 and 50 μg/m^3 for unstable, neutral and stable conditions respectively were observed (Figs. 27.2-27.4). The spreading of concentration is predominant from SSE to WSW through SW. The residential areas near these industries like Chukkavanipalem, Srihari Puram, Gajuwaka, Malkapuram and Mindi are worst affected in this season. Further, these pollutants are collected and stagnated in the residential areas near the foot hills of Yarada ranges. Also when the concentrations of different industries are analysed individually it is found that in January SO_2 concentration is more due to HPCL in comparison to that of HZL and CFL.

April

In this pre-monsoon season (Figs. 27.5-27.7) the pollution spread is very much isolated in the North-East region and the rest of the area is not much affected. As the prevailing winds are West and South-West, this type of spreading is expected in this season. The central maximum concentrations in this season in different stability conditions of unstable, neutral and stability respectively are 548, 335 and 223 μg/m^3, respectively.

The reason for a stream line transport and concentration in North-East residential sector is due to channelling effect of the two hill ranges of Kailasa and Yarada in the North and South of the city.

August

Like pre-monsoon season (Figs. 27.8-27.10) the pollution spread is more in North-East sector due to South-West monsoon winds. The maximum values of 322, 225 and 130 μg/m^3 are the respective concentrations, in three different stability conditions. However, due to rainfall some pollutants are likely to be washed out. So the concentration levels are less compared to pre-monsoon season.

October

During post monsoon season winds are not in any single direction. So only hills are controlling and spreading of pollutants. So in this season the spread is mainly South to North with maximum concentrations 97, 63 and 51 μg/m^3, in different stabilities respectively.

Conclusion

1. It is clearly evident that North-East sector during premonsoon and monsoon seasons is the worst affected.
2. The pollution concentrations are relatively less in North-West sector and so it is best suited for residential development.
3. Future industries may be advised to be set up on the North-East beach areas rather than the present existing industrial areas of Gajuwaka and towards Anakapalli.

Acknowledgment

The authors wish to thank the Head, Department of Marine Sciences, Berhampur University for providing necessary facilities and encouragement and CSIR.

References

Aggarwal, A.L., Patel, T.S., Vitthala Murty K.P.R. (1976). Incidence of bronchitis and asthma as a function of ambient pollution of sulphur oxides and bioclimatic parameters: a case study of industrial population of Ahmedabad. *Indian J. Ecol.*, 3: 1-6.

Laikhtman, D.L. (1970), Physika programitshnova sloya (in Russian), Hydrometrological publications, Leningrad, 288-290.

Prasad, N.S. (1995). Theoretical studies of air pollution with special reference to Vishakhapatnam: a meteorological approach, Ph.D. thesis awarded by the Berhampur University, 206 p.

Vitthala Murty, K.P.R. (1972). Air pollution studies over Ahmedabad. *Proceedings of the seminar on "Environmental health problems of Ahmedabad"*. Ahmedabad D 3-1 to D3-9.

28

Disease in Aqua-farming can be Challenged Through Probiotics: A New Era

Bidhan C. Patra and Partha Bandyopadhyay

Aquaculture Research Unit, Department of Zoology, Vidyasagar University Midnapore - 721 102 (West Bengal)

ABSTRACT

Probiotics are the live microbial feed supplement, which beneficially affects the host animal nutrition and health by improving its intestinal microbial balance. For sustainable aquaculture in all types of aquatic area, the probiotics will help to improve growth, immunity and may also to be disease resistant. The inoculate may be from external probiotic like *Pseudomonas, Lactobacillus* and Yeast etc or from intestinal microflora. The application of probiotics in aquaculture, starting from growth promotion to disease resistance and larvae to adult shows promising results but more considerable attempt and dependable research is needed by using suitable immunological and molecular probes.

Key Words: Probiotics, Growth promotion, Disease resistance, Intestinal microflora.

Introduction

'Pro' means favour and 'bios' means life. "*Probiotics*" literally means "for life" (Abraham and Mohna, 2000) and an antonym to antibiotics, which involves multiplying few useful microbes to compete with the harmful ones, thus suppressing their growth (Mishra *et al.*, 2001). Antibiotics have been used profusely as therapeutic agents and as growth promoter from the year 1950 onwards. However, the excessive use of these antibiotics in the production system has led to the development of resistant to bacterial pathogens, which lead to the problem in the effective use of antibiotics as therapeutics. In the year 1969, Swan Committee restricted a list of antibiotics used in treatment of disease. The therapeutic use of antibiotics result in intestinal disorders. This is believed to be due to the different effects of antibiotics, which tends to remove indigenous gut flora along with disease causing agents. This situation is avoided by an alternative situation in the production system through the use of beneficial bacteria to fight against pathogenic bacteria *i.e.,* probiotics that is an acceptable practice in aquaculture. The health of the animal thus improved by the elimination of pathogens or at least minimizing the effects, which benefits aquaculture.

The first application of probiotics in aquaculture seems relatively recent (Kozasa, 1986), but the interest in such environment friendly treatment is increasing rapidly. Scientific evaluation corroborated seldom the first empirical trials, and the information was mainly spread by "gray literature" (Gatesoupe, 1999). However, a growing number of scientific papers have dealt except with probiotics, and it is now possible to survey the state of the art, from the empirical use to the scientific approach.

Definition of Probiotics

Biological control has been described as the utilization of natural enemies to reduce the damage caused by noxious organisms to tolerable levels (Debach and Rosen, 1991) or more precisely, the control or regulation of pest populations by natural enemies (Smith, 1919). Strictly speaking, a probiotic ought not be classified as a biological agent, since a probiotic micro-organism doesn't necessarily attack the noxious agent (pathogen).

Elie Metchnikoff's work at the beginning of this century is regarded as the first research conducted on probiotics (Fuller, 1992). He described them as "microbes ingested with the aim of promoting good health". This same definition was modified to "organisms and substances which contribute to intestinal microbial balance" (Parker, 1974) and latter by Fuller (1989), to "a live microbial feed supplement which beneficially affects the host animal by improving its intestinal microbial balance". Gibson and Roberfroid (1995) introduce the concept of 'prebiotics'. It is defined as non-digestable food ingredients that beneficially affect the host by selectively stimulating the growth and the activity of one or a limited number of bacteria in the colon, and thus improves host health. Ruiz *et al.* (1998) defined it as beneficial micro-organisms which can protect organisms against pathogens or enhance their growth.

Probiotics are now also being used in aquaculture and therefore, the definition may have to be modified. In aquatic animals, not only the digestive tract is important but also the surrounding water. Jory (1998) defined probiotics from aquaculture point of view as culture (single or mixed) of selected strains of bacteria that are used in culture and production systems (tanks, ponds and others) to modify or manipulate the microbial communities in water and sediment reduce or eliminate selected pathogenic species of micro-organisms and generally improve growth and survival of the targeted species. Gatesoupe (1999) defines probiotics as "microbial cells that are administered in such a way as to enter the gastrointestinal tract and to be kept alive, with the aim of improving health". Gram *et al.* (1999) broadened the definition by removing the restriction to the improvement of the intestine: "a live microbial supplement which beneficially affects the host animals by improving its microbial balance". Neither should probiotics be classified as growth promoters, since their action is not confined to improve growth but is associated with a general improvement in health.

Probiotics for Sustainable Aquaculture

Microbes are very important and have critical roles in aquaculture systems, including shrimp farming at both the hatchery and the grow out level, because water quality and disease control are directly related and closely affected by microbial activity. Probiotic protection can be due to different mechanisms, such as nutritional competition or production of antibacterial substances. In aquaculture, the current trend is to replace antibiotics by probiotics for ecological considerations as well. The transience of aquatic microbes may legitimate the extension of the probiotic concept to living microbial preparations used to treat aquaculture ponds. Moirarty, 1998 proposed to extend as "water additives" but Tannock (1997) described this extension would make too vague.

Improved hygiene and bio-security, development of probiotics and immune-stimulants, and improvement of artificial feeds promise better post larval fitness while reducing cost of reliability of production (Borwdy, 1998).

The probiotic treatments may be considered as methods of biological control, the so-called 'bio-control' that termed the limitation or the elimination of pests by the introduction of adverse organisms, likes parasites or specific pathogens. Maeda *et al.*, 1997 proposed to designate as bio-control, the methods of treatment using "the antagonism among microbes through which pathogen can be killed or reduced in number in the aquaculture environment". In this acceptation, the pond treatment proposed by Moriarty (1998) is indisputably relevant to bio-control as well as many other microbial treatments, including those whose target organisms are not animals, but micro algae.

During growth of fish spawn and prawn larvae in aquaculture, the first food supplied is usually the rotifer: *Brachionus plicatilis* and algae are commonly included in the system as food for the rotifers, thereby maintaining their nutrient quality. The addition of lactic acid bacteria to rotifers fed to turbot larvae, *Scophthalmus maximus*, was found to improve growth (Gatesoupe, 1991a) and increase resistance against pathogenic *Vibrios*, (Gatesoupe, 1994). The bacterial species composition was different in the pond upon use of probiotics that demonstrated the possibilities to change bacterial species composition and improve prawn production in large water bodies (Moriarty, 1998). Nutrients such as nitrate, phosphate and silicate levels are also found to be more in *Penaeus indicus* experimental tanks compare to control on provision of probiotics (Ravi *et al.*, 1998).

Aquatic animals are quite different from the land animals for which the probiotic concept was developed, and a preliminary question is the pertinence of probitic application to aquaculture. As fish are poikilothermic, their gut flora is temperature dependent, *i.e.*, there is an increase in total counts and in the proportion of anaerobic forms with temperature. The state-of-the-art concerning probiotics is not as advanced in fish as it is in homeothermic vertebrates, and further studies are thus required in the field to promote the environmental friendly development of fish culture (Mishra *et al.*, 2001).

Application of Probiotics in Aquaculture

Generally, the best approach for screening of a potential probiotic is to isolate the microbe from the culture and enhance its natural effects favouring their growth by adding some nutrients or distributing large quantities of them in the culture. Interesting results have already been obtained in culture of fin fishes, shrimps, crabs and bivalves by using probiotics are discussing below:

Probiotics in Fin Fishes

The digestive tract of fish contains a much higher number of micro-organisms than the surrounding water, as many as 10^8 cells g^{-1} (RingØ *et al.*, 1995). Reports on the presence of lactic acid bacteria in the intestinal microbiota of fish (Schroder *et al.*, 1980, Strom, 1988), suggest that there exist lactic acid bacteria that constitute non-pathogenic members of the indigenous intestinal micro-flora of healthy fish. After hatching, the gastrointestinal tract of Atlantic cod (*Gadus morhua*) larvae is colonised by almost the same bacterial genera as found in the eggs (Hansan and Olafsen, 1989). Dry feed containing lactic acid bacteria (*Carnobacterium divergens*) isolated from Atlantic cod (*Gadus morhua*) intestines improved disease resistance of cod fry exposed to a virulent stain of *Vibrio anguillarum* (Gildberg *et al.*, 1997).

It has been demonstrated that adult marine flat fish, turbot (*S. maximus*) and dab (*Limanda limanda*), harbour bacteria capable of suppressing the growth of *Vibrio anguillarum* (Olsson *et al.*, 1992), that might therefore act as probionts against such pathogens. Some of the bottleneck is ensuring long-term benefits of probiotic-feeding are, competition for attachment size, *i.e.*, colonization, and antagonism shown by these organisms in terms of reduction in pH, production of secondary metabolites, and production of bacteriocins etc.

Austin *et al.* (1995) reported that a strain of *Vibrio alginolyticus* was effective in reducing disease caused by *Aeromonas salmonicida* and two pathogenic *Vibrio* species in Atlantic salmon.

Some lactic acid bacteria are pathogenic to fish (Cone, 1982). Pathogenic lactic acid bacteria such as Streptococcus, Enterococcus, Leuconostoc, Lactobacillus etc. belong to be normal microbiota of the gastrointestinal tract in healthy fish and have been detected from kidney, liver, spleen and heart (RingØ and Gatesoupe, 1998).

Andlid *et al.* (1994) reintroduced and successfully colonized the isolated-yeast in fish intestine in a higher number, without visible negative effects on the fish. A non-virulent *Carnobacterium* sp. isolated from the GI tract of Atlantic salmon, *Salmo salar*, produced inhibitory substances against bacterial fish pathogens. Production of growth inhibitors *in vitro*, in mucus and feacal extracts, against *Vibrio anguillarum* and *Aeromonas salmonicida* was demonstrated (Joborn *et al.*, 1997).

Gram *et al.* (1999) reported that a *Pseudomonas fluorescens* stain (AH_2) reduced the mortality of 40 g rainbow trout (*Oncorhynchus mykiss*) infected with *Vibrio anguillarum* (90-11-287, serotype 01) strain. Controls inoculated with the pathogenic bacterium had a cumulative mortality of 47 per cent after 7 days, where as in treated fish the mortality was 32 per cent.

Yeast adheres to the intestinal mucus of rainbow trout (Vazquez-Juarez *et al.*, 1997) and the involvement of specific adhesions demonstrated by Vazquez-Juarez *et al.*, 1996. Yeasts have, therefore, a great potential to adhere and to colonise the intestine of fish and their application, as probionts in aquaculture deserves more attention (Gatesoupe, 1999).

Probiotics in Shell Fishes

There is almost no scientific information on the use of bacteria as probiotics in shrimp larvae culture systems, although some work has been done in South America (Gomez-Gill *et al.*, 2000). In Asia, a report of several species of bacteria being used in the larvae culture of *Peneaus monodon* and *P. penicillatus* with promising results (Anonymous, 1991). Early experiment started when hatchery production managers wanted to improve the levels of good *Vibrios* namely the sucrose fermentors (Garriques and Wayban, 1993) in this system. Latter, these bacteria were culture in a similar manner to microalgae and then added to the larval rearing tanks (Garriques and Wayban, 1993, Daniels, 1993). Introduction of probiotics in Ecuador in 1992, hatchery down-time between batches were reduced from 7 days per month to 21 days annually, production volume increased by 35 per cent and overall antimicrobial use decreased by 94 per cent (Griffith, 1995). Garriques and Arevalo (1995) tested a strain of *V. alginolyticus* isolated from the seawater next to their hatchery on *Litopenaeus vannamei* larvae. No mortalities were observed in a bath challenged pathgenecity test, whereas 100 per cent mortality was obtained with *V. parahaemolyticus* strain after 96 hours. Austin *et al.* (1995) and by Garriques and Arevalo (1995) suggest that *V. alginolyticus* may have characteristics capable of conferring some degree of protection against disease. Gatesoupe (1990) also detected the same and established a positive correlation between the survival rate of turbot larvae and the proportion of *V. alginolyticus* in the rearing environment.

Yeast and fungi have also been used to promote the growth rate and performance of *L. vannamei* larvae (Intriago *et al.*, 1998). A red pigmented yeast and a chitin-degrading fungus isolated from the marine environment was used, but the information was little.

Nagami and Maeda (1992) and Maeda (1992a) isolated a bacterial strain, coded PM-4 and identified as *Thalassobacter utilis* (Nagami *et al.*, 1997), from seaweeds and inoculated it into blue crab (*P. trituberculatus*) larval rearing the tank at concentration of 10^6 cells ml^{-1}. They obtained a survival of 27.2 per cent in the test tank compared with 6.8 per cent in the controlled tank with no bacteria inoculated. Gomez-Gill 1998 isolated a strain of *A. media* (A 199) capable of producing bacteriocin-like inhibitory substances. This strain displayed antagonistic activity against several potentially pathogenic bacteria and was used as a probionts to control interaction of *C. gigas* larvae with a *Vibrio tubiashii* strain, significant differences were observed between treatments, demonstrating the probiotic characteristic of this strain. Bacteria specially *Vibrio* sp. are important pathogens in the culture of the Scallops, *P. maximus* (Nicolas *et al.*, 1996), *A. purpuratus* (Requelme *et al.*, 1996) and oyster larvae (Elston *et al.*, 1981), and probiotics have been used in an attempt to control them. For the scallop *P. maximus*, a *Reseobacter* sp. was found to confer protection only bacterial cell extracts were inoculated into larval cultures (Ruiz-Ponte *et al.*, 1999).

Presently there are only very few commercial probiotic preparations that are available for application in shrimp farms. Further research is needed to isolate and characterised probiotic strains of bacteria from shrimp ponds, which have got antagonistic properties against common shrimp pathogens.

In Vivo Challenges of Pathogens by Using Probiotics

Many scientists challenged the pathogens of aquatic animals by treating with probiotic supplements. Such tests were particularly useful to point out the peculiarity of the application of probiotics of bivalve larvae (Gatesoupe, 1999). In some studies, the mortality was only delayed, in comparison with the control without probiotic treatments (Gatesoupe, 1994, 1997). Some of the *in vivo* challenge studies are given in Table 28.1.

Table 28.1: *In Vivo* Challenge Study of Pathogens of Fin-fishes Treated with Probiotics

Host	*Probiotic*	*Pathogen*	*Reference*
S. salar (j)	*P. fluorescens*	*A. salmonicida*	Smith and Davey, 1993
S. maximus (l)	*Camobacterium* sp.	*V. splendidus*	Gatesoupe, 1994
S. salar (j)	*V. alginolyticus*	*A.S., A. an., V.o.*	Austin *et al.*, 1995
A. purpuratus (l)	*A. haloplanktis*	*V. anguillarum*	Requelme *et al.*, 1996
S. maximus (l)	*Vibrio* sp.	*V. splendidus*	Gatesoupe, 1997
G. morhua (j)	*C. divergens*	*V. anguillarum*	Gildberg *et al.*, 1997
A. purpuratus (l)	*Vibrio* sp.	*V. anguillarum*	Requelme *et al.*, 1997
C. gigas (l)	*A. media*	*V. tubiashii*	Gibsin *et al.*, 1998
S. maximus (l)	*V. pelagicus*	*A. caviae*	RingØ and Vadstein, 1998
P. monodon (j)	*Bacillus* sp.	*V. harveyi*	Rengpipat *et al.*, 1998
B. plicatilis	*L. lactis*	*V. anguillarum*	Shiri Harzevili *et al.*, 1998
O. myskiss (j)	*P. fluorescens*	*V. anguillarum*	Gram *et al.*, 1999

(j): Juveniles, (l): Larvae.

Potential Benefits of Probiotics in Aquaculture

1. Lower incidence of disease and greater survival.
2. Better growth and development.
3. Enhance decomposition of organic matter.
4. Control of built up of stress causing chemicals like NH_3, NO_2 and H_2S.
5. Greater availability of dissolved oxygen.
6. Better algal growth.
7. Reduction in nitrogen and phosphorus concentration.
8. New dietary alternatives for immuno-therapy to counteract local immunological dysfunctions and to stabilise the natural gut mucosal barrier mechanisms.

Selection Criteria of Probiotics in Aquaculture

Selection of probiotic bacteria has usually been an empirical process based on limited scientific evidence and many of the failures in porbiotic research can be attributed to the selection of inappropriate micro-organisms (Gomez-Gill *et al.*, 2000), but it is dependent upto the host and environment.

It is essential to understand the mechanism of probiotic action and to define selection criteria for potential probiotics (PP) (Huis in 't veld *et al.*, 1994). Selection criteria are determined by biosafety considerations, methods of production and processing, the method of administration and the location in the body where the micro-organisms are expected to be active (Huis in 't veld *et al.*, 1994). According to Gomez-Gill *et al.* (2000), selection of probiotic bacteria for the use in the larvae culture of aquatic animals are, I) collection of background information, II) acquisition of PP, III) evaluation of the ability of PP out-complete pathogenic strain, IV) assessment of the pathogenecity of the PP, V) evaluation of the effect in the PP in larvae, and VI) an economic cost-benefit analysis.

Selection of probiotic bacteria for aquaculture should be introduced by the following steps:

Origin of the Probiotic

↓

Collection of Information About the Biosafety

↓

Activity and the Viability in that Environment

↓

Adherence to Intestinal Tissue of the Host

↓

Antimicrobial Activity to Pathogens

↓

Immune Response

↓

Harmless to other Aquatic Organisms

↓

Economics Benefits

Future Prospects

The application of probiotics in aquaculture, starting from growth promotion to disease resistance and larvae to adult shows promising results but more considerable attempt and dependable research is needful. Immunological and molecular probes will be useful tools to trace the probiotic cells (RingØ *et al.*, 1996, Austin, 1998, O'Sullivan, 1999). It is essential to investigate the best way of introduction and the optimal dose, and technical solutions are required, especially to keep the probiotic alive in dry pellets (Gayesoupe, 1999). The influence of probiotics on gastrointestinal microbiota remains poorly described, but further investigation may be expected with the propagation of molecular approaches to analyse bacterial communities (Raskin *et al.*, 1997, Wallner *et al.*, 1997, Hugenholtz *et al.*, 1998).

The bacteria normally dominant in healthy animals may be source of probiotics, but there are many potential pathogens among Vibrionaceae and Pseudomonads which may be wise to carry out long-term surveys to make sure that the bacteria keep innocuous, without risk of application of potentially detrimental mutants (Gatesoupe, 1999). Lactic acid bacteria have good characteristics of probiotics, is already proved by several scientists but further research on yeasts as probiotics is necessary because it adheres and colonizes the intestine of fish (Gatesoupe, 1999). Finally, the research should be optimized about the cost-benefit analysis of those applications.

Conclusion

Most probiotics have been designated as 'generally recognized as safe' based on their long history. Probiotics have been successfully used in aquaculture to enhance both external and internal microbial environments. The role of bacteria in fish culture is important for fish health and also environment conservation. Presently, there are only few commercial probiotic preparations are available for application in aquaculture farms. Further research is needed to isolate the probiotic strains of microflora from aquatic pond, which have potential antagonistic properties against common aquatic pathogens. It is clear from the above clarification that probiotics have immense potential for use in aquaculture. The advantages are varied and manifold. More research on its use in aquaculture should help to achieve better performances and yeast may plays a vital role.

Acknowledgements

The authors are highly acknowledge to the Head, Department of Zoology, Vidyasagar University, India for necessary facilities and the research scholars of the Aquaculture Research Unit for their valuable co-operation and for preparation of this article and University Central Library for providing us the Internet facilities.

References

Abraham, J. and Mohan, B.K. (2000). Probiotics: A safe alternative to antibiotics for sustainable shrimp culture. *Fishing chimes*. 20(2): 35-36.

Anonymous (1991). Disease control in hatchery by microbiological techniques. *Asian shrimp news*. 8(4) 4th quarter.

Andlid, T., Vazquezjuarez, R. and Gustafsson, L. (1994). The use of yeast in aquaculture. 3rd Intl. Mar. Biotech. Conf., Intl. Advisory Comm. Tromsoe Univ., Tromsoe, Norway: 115.

Austin, B., Stuckey, L.F., Robertson, P.A.W., Effendi, I. and Griffith, D.R.W. (1995). A probiotic strain of *Vibrio algilolyticus* effective in reducing diseases caused by *Aeromonas salmonicida*, *Vibrio anguillarum* and *Vibrio ordalii*. *J. Fish Dis.*, 18: 93-96.

Austin, B. (1988). Biotechnology and diagnosis and control of fish pathogens. *J. Mar. Biotecnol.* 6: 1-2.

Borwdy, C.L. (1998). Recent development in penaeid broodstock and seed production technologies: improving the outlook for superior captive stocks. *Aquaculture*, 164: 3-21.

Cone, D.K. (1982). A *Lactobacillus* sp. from diseased female rainbow trout, *Salmo gairdneri* (Richardson), in Newfoundland, Canada. *J. Fish Dis.*, 5: 479-485.

Daniels, H.V. (1993). Disease control in shrimp ponds and hatcheries in Ecuador. Associacao Brasileria de aquicultira. IV simposiabrasileiro sobre de camarao, 22-27 November, Brasil, pp. 175-84.

Debach, P. and Rosen, D. (1991). Biological control by natural enemies. Cambridge University Press, Cambridge, 440 p.

Elston, R., Leibovitz, L., Relyea, D. and Zatila, J. (1981). Diagnosis of vibriosis in a commercial oyster hatchery epizootic: diagnostic tools and management features. *Aquaculture*, 24: 53-62.

Fuller, R. (1989). Probiotics in man and animals. *J. Appl. Bacteriol.*, 66: 365-378.

Fuller, R. (1992). History of development of probiotics. In. Fuller, R. (Ed.), Probiotics: the Scientific Basis. Chapman and Hall, New York, pp. 1-8.

Garriques, D. and Wyban, J. (1993). Up to date advances on *Penaeus vannamei* mutation, naupli and post larvae production. Associacao Brasileira de Aquicultera. IV simposio brasileiro sobre cultivo de camarao, 22-27, November, Brasil, pp. 217-235.

Garriques, D. and Arevalo, G. (1995). An evaluation of the production and the use of a live bacterial isolate to manipulate the microbial flora in the commercial production of *Penaeus vannamei* post larvae in Ecuador. In. Browdy, C.L., Hopkins, J. S. (Eds.), Swimming Through Troubled Water. Proceedings of the special session on shrimp farming, *Aquaculture*, 1995. World Aquaculture Society, Baton Rouge, pp. 53-59.

Gatesoupe, F.J. (1990). The continuous feeding of turbot larvae, *Scophthalmus maximus*, and control of the bacterial environment of rotifers. *Aquaculture*, 89: 139-148.

Gatesoupe, F.J. (1991a). The effect of three strain of lactic acid bacteria on the production rate of rotifers, *Brachionus plicatilis*, and their dietary value for larval turbot, *Scophthalmus maximus. Aquaculture*, 96: 335-342.

Gatesoupe, F.J. (1994). Lactic acid bacteria increased the resistance of turbot larvae, *Scophthalmus maximus*, against pathogenic vibrio. *Aquat. Living Resour.* 7: 277-282.

Gatesoupe, F.J. (1997). Siderophore production and probiotic effect of *Vibrio* sp. associated with turbot larvae, *Scophthalmus maximus*. *Aquat. Living Resour.* 10: 239-246.

Gatesoupe, F.J. (1999). The use of probiotics in aquaculture. *Aquaculture*, 180: 147- 165.

Gibson, G.R., Roberfroid, B. (1995). Dietary modulation of the human colonic microbiota: introducing the concept of prebiotics. *J. Nutr.*, 125: 1401-1412.

Gibson, L.F., Woodworth, J., George, A.M. (1998). Probiotic activity of *Aeromonas media* on the Pacific oysrer, *Crassostrea gigus*, when challenged with *Vibrio tubiashii. Aquaculture*, 169: 111-120.

Gildberg, A., Mikkelsen, H., Sandaker, E and RingØ, E. (1997). Porbiotic effects of lactic , acid bacteria in the feed and growth and survival of fry of Atlantic cod (*Gadus morhua*). *Hydrobiologia*, 352: 279-285.

Gildberg, A., Mikkelsen, H. (1998). Effects of supplementing the feed to Atlantic cod (*Gadus morhua*) fry with lactic acid bacteria and immuno-stimulating peptides during a challenge trial with *Vibrio anguillarum*. *Aquaculture*, 167: 103-113.

Gildberg, A., Mikkelsen, H., Sandaker, E. and Riogo, E. ((1997). Probiotic effect of lactic acid bacteria in the feed on growth and survival of fry of Atlantic cod (*Gadus morhua*). *Hydrobiologia*, 352: 279-285.

Gomez-Gil, B. (1998). Evaluation of potential probionts for use in penaeid shrimp larvae culture. Ph.D. Thesis. University of Starling, 269 pp.

Gomez-Gil, B., Roque, A. and Turnbull, J.F. (2000). The use and selection of probiotic bacteria for use in the culture of larval organisms. *Aquaculture*, 191: 259- 270.

Gram, L., Melchiorsen, J., Spanggaard, B., Huber, I. and Nielsen, T.F. (1999). Inhibition of *Vibrio anguillarum* by *Pseudomonas fluorescens* AH2, a possible probiotic treatment of fish. *Appl. Environ. Microbiol.* 65: 969-973.

Griffith, D.R.W. (1995). Microbiology and the role of probiotics in Ecuadorium shrimp hatcheries. In: Lavens, P., Jaspers, E., Roelants, I. (Eds), Larvi 1995-Fish and Shellfish Larvae culture Symposium. European Aquaculture Society, Special Publication, Vol. 24, Gent, Belgium, p. 478.

Hansan, G. H. and Olafsen, J.A. (1989). Bacterial colonization of cod (*Gadus morhua* L.) and halibut (*Hippoglossus hippoglossus*) eggs in marine aquaculture. *Apll. Environ. Microbiol.* 55: 1435-1446.

Huis in 't Veld, J.H.J., Havenaar, R. and Marteau, Ph. (1994). Establishing a scientific basis for probiotics R & D. *Tibtech*. 12: 6-8.

Hugenholtz, P., Pitulle, C., Hershberger, K.L., Pace, N.R. (1998). Novel division level bacterial diversity in a Yellowstone hit spring. *J. Bacteriol.* 180: 366-376.

Intriago, P., Krauss, E. and Barmiol, R. (1998). The use of yeast and fungi as probiotics in *Penaeus vannameri* larviculture. Aquaculuture 98. World Aquaculture Society, Baton Rouge, p. 263.

Joborn, A., Olsson, C., Westerdahl, A., Conway, P.L., Kjellberg, S. (1997). Colonization in the fish intestinal tract and production of inhibitory substances in intestinal mucus and faecal extract by *Carnobacterium* sp. strain K. *J. Fish Dis.*, 20: 383-392.

Jory, D.E. (1988). Use of probiotic in penaeid shrimp growout. *Aqcuacult. Mag.*, 24(1): 62-67.

Kozasa, M. (1986). Touocerin (*Bacillus toyoi*) as growth promoter for animal feeding. *Microbial. Aliment. Nutr.*, 4: 121-135.

Maeda, M. (1992a). Fry production with biocontrol. Isr. J. Aquacult-Bamidgeh, 44: 142-143.

Maeda, M., Nogami, K., Kanematsu, M., Hirayama, K. (1997). The concept of biological control methods in aquaculture. *Hydrobiologia*. 358: 285-290.

Mishra, S., Mohanty, S., Pattnaik, P. and Ayyappan, A. (2001). Probiotics: possible application in aquaculture. *Fishing Chimes*, 21(1): 31-37.

Moriarty, D.J.W. (1998). Control of luminous *Vibrio* species in penaeid aquaculture ponds. *Aquaculture*, 164: 351-358.

Nicolas, J.L., Corre, S., Gauthier, G., Robert, R. and Ansquer, D. (1996). Bacterial problems associated with scallop *Pecten maximus* larval culture. Dis. Aquacult. Org., 27: 67-76.

Nogami, K. and Maeda, M. (1992). Bacteria as a biocontrol agents for rearing larvae for the crab, *Portunus trituberculatus*. *Can. J. Fish. Aquat. Sci.*, 49: 2373-2376.

Nogami, K., Hamasaki, K., Maeda, M. and Hirayama, K. (1997). Biocontrol method in aquaculture for rearing the swimming crab larvae, *Portunus trituberculatus. Hydrobiologia*. 358: 291-295.

O'Sullivan, D.J. (1999). Methods of analysis of the intestinal microflora. In: Tannock, G.W. (Ed.). Probiotics: A Critical Review. Horizon Scientific Press, Wymondham, England, pp. 23-44.

Olsson, J.C. Westerdahl, A., Conway, P.L. and Kjelleberg, S. (1992). Intestinal colonization potential of turbot (*Scophthalmus maximus*) and Dab (*Limada limada*): associated bacteria with inhibitory against *Vibrio anguillarium. Appl. Environ. Microbiol*. 58: 551 -556.

Parker, R.B. (1974). Probiotics: The other half of the antibiotics story. *Anim. Nutr. Health,* 29: 4-8.

Raskin, L., Capman, W.C., Sharp, R., Poulsen, L.K. and Stahl, D.A. (1997). Molecular ecology of gastrointestinal ecosystem. In: Mackie, R.I., With, B. A., Isaacson, R. E. (Eds.), Gastrointestinal Microbiology, Vol. 2, Gastrointestinal Microbes and Host Interactions. Chapman and Hall Microbiology Series, International Thomson Publishing, New York, pp. 243-298.

Ravi, V., Khan, S.A. and Rajagopal, S. (1998). Influence on probiotics on growth of Indian white prawn, *Penaeus indicus. J. Sci. Ind. Res.,* 57(10-11): 752- 756.

Rengpipat, S., Phianphak, W., Piyatiratitivorakul, S., Menasveta, P. (1998). Effects of a probiotic bacterium on black tiger shrimp *Penaeus monodon* survival and growth. *Aquaculture,* 167: 301-313.

Rico-Mora, R., Voltolina, D., Villaescusa-Celaya, J.A. (1998). Biological control of *Vibrio alginolyticus* in Skeletonema costatum (Bacillariophyceae) cultures. *Aquacult. Eng.,* 19: 1-6.

RingØ, E., Gatesoupe, F.J. (1998). Lactic acid bacteria in fish: a review. *Aquaculture,* 160: 177-203.

RingØ, E., StrØm, E., Tabachek, J.A. (1995). Intestinal microflora of salmonids: a review. *Aquacult. Res.* 26: 773-789.

RingØ, E., Birkbeck, T.H., Munro, P.D., Vadstein, O. and Hjelmeland, K. (1996). The effects of early exposure to *Vibrio pelagius* on the aerobic bacterial flora of turbot, *Scophthalmus maximus* (L) larvae. *J. Appl. Bacterol.* 81: 207-211.

RingØ, E., Bendiksen, H.R., Gausen, S.J., Sundsfjord, A. and Olsen, R.E., (1998). The affect of dietary fatty acids on lactic acid bacteria associated with the epithelial mucosa and from faecalis of Arctic charr, *Salvelinus alpinus* (L). *J. Appl. Microbiol.* 85: 855-864.

Requelme, C., Araya, R., Uchida, A., Satomi, M. and Ishida, Y. (1996). Isolation of a native bacterial strain from the scallop *Argopecten purpuratus* with inhibitory effects against pathogenic vibros. *J. Shellfish Res.,* 15: 369-374.

Requelme, C., Araya, R., Vergare, N., Rajos, A., Guaita, M. and Candia, M. (1997). Potential pro-biotic strains in the culture of the Chilean scallop, *Argopecten purpuratus* (Lamark, 1819). *Aquaculture,* 154: 17-26.

Ruiz-Ponte, C., Samain, J.F., Nicolas, J.L. (1998). Antibacterial bacterial activity exhibited by the marine strain *Roseobactor* sp. In: Le Gal, Y., Muller-Feuga, A. (Eds.), Marine Microorganisms for industry. Proceedings of a Meeting, 17-19 September, 1997, Brest, France. Actes Colloq. IFREMER No. 21, IFREMR, Plouzana, France, pp. 166-168.

Ruiz-Ponte, C., Samain, J.F., Nicolas, J.L. (1999). The benefit of a *Roseobacter* species on the survival of scallop larvae. *Mar. Biot.,* 1: 52-59.

Schoder, K., Clausen, E., Sandberg, A.M. and Raa, J. (1980). Phychrotriphic *Lactobacillus plantarum* from fish and its ability to produce antibiotic substances. In: J.J. Connell (Eds.) Advances in fish science and technology, Fishing New Books Ltd. pp. 480-483.

Shiri, Harzevili, A.R., Van Duffel, H., Dhert, P., Swings, J. and Sorgeloos, P. (1998). Use of a potential probiotic *Lactociccus lactis* AR 21 strain for the enhancement of growth in the rotifer *Brachionus plicatilis* (Muller). *Aquacult. Res.*, 29: 411-417.

Smith, H.S. (1919). On some phase of insect control by the biological method. *J. Ecin. Entomol.*, 12: 288-292.

Smith, P. and Davey, S. (1993). Evidences for the competitive exclusion of Aeromonas salmonicida from fish with stress-inducible furunculosis by a flurscent pseudomonad. *J. Fish Dis.*, 16: 521-524.

Tannock, G.W. (1997). Modification of a normal microbiota by diet, stress, antimicrobial agents, and probiotics. In: Mackie R.I., With, B.A., Isaacson, R.E. (Eds.), Gastrointestinal Microbiology, Vol. 2, Gastrointestinal Microbes and Host Interactions. Chapman and Hall Microbiology Series, International Thomson Publishing, New York, pp. 434-465.

Vazquez-Juarez, R. (1996). Factors involves in the colonization of fish intestine by yeast. Ph.D. Thesis, Goteburg University, Sweden, 134 pp.

Vazquez-Juarez, R., Andlid, T. and Gustafsson, L. (1997). Adhesion of yeast isolated from fish gut to crude intectinal mucus of rainbow trout, *Salmo gairdneri. Molec. Mar. Biol. Biotecnol.*, 6: 64-71.

Wallner, G., Fuchs, B., Spring, S., Beisker, W. and Amann, R. (1997). Flow sorting of micro-organisms for molecular analysis. *Appl. Environ. Microbiol.*, 63: 4223- 4231.

29

Assessment of Water Quality and Self-Purification Process in River Sai at Bela Pratapgarh

Tapan Routh and R.K. Gupta

National Environmental Engineering Research Institute, Kanpur Zonal Laboratory, 6/33, Civil Lines, Kanpur - 208 002

ABSTRACT

The objective of this study is to assess the water quality and self purification process of river Sai in Bela Pratapgarh, receiving untreated domestic sewage from the town. Five sampling stations spread over 6 km stretch of river were selected to collect grab samples during lean flow period of summer and winter seasons. The untreated sewage of Bela Pratagarh town is discharged into river through two drains. The flow measurement and composited samples of wastewater were collected in summer and winter seasons for 24 hours on two consecutive days. The wastewater analyses show the average flow of 1368 and 2442 m^3/d in drains W1 and W2, with average BOD load of 93 and 214.9 kg/d respectively. The quality of river water reveals that the pollution is within the prescribed limit except for BOD and Total coliforms at few monitoring stations. Though the sewage is of weak nature, it has BOD value higher than 30 mg/L which is the prescribed limit for its discharge in streams. The total coliforms counts are also high which necessitate sewage treatment before disposal. The self purification constant calculated from the river data is 2, and this value is for large streams with low to moderate velocity.

Key Words: Bela Pratapgarh, Sai River, River Water Quality, Sewage, Self-purification.

Introduction

Studies undertaken to assess the pollution load and its mitigation measures are mostly centred around wastewater discharges from major cities to adjoining water courses. The cities/towns covered for such studies are categorised as class-I, with population exceeding one lakh. With the growing concern for pollution control in relatively smaller towns, class-II (population between fifty thousand and one lakh) and class-III, (population between twenty thousand to fifty thousand) towns are also considered for such studies. The present study is carried out to assess the water quality and self-purification process of river Sai at Bela Pratapgarh, which receives untreated wastewater from the town, Bela Pratapgarh (class-II) situated at its bank.

The Sai is a tributary of river Gomti joining its right bank, which in turn is a tributary of the main river Ganga. It originates as a small channel from Uchaulia (27°38 N 80°12 E) in district Hardoi. Subsequently, it passes through the districts of Unnao, Rae Bareli, Pratapgarh and finally confluences with river Gomti at Rajapur-Tirmuhani in the district Jaunpur, covering a distance of about 450 km. Town Hardoi lies on western side, Rae Bareli on north eastern side, Pratapgarh on its southern side and Jaunpur on northern side of the river stretch. The entire course of the river Sai is depicted in Fig. 29.1.

In the dry seasons, the river Sai is narrow, shallow and easily fordable, while its tributaries are mere ravines. However, during rainy season the tributaries carry a large volume of water into the river, the water level of which rises high with significant velocity. The banks of the river are high and generally well defined. The river enters Bela Pratapgarh from eastern side near Avas Vikas township and traverses through north-western side and leaves the town at Kadipur in south-west direction. In the north outskirts of the town, famous temple of Bela Bhawani is situated on the right bank of river Sai. Two open drains carrying untreated wastewater of the town discharge into the river on the left and right side of the temple. The drain on the left side is having natural course through fields, while the drain on right side is cemented upto 50 m from discharge point. Large number of devotees take holy dip in the river specially on Mondays. On eighth day of Navratra festival, a mass bathing is performed in the river by devotees exceeding one lakh.

The objective of the present study is to assess the quantum of pollution in river Sai at Bela Pratapgarh due to wastewater out-falls during low flow periods and also the rate of self-purification of the river and allowable BOD load.

Materials and Methods

Sampling Stations

Five water sampling stations were selected over 6 km stretch of the river Sai within Bela Pratapgarh municipal area. The first sampling station (R1) is located in the upstream of the town. The second station (R2) is located at Belha Devi Mandir ghat at 2.5 km downstream of R1. The third station (R3) is located below the railway bridge. The city sewage is discharged through concrete drain, about 200m upstream to this point. The fourth sampling station (R4) is located near Ziriamau village. The fifth sampling station (R5) is located at Kadighat from where the river leaves Bela Pratapgarh town area. The two major drains were selected for sampling of wastewater. The sampling station (W1) is located at the outfall having natural course. The second wastewater sampling station (W2) is located at concrete drain. The sampling stations in river and drains are presented in Table 29.1 and shown in Fig. 29.2.

Sample Collection and Analyses

Two sets of grab samples of river water were collected on two consecutive days in winter (December, 1995) and summer (May, 1996) seasons, from the midstream. Samples were collected in 1 liter polypropylene bottles for physico-chemical analysis. Samples for bacteriological examination were

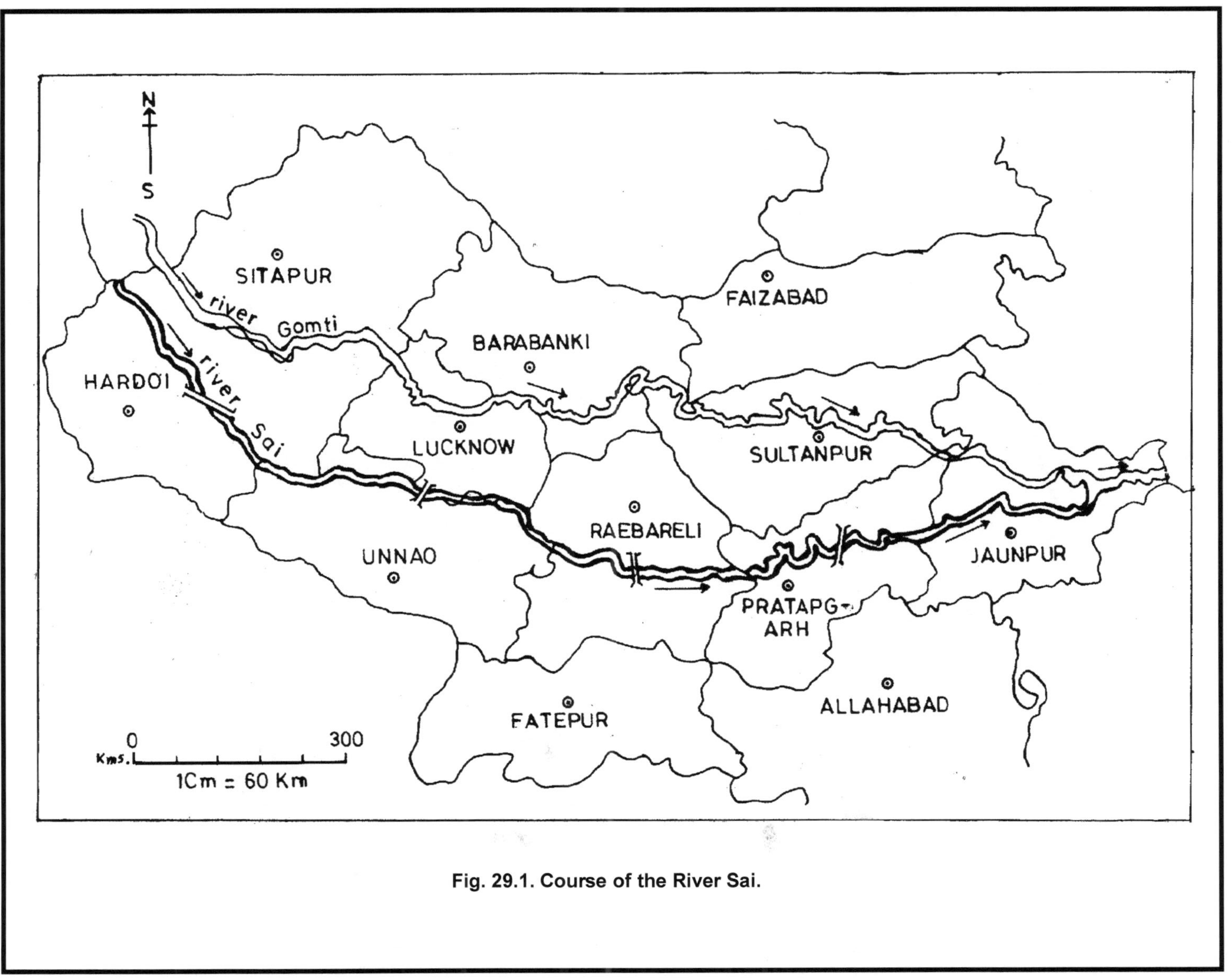

Fig. 29.1. Course of the River Sai.

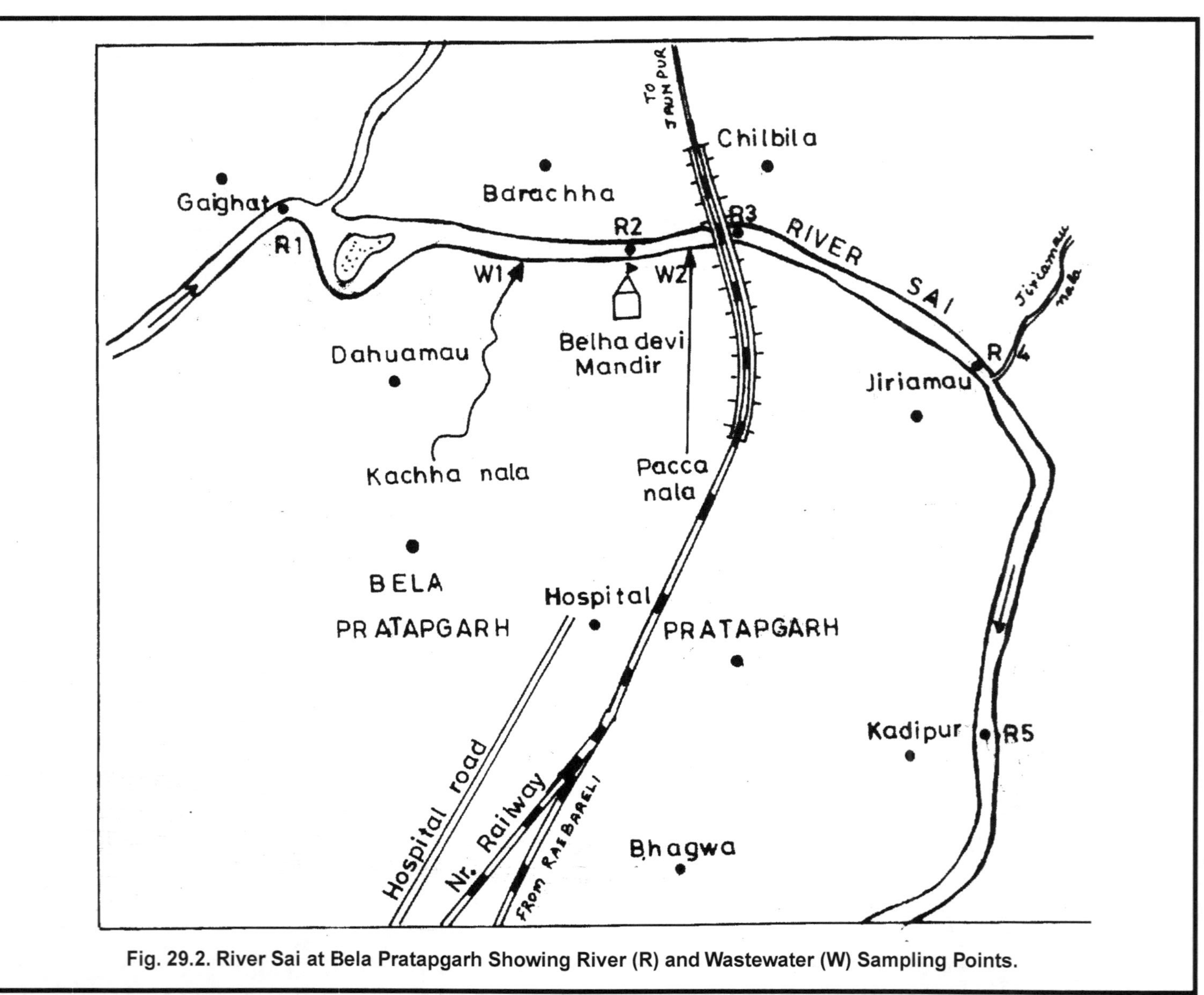

Fig. 29.2. River Sai at Bela Pratapgarh Showing River (R) and Wastewater (W) Sampling Points.

collected in 100 mL sterilized glass bottles. Samples for heavy metals were collected in 125 mL polypropylene bottles and fixed with conc. HNO_3 (pH 2.0). Biological samples for zooplankton after concentration of 50 L water, and phytoplankton, were collected in 100 mL bottles and preserved with formalin and lugol's solution respectively. Samples for biochemical oxygen demand (BOD) and dissolved oxygen (DO) were collected in BOD bottles. DO was fixed on the site itself. Physical parameters (temperature, turbidity, specific conductivity, suspended solids), chemical parameters (pH, alkalinity, hardness, dissolved oxygen (DO), chemical oxygen demand (COD), total phosphorus, total kjeldahl nitrogen, chloride, sulfate) and biological parameters (BOD, total coliform, fecal coliform, zooplankton, phytoplankton) were selected for analyses to determine the existing water quality. Heavy metals (Cd, Cr, Cu, Pb, Zn, Ni, Mn and Fe) were also determined in water samples. Temperature, pH and DO were measured at the site. The discharge of river at downstream stations, R3 and R4 was determined by measuring the depth and velocity in the cross-section of the river.

Table 29.1. Description of Sampling Stations at the Study Area

Station	Distance (Km)	Location	Particulars
River			
R1	0.00	Gaighat	Upstream of town, representing baseline water quality
R2	2.50	Belha Mandir Ghat	Famous bathing ghat of Belha Devi temple
R3	2.60	Railway Bridge	Faizabad-Lucknow Railway line
R4	4.00	Ziriamau Village	Small Ziriamau drain opens here
R5	6.00	Kadighat	The river leaving Bela Pratapgarh municipal area from this point
Wastewater Outfalls			
W1	2.00	500 m Upstream of Bela Mandir ghat	The drain has natural course through fields upto the confluence point
W2	2.55	50 m Downstream of Bela Mandir ghat	The drain has open cemented channel near confluence point

Wastewater samples from two drains were monitored for 24 hour on two consecutive days in December, 1995 (winter season) and in May, 1996 (summer season). Flow was measured by fixing V-notch in the drains and samples were collected. The 24 hour flow-weighted composited samples were prepared from hourly grab samples for both the drains. The samples were analysed for physical (temperature, specific conductivity, total dissolved solids, total suspended solids), chemical (pH, alkalinity, COD, chlorides, sulfates, ammonical nitrogen, total kjeldahl nitrogen, total phosphorus) and biological (BOD, total coliform, fecal coliform) parameters. Heavy metals (Cd, Zn, Ni, Mn, Fe, Pb, Cu, Cr) were also analysed. The Standard Methods for the Examination of Water and Wastewater (APHA, AWWA, WEF, 1995, 19th Edition) was followed for sample handling, preservation and analytical procedures.

Results and Discussion

Wastewater Quality

The flow measurement of wastewater discharged through drains, W1 and W2 was carried out hourly and continuously for 48 hours on two consecutive days in both winter and summer seasons. Average hourly flow variation in drains, W1 and W2 is shown in Fig. 29.3. The total average flow is

1337 and 1399 m^3/day in W1 and 2374 and 2509 m^3/day in W2 during winter and summer seasons respectively. The characteristics of 24 hour composited wastewater sample from drains W1 and W2 are presented in Table 29.2. The BOD value ranged from 64 to 98 mg/L in the samples. The higher values observed in W2 may be attributed to the fact that it receives wastewater from area having mostly service type toilets whereas, W1 receives the wastewater from areas with overflow from septic tanks. The observation reveal that all the parameters, except BOD, of the wastewater are within the permissible limit for discharge into inland surface water as laid down by Indian Standard (CPCB, 1997). The standard for BOD is 30 mg/L maximum for discharge into inland surface water. The wastewater in both the drains is domestic in nature as evident from COD: BOD ratio varying between 1.5 and 1.79. The pollution load exerted by wastewater in drains, W1 and W2 is presented in Table 29.3. Based on average flow of 1368 and 2442 m^3/day, the average BOD load of 93 and 214.9 Kg/day are exerted into river Sai at Bela Pratapgarh through the drains, W1 and W2, respectively.

Table 29.2. Average Characteristics of 24 Hourly Composited Wastewater Samples

Parameters	*Drain - W1*		*Drain -W2*	
	Winter	*Summer*	*Winter*	*Summer*
Temperature (°C)	18-20	30-31	16-20	30-31
pH	7.8	7.7	7.7	7.8
Conductivity (umhos/cm)	3400	3100	3420	2912
$BOD_5$20°C (mg/L)	64	72	78	98
COD (mg/L)	96	106	131	176
Alkalinity (mg/L)	269	250	246	252
Chlorides (mg/L)	610	584	585	595
Sulfates (mg/L)	14	12	12	14
NH_4-N (mg/L)	6.8	6.4	6.6	5.8
NO_3-N (mg/L)	0.01	0.40	0.60	0.55
NO_2-N (mg/L)	0.003	0.003	0.004	0.003
TKN (mg/L)	10	11	12	11
Total Phosphorus (mg/L)	1.4	1.2	1.6	1.6
Total Dissolved Solids (mg/L)	1253	1416	1364	1574
Total Suspanded Solids (mg/L)	62	68	70	82
Total Coliforms (CFU/100 mL)	2.2×10^5	4.0×10^6	4.3×10^6	3.8×10^6
Fecal Coliforms (CFU/100 mL)	1.4×10^4	2.8×10^4	1.2×10^5	2.6×10^5
Cd (mg/L)	BDL	BDL	BDL	BDL
Zn (mg/L)	0.008	0.010	0.012	0.011
Ni (mg/L)	BDL	BDL	BDL	BDL
Mn (mg/L)	0.046	0.048	0.048	0.050
Fe (mg/L)	0.384	0.378	0.456	0.512
Pb (mg/L)	0.046	0.051	0.062	0.058
Cu (mg/L)	BDL	BDL	BDL	BDL
Cr (mg/L)	0.028	0.032	0.046	0.036

BDL = Below detectable limit

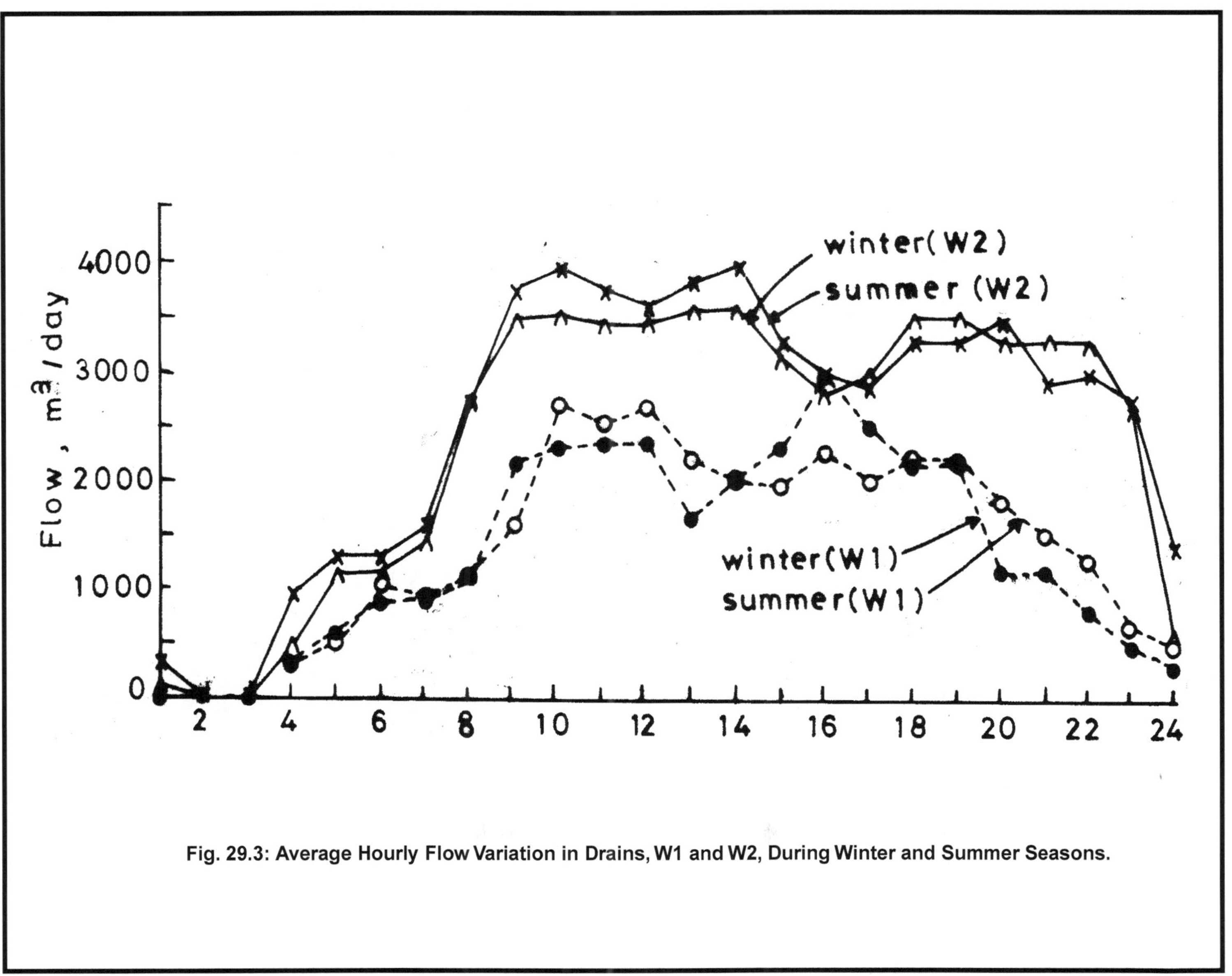

Fig. 29.3: Average Hourly Flow Variation in Drains, W1 and W2, During Winter and Summer Seasons.

Table 29.3: Discharge and Pollution Load of Wastewater in Drains

Particulars	*Drain W1*	*Drain W2*
Average Flow (m^3/day)	1368	2442
Pollution Load (Kg/day) :		
BOD	93.0	214.9
COD	138	376.1
Suspended Solids	88.9	185.6
Nitrogen	18.3	28.1
Phosphorus	1.8	3.9

River Water Quality

The average physico-chemical and bacteriological quality of river Sai at five sampling stations in Bela Pratapgarh during winter and summer seasons is presented in Table 29.4 and Fig. 29.4.

Table 29.4: Physico-chemical Quality of Water in River Sai at Bela Pratapgarh

Parameters	*Winter Station No.*					*Summer Station No.*				
	R1	*R2*	*R3*	*R4*	*R5*	*R1*	*R2*	*R3*	*R4*	*R5*
Conductivity (μmhos/cm)	310	316	417	318	317	322	740	326	322	320
Total dissolved Solids (mg/L)	140	158	189	140	170	156	513	180	162	168
Chlorides (mg/L)	15	15	17	16	16	14	16	26	16	15
Sulphates (mg/L)	9	12	11	11	11	12	19	20	11	14
Nitrates (mg/L)	0.33	0.42	0.46	0.24	0.21	0.46	0.30	0.42	0.15	0.39
Iron (mg/L)	0.31	0.28	0.19	0.31	0.32	0.28	0.28	0.24	0.32	0.30
Lead (mg/L)	0.053	0.058	0.064	0.054	0.036	0.033	0.038	0.044	0.044	0.040
Copper (mg/L)	BDL	BDL	BDL	BDL	BDL	BDL	BDL	BDL	BDL	BDL
Zinc (mg/L)	0.016	BDL	BDL	BDL	BDL	0.011	BDL	BDL	BDL	BDL

BDL = Below Detectable Limit.

pH

The pH value ranges between 8.5 and 8.6 in winter and 8.4 and 8.6 in summer seasons (Fig. 29.4a). These are observed to be within the permissible range of 6.5 to 8.5, as per primary water quality criteria for various uses of fresh waters, category 'A' to 'D', laid down by Central Pollution Control Board (CPCB). All the values, however, exceed the permissible limit of 6.5-8.0, for category 'E' which is desired for designated use in irrigation, industrial cooling and controlled waste disposal (CPCB, 1978-79). Freshwater with pH higher than 8.5 indicates presence of phenolphthalein alkalinity which does not harm normal aquatic life of the stream.

Alkalinity

The total alkalinity varies from 110 to 122 mg/L in winter and 110 to 180 mg/L in summer seasons. Phenolphthelin alkalinity is also present in all the stations which varies from 9.8 to 10.5 mg/

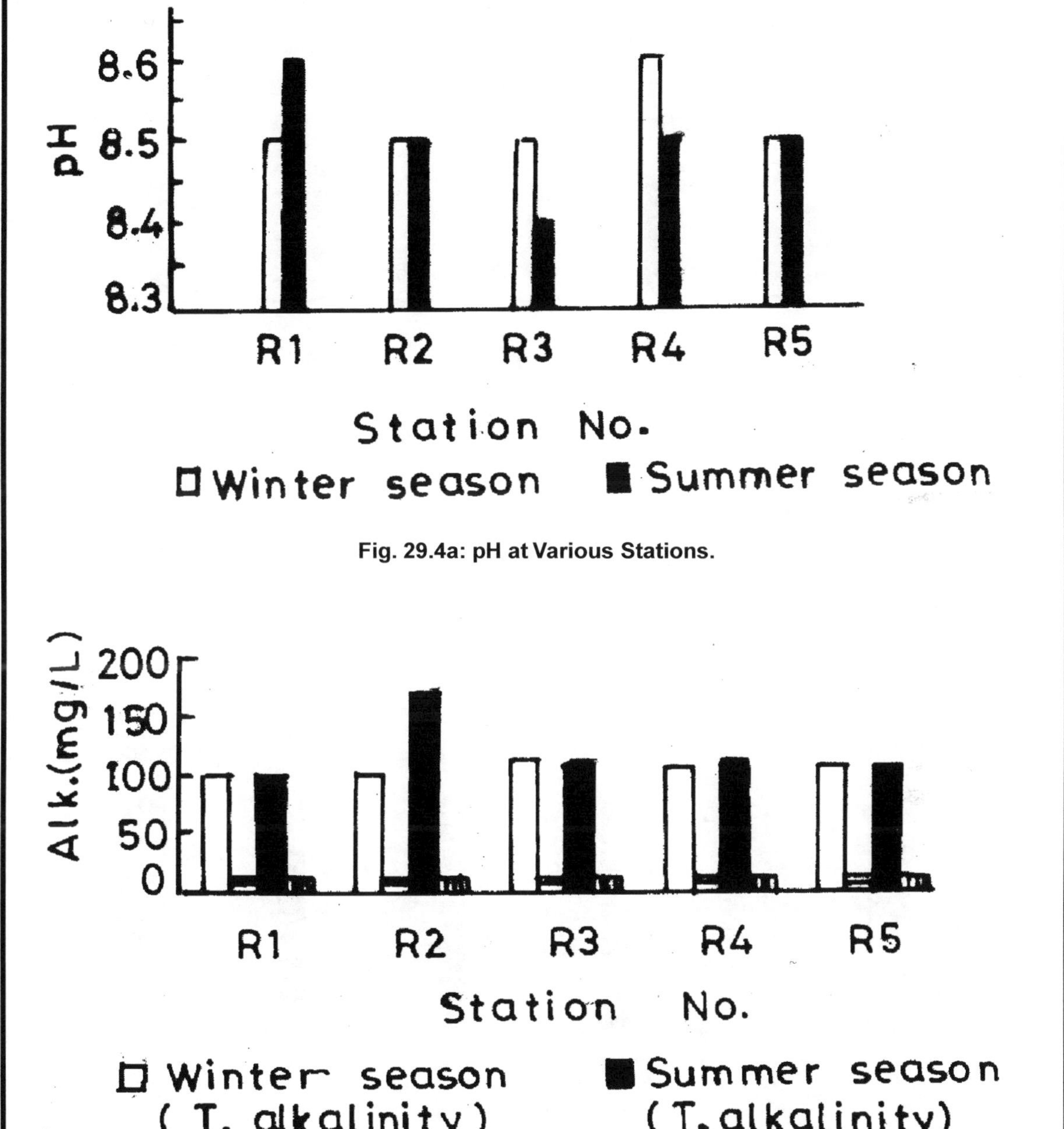

Fig. 29.4a: pH at Various Stations.

Fig. 29.4b: Alkalinity at Various Stations.

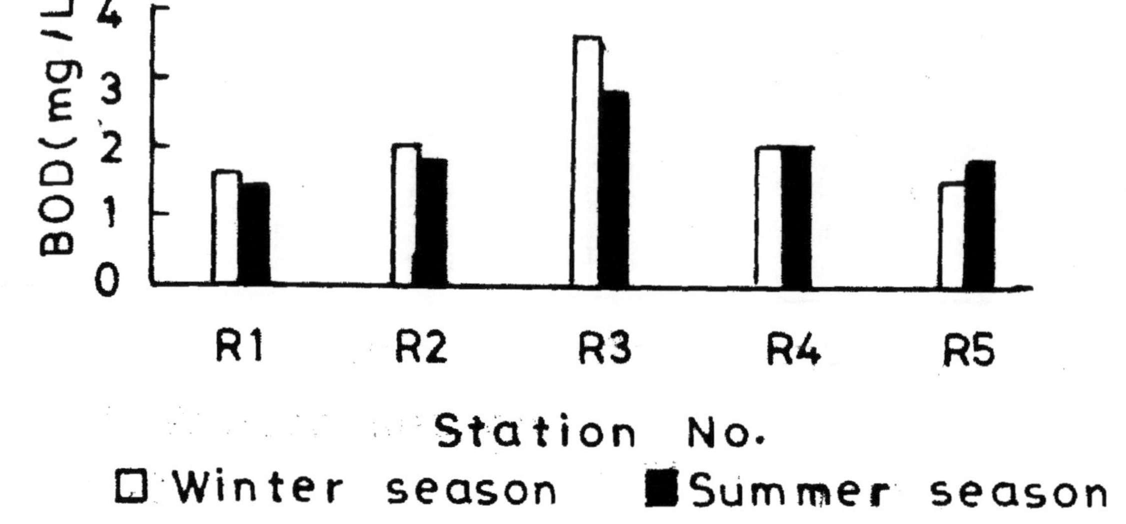

Fig. 29.4c: BOD at Various Stations.

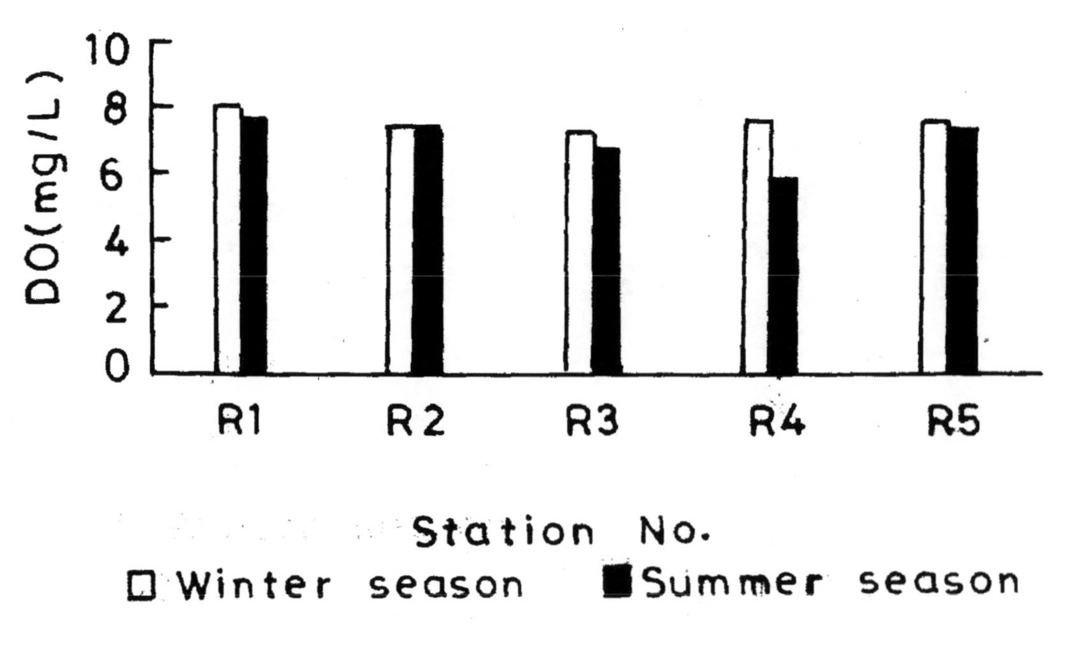

Fig. 29.4d: DO at Various Stations.

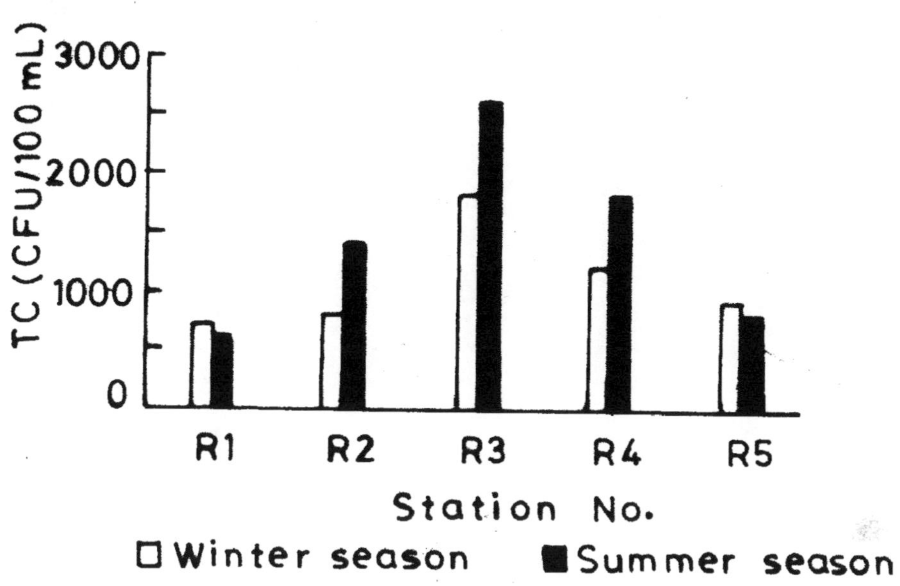

Fig. 29.4e: Total Coliforms at Various Stations.

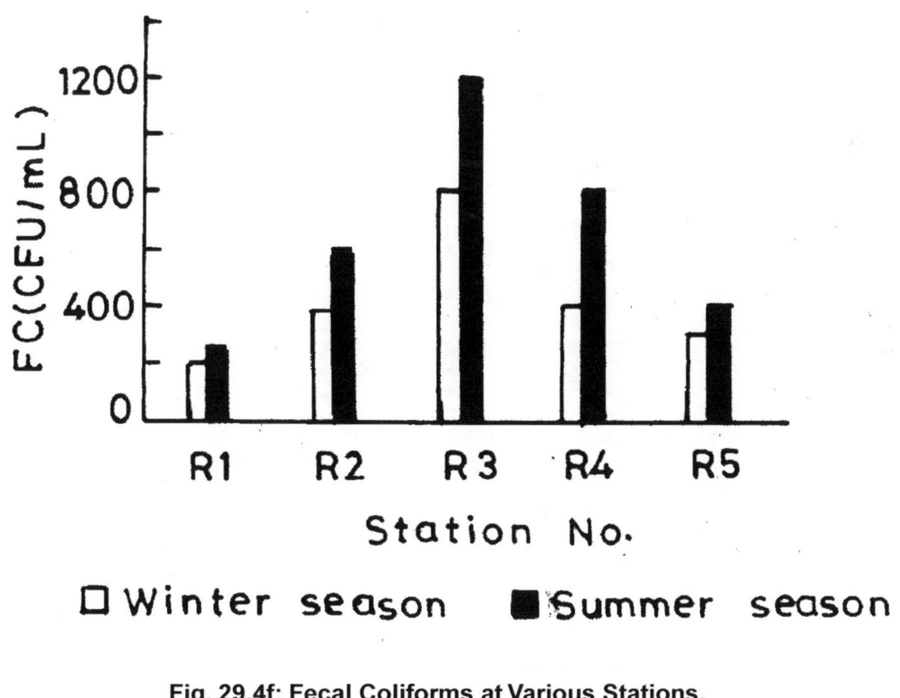

Fig. 29.4f: Fecal Coliforms at Various Stations.

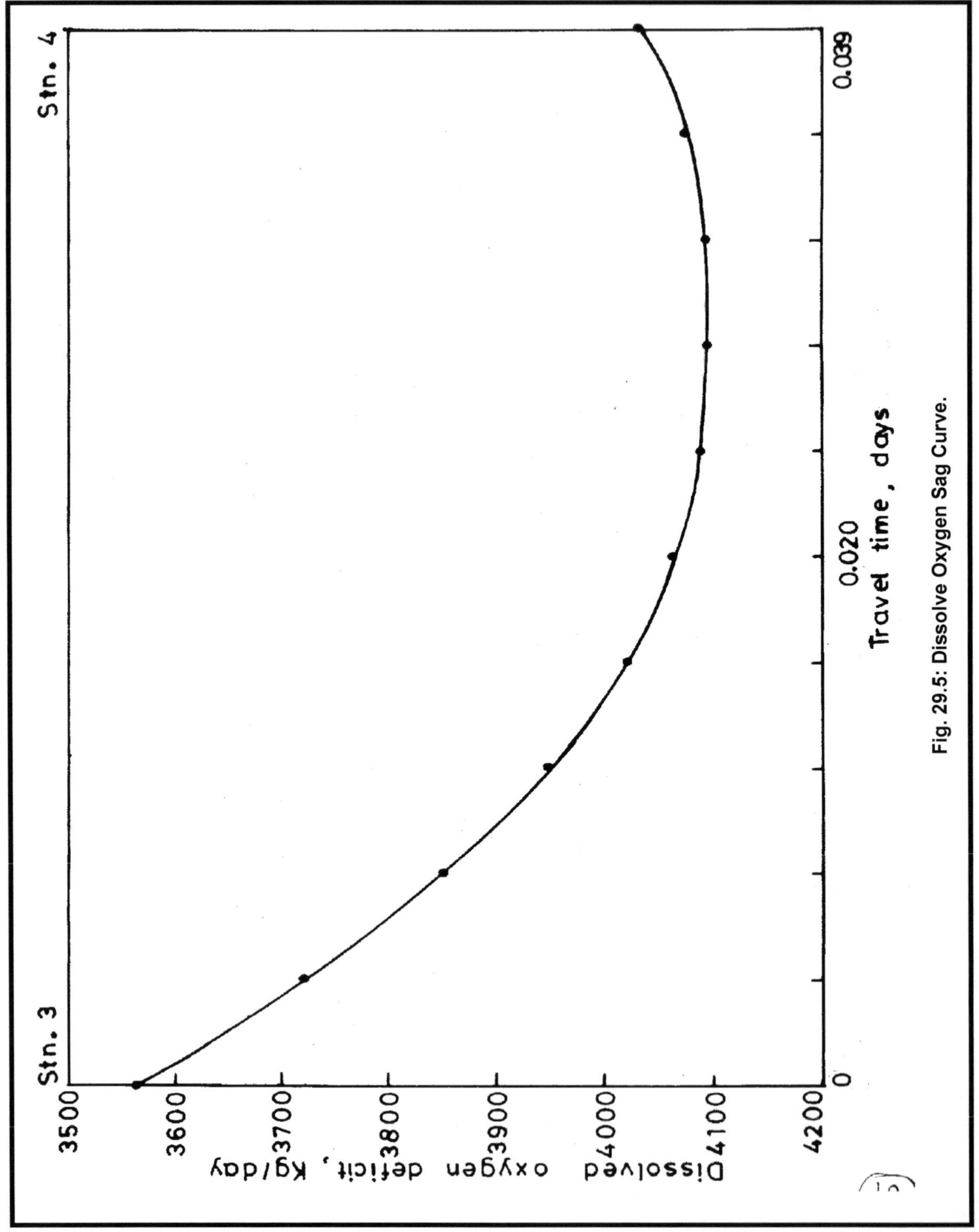

Fig. 29.5: Dissolve Oxygen Sag Curve.

L in winter and 9.8 to 10.0 mg/L in summer (Fig. 29.4 b). There is no standard for alkalinity, but its presence is desirable because of its buffering action in the receiving streams which is required for safe disposal of wastewater.

Biochemical Oxygen Demand (BOD)

The BOD of the river water varies from 2.0 to 3.6 mg/L in winter and from 2.0 to 2.8 mg/L in summer seasons (Fig. 29.4c). The BOD values generally lie within the permissible limit of Indian Standard classification of fresh water sources under category 'A' which may be used as drinking water source without treatment but after disinfection (CPCB, 1978-79). The BOD value of 3.6 mg/L, as observed at station, R3 in winter keeps the stream under category 'B' and 'C', designated for outdoor bathing (organized) and drinking water source with conventional treatment followed by disinfection, respectively.

Dissolved Oxygen (DO)

The concentration of DO was found in the range of 7.2-7.6 mg/L during winter and 5.8-6.7 mg/L during summer (Fig. 29.4d). The DO level at all the stations, except at R4 are well within the standard limit of 6 mg/L required for category 'A' streams. The DO values at R4 during summer was however, lower but more than the stipulated standard of 5 mg/L, (min.) for category 'B' streams. The strength of domestic wastewater is too low to cause any significant variation in DO concentration in river water due to dilution. A little impact has however, been observed at downstream points, R3 and R4, after confluence of wastewater with river, upstream to R2 and R3 during summer. The water temperature in summer rises to 30°C, resulting into increased microbial activity associated with more demand for DO, leading to its depletion.

The other physico-chemical parameters *viz.* conductivity, total dissolved solids, chlorides, sulphates and nitrates, presented in (Table 29.4) are well within the limits of Indian Standard laid down for various use of surface water (BSI, 1979). Heavy metals are also within the prescribed limit. Zinc is present below detectable limit (BDL) except at Station R1 where it has been found to 0.016 mg/L in winter and 0.011 mg/L in summer. The average concentration of iron is found in the range of 0.19 to 0.32 mg/L whereas lead is present in the range of 0.033 to 0.064 mg/L.

Total and Fecal Coliform Organisms

The total coliform shows higher count both in winter and summer seasons at all the stations, exceeding the standards laid down for category 'A' and 'B' of inland surface water (Fig. 29.4e). The values are however, within 5000 CFU/100 mL (max.), standard for category 'C' streams meant for best designated use as drinking water source with conventional treatment followed by disinfection. The fecal coliform counts also exceed the Indian Standard of more than twenty per cent of total coliform present at all the stations (BSI: 1979) (Fig.29.4f).

Biological Quality

The biological quality of river water was examined by counting phyto- and zooplankton. The phyto-plankton community structure and observed algal species are presented in Table 29.5 and 29.6. The dominance of diatoms in river Sai at Bela Pratapgarh has been observed which is indicative of no pollution. The low Palmer's Pollution Index also confirms the same fact. However, the river shows slight organic pollution as indicated by the presence of euglenophyceae which is indicative of organic pollution. The zoo-plankton species and density measured at different sampling stations are presented in Table 29.7. The zoo-plankton count of 120-240/ml was found at different sampling stations with occurrence of four groups. The Shannon-Weaver Index ranged from 0.92 to 1.26, which indicates low pollution in the river.

Table 29.5: Phytoplankton Community Structure of River Sai at Bela Pratapgarh

Station No.	*Number of Species in Sample*	*Total Algal Count/mL*	*Per cent Composition of Algal Groups*					*Palmer's Pollution Index (PI)*
			Cyano-phyceae	*Bacill-ariophyceae*	*Chloro-phyceae*	*Eugleno-phyceae*	*Pyrrho-phyceae*	
R1	11	30	0.0	93.5	6.5	0.0	0.0	12
R2	10	20	0.0	82.3	6.6	11.1	0.0	15
R3	8	20	0.0	80.4	10.3	9.3	0.0	13
R4	-	-	-	-	-	-	-	-
R5	13	50	2.8	87.7	9.5	0.0	0.0	8

Table 29.6: Algal Species in River Sai at Bela Pratapgarh

Sl.No.	*Algal Group/Species*	*Station No.*			
		R1	*R2*	*R3*	*R5*
1.	*Cyanophyceae*				
	Oscillatoria sp.	-	-	-	-
	Phormidium sp.	-	-	-	-
	Aphanocopsa sp.	-	-	-	-
2.	*Bacillariophyceae*				
	Navicula sp.	+	+	+	+
	Gyrosigma sp.	-	+	-	-
	Synedra sp.	+	+	+	+
	Stephanodiscus sp.	+	+	+	+
	Fragillaria sp.	-	-	-	-
	Ceratoneis sp.	+	+	-	+
	Nitzschia sp.	+	+	+	+
	Molosira sp.	-	+	-	-
	Cymbella sp.	+	-	-	-
	Stauroneis sp.	+	-	-	-
	Diatonia sp.	+	-	-	-
	Gomphonema sp.	-	-	-	-
	Pinnularia sp.	-	-	-	-
3.	*Chlorophyceae*				
	Chlorococcus sp.	-	-	-	+
	Chlamydomonas sp.	-	-	-	-
	Chroomonas sp.	-	-	-	-
	Chlorella sp.	-	+	-	-
	Sirocladium sp.	+	-	-	+
	Spirogyra sp.	-	-	-	+
	Phacotus sp.	+	-	-	-
	Selenastrum sp.	+	-	-	-
	Actinastrum sp.	-	-	-	+
	Pithophora sp.	+	-	-	-
	Scenedsmus sp.	-	-	-	-
	Ankistrodesmum sp.	-	-	+	-
	Schroderia sp.	-	-	-	-
	Closterium sp.	-	-	-	-

Contd...

Tab;e 29.6-Contd...

Sl.No.	*Algal Group/Species*	*Station No.*			
		R1	*R2*	*R3*	*R5*
4.	*Euglenophyceae*				
	Euglena sp.	-	+	+	-
5.	*Pyrrhophyceae*				
	Peridinium sp.	-	-	-	-
	Ceratium sp.	-	-	-	-
	Dinobryon sp.	-	-	-	-

+ = Present, - = Absent.

Table 29.7: Density Dominance and Diversity of Zooplankton in River Sai at Bela Pratapgarh

Station No.	*Total Zooplankton per ml*	*Per cent Contribution of Different Groups*					*Shannon Weaver Index (SWI)*
		Protozoa	*Rotifera*	*Copepoda*	*Diptera*	*Nematoda*	
R1	240	16	66	18	-	-	1.26
R2	120	34	66	-	-	-	0.92
R3	160	-	50	-	50	-	1.00
R5	120	34	66	-	-	-	0.92
Species Present :							
R1			Difflugia	Lecane	Cyclops	-	-
R2			Difflugia	Lecane	-	-	-
R3			-	Nothoica	-	Ceratophogon Larva	
R5			Vorticella	Brachinus calcifiorus	-	-	-

Self-Purification Process

The rate of self-purification in river Sai at Bela Pratapgarh is computed from stream data collected at sampling stations, R3 and R4, and presented in Table 29.8. This is the downstream stretch of the river after receiving domestic wastewater from the town. The stretch represents the stream receiving wastewater prior to any treatment. Streter and Phelps formulations have been applied for the computation of self-purification rate constant and allowable BOD loading (Nemerow and Dasgupta, 1991). The following formulae are used to determine the deoxygenation rate (K_1), re-aeration rate (K_2) and the dissolved oxygen-deficit (Dt) at downstream location:

$$K_1 = \frac{1}{t} \log \frac{L_A}{L_B} \quad = \text{de-oxygenation rate} \qquad (1)$$

$$K_2 = K_1 \frac{L}{\quad} - \frac{D}{\quad} \quad = \text{re-aeration rate} \qquad (2)$$

D 2.3 tD

$$Dt = \frac{K_1 L_A}{K_2 - K_1} [10^{-K_1 t} - 10^{-K_2 t}] + D_A \cdot 10^{-K_2 t}$$

= dissolved oxygen deficit downstream (3)

Where,

K_1 = de-oxygenation rate/day

t = time of travel (days)

L_A = upstream ultimate BOD (Kg)

L_B = downstream ultimate BOD (Kg)

K_2 = re-aeration rate/day

L = average ultimate oxygen demand in stream section (Kg)

D = average oxygen deficit in stream section (Kg)

D = change in oxygen deficit from upstream to downstream sampling stations (Kg)

t = time of stream flow from upstream to downstream sampling stations (days)

D_t = dissolved oxygen deficit downstream (Kg) at time t

D_A = dissolved oxygen deficit upstream (Kg)

Table 29.8: River Sai Data Collected on Two Seasons

Station No.	*Temperature*	*Flow*		*BOD_5 20°C*		*Dissolved Oxygen*			
	°C	*(m^3/sec)*	*(m^3/day)*	*(mg/L)*	*(Kg/day)*	*Kg/day*	*Saturated (mg/L)*	*(mg/L)*	*Deficit (mg/L)*
R3	20	30.03	2594592	3.6	9341	18681	9.2	7.2	2.0
	30	25.00	2160000	2.8	6048	14472	7.6	6.7	0.9
				Total	15389				
				Average	7695				
R4	20	33.92	2930688	2.0	5861	22273	9.2	7.6	1.6
	30	23.62	2040768	2.0	4082	11836	7.6	5.8	1.8
				Total	9943				
				Average	4972				

Distance Between Stations 3 and 4 = 1.4 Km

Velocity of River Water = 1.494 Km/sec

The following quantitative results have been obtained using the stream data (Table 29.8):

$$K_1 = \frac{1}{(1.4/1.4944)/24} \cdot \log \frac{7695 \times 1.46^*}{4972 \times 1.46^*}$$

* multiplier used to convert 5 day 20°C BOD to ultimate first stage BOD assuming normal domestic sewage de-oxygenation rate.

$$= 25.64 \times 0.19$$

$$= 4.86;$$

$$L = \frac{\text{BOD (Station 3)} + \text{BOD (Station 4)}}{2},$$

$$= \frac{7695 \times 1.46 + 4972 \times 1.46}{2} = 9247 \text{ Kg/day};$$

$$D = \frac{(1.6 \times 86.4 \times 33.92) + (2.0 \times 86.4 \times 30.03) + (1.8 \times 86.4 \times 23.62) + (0.9 \times 86.4 \times 25.0)}{4},$$

$$= \frac{4689 + 5184 + 3673 + 1944}{4} = 3873;$$

$$D = ½ (1.6 \times 86.4 \times 33.92 + 1.8 \times 86.4 \times 23.62) - ½ (2.0 \times 86.4 \times 30.03 + 0.9 \times 86.4 \times 25.0),$$

$$= 4181 - 3564 = 625 \text{ Kg/day};$$

$$t = \frac{1.4/1.494}{24} = 0.039 \text{ days};$$

$$K_2 = 4.86 \frac{9247}{3873} - \frac{625}{2.3(0.039)\ 3873};$$

$$= 11.60 - 1.799 = 9.8$$

Using Fair's formula, self purification constant, 'f', of river Sai at Bela Pratapgarh is,

$$f = \frac{K_2}{K_1} = \frac{9.8}{4.86} = 2.0$$

According to Fair's classification, the river is falling between category 'C' and 'D' which is characterized as large streams with low to moderate velocity (Fair and Geyer, 1961).

Dissolved Oxygen Sag

The dissolved oxygen-sag curve is plotted by using the values of K_1 and K_2 and the initial condition at station, R3 (D_A and L_A) in Eq. 3. The dissolved oxygen deficit is obtained at any point of the river stretch between stations, R3 and R4. Assuming the reaction rates remain constant in the stretch considered, the dissolved oxygen-sag curve obtained is shown in Figure 29.5. The minimum deficit of DO occurs after 0.031 days time.

Allowable BOD Load

The BOD load exerted to river Sai at Bela Pratapgarh due to the discharge of untreated sewage from the town causes DO deficit, from saturation point, during winter and summer, equivalent to 1.89 and 2.27 mg/L respectively. The actual DO found at station R3 and R4 is above 5 mg/L, which is within the limit prescribed for 'A', 'B' and 'C' category of inland surface water. The BOD values observed in the river after confluence with sewage were also within the standards for B and C category, except at sampling point R3 where the value exceed to 3 mg/L (max.) during winter. The total coliform organisms were found in high concentration in river after confluence with untreated sewage, exceeding the standards laid down for category 'A' and 'B'. The BOD value of the sewage also exceeds the Indian standard of 30 mg/L max for discharge into inland surface water.

Although, the present BOD load can be allowed with the existing river situation, it is recommended to treat the wastewater to the extent that its BOD value comes down to less than 30 mg/L and coliform count is reduced. In order to achieve the twin goals a low-cost treatment through of Pond system has been suggested for the combined treatment of wastewater from Kachha and Pucca drains. The treatment system comprises of facultative and maturation ponds which are with an expected to generate effluent with BOD of 6 mg/L and fecal coliform count of 54 MPN/100 mL.

Conclusion

The quality of water in river Sai is within the acceptable standards except for total and fecal coliforms, which exceed the limit specified by Central Pollution Control Board (CPCB). The wastewater from two out-falls has impact on river water quality as evident by existing pollution load. The stretch of river Sai at Bela Pratapgarh falls under 'C' classification of Indian rivers which is categorised for use for drinking water supply after conventional treatment followed by disinfection.

The river with low to moderate flow as per Fair's classification having self- purification constant of 2.0, is able to retain its DO concentration sufficient to maintain fish life. The DO level between upstream and downstream points in 6 km distance is almost levelled off in both the seasons which indicates that self-purification process is going on simultaneously. The heavy metals, both in wastewater and river water are present in low concentrations or below detectable level. The biological parameters indicate negligible to slight organic pollution in the river.

Acknowledgement

The authors are thankful to Director, NEERI for giving permission to publish this paper.

References

APHA, AWWA, WEF (1995). *Standard Methods for the Examination of Water and Wastewater*, 19th Ed. American Public Health Association, Washington, D.C.

CPCB (1997). *Standards for Liquid Effluents, Gaseous Emissions Automobile Exhaust, Noise and Ambient Air Quality, Pollution Control Law Series*: PCL 14/1995-96.

CPCB (1978-79). *Scheme for Zoning and Classification of Indian Rivers, Estuaries and Coastal Waters* (Pt. One sweet water: ADSORBS/3).

BSI (1979). *Surface Water Standards for various use.*

Fair, G.M. and Geyer, J.C. (1961), *Water Supply and Wastewater Disposal*, John Wiley and Sons, pp. 846-860.

Nemerow, N.L. and Dasgupta, A. (1991). *Industrial and Hazardous Waste Treatment*, Van Nostrand Reinhold, New York, pp. 57-60.

30

Impact of Degradation of Aquatic Ecosystems on Fisheries: A Case Study of Midnapore District, West Bengal

Tapas Paria and Sushil Kanta Konar

Fisheries Laboratory, Department of Zoology, University of Kalyani, Kalyani - 741 235

ABSTRACT

The study was made to assess the impact of degradation of aquatic ecosystems on fisheries in Midnapore district, West Bengal. The field studies were made in 20 blocks and a total of 167 ponds were selected at random. The study was conducted from August 1994 to July 1997. The physico-chemical parameters such as pH, temperature, transparency, alkalinity, hardness, dissolved oxygen, nitrate and phosphate of water and pH, available nitrogen and available phosphorus of soil were evaluated. Different groups of pesticides such as organophosphate, chlorinated hydrocarbon and carbamate and different types of fertilizers such as nitrogenous, potash, phosphorus and mixed types drained from adjacent crop fields to water bodies. Ponds were to be highly silted. Stocking density was high, 15,984 to 56,803 numbers/hectare, which was 2.45 to 8.73 times more than the normal recommended rate of 6,500 fish/hectare. Fishes in most of the ponds were affected by different types of diseases, such as epizootic ulcerative syndrome (EUS), gill-rot, dropsy, malnutrition, tail and fin-rot, argulosis, tumor and fungal diseases. Several factors were found to be responsible for low productivity of these impounded waters. Fish production ranged from 1,027.77 to 2,362.50 kg/ha per year, which was 4.37 to 1.90 time less than the normal predicted rate of 4,500 kg/ha per year. Therefore, rational eco-friendly management methods are to be adopted for sustainable yield not only for economic point of view but also from the ecological aspect of the management.

Introduction

Fresh water fish production greatly depends on culture-based impounded water resources. Almost all the impounded water resources were gradually become degraded by contamination with foreign materials and antropogenic activities (Konar *et al.*, 1997). Aquaculture and agriculture can go very well together side by side balancing the economy of natural resources and they could also complement each other (Thakur and Thakur, 1991). Water pollution manifests through changes of physical, chemical and biological conditions (Rao *et al.*, 1996). Anthropogenic activities are subjected to various physico-chemical stress either of short or prolonged duration (Das *et al.*, 1994). Good pond management schedules such as liming, manuring, and application of supplementary feed are not followed by fishermen in West Bengal (Mahapatra *et al.*, 1992). When population densities are increased in a particular ecosystem there is corresponding increase the risk of disease outbreak (Kumar and Dey, 1991). According to Ayyappan and Bhowmik (1999), scientists feel that systematic and comprehensive research is needed on diseases and their causative factors to prevent fish mortalities and increase the fish yield. Therefore, this research programme was undertaken in Midnapore district of West Bengal for generating scientific data on the present status of impounded these water bodies and to formulate a scientific management plan for sustainable fish yield.

Materials and Methods

Midnapore district lies in between latitude 21°36′ and 22°57′ N and longitude 86°33′ and 88°11′E. The study was conducted for two consecutive years (August 1994 to July 1996). The study made in 20 blocks such as Daton (8), Debra (8), Garbeta (11), Keshpur (8), Midnapore Sadar (9), Chandrakona (8), Ghatal (8), Kharagpur (11), Sabong (8), Tamluk (9), Panskura (8), Nandigram (6), Mahishadal (8), Contai (7), Egra (9), Ramnagar, (11), Khejuri (7), Binpur (9), Gopiballavpur (8) and Nayagram (9) and total 170 number of ponds were selected for investigation (Fig. 30.1) (Number of ponds studied are

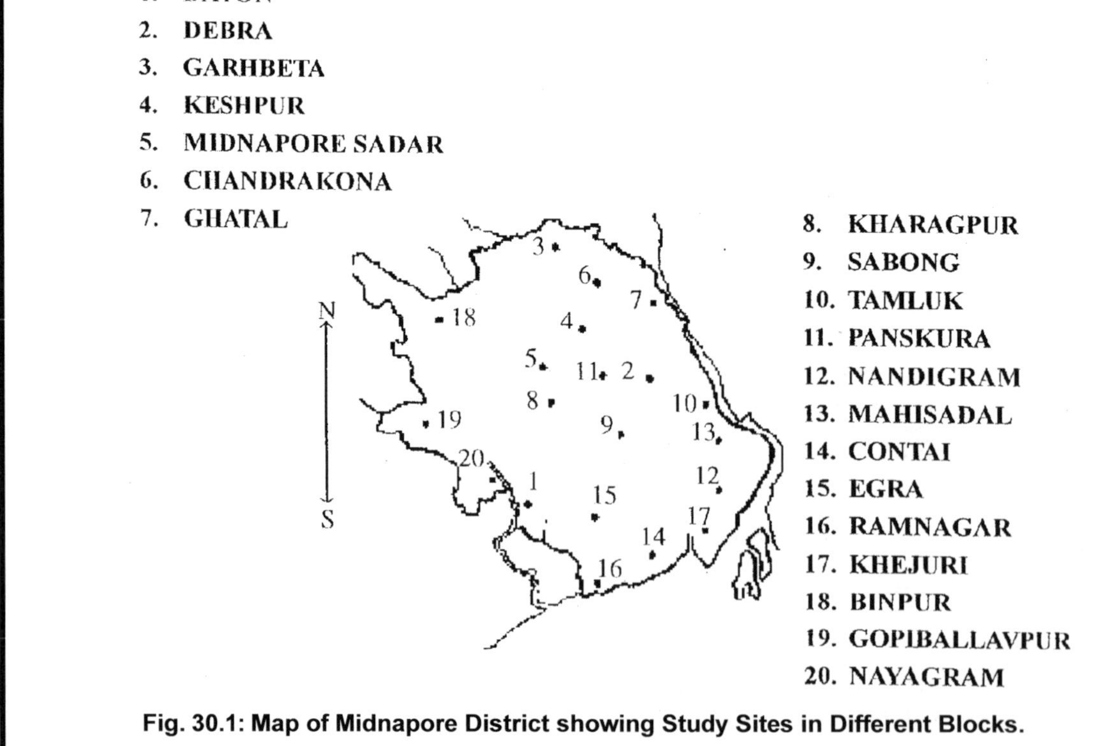

Fig. 30.1: Map of Midnapore District showing Study Sites in Different Blocks.

given in parentheses). During the time of study, two methods were followed: direct investigation on randomly selected ponds for evaluating different physico-chemical parameters (temperature, water depth, transparency, water pH, dissolved oxygen, alkalinity, hardness, nitrate and phosphate of water and soil pH, available nitrogen, and available phosphorus and depth of siltation and secondly, by taking interviews with the fish farmers using a pre-tested structural schedule of questionnaires. A total 170 number of farmers were interviewed on various aspects of pond management.

The information of methods on pond management, fish stocking density, fish production, drainage of fertilizers and pesticides from adjacent crop fields and incidence of fish diseases throughout the year were collected by help of pre-tested structured schedule from different culture impounded waters.

Assessment of the physico-chemical parameters followed by the standard methods given in APHA *et al.* (1980) and Jackson (1973). Studies were made using field kit for most of the parameters except nitrate and phosphate of water and available nitrogen and available phosphorus of soil, which were done on the laboratory on sample collected from various study sites. Statistical analysis followed by the methods of Snedecor and Cochran, (1967), Downie and Heath, (1970), Elhance and Elhance (1992). Climate conditions of Midnapore district such as rainfall, temperature, pressure and humidity for entire period of present investigation were recorded from Meterology Department of Alipur, Calcutta (Fig. 30.2).

Results and Discussion

The data on physico-chemical parameters of water and soil, present status of fish farming, practices, drainage of pollutants into ponds, incidence of fish diseases and fish production, in Midnapore district of West Bengal were recorded.

Physico-chemical Parameters

The physico-chemical parameters of water and soil of different blocks in different impounded waters of Midnapore district are shown in Table 30.1 and 30.2.

The depth of water ranged from 1.28 to 2.17 m and average was 1.83 m, water pH and soil pH was found to be neutral, it ranged from 6.5 to 9.1 and 6.83 to 7.45 respectively. The fish production is quite normal when the pH value ranged between 7.5 to 8.5 in culture ponds (Swingle, 1969) and a depth little above 2 m can be considered good enough from the point of biological activity of a water body and normal fish growth was observed at pH 7.5 - 8.5 (Jhingran, 1982). In the present study dissolved oxygen was ranged from 3.6 to 8.4 ppm. The suitable environment for fish culture seeds 5.0 ppm of dissolved oxygen (Ghosh, 1978, Boyd, 1982). The observed minimum alkalinity was 18.0 ppm and maximum 244.0 ppm and total hardness was minimum 10.0 ppm and maximum 164.0 ppm. Alikunhi, (1957) reported that a highly productive ponds, the alkalinity ought to be over 100.0 ppm and Boyd, (1982) reported that above 20.0 ppm it become satisfactory for productivity and for protecting fish against harmful effects. Water transparency caused by zooplankton and phytoplankton is generally desirable in fish pond. Transparency of water varied from 6.5 to 19.0 cm and average 13.12 cm. Boyd, (1982) observed, the normal fish growth at 30-80 cm of transparency but Mahajan and Kanhere, (1995) reported that transparency ranged of 8.2-3.2 cm is suitable for fish culture. Nitrate and phosphate of water ranged from 0.84 to 7.70 ppm and 0.86 to 7.43 ppm respectively. Banerjea, (1967) and Jhingran, (1982) reported that the nitrate and phosphate of water above 0.20 ppm is characteristic of productive waters. Available nitrogen and available phosphorus showed wide fluctuation which varied from 94.0 to 676.0 mg/100g of soil and 1.09 to 9.41 mg/100 g of soil

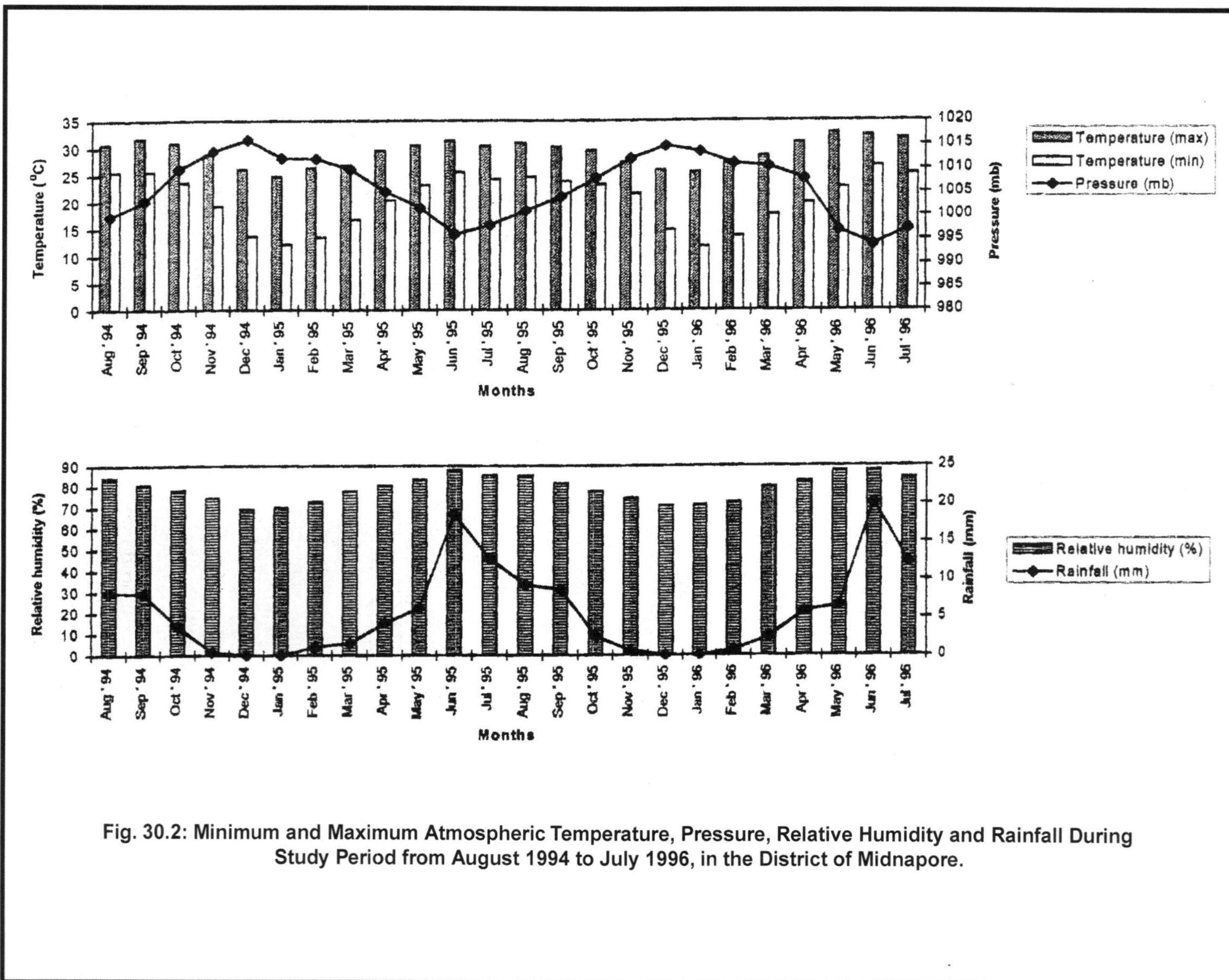

Fig. 30.2: Minimum and Maximum Atmospheric Temperature, Pressure, Relative Humidity and Rainfall During Study Period from August 1994 to July 1996, in the District of Midnapore.

Table 30.1: Estimation of Physico-chemical Parameters of Pond Water in Different Blocks of Midnapore District of West Bengal

Sl. No.	Name of Blocks	Depth of Water (m)	Water pH	Dissolved Oxygen (ppm)	Alkalinity (ppm)	Hardness (ppm)	Transparency (cm)	Nitrate (ppm)	Phosphate (ppm)
1.	Daton	2.01	7.38 (6.9 - 7.9)	5.40 (3.6 - 6.8)	35.25 (26 - 64)	55.25 (33 - 92)	13.43 (10.5 - 15.5)	1.36	1.16
2.	Debra	1.94	7.78 (7.3 - 8.2)	6.15 (4.0 - 7.6)	30.25 (24 - 40)	31.50 (18 - 48)	12.93 (10.5 - 14.5)	1.38	1.32
3.	Garbeta	1.86	7.78 (7.1 - 9.0)	5.74 (5.2 - 6.4)	95.81 (38 - 150)	73.27 (10 - 146)	12.36 (10.5 - 16.5)	2.44	1.58
4.	Keshpur	2.11	7.53 (7.1 - 8.2)	5.55 (4.0 - 6.8)	37.50 (26 - 46)	53.50 (26 - 72)	13.12 (11.0 - 14.5)	2.39	1.98
5.	Midnapore-Sadar	2.17	7.81 (7.0 - 9.1)	5.00 (4.8 - 6.4)	34.22 (20 - 46)	29.77 (14 - 48)	12.94 (9.5 - 16.0)	6.35	7.43
6.	Chandrakona	2.05	7.58 (7.1 - 8.3)	6.15 (4.4 - 7.6)	37.25 (20 - 54)	77.00 (46 - 104)	14.31 (11.5 - 16.5)	2.86	1.45
7.	Ghatal	1.91	7.26 (6.7 - 8.1)	5.55 (4.4 - 6.4)	46.50 (26 - 72)	40.25 (20 - 52)	12.62 (9.5 - 15.5)	7.70	2.95
8.	Khargapur	1.67	7.48 (6.8 - 8.5)	5.65 (4.4 - 6.4)	32.85 (12 - 68)	32.57 (14 - 82)	13.71 (11.5 - 19.0)	2.69	1.60
9.	Sabong	2.17	7.56 (6.9 - 8.1)	5.15 (4.0 - 6.0)	34.75 (24 - 42)	33.00 (18 - 50)	13.37 (10.5 - 16.5)	2.18	1.58
10.	Tamluk	1.67	7.57 (6.8 - 9.4)	5.42 (4.0 - 6.8)	58.89 (20 - 160)	44.00 (24 - 68)	12.39 (9.0 - 16.5)	1.48	1.32
11.	Panskura	1.90	7.91 (7.0 - 9.1)	6.00 (5.2 - 7.6)	86.25 (34 - 244)	42.25 (16 - 56)	11.87 (10.0 - 13.0)	1.60	1.12
12.	Nandigram	1.86	7.25 (6.7 - 7.6)	5.53 (4.8 - 5.6)	57.33 (28 - 142)	87.67 (16 - 164)	14.17 (12.0 - 18.0)	0.97	1.20
13.	Mahisadal	2.09	7.40 (7.1 - 7.8)	6.15 (4.8 - 8.4)	75.50 (28 - 162)	64.50 (30 - 124)	13.37 (10.5 - 17.5)	1.87	2.88
14.	Contai	1.56	7.38 (6.6 - 7.9)	5.37 (4.0 - 7.2)	41.71 (18 - 96)	49.42 (16 - 86)	13.85 (8.5 - 16.5)	0.84	1.62
15.	Egra	1.29	7.42 (6.9 - 8.9)	5.33 (3.6 - 6.8)	37.78 (24 - 56)	40.00 (10 - 86)	12.83 (9.5 - 17.5)	1.58	1.51
16.	Ramnagar	1.28	7.23 (6.5 - 7.9)	5.30 (4.0 - 6.0)	38.67 (20 - 64)	44.17 (16 - 84)	13.67 (8.5 - 18.5)	1.23	1.81
17.	Khejuri	1.40	7.20 (6.9 - 7.5)	5.26 (4.0 - 6.4)	45.14 (40 - 58)	45.14 (34 - 58)	14.14 (10.5 - 18.5)	2.08	1.15
18.	Binpur	1.60	7.36 (6.7 - 8.0)	4.84 (4.0 - 6.0)	47.77 (26 - 88)	70.22 (20 - 128)	13.16 (9.5 - 15.5)	3.24	0.86
19.	Gopiballavpur	2.05	7.21 (6.8 - 7.8)	4.80 (3.6 - 5.6)	43.50 (22 - 126)	36.00 (18 - 52)	11.31 (6.5 - 13.5)	2.32	1.23
20.	Nayagam	2.05	7.48 (7.1 - 8.1)	5.22 (4.4 - 6.4)	27.77 (18 - 38)	41.77 (26 - 54)	13.00 (11.0 - 14.5)	3.09	2.19
	Average	1.83	7.47	5.47	47.23	49.53	13.12	2.48	1.89

respectively. Banerjea, (1967) reported that available nitrogen above 75.0 mg/100g of soil and available phosphorus above 6.0 mg/100g of soil indicate ponds are highly productive. Water temperature fluctuation was recorded to be 27.21°C. Fish growth was also recorded better at temperature range of 14.5 to 38.6°C (Sharma and Gupta, 1994).

Table 30.2: Estimation of Siltation, Soil pH, Available Nitrogen and Available Phosphorus in Different Blocks Under Midnapore District of West Bengal

Sl. No.	*Name of Blocks*	*Siltation (m)*	*Soil pH*	*Available Nitrogen (mg/100 g of soil)*	*Available Phosphorus (mg/100 g of soil)*
1.	Daton	0.457	7.06	228.0	4.55
2.	Debra	0.353	7.12	291.0	2.47
3.	Garbeta	0.658	7.45	587.0	4.45
4.	Keshpur	0.874	7.00	291.0	3.56
5.	Midnapore-Sadar	0.816	7.22	127.0	1.09
6.	Chandrakona	0.557	7.18	117.0	5.49
7.	Ghatal	0.627	7.06	324.0	2.87
8.	Khargapur	0.341	6.92	676.0	3.46
9.	Sabong	0.348	7.06	183.0	2.57
10.	Tamluk	0.417	7.11	648.0	9.41
11.	Panskura	0.405	7.24	226.0	2.67
12.	Nandigram	0.533	7.17	390.0	3.46
13.	Mahisadal	0.493	7.12	117.0	2.57
14.	Contai	0.417	6.92	353.0	4.75
15.	Egra	0.435	6.83	620.0	2.09
16.	Ramnagar	0.777	6.92	259.0	4.75
17.	Khejuri	0.353	6.93	653.0	2.77
18.	Binpur	0.304	6.88	94.0	3.66
19.	Gopiballavpur	0.285	6.93	259.0	6.24
20.	Nayagram	0.341	7.16	554.0	1.18
	Average	0.489	7.06	349.8	3.70

Siltation

Siltation was found to be common problem in almost all the ponds in Midnapore district, West Bengal. Investigated ponds were found to be highly silted, it ranged from 0.285 to 0.874 m (Table 30.2). Nitrate originates from the reduction of nitrate by bacteria in anaerobic mud or water (Boyd, 1980). The abundance of bottom fauna adversely affected by siltation (Jhingran, 1991). Organic matters produced different types of noxious gases such as hydrogen sulphide and ammonia. The ammonia produce toxic nitrite, causing respiratory trouble in fishes even though enough dissolved oxygen present in pond water. High siltation cause adverse effect to fish population (Jhingran, 1982).

Source of Pollution

Fertilizers drainage

Several types of fertilizers were used in crop fields such as nitrogen, potash, phosphorus and mixed types. These fertilizers may contaminated with pond water by run-off. Fertilizers increase the pond fertility. However, excess of fertilizers reduces the pond productivity (Colt and Armstrong, 1981). Nitrogen fertilizers was maximum used in Binpur, Chandrakona and Daton, Potash was maximum used in Binpur, Gopiballavpur, Phosphorus was maximum used in Chandrakona, Binpur, Ghatal and Sabong and mixed types was maximum used in Panskura and Mahishadal blocks (Fig. 30.3). Sexena and Chauhan, (1993) reported the excess fertilizers are harmful to fish growth. The frequent addition of fertilizers into fish ponds may affect the feeding rate of fish and fish yield (Sarkar and Konar, 1983).

Pesticides drainage

Different groups of pesticides such as organophosphate, chlorinated hydrocarbon and carbamates were used in pond adjacent crop fields. The pesticides of organophosphate group was maximum used in Ghatal, Binpur, Panskura, Nandigram, Chandrakona, Daton and Midnapore Sadar, chlorinated hydrocarbon group was maximum used in Khejuri and Ramnagar and carbamate group was maximum used in Mahishadal and Binpur and moderate doses was used in rest blocks (Fig. 30.4). Pesticides reduce the survival rate, growth and reproduction of fishes and it may also induce histological and physiological disorders (Chakraborty and Konar, 1974, Das and Kona, 1974). Blakely and Hrusa, (1989), Paria and Konar, (1999a, 1999b, 1999c, 1999d, 2000) reported that pollution is a problem for fish pond managers in areas where agricultural pesticides may be carried into the pond water by storm run-off from surroundings fields.

Inputs used

Several types of inputs used by fish farmers into their ponds during fish culture. These inputs generally include manure and fertilizers, lime, supplementary feed and fish seed. The observation were made both by direct methods and assessment through personal interview with fish farmers using pre-tested structured schedule.

Application of fertilizer and manure

In present investigation maximum amount of manures and fertilizers was found to be 7,590.94 kg/ha per year in Chadrakona block and minimum amount was 693.75 kg/ha per year in Sabong block of Midnapore district (Table 30.3). Some farmers applied less quantity than the required amount while others applied higher them at rates. Deficiency of nutrients is usually corrected through application of fertilizers or manures. But fish farmers use the fertilizers and manures irregularly and indiscriminately. Therefore, instead of improvement, production is reduced due to degradation of the pond environment. According to Jhingran (1982), the natural productivity of a pond can be greatly enhanced by use of fertilizers. But, Chatterjee (1996) reported that excess amount of fertilizer create adverse effect to fishes.

Application of lime

Lime usually applied to raise the soil pH to a desired level and to reduce the toxicity of harmful compounds besides disinfecting the environment against fish diseases. The application of lime in culture ponds varied from 117.18 to 679.10 kg/ha per year (Table 30.3) Jhingran (1982) reported that the rate of lime application as 500 kg/ha at pH 5.5-6.5 and 200 kg/ha at pH 6.5-7.5.

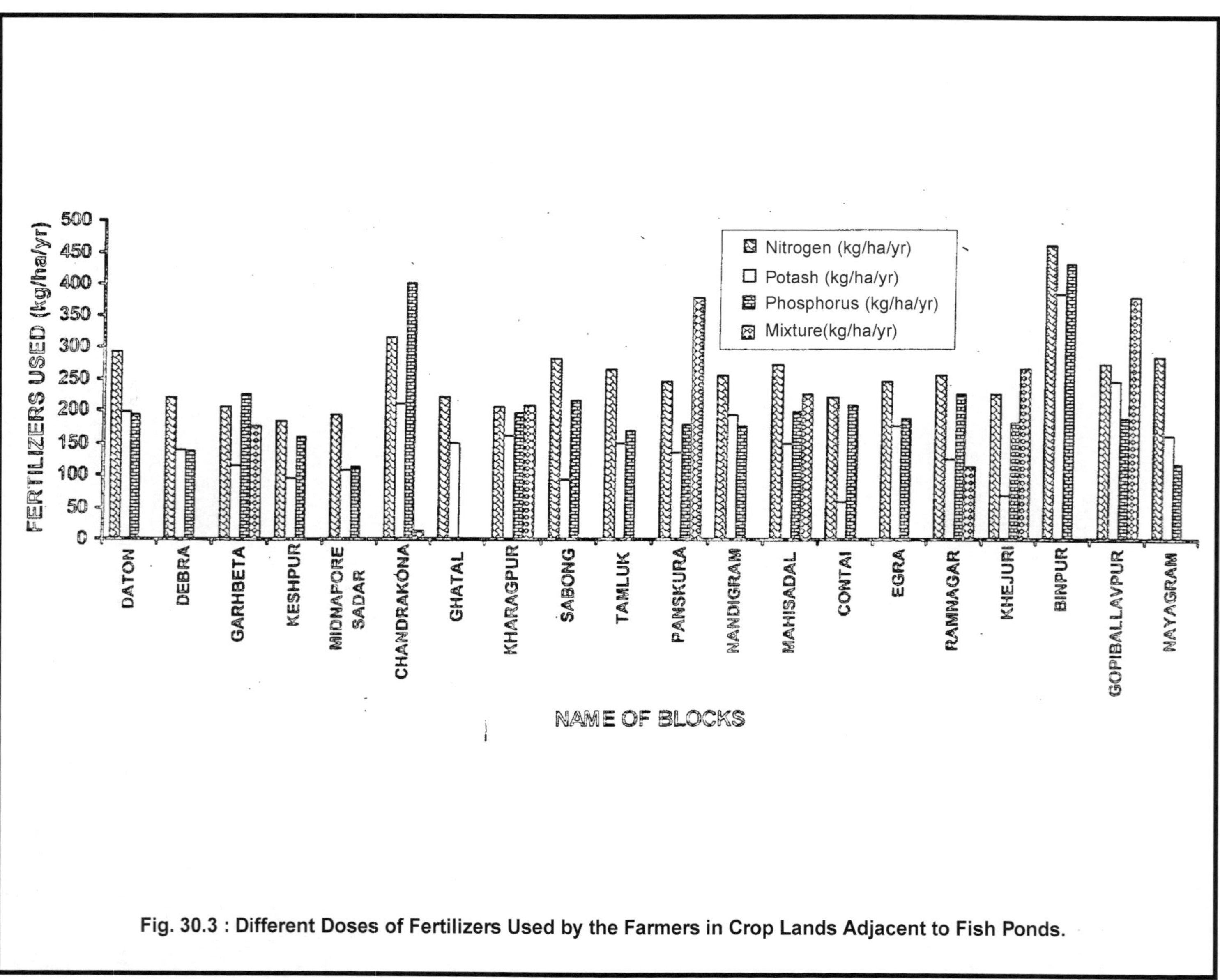

Fig. 30.3 : Different Doses of Fertilizers Used by the Farmers in Crop Lands Adjacent to Fish Ponds.

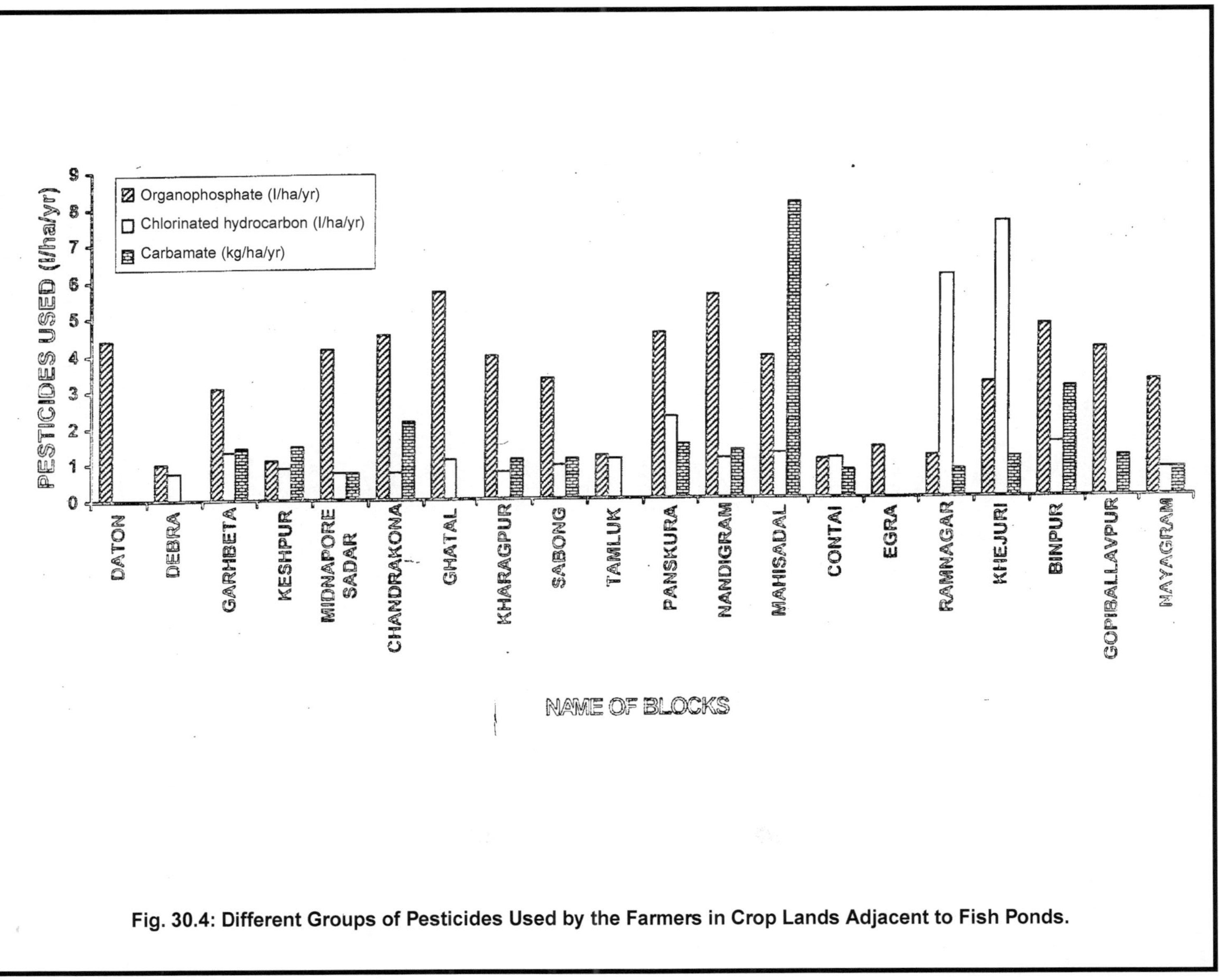

Fig. 30.4: Different Groups of Pesticides Used by the Farmers in Crop Lands Adjacent to Fish Ponds.

Application of feed

The fish farmers in Midnapore district of West Bengal used rice bran and oil-cake as supplementary feed. The quantity of feed is also negligible as most of the feed are added together in one corner of the pond in one instalment. In present study application of supplementary feed ranged from 595.31 to 2,643.75 kg/ha per year (Table 30.3). According to Patra (1990) the organic manure and supplementary feeding enhance the production of Indian Major Crops (IMC) in rural areas of West Bengal.

Stocking density

The fish production has been directly depends on the stocking, density. Fish farmers were stocked high density of fingerlings, in the hope of getting higher yield from these fish culture ponds. The stocking density of fry and fingerlings were generally very high, it was ranged from 15,984 to 56,803/ha, which was 2.45 to 8.73 times more than the normal recommended rate 6,500/ha (Table 30.4). The fry and fingerlings did not sterilised generally in the time of stocking. The fish growth and fish mortality is directly depends with density (Krebs, 1972).

Table 30.3: Different Amount of Manure, Fertilizer, Lime and Feed Applied by the Farmers in Different Blocks Under Midnapore District of West Bengal

Sl. No.	*Name of Blocks*	*Manure and fertilizer (kg/ha/yr)*	*Lime (kg/ha/yr)*	*Supplementary Fed (kg/ha/yr)*
1.	Daton	2681.25	178.12	1763.31
2.	Debra	1610.52	267.18	1358.14
3.	Garbeta	1777.23	206.25	1084.68
4.	Keshpur	1884.37	210.00	775.78
5.	Midnapore-Sadar	3095.21	196.87	1158.60
6.	Chandrakona	7590.94	679.10	1665.62
7.	Ghatal	1875.00	243.75	1443.34
8.	Khargapur	1085.93	342.18	1139.06
9.	Sabong	693.75	271.87	1633.59
10.	Tamluk	907.50	300.00	595.31
11.	Panskura	1106.25	318.75	1987.50
12.	Nandigram	2554.09	192.18	750.00
13.	Mahisadal	1653.12	355.48	1400.00
14.	Contai	984.37	117.18	685.00
15.	Egra	1307.81	296.25	1412.50
16.	Ramnagar	1183.12	158.48	2643.75
17.	Khejuri	743.75	625.00	890.00
18.	Binpur	1574.05	215.62	1122.06
19.	Gopiballavpur	1196.62	120.31	862.50
20.	Nayagram	1185.00	135.00	670.20
	Average	1834.79	271.47	1252.05

Diseases

The aquatic environmental factors changes, pathogenicity of pathogens become high and fish with poor health and weak resistance will be infected. Fishes were suffered from various types of diseases such as epizootic ulcerative syndrome (EUS), gill-rot, tail and fin rot, dropsy, argulosis, fungal diseases and tumour (Fig. 30.5). Under stressed condition of agrochemicals, high densities and mismanagement, fishes lose appetite and thus they suffered from malnutrition. EUS was most prevalent than all other diseases. The outbreak of EUS found in agrochemical contaminated water (Das and Das, 1993). Das *et al.* (1994) reported that the unfavourable water quality is responsible for different types of fish diseases.

Production

Production of culture impounded waters directly depends upon the surrounding environmental factors and management procedures. Production was reduced 4.37 to 1.90 time from the normal predicted rate 4,500 kg/ha per year, ranged from 1,027.77 to 2,362.50 kg/ha per year (Table 30.4). Production was positively correlated with hardness (r = 0.039), transparency (r = 0.238) and negatively correlated with water pH (r = 0.211), dissolved oxygen (r = - 0.056), alkalinity (r = - 1.56) and soil pH (r = - 0.141).

Table 30.4: Estimation of Stocking Density and Production in the District of Midnapore, West Bengal

Sl. No.	*Name of block*	*Stocking density* (no./ha/yr)	*Production* (kg/ha/yr)
1.	Daton	22,699.00	1,590.62
2.	Debra	38,275.00	1,681.25
3.	Garbeta	74,924.00	2,362.50
4.	Keshpur	15,984.00	2,085.93
5.	Midnapore-Sadar	18,945.00	1,230.70
6.	Chandrakona	22,375.00	1,781.25
7.	Ghatal	31,536.00	1,497.00
8.	Khargapur	31,157.00	2,025.56
9.	Sabong	25,500.00	2,156.25
10.	Tamluk	29,970.00	1,921.87
11.	Panskura	33,955.00	1,135.68
12.	Nandigram	38,768.00	1,546.87
13.	Mahisadal	46,375.00	1,937.50
14.	Contai	33,633.00	1,916.66
15.	Egra	33,867.00	1,950.00
16.	Ramnagar	23,503.00	1,847.50
17.	Khejuri	23,624.00	1,027.77
18.	Binpur	56,804.00	1,950.00
19.	Gopiballavpur	29,859.00	1,168.75
20.	Nayagram	23,472.00	1,634.37

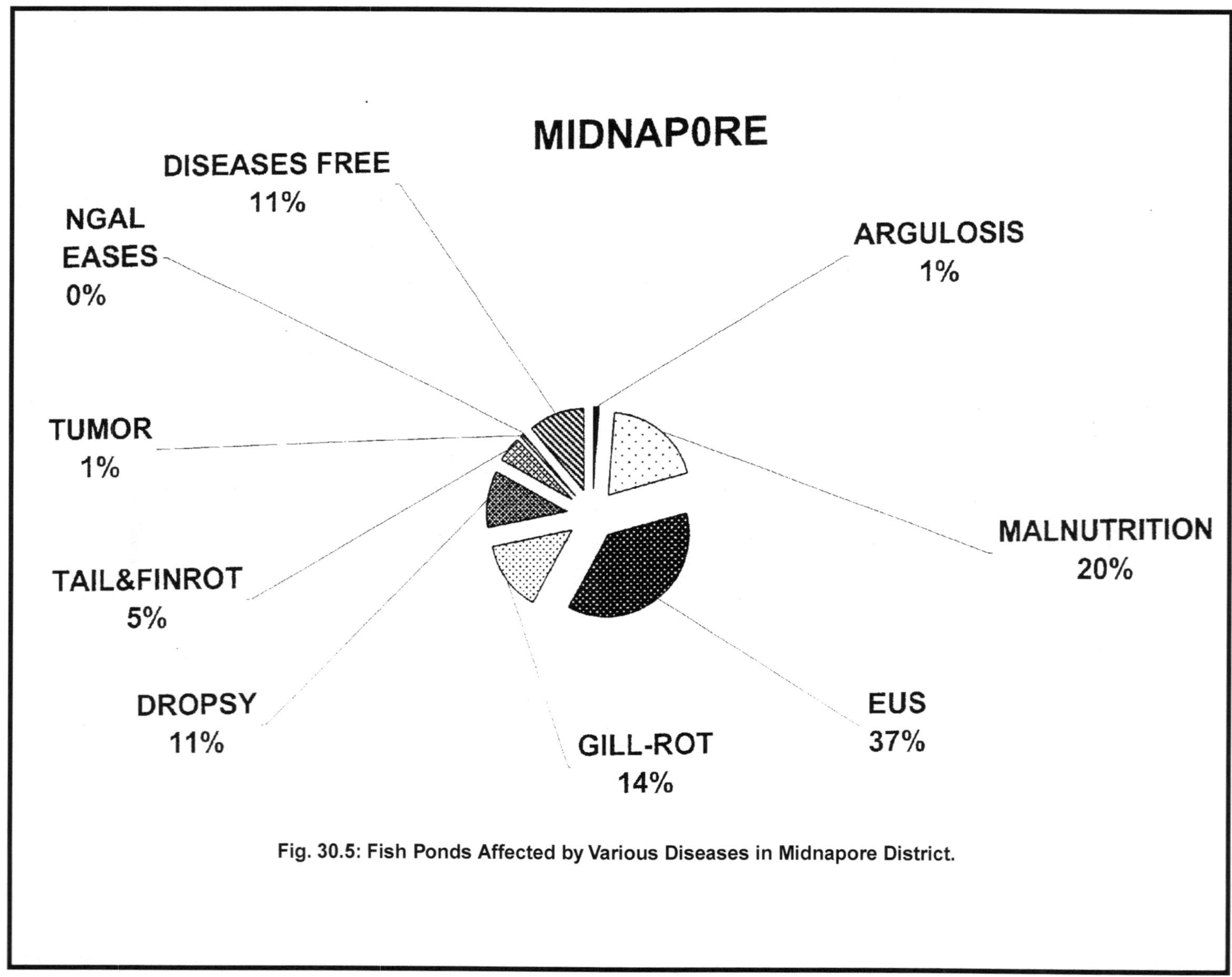

Fig. 30.5: Fish Ponds Affected by Various Diseases in Midnapore District.

Conclusion

The multiplicity of factors were influencing the productivity of ponds. Physico-chemical parameters were influenced the normal, fish growth. Drainage of pesticides and fertilizers from adjacent crop fields, heavy siltation, high density of fry and fingerlings stocking and poor management were may be the caused for low productivity of the impounded water bodies. Epizootic ulcerative syndrome was a major thrust on the economy of aquaculture caused means moralities of pond fishes. Excessive fertilizer can also reduced the fish production. Rational management methods are to be adopted for sustainable production of fishes from these ponds and prevent the rural economic loss.

Acknowledgements

We thank Head of Zoology Department, Kalyani for providing facilities for the research. We are thankful to Department of Science and Technology and Non-Conventional Energy Sources, Government of West Bengal for financial help of this work. We also thankful the Gram Panchyats of Midnapore district of West Bengal for rendering a helping hand during the course of this study.

References

Alikunhi, K.H. (1957). Fish culture in India. Fm. Ball Indian coun. *Agric. Res.*, 20: 144.

APHA (American Public Health Association), American Water Works Association and Water Pollution Control Federation (1980). Standard methods for the examination of water and waste water. 14th edition. *Am. Publ. Hlth. Assoc.*, Washington, USA.

Ayyapan, S. and Bhowmik, M.L. (1999). Aquaculture development in India with particular reference to West Bengal. *In: Workshop on Environmental friendly Inland Fisheries and Aquaculture for 21st Century in West Bengal* organized by the retired Govt. Fishery Officers and Staff Association, Calcutta in March 10, 1999 at Bangla Academy, Calcutta: pp. 1-15.

Banerjea, S.M. (1967). Water quality and soil condition of fish ponds in some states of India in relation to fish production. *Indian. J. Fish.*, 14: 115-144.

Blakely, D.R. and Hrusn, C.T. (1989). *Inland Aquaculture Development Handbook.* Fishing News Book: pp. 44-83.

Boyd, C.E. (1980). Reliability of water analysis kits. *Trans. Amer Fish Soc.*, 109: 239-243.

Boyd, C.E. (1982). *Water quality management for pond fish culture.* Elsevier Publication. Netherland: 318 p.

Chakraborty, G. and Konar, S.K. (1974). Chronic effects of sublethal levels of pesticides on fish. *Proc. Nat. Acad. Sci. India.* 44: 241-426.

Chatterjee, N.R. (1996). Influence of fertilizer combination on the productivity of water body. *Poll. Res.*, 15(2): 99-101.

Colt, J. E. and Aarmstrong, D.A. (1981). Nitrogen toxicity to crustarcean, fish and molluscs. *Proc. of the bioengineering symp. for fish culture.* Fish culture sections. *Ans. Fish Soc. Bethesda,* MD, USA: pp 34-37.

Das, M.K. and Das, R.K. (1993). A review of the fish disease epizootic ulcerative syndrome in India. *Environ. Ecol.*, 11: 134-145.

Das, M.K. and Konar, S.K. (1974). Effects of sublethal levels of pesticides on the feeling, behaviour, survival and growth of fish. *Proc. Nat. Acad. Sci. India.* 44: 235-240.

Das, M.K., Das, R.K., Ghosh, S.P. and Bhowmik, S. (1994). Fish disease in relation with environmental factors in a sewage fed wetland. *J. Inland Fish. Soc., India,* 26: 100-105.

Downie, N.M. and Health, R.W. (1970). *Basic Statistical Methods.* Harper and Row Publishers, New York, 356 p.

Elhance, D.N. and Elhance, V. (1992). Fundamentals of Statistics. Kitab Mahal, Allahabad: pp. 11.1 - 11.7.

Ghosh, S.R. (1978) Role of soil and water quality study in fishpond. *Summer Institute on Inland Aquaculture CIFRI (ICAR),* Barrackpore, 6 p.

Jackson, M.L. (1973). *Soil chemical analysis.* Prentice Hall of India Pvt. Ltd. New Delhi, India.

Jhingran, A.G. (1991). Impact of Environmental stress on freshwater fisheries resources. *J. Inland Fish. Soc., India,* 23: 20-32.

Jhingran, V.G. (1982). *Fish and Fisheries of India.* Hindustan Publishing Corporation (India). New Delhi, India.

Konar, S.K. Paria, T. and Das, M.C. (1997). Habitat degradation as threat to piscine abundance. In: *Changing Perspectives of Inland Fisheries,* (Eds. Vass, K.K. and Sinha, M), pp. 8-14, Proc. of the National Seminar. March 16-17. 1997. IFSI Barrackpore, India.

Kumar, D. and Dey, R.K. (1991). Fish diseases in India. In: *Aquaculture Productivity* (Eds. Sinha, V.R.P. and Srivastava, H.C.), pp. 316-343.

Krebs, C.J. (1972). *Ecology,* Harper and Row publishers, New York.

Mahajan, A. and Kanhere, R.R. (1995). Seasonal variations of abiotic factors of a fresh water pond at Barwani (M.P.) *Poll. Res.,* 14: 347-350.

Mahapatra, B.K., Dutta, N.C., Chatterjee, P. and Saha, D. (1992). Adoption of composite fish culture technology by the rural fisherfolks of Sundarbans. *J. Fresh Biol.,* 5: 197-202.

Paria, T. and Konar, S.K. (1999a). Management of pond resources for fish production in Bankura district, West Bengal. *Environ. Ecol.,* 17: 306-309.

Paria, T. and Konar, S.K. (1999b). Ecological status of impounded waters in the districts of Cooch Behar and Dajeeling for fish production. *Indian J. Environ. of Ecoplan.,* 2: 269-274.

Paria, T. and Konar, S.K. (1999c). Present status of the impounded waters of district Birbhum (West Bengal) far sustainable fish production. *Environ. Ecol.,* 17: 931-935.

Paria, T. and Konar, S.K. (1999d). Management of fish ponds and its relation to fish diseased in West Bengal, India, *Environ. Ecol.,* 17: 962-970.

Paria, T. and Konar, S.K. (2000). Management practices for fish production in impounded waters of Howrah (West Bengal, India). In water Recycling and Resource Management in the Developing World, Eds, B.B. Jana, R.D. Banerjee, B. Guterram and J. Heeb, Univ. Kalyani, India & International Ecological Engineering Society, Switzerland: pp. 141-147.

Patra, B.C. (1990). A report on the effect of organic manures and supplementary feeding on the production of Indian Major Carps in rural areas of West Bengal. *J. Freshwater Biol.,* 5(4): 347-357.

Rao, M.M., Rao, A.V.N., Mahmood, S.K. (1996). Assessment of water quality and pollution of Narsingi pond. *Ecol. Environ. and Cons.*, 2: 45-49.

Sarkar, S.K. and Konar, S.K. (1983). Impact of muriate of potash on the feeding behaviour, survival growth and reproduction of fish. *Environ. Ecol.*, 1: 259-261.

Sexena, K.K. and Chanhan, R.R.S. (1993). Acute toxicity of agriculture fertilizers complex on food composition and growth in *Heteropneustes fossils* (BL). *Poll. Res.*, 12: 135-138.

Snedecor, G.W. and Cochran. W.G. (1967). *Statistical Method.* Oxford and IBH Publishing Co. New Delhi, 539 pp.

Sharma, L.L. and Gupta, M.C. (1994). Some aspects of limnology of Awarchand Reservoir Rajsamend, Rajasthan: Physical Parameters. *Poll. Res.*, 13: 169-179.

Swingle, H.S. (1969). Standardization of chemical analysis for water and pond muds. *FAO Fish Rep.*, 4: 397-421.

31

Benthic Foraminifera in Evaluating Environmental Stresses in Marginal Marine Environment: A Case Study

Sabyasachi Majumdar, Abhijit Mitra*, U.C. Panda and Amalesh Choudhury**

S.D. Marine Biological Research Institute, Sagar Island, Sundarbans, West Bengal
** Department of Marine Science, University of Calcutta*
*** DOD Unit, Regional Research Laboratory, Bhubaneshwar - 751 013*

ABSTRACT

Coastal, estuarine and other marginal marine environment act as recipient for various kinds of anthropogenie wastes thereby resulting in severe negative impact on local biota. Due to their abundance and better preservative potential benthic foraminifera serve as one of the most sensitive and inexpensive tracers in evaluating environmental stresses in marginal marine environment.

Recent benthic foraminiferal studies were carried out in eight stations across a sector connecting Sewri mudflat and Shivagi Nagar, Mumbai. Two stations each were choosen from Sewri and Shivagi Nagar whereas four stations were sampled along the bay connecting the two sides (Arabian sea). Assessment of hydrobiological and sedimentological characteristics by employing standard analytical techniques revealed that the study area was grossly polluted by heavy metals, oils, tars and other organics. A total of 24 species of benthic foraminifera were identified from the study site. In general, Foraminiferal diversity and density were poor. The faunal health also shows impact of severe pollution. Stunted growth and deformation of tests were noticed. Total absence of agglutinated foraminifera and abundance of polychaete worm were also observed, indicating severe environmental stresses.

Key Words*:* Benthic foraminifera, stress, marginal marine.

Introduction

It is well established that anthropogenic habitat disturbances and destructions have severe hazardous impact on marine biodiversity. (Connell 1978, Ehrlich 1986, Richmond 1993). Marginal marine habitats, are being under constant threat of destruction by various developmental activities including mining, dredging, filling, tourism, aquafarming etc. (Leatherman 1991), beside, such environments are the recipient of various anthropogenic wastes (contamination through heavy metals, petroleum, pesticides, organics etc.) increasing environmental stresses may well lead to disruption and dynamic reconstruction of communities, localised extinction of both rare and abundant species and total extinction of sensitive species leading to eventual loss of genetic diversity (Buzas and Culver 1998, Culver *et al.*, 1995). Benthic foraminifera play useful role towards identification and depiction of different environmental processes from the marginal marine environment. Benthic foraminiferal distribution in the polluted marine environment had been investigated for the last few decades (Alve 1995) and with the advancement of knowledge in foraminiferal ecology it is now clear that they provide one of the most sensitive and inexpensive invertebrate marker for evaluating deterioration of marginal marine environment (Scott *et al.*, 1995). Pioneering work on pollution studies using benthic foraminifera as proxy indicator was carried out by Resig 1960 & Watkins 1961. The role of benthic foraminifera in the identification of polluted Indian marine environment was reported by Setty 1976, Kameswara Rao 1979, Naidu 1985, Banerji 1992 and Majumdar 1997. Despite having its own limitations, among the various group of invertebrates (viz. polychaetes, ostracods etc.) forams are the best possible marker for depicting polluted and quasipolluted nearshore marine environment. High diversity, wide ecological amplitude, relative ease of collection, small size, shelled morphology etc., have rendered them as unique proxy indicator. The most common response of benthic foraminifera is a decrease in diversity with increasing contaminant concentration even with ultimate total loss of foraminiferal species or persistence of one or two opportunistic species.

The combined waste water discharge from various domestic and industrial establishments in and around Bombay (lat, 18°55′ N, long. 72°49′E) imposes heavy strain on the receiving marine environment and deteriorates the environmental quality (Zingde *et al.*, 1979, NIO, 1975).

Study Site

Recent benthic foraminiferal studies were carried out in eight selected stations under 3 sectors across a transact connecting Sewri Mudflat and Shibagi Nagar, Mumbai.

Sector - I

Sewri Mudflat: The Sewri Mudflat on Mumbai falls under CRZ-1 classification. The mudflat gets exposed to about 3 km. stretch during the lowtide. The presence of only few stunted *Avicennia marina* reflects the severe stress already existing in the system. A large number of dead *Avicennia marina* due to chemical pollution also corroborate the severe stress operating in this mudflat. The mudflat receives a wide range of pollutants from the fishing vessels and trawlers that often use antifouling paints for conditioning purpose and thereby introducing Cu and Zn in the system well in excess amount. The floating, old, stranded rusty barges are the source of Fe. The bilge water released from the vessels and trawlers together with the drilling activity in the Bombay High area and oil refineries of a petroleum company have made the area potently polluted with oil. Conspicuous black marks of coal tar are evident throughout the flat. The mudflat was also highly polluted with organic matter and microbes. The abnormally high values of organic Carbon, Fe, Zn and microbes (vide Table 31.2 and Table 31.5) may well be related to poor biological diversity.

Sector - II

Bay region: It falls within the Arabian Sea. The bay receives mainly the industrial discharges as evident from the significantly high amount of dissolved Pb concentration in the aquatic phase (vide Table 31.2). As it falls under major shipping channel, the high oil and grease concentration can be correlated with the routine operations and bilgewater from shipping activities as well as from the nearby Bombay high operations. The industrial effluents discharged from the major industries contribute significant organic load to the Bay region (vide Table 31.6).

Sector - III

Shivaginagar Mudflat: It is located towards the Nabha end of the transect. This mudflat fails under CRZ-II classification, Congenial growth of mangrove vegetation dominated by *Avicennia marina* and a few species of *Sonneratia* were noticed. The existing road has obstructed the periodic inundation of mudflats by the tidal water to a considerable extent which probably inhibited the plankton and nutrient supply to the mudflat. Although organic carbon values is low, yet presence of high values of Pb, Zn and Cu in the soil samples stand as a proof of deterioration of the quality of mudflat (*vide* Table 31.2 and Table 31.5).

Materials and Methods

Foraminfera were studied following standard wet sieving and floatation technique (Schroder Adams 1987). Rose Bengal solution was used to differentiate the living forams from the dead ones. (Walton 1955, Corliss 1991, Barmawidjaja *et al.,* 1992). 50 gms, of dry sediments were used for standardising all the foraminiferal parameters. Faunal diversity studies were performed using the following parameters - (i) Species number, (ii) Absolute abundance, (iii) Relative abundance, (iv) Shannon-Weiner Index (Compound diversity).

The physico-chemical variables of the aquatic medium were analysed as per the standard method of APHA (1985). Sediment analyses were done as per the standard techniques proposed by Agimian and Chau 1976 & Malo 1977.

Microbial analysis were carried out using the membrane filter technique as proposed by APHA (1985).

Results

Level of Stress

Sector - I: Sewri Mudflat

Due to the presence of huge organic load in the Sewri Mudflat area, the DO (Dissolved Oxygen, level is very poor (vide Table 31.1) as most of the oxygen is used up for oxidation of the organic matter. The nitrate level is also of significantly high values (13.5 mg/l and 16.43 mg/l respectively) suggesting the direct contamination of the area by raw sewage. The presence of maximum Coliform load confirms the contamination source (*vide* Table 31.1).

The presence of heavy metals Zn, Cu, Mn and Fe in maximum concentration in dissolved state is due to the presence of large number of vessels, trawler and rusty barges. The relatively low pH value of the aquatic medium might well enhance the change of dissolution of the metal from the sediment bed to the water column. The BOD and COD value is much above the permisible level (Table 31.1) indicating the presence of organic load in the system.

Table 31.1: Hydrology of the Study Site

Variables	*Sector-I*		*Sector-II*				*Sector-III*	
	Station 1	*Station 2*	*Station 3*	*Station 4*	*Station 5*	*Station 6*	*Station 7*	*Station 8*
Water Temp. (ºC)	28.5	28.0	28.5	28.3	28.5	28.5	29.2	29.5
pH	7.2	7.3	8.1	8.1	8.0	8.2	8.3	8.2
Salinity % 0	34.6	35.0	36.2	36.0	36.2	36.2	35.3	35.9
Dissolved Oxygen (mg/l)	3.0	3.5	5.0	5.6	5.0	4.9	5.8	6.2
Oil & Grease (mg/l)	19.37	19.25	15.20	15.58	14.96	14.82	13.49	14.86
BOD	135	139.5	87.0	82.5	75.0	80.2	129.2	127.5
COD	398.0	335.0	279.5	251.5	258.0	239.0	346.0	340.5
Nitrate (mg/l)	13.5	14.5	0.09	0.06	0.05	0.05	0.42	0.40
Phosphate (mg/l)	0.15	0.20	0.08	0.07	0.08	0.08	0.03	0.03
MPN (Coliform/100 ml)	624	587	179	168	182	174	543	529

Table 31.2: Trace metal level in dissolved State in the Aquatic Medium of the Sampling Stations

	Sector-I		*Sector-II*				*Sector-III*	
Dissolved Trace Metals (mg/l)	*Station 1*	*Station 2*	*Station 3*	*Station 4*	*Station 5*	*Station 6*	*Station 7*	*Station 8*
Zn	327.8	419.22	117.52	189.26	172.30	185.46	125.53	196.50
Cu	562.6	549.72	462.00	389.82	318.21	335.00	428.30	372.60
Mn	487.53	459.20	405.00	326.32	311.00	328.25	330.45	327.72
Fe	658.50	515.00	179.25	182.41	170.00	176.20	329.00	315.20
Cr	20.10	16.59	12.32	14.40	13.36	13.32	12.80	10.00
Cd	BDL	BDL	BDL	BDL	BDL	BDL	BDL	BDL
Pb.	2172.30	2105.15	2650.00	2739.22	2400.00	2372.12	1987.00	1675.00

BDL: Below Detectable Limit.

Moreover, the total degradation of the environment was visually evident from the complete blackening of the sediments with marks of oils signifying acute contamination by coal tar.

The synergistic impact of all the above stated stress factors was tremendous on the existing *Avicennia marina* species, the only mangrove thriving in the area with the acute stunted growth and even many trees were found in dead condition. Low species diversity and density of zooplankton and almost total absence of phytoplankton excepting *Rhizosolenia* sp. corroborated the stress factor (*vide* Table 31.3 and 31.4).

Sector - II: Bay Region

The bay waters depict clear sign of deterioration particularly regarding Pb which has exhibited the maximum concentration level (*vide* Table 31.2) among all the three sectors. This probably is due to the fact that the region is a recipient of a wide variety of pollutants from the various industries (Chloro alkali, Rayons, Pigments, Thermal Power Plants, Oil refineries, Electroplating etc.).

Table 31.3: Zooplankton Status

Zooplankton Species	*Sector-I*	*Sector-II*	*Sector-III*
Mysis	Absent	Medium	Dominant
Calanold Copepod	Medium	Dominant	Less
Cyclopold Copepod	Dominant	Medium	Less
Cyelops	Medium	Less	Dominant
Naupllus	Absent	Medium	Dominant
Fish eggs	Less	Dominant	Dominant
Prawn larvae	Less	Dominant	Medium
Clenophore	Absent	Dominant	Medium

Table 31.4: Phytoplankton Status

Phytoplankton Species	*Sector-I*	*Sector-II*	*Sector-III*
Rhizosolenla sp.	Dominant	Dominant	Dominant
Chaetoceros sp.	Absent	Medium	Less
Leptocyllndus sp.	Absent	Medium	Medium
Nitzchia sp.	Absent	Less	Less
Asterionella sp.	Absent	Absent	Less
Cosclnodiscus sp.	Absent	Less	Less
Biddulpphla sp.	Absent	Less	Absent

Table 31.5: Sedimentology of Study Site

Variables Texture	*Sector-I*		*Sector-II*				*Sector-III*	
	Station 1	*Station 2*	*Station 3*	*Station 4*	*Station 5*	*Station 6*	*Station 7*	*Station 8*
Sand (%)	42.53	51.00	73.20	63.50	68.00	56.72	69.30	58.25
Silt (%)	27.60	32.24	21.14	24.42	27.35	31.40	26.69	26.36
Clay (%)	29.87	16.76	5.66	12.68	4.65	11.88	4.01	15.39
Organic Carbon (%)	0.96	0.87	0.12	0.17	0.10	0.10	0.21	0.17
Zn (ppm)	897.0	783.2	652.5	618.20	583.0	559.25	912.5	829.32
Fe (ppm)	1,31,800	1,39,000	17,436	1,13,000	15,215	1,19,000	1,68,320	1,47,419
Cu (ppm)	94.0	79.0	72.0	68.0	72.2	85.45	66.2	58.55

The zooplankton and phytoplankton components (both diversity and density) were relatively poor indicating environmental inhospitability though the condition was comparatively better than

Sector-I.

Table 31.6: Organic Pollution Load Generation from the Industries

Zones	Major Industries	Pollution Load Organic (Kg/d)
Island city	Bombay Dying and Manufacturing	519
	Govt. Milk Dairy	330
	Century Textiles	268
	Hindustan Spinning and Weaving	142
South-Eastern	Standard Mills	19
	B.P.C.L.	4251
	H.P.C.L.	2893
	R.C.F.	4717
	Oswal Petrochemicals	314
	I.L.A.C.	10
North-Eastern	Rhone-Pouleue	56
	Hoechst (I) Ltd.	101
	Guest Keen Williams	432
	Valson Dying & Painting	21

Sector - III: Shivagi Nagar Mudflat

The mangrove vegetations in this zone is dominated by *Avicennia marina,* few trees belonging to *Sonneratia* sp. were also observed. Associate species like *Porteracea coarctata, Acanthus ilicifolius, Sweda maritima, Salicornia* sp., *Enteromorpha* sp., and *Catenella* sp. were also noticed. The mangrove vegetations exhibited stunted growth. The level of chemical stress is not very prominent in this sector excepting for BOD, COD, dissolved Pb and Oil values. (*vide* Table 31.1 and 31.2), signs of Microbial pollution is evident (*vide* Table 31.1). The poor nutrient content of aquatic phase and oil pollution might have degraded the mangrove vegetation and associated mangrove fauna. Poor density and diversity of phytoplankton and zooplankton also signifies stress environment.

Table 31.7: Taxonomic Checklist of Benthic Foraminifera from the study site

Ammonia beccarii, A. tepida, A. sobrina, Asterorotalia dentata, Bolivina striatula, Elphidium advenum, Elphidium crispum, E. craticulatum, E. discoidale, Eponides repandus, Florilus schapha, Hanzawaia concentrica, Nonion asterizans, N. boeuanum, Nonionellina turgida, Poroeponides lateraílis, Quinqueloculina laevigata, Q. seminulum, Q. lamarckiana, Q. vulgaris, Rupertinella rupertiana, Spiroloculina excavata, Triloculina tricarinata, T. trigonula.

Forminiferal Abundance

Sector-I

Total foraminiferal number (Living + Dead) completely outnumbered the living foraminiferal number at this sector. Only 2 species (*Ammonia beccarii* and *A. tepida*) were encountered in living condition. 13 species were found in deal condition. They include, *Asterorotalia dentata, Elphidium crispum, E. craticulatum, Florilus schapha, Nonion asterizans, N. boeuanum, Eponides repandus, Poreponides lateralis, Hanzawaia concentrica, Quinqueloculina seminulum, Q. lamarckina, Triloculina trigonula* and

Spiroloculina excavata.

In general, the health of the fauna were quite poor. Pyritization was often observed in certain forams. Complete blackening of shells were also frequently noticed. Forams with distorted chambes were found specially among the rotallids. Total absence of agglutinated forams were conspicuously marked from the study site. Abundant occurence of polychaete worm from this sector also suggested for environment overloaded with organic matters and corroborated with the similar findings of Schafer *et al.*, 1995.

Sector - II

In this sector also, the total foraminiferal number overwhelmed the living foraminiferal number, *Ammonia beccarii, A. tepida, Asterorotalia, dentata, Elphidium advenum, Nonion boeuanum, Nonionellina turgida, Q. seminulum, Q. vulgaris, Q. lamarckiana, T. trigonula* were present in living condition.

Ammonia sobrina, Bolivina striatula, E. discoidale, F. schapha, Q. laevigata, Triloculina tricarinata, Rupertinella rupertiana, S. excavata were observed only in dead condition. The environment was absolutely devoid of agglutinated foraminifera.

Sector - III

Like the other two sectors the total foraminiferal abundance outnumbered the living abundance, in this sector also.

A. beccarii, A. tepida, E.advenum, Q. lamarckiana were encountered in living condition. The dead species includes - *A. sobrina, E. crispum, E. discoidale, Q. seminulum, Q. vulgaris, T. trigonula.* The fauna in general exhibited stunted growth, No. aberrant forms were however noticed.

Conclusions

A total of 24 species of recent benthic foraminifera were identified from the study stretch (Table 31.7). From the hydrological and sedimentological studies it is clear that the Sewri Mudflat was the most organically polluted environment coupled with significant stress from oil and heavy metals. The Shivagi Nagar Mudflat area on the other hand was short of nutrients and planktons. Considerable signs of oil contamination is also evident from this site. The bay area is the recipient of huge industrial pollutants, oil and metal contaminants.

The presence of *A. baccerii* and *A. tepida* as living member in the Sewri Mudflat shows their afinity towards organic matter and thus corroborated with the findings of Setty *et al.*, 1986. This also supports the findings of Yanko *et al.*, 1992 from similar environment. However the absence of other organic loving species *Ammobaculites* supports the findings of Alve 1995 regarding total absence of agglutinated foraminifera from the environment degraded with oil and tar.

The poor foraminiferal diversity and density in the bay area differs from the findings of foraminiferal diversity and density of Setty *et al.* 1986 from the self regime off Mumbai thereby clearly suggesting for a strong stress gradient operating in the study area.

Acknowledgement

The authors wish to express their thankfulness to Prof. (Rtd.) M.S. Rao, Dr. T.Y. Naidu and Dr. V.V. Shyamsundar of Andhra University and Dr. A.B. Bhunia of CPCB, Kolkata. S.M. is also thankful to the Asstt. Commissioner, KVS, Kolkata Region.

References

Agmeian, L L. and Chau, A.S.Y., (1976). Evaluation of extraction techniques for the determinaton of metals in aquatic sediments, *Analyst* 101: 761-767.

Alve, E. (1995). Benthic Foraminiferal responses to estuarine pollution: A review, *Journal of Forminiferal Research*. 25(3): 190-203.

APHA. (1995). *Standard methods for water analysis.*

Banerji, R.K. (1992). *Heavy metals and benthic foraminiferal distribution along Bombay Coast, India*: Benthos' 90, Sendai, Studies in Benthic Foraminifera, Tokai University Press: 151-157.

Barmawidjaja, D.M., Jorissen, F.J., Puskarie, S., and Zwaan Vander (1992). Microhabitat Selection by benthic foraminifera in the Northern Adriafic Sea. *Journal of foraminiferal research* 22: 197-317.

Buzas M.A. and Culver S.J. (1986). Geographic origin of benthic foraminiferal species: *Science*. 232: 775-776.

Connell, J.H. (1978). Diversity in tropical rain forests and coral reefs: *Science*, 199: 1302-1310.

Corilss, B.H. (1991). Morphology and microhabitat preferences of benthic foraminifera from the northwest Atlantic Ocean, *Marine Micropaleontology*. 17: 135-236.

Culver, S.J. and Buzas M.A. (1995). The effects of Anthropogneic Habitat Disturbance, Habitat Destruction and Global Warning on Shallow Marine Benthic Foraminifera. *Journal of Foraminiferal Research*, 25(3): 204-211.

Ehriich, P.R. (1986). Extinctions: What is happening now and what needs to be done. In Elliot, D.K. (Eds.) *Dynamics of Extinction*. Wiley and Sons. New York: 157-164.

Leatherman, S.P. (1991). Impact of Climate-Induced Sea level rise on coastal areas. In: Wyman, R.P. (Eds.). *Global Climate Change and Life on Earth*, Routledge, C. Chapman and Hall, New York: 170-179.

Majumdar, S. (1997). Study of Recent Benthic Foraminifera for Evaluating Environmental Stress in and around Alang Ship-Breaking Yard (Ghogha - Gopnath Sector) of Saurashtra Coast, Gujarat: A Case Report, (Unpublished report submitted to G.E.C., Vadodara).

Malo, B.A. (1977). Partial extraction of metals from aquatic sediments. *Environ. Sci. Technol*. 11: 277-288.

Naidu, T.Y., Rao, D.C., and Rao, M.S. (1985). *Bull. Geol. in. Met. Soc. India*, 52: 88-96.

Pollution Survey in Thana Creek and Bombay Harbour Region: (Reports submitted to the Maharashtra Prevention of Water Pollution Board) by Bombay Regional Centre NIO: (1975).

Rao, K.K. and Rao, T.S.S. (1979). *Indian Journal of Marine Sciences*, 8: 31-35.

Resig, J.M. (1960). *Foraminiferal ecology around ocean outfalls off Southern California in Waste Disposal in the Marine Environment*, Pergamon Press, London: 104-121.

Richmond, R.H. (1993). Coral reefs: Present problems and future concerns resulting from anthropogenic disturbance: *American Zoologist*, 33: 524-536.

Scott, D.B., Schafer, C.T. Honig, C., and Younger, D.C. (1995). Temporal variations of benthic foraminiferal assemblages under or near aquaculture operations: Documentation of impact history, *Journal of Foraminiferal Research*. 25(3): 224-235.

Schroder - Adams C.J., Cole F.F., Medioli E.S., Mudle P.J., Scott DB., and Dobbion I. (1990). Recent Arctic shelf Forminifera: Seasonally ice covered Vs perenially ice covered area. *Journal of Formainiferal, Research*, 20: 8-36.

Setty, M.G.A.P. and Nigam, R. (1986). Benthic Foraminifers as indices of Diversity and Hyposalinity in a Modern Clastic Shelf Regime, of Bombay, India in Thompson, M.F. *et al.* (Eds.). *Indian Ocean - Biology of Benthic Marine Organisms*: Oxford and IBH Publishing Co., New Delhi.

Setty, M.G., Anantha Padmanabha, (1976). The relative sensitivity of Benthonic Foraminifera in the polluted marine environment of Cola Bay. Goa, *Proc. VI Indian Coll. Micropal, Strat.*, 255-231.

Walton, W.R. (1955). Techniques for recognition of living foraminifera. *Contr. Cushman Found. Foram. Res.* 3: 56-60.

Watkins, J.G. (1961). Foraminiferal ecology around the Orange County, California Ocean sewer outfall, *Micropaleontology*. 7: 1999-206.

Yanko, V., Flexer, A., Kress, N., Hornung, H., and Kronfeld, J. (1992). Benthic foraminifera as indicators of heavy metal pollution in Israel's eastern Mediterranean margin, French-Isreali Symposium on the Continental Margin of the Mediterranean Sea. May 11-13, 1992, Haifa, N.C.R.D. 4-92, *Scientific Program and Abstracts*, 73-79.

Zingde, M.D., Trivedi, S.K. and Desai, B.N. (1979). Physiochemical Studies on Coastal Pollution of Bombay. *Indian Journal of Marine Sciences*, 8: 271-277.

32

Regression Models for Assessment of Pollutant Loads in a River

M. Chandra Sekhar, N.V. Umamahesh and M.V.K. Satya Prasad

Department of Civil Engineering, Water and Environment Division,
Regional Engineering College, Warangal- 506 004

ABSTRACT

The river system under study is a very large basin, with a number of tributaries. Monitoring all sources of pollution to assess the loads contributed by these sources is rather difficult and/or impossible and expensive and subjected to analytical errors. Hence, modeling which is relatively cheaper and less time consuming allows for estimation of loadings which otherwise could not be measured (*i.e.* rare events). However, modeling is also subjected to analytical errors, methodological errors (if model is to reflect all processes and interactions) and calculation errors (if the procedure is wrong). Regression models can provide an alternative means and is the main focus of the work presented in this paper. These models are developed using upstream and downstream flows and water quality data and tested successfully for the river system under study. As the catchment is very large, the loads estimated using these models can serve to separate non-point sources contributions, which predominantly include pollution due to agricultural practices and activities, soil erosion, dissolution of soil minerals or combination of these sources. The estimated loads can also serve as controls for abatement of pollution of the river water.

Introduction

Water quality management studies to achieve goals of any water quality management programme must contain evaluation of pollution loads from various sources. Catchment management plans require to recognize that total pollutional load to a water body consists of three components:

1. Direct/point wastewater discharges,
2. Diffuse/non-point contribution in seepage and runoff water from the catchment manipulated/managed by man, and

3. A background contribution from natural sources (possibly due to scouring/erosion from catchment surfaces and in stream secondary pollution).

Maintenance or improvement of water quality and control water quality degradation may require control of both point and non-point sources. An assessment of non-point source pollution has, to certain extent been overshadowed by the urgent need for treatment of domestic and industrial wastewaters. Consequently non-point source pollution has not been adequately studied, and its role in water quality degradation is poorly understood. Now, it is becoming evident that to establish the goals of water quality management programme, regulating and controlling only point sources is not sufficient. In fact, in many cases, the pollutants emanating from non-point source comprised major contribution of pollutional load into the receiving water bodies.

Indirect methods to study source contributions of pollutant loads are essential to control water quality degradation in rivers. Especially in the case of rivers draining large basins, the application of direct methods/models becomes difficult as the availability/collection of data will be a major constraint (Sekhar, 2001). Regression models are useful especially when only limited data. i.e., receiving water quality and low data are available in the developing countries like India. Mass balance studies using water quality and flow data are extensively used during recent years to study the in-stream reactions and pollution loading patterns (Plummer and Back, 1980, Yuretich and Batchelder, 1988, Jain, 1996). The study on the river Hoje in Sweden has indicated that an essential pollution build up seems to take place in stream sediments which receive rural and urban runoff (Berndtsson, 1990). Development of regression models for prediction of pollutant loads is the objective of the work presented in the present paper. The available data is used for development and testing the models.

Description of the Study Area

Krishna river is one of the major perennial rivers, which drains three important States of South India. The river basin is the second largest river basin which is situated in the Deccan plateau. The river Krishna drains an area of 258,948 km^2, which is nearly 8 per cent of the total geo-graphical area of the country. The total population in the basin as per 1991 causes has been estimated as 60.78 million. There are about 25 towns within the basin with the population more than hundred thousands. The river and its tributaries flow through different terrain having varied land use activities, soil conditions, vegetation and agricultural practices. The water potential of the River Krishna and its tributaries are mainly used for drinking, industries, irrigation and power generation. The study area presented in Fig. 32.1, in particular is part of the Krishna river reach between two monitoring stations: Pondugala (upstream) and Wadenapalle (downstream). The river reach between the monitoring stations is approximately 80 km long along the river. In addition to other districts, major parts of Nalgonda and Guntur districts drain into this part of the Krishna river reach in Andhra Pradesh. Typical tropical climate prevails in the basin for better part of the year. For practical considerations two seasons: dry (December-May) and wet (June-November) seasons exist in the area.

Methodology

The Central Water Commission (CWC), Government of India collects hydrological (gauge and discharge observations) and water quality data from 57 monitoring stations on the River Krishna and it's tributaries. The Seba current meter is used to measure velocity and discharges are estimated by area velocity method. The discharge and water quality are monitored at an interval of 10 days CWC Manual, 1995). All the samples were collected from mid-stream at about 15 cm depth and stored in pre-cleaned polyethylene bottles. After measuring some of the parameters like conductivity, pH,

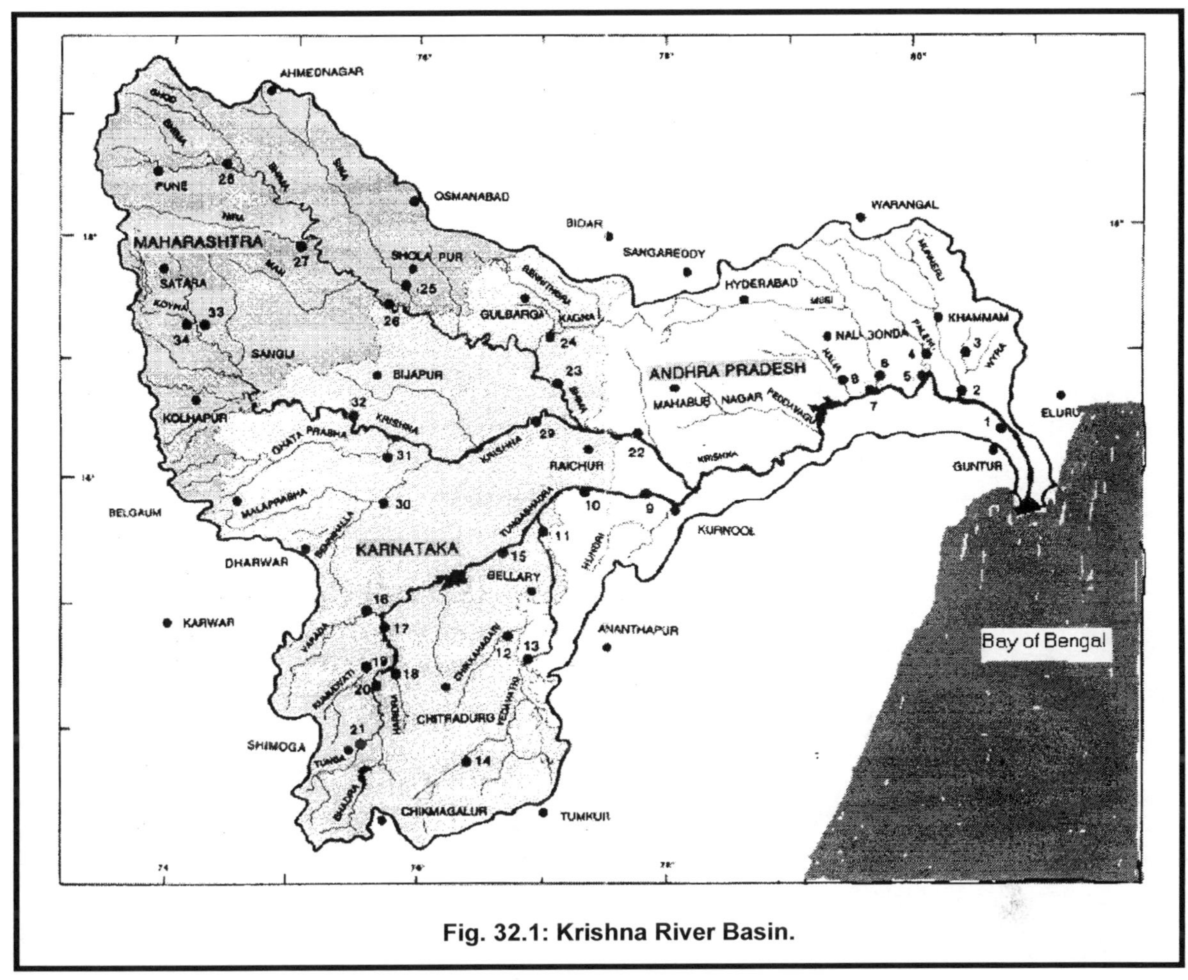

Fig. 32.1: Krishna River Basin.

temperature, etc., on the spot, appropriate reagents were added for preservation of samples before taking them to the laboratory for further analyses. The analyses for water quality parameters are carried out as per the "Standard Methods for Examination of water and wastewater" (APHA 1992).

Data and preliminary information are obtained from Central Water Commission, National Remote Sensing Agency (NRSA), Andhra Pradesh State Remote Sensing Application Center (APSRAC), National Hydrological Project, etc. Time series analysis is carried out for the discharge and loading of water quality parameters to understand the seasonal variations. Regression models are attempted for establishing load-discharge relationships for dissolved conservative pollutants namely, Na^+, Ca^{++}, Mg^{++}, HCO_3^-, Cl^-, F^-, SO_4^{--}, NO_3^- and SiO_3^{--}. The data for the years 1989-1994 is used for developing the models and 1994-1997 data is used for verification of the models.

Results and Discussions

Flow Characteristics

The discharge in the river under study ranged from 100 m^3/sec to 4000 m^3/sec approximately at both upstream and downstream monitoring stations. The existing seasonal variation in rainfall during the dry (December-May) and wet (June-November) seasons suggests such a variation in the stream discharges. The stream hydrographs presented in Fig. 32.2 is characterized by high discharges from June to November, followed by a decline, until a low is reached in May which is the end of the dry period. The temperature variation pattern in the study area supports the variation in flows. The maximum temperature 40°-42°C in the study area is recorded during the month of May and the minimum 10°-12°C during the month of December. Occasional summer storms/cyclones caused small peaks during dry seasons. In addition to the overall seasonal trends, the river hydrographs are characterized by rapid rises and falls in response to individual rainfall events. This rapid response is probably related to the ease with which water can travel through the porous, shallow and loamy soils of the study area.

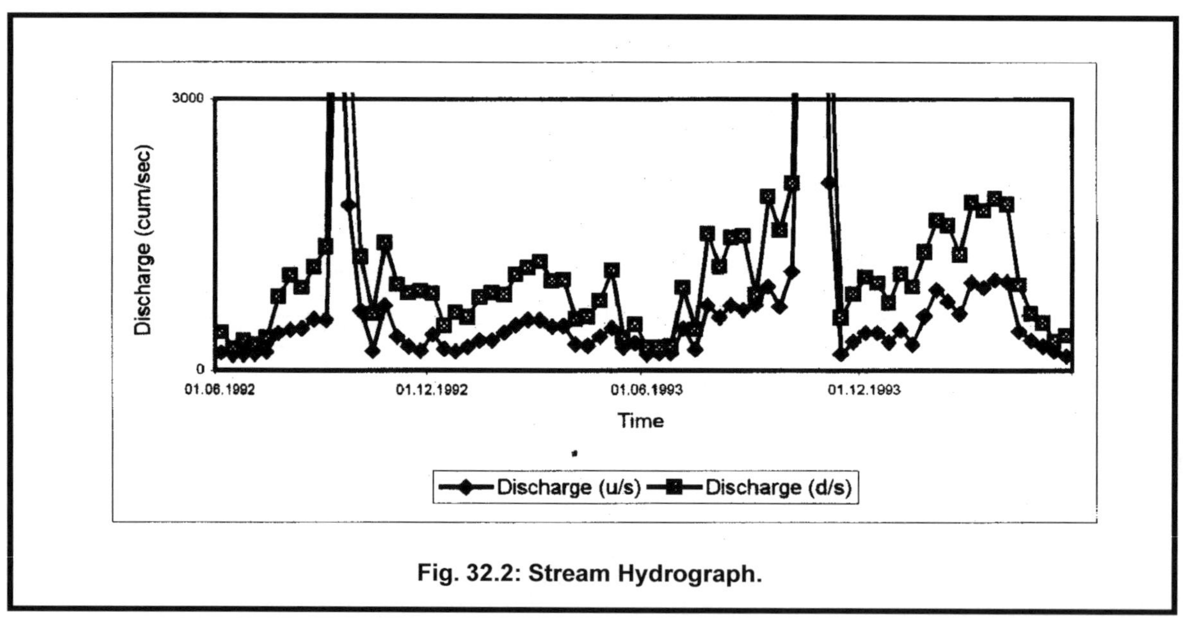

Fig. 32.2: Stream Hydrograph.

Seasonal Variation

Seasonal variations in quantity and quality of river water are mainly due to the non-uniform distribution of rainfall. During the dry season, the river water is polluted due to discharge of treated/untreated domestic and industrial wastewater, whereas during the wet season, the water quality of the river is degraded by both point and non-point sources. This pattern is not reflected when only concentrations of pollutants are taken into consideration due to the dilutional effects of large flows during the wet season. Though dilution also occurs during dry season, the river flow is so small to reflect the effect of dilution significantly.

Time series plots suggest that there is a strong seasonal dependence among the pollution load with higher loads during the wet season and lower loads during the dry season. Also, when pollutant concentrations are taken into account, the concentrations during the wet season are marginally higher than the concentrations during the dry season. Such an observation is possible only when the flow variations are very large (presented in Fig. 32.2) as in the present case. In terms of concentrations, the dilutional effects of the river on point sources during the dry season are not significant due to lean flows. However, when total pollution loads are considered, it is obvious that during the wet season, the pollution loads are significant (though the concentrations are marginally higher) due to considerably greater flows in the river. Secondly, the pollution load among the wet season is contributed by both point and non-point sources of pollution. With the data analyzed during the present study, year-wise trends could not be established.

Development of Modelling Equations

Seasonal variations in quantity and quality of river water are significant due to non-uniform distribution of rainfall and hence the discharge in the river. During the dry season the river water polluted due to discharge of treated/partially treated/untreated domestic and industrial wastewaters. In winter, the river receives pollutants from non-point sources in addition to pollutants from point sources. As the flow variations in the river are very large, flows are classified as wet flows (June-November) and dry flows (December-May) for developing modeling equations. The regression equations obtained during the present study are of the form

$$Y = a\,X^b$$

where,

Y is the pollutant load (dependent variable),
X is the river discharge (independent variable) and
a, b are the constants.

The constants of regression equations are estimated using the water quality data for the years 1989-1994. In some of the equations presented in Table. 32.1, the constant 'b' approached unity indicating that the relationship is linear. However, to have generalized type of equation the non-linear regression is adopted. Significant correlation coefficients indicates good applicability of the modeling equations for the river under study. For many parameters the constants in the modeling equations are different for upstream and downstream monitoring stations. This is probably due to the variations in many of the basin characteristics such as land use, soil, geomorphology, etc. Hence, no attempt is made to arrive at generalized equations for upstream and downstream stations.

Model Verification

The models are verified using linear regression. The water quality and flow data for the years 1994-1997 is used for verification. The model results are in agreement with calculated pollutant loads within the 10 per cent error. The models provide a quantitative description of the loads in river water over a year. The time sequence of load is thus qualitatively related to changes in pollutant loads reaching the river. Hence, the model equations developed during the present study can be used for prediction of pollutant loads and thus for water quality management programs in the river basin.

Conclusions

The Krishna river basin under study is a very large basin, with number of tributaries. The river system is polluted by numerous point and non-point sources. The pollutant loadings obviously reduce the self purification capacity of the receiving water body. The present investigation provides an account of the advantages of using u/s and d/s river water quality data to estimate pollutant loads. Monitoring all individual sources of pollution to assess the river pollution loads is rather difficult and/ or impossible and expensive and further subjected to analytical errors. Hence, modeling which is relatively cheaper and less time consuming, allows for estimation of loads which otherwise could not be measured (*i.e.*, rare events). However, it is to be kept in mind that modeling is also subjected to analytical errors, methodological errors and calculation errors. Time series analysis indicates a strong seasonal dependence of pollution loadings to the river. Regression models based on observed in-stream water quality parameters provide general indication of the quality and quantity of pollution loads on relative magnitude of loads. This approach is also useful to detect changes in water quality constituents within the system. Indirect modelling of point and non-point sources carried out in the present work provides a better alternative for a systematic study over the conventional techniques. However, the limitations such as short time of travel, changes/losses with respect to time and distance, source classification, data limitations, etc., have to be considered carefully. Comparisons between upstream and downstream monitoring stations can be further attempted to estimate loads contributed by different sources to the river between the monitoring stations. The resulting differential loads may be of help to indicate the non-point source loads in the river reach under study. The particular advantage of this approach discussed in this study is that it requires only flow and water quality data of the river which is generally available. However, sediment and paired sampling can result in more accurate results.

Table 32.1: Modelling Equations for Pollutant Loads

Monitoring Station	*Season*	*Parameter*	*Regression Equation*	*Correlation Coefficient*
Wadenappli	Monsoon	Na^{+}	$Y = 0.11 \times X^{1.3}$	0.8673
		Ca^{+2}	$Y = 0.93 \times X^{0.99}$	0.9811
		Mg^{+2}	$Y = 14.74 \times X^{0.58}$	0.9608
		HCO^{-3}	$Y = 1.89 \times X^{0.91}$	0.9399
		NO^{-3}	$Y = 3.50 \times X^{0.77}$	0.9449
		Cl^{-}	$Y = 0.65 \times X^{1.03}$	0.9729
		SO_4^{-3}	$Y = 0.40 \times X^{1.1}$	0.8872
		SiO_3^{-2}	$Y = 4.89 \times X^{0.64}$	0.9526
Pondugala	Monsoon	Na^{+}	$Y = 1.57 \times X^{1.09}$	0.9240
		Ca^{+2}	$Y = 2.25 \times X^{0.87}$	0.9382
		Mg^{+2}	$Y = 0.98 \times X^{0.99}$	0.8382
		HCO^{-3}	$Y = 1.22 \times X^{0.97}$	0.9540
		NO^{-3}	$Y = 0.58 \times X^{1.05}$	0.9674
		Cl^{-}	$Y = 14.51 \times X^{0.54}$	0.8947
		SO_4^{-3}	$Y = 0.08 \times X^{1.23}$	0.9305
		SiO_3^{-2}	$Y = 0.16 \times X^{1.09}$	0.9899
Wadenappli	Dry	Na^{+}	$Y = 0.30 \times X^{1.13}$	0.6858
		Ca^{+2}	$Y = 6.92 \times X^{0.71}$	0.9506
		Mg^{+2}	$Y = 3.81 \times X^{0.75}$	0.5710
		HCO^{-3}	$Y = 9.01 \times X^{0.72}$	0.7987
		NO^{-3}	$Y = 14.09 \times X^{0.40}$	0.6363
		Cl^{-}	$Y = 3.41 \times X^{0.82}$	0.6340
		SO_4^{-3}	$Y = 0.29 \times X^{1.18}$	0.8729
Pondugala	Dry	Na^{+}	$Y = 1.03 \times X^{0.98}$	0.9262
		Ca^{+2}	$Y = 4.46 \times X^{0.78}$	0.9339
		Mg^{+2}	$Y = 6.88 \times X^{0.63}$	0.7538
		HCO^{-3}	$Y = 1.76 \times X^{0.91}$	0.8539
		NO^{-3}	$Y = 12.49 \times X^{0.28}$	0.7054
		Cl^{-}	$Y = 3.69 \times X^{0.79}$	0.7479
		SO_4^{-3}	$Y = 0.59 \times X^{1.01}$	0.7092

References

APHA. (1992). *Standard Methods for the Examination of Water and Wastewater* (18th edn), APHA, New York.

Berndtsson, R. (1990). Transport and Sedimentation of Pollutants in a river reach: A chemical Mass Balance Approach, *Journal of Water Resources Research*, 26(7): 1549-1558.

CWC Manual. (1995). *Water Quality monitoring in Krishna River Basin*, Central Water Commission, Hyderabad.

Jain, C. K. (1996). *Application of Chemical Mass Balance to Upstream/Downstream River Monitoring Data*, 182: 105-115.

Plummer, L.N. and W. Back. (1980). The mass balance Approach: Application to Interpreting the Chemical Evaluation of hydrologic Systems, *American Journal of Science*, 280: 130-142.

Sekhar, M.C. (2001). Immission Approach for Modelling Dissolved Solids in a River and Separation, of Point and Non-point Loads, Ph. D. Thesis, Regional Engineering College, Warangal, India.

Yuretich, R.F. and Batchelder, G.L. (1988). Hydro chemical Cycling and Chemical Denudation in the Fort River Watershed, Central Massachussetts: An Approach of Mass Balance Studies, Water Resources Research, 24(1): 105-114.

Section II

TOXICOLOGY

33

Toxicological Effects of Cadmium Chloride on a Fresh Water Fish, *Tilapia mossambica*, Peters and its Ecological Significance

Nilalohit Harichandan, Bibekananda Beura and A.K. Panigrahi

Laboratory of Environmental Toxicology, Environmental Science Research Centre, Berhampur University, Berhampur -760 007, Orissa

ABSTRACT

Fishes were exposed to graded series of concentrations of cadmium chloride for acute toxicity studies. The MAC value in 30 days was 1.25 ppm of cadmium chloride. A safety concentration of 1.2 ppm of Cadmium chloride l^{-50} was selected for this study. The LC_{10}, LC_{50}, LC_{90}, LC_{100}, determined for *Tilapia* fish was found to be 1.32, 1.63, 1.91 and 2.12 ppm respectively after 30 days of exposure. Exposed fishes appeared lethargic showed erratic movements, loss of equilibrium, gradual onset of inactivity etc. when compared to the control fish. Inappetence and ataxia was observed in the exposed fish. The body weight of the $CdCl_2$ exposed fish decreased (6.17 per cent) significantly, when compared to the control fish, where a significant increase in body weight was marked. When the exposed fish was transferred to toxicant free medium, no recovery was noted. A maximum of 22.32 per cent decrease was noted in BSI of $CdCl_2$ exposed fish, where 17.94 per cent recovery was noted after 28 days of recovery. The HSI decreased by 71.72 per cent after 28 days of exposure and 27 per cent recovery was noted after 28 days of recovery. The KSI increased by 68 per cent after 21 days of exposure and by 24 per cent on 28d of exposure and 27 per cent recovery was noted after 28 days of recovery. The exposed fish could not recover to its pre-test activity.

Introduction

The purpose of acute toxicity test is to assess various abnormalities caused due to administration of a chemical to animals on occasion or other and to determine the order of lethality of the chemical. Considerable efforts have been directed towards the establishment of short-cut procedure. In aquatic toxicology, acute lethal toxicity tests with fish or invertebrates are usually intended to assess the numerical value of toxicity, to compare potencies of toxicants, to assess the effect of environmental variables on toxicity. Besides, statistical analysis of toxicity is essentially a tool in all these studies. Fathead minnows when exposed to low concentrations of methyl mercuric chloride (Olson *et al.*, 1973) showed no effect on survival at concentrations causing an acceptable residue. Panigrahi and Misra (1980) found severe poisoning symptoms in *Tilapia mossambica* exposed to inorganic mercury. Relative toxicity of Dichlorovos (DDVP) was studied by Rath and Misra (1979) on *Tilapia mossambica*, Peters belonging to 3 different age groups. Gouda *et al.* (1981) tested the toxicity of dimecron, sevin and lindax to *Anabas scandens* and *Heteropneustees fossilis*.

Cadmium toxicity of gold fish, *Carassius auratus* by McCarty *et al.* (1978), lethal and sub-lethal effects of Binary mixtures of cyanide and hexavalent chromium, zinc or ammonia to the fathead minnow (*Pimephales promelas*), all these add to the knowledge of heavy metal toxicity on fishes. McCarty *et al.* (1978) studied the toxicity of cadmium to gold fish, *Carassius auratus* in hard and soft water. Panigrahi and Misra (1978, 1980) and Panigrahi (1980) reported the behavioral changes observed in inorganic mercury exposed fresh water fishes. Macleod and Passah (1973) reported the behavioural changes induced by mercury in Rainbow trout. Larson and Lewander (1973) reported metabolic effects of starvation in the eel, *Anguilla anguilla* L., Gibilin and Massaro (1973) observed behavioural changes induced by methyl mercury in Rainbow trout. Bhatia *et al.* (1973) reported behavioral changes induced by dieldrin in albino rats. Although adequate literature on acute toxicity of different pesticides and mercury based chemicals, mercury contained effluents and solid wastes to fish is available, no adeqaute information is available regarding this metal, cadmium and its acute toxicity. This study was designed to find out the lethal concentration values and behavioural changes induced by Cadmium chloride on *Oreochromis mossambicus* sy. *Tilapia mossambica*, Peters.

Materials and Methods

Test fish, *Tilapia* were collected, acclimatized in the laboratory. A graded series of concentrations of Cadmium chloride ranging from 1 mg l^{-1} to 20 mg l^{-1} were prepared. Ten healthy fish were exposed to each concentration in 10 litre glass jars. The experiments were conducted in chlorine-free tap water.

The mortality rate of test-fish was studied following the method described by Finney (1971). Observation on the toxicity of Cadmium chloride was made at 24, 48, 72, 96 hours and 28 days after the experimental animals were first exposed. Individuals showing no respiratory movements, no opercular movements and no response to a tactile stimulus were recorded as dead, and were immediately removed. The test fish exposed to lower range of the heavy metal was exposed for a period of five weeks to find out the maximum allowable concentration (MAC), where no mortality was noticed and this was expressed as mg l^{-1}, LC_{50}. Different values such as LC_{10}, LC_{50}, LC_{100} were deduced from graphical interpolation (Finney, 1971).

Fishes were observed for behavioral changes influenced by Cadmium chloride. Experimental fishes were sacrificed, brain, liver, kidney and gill were dissected out carefully and washed thoroughly with distilled water. The sacrificed fish were weighed. Brain, liver and kidney were separated and weighed carefully in a single pan electric balance and the different somatic indices were calculated.

All the parameters were noted at 7-day interval and the per cent change in parameters were computed.

Results

The MAC value of Cadmium chloride was found to be 1.25 ppm for 30 days and to be on the safe side 1.20 ppm was considered for 28 days of the exposure for sub-lethal toxicological studies. With the increase in exposure period, the concentration of the chemical decreased showing the existence of a negative correlation. No mortality was observed in the control set. The following Table 33.1 indicates the lethal concentration values obtained from statistical interpolations.

Table 33.1: Lethal Concentration Values

Lethal Concentration	*Cadmium Chloride Toxicant Concentration (ppm)*	
Value	24 h	48 h
LC_{10}	4.5 ppm	4.2
LC_{50}	9.5 ppm	9.2
LC_{100}	14.6 ppm	14.4
Exposure Period 30 days		
MAC	1.25	
Used Concentration	1.20	
LC Value After 30 Days of Exposure		
LC_{10}	1.32	
LC_{50}	1.63	
LC_{90}	1.91	
LC_{100}	2.12	

All the exposed fish appeared lethargic after exposure to Cadmium chloride. The major clinical symptoms such as inappetance and ataxia appeared after 2 to 3 days exposure. At higher concentration of the Cadmium chloride, the exposed fish showed erratic movements. The other signs of toxicity such as loss of equilibrium, gradual onset of inactivity, erratic swimming with irregular collision to the inner glass walls of the aquarium were observed. Circular movement of the fish was recorded. The exposed fish shed fins and huge destruction of dermal mucous layer was marked. Scar and ulceration of the gill lamellae was observed in the cadmium chloride exposed fish. Some symptoms were similar to the preliminary symptoms of Epizootic Ulcerative Syndrome (EUS) and was marked in the exposed fish. At higher exposure periods, profuse bleeding was marked in the gills. Infection of eyes of test fish was not observed. In the present investigation, peculiar symptoms were marked. The cadmium chloride exposed fish showed initial symptoms like mercury poisoning but later on the exposed fish became over excited and periodic outbursts in irratic swimming was noted, ultimately the excited fish died of suffocation. When the excited fishes were heavily oxygenated and kept in undisturbed water, survived. The exposed fishes showed high level of sensitivity. Any type of external stimuli like a flash of light, some one coming closer to the aquarium, any type of movement, even any sound, the sluggish fish instantly starts vigorous movement and many a times colloid with the glass of the aquarium frequently and showed circular movements. After 1/2 minutes, the movements slow down and the fish becomes senseless, ventillation rate almost stops. If heavily oxygenated, the fish survives otherwise the exposed fish dies. During this later period of slowing down the movement, the colour of the fish becomes black due to dispersion of melanin pigment. This black colour disappears, when the system was heavily oxygenated. At times, the shocked fish automatically comes back to normal, and the black colour disappears. This type of self recovery was only in 30 per cent of the cases. Heavy oxygenation can help to around 65 per cent fishes to survive but the rest of the excited fish die instantly and never recover from the shock. The black colour persists for some time and after death the black colour again disappears. In cadmium exposed fish lesions develope on the fins and later the skin erodes and the bones were clearly visible and becomes naked and shedding of fins were noted in few exposed fish towards later period of exposure or at higher exposure period.

Body weight of the control fish increased by 3.96 per cent after 28 days and by 7.86 per cent after 56 days of exposure. Whereas, the body weight of the exposed fish gradually decreased during the exposure period and a maximum of 6.17 per cent decrease was observed after 28 days of exposure (Fig. 33.1), when compared to control fish. No recovery in body weight was marked, when the exposed fish was transferred to toxicant free normal medium. Rather further depletion upto 8.92 per cent over the 28 d exposure value was marked, on 28th day of recovery (Fig. 33.1). Recovery studies clearly indicated that transferring the exposed fish to normal water medium has no effect. Rather higher depletion in body weight was marked. The body weight showed a linear increase showing the existence of significant positive correlation ($r = 0.984$, $P \leq 0.01$) with the exposure period in the control set. Whereas, a significant negative correlation was observed between the exposure period and body weight. With the increase in exposure period, the body weight significantly ($r = -0.898$, $p \leq 0.005$) declined.

Autopsy studies revealed that the liver and brain of exposed fish were congested and tender. The brain somatic index initially increased from 1.13 to 1.22 after 7 days of exposure. Than, the BSI decreased gradually with the increase in exposure period. On 28th day of exposure, the BSI decreased from 1.12 to 0.87 (Fig. 33.2). After 28d of exposure to cadmium chloride, the exposed fish was transferred to toxicant free medium for recovery studies. The BSI further decreased from 0.87 to 0.72 on 14th day of recovery. Interestingly, on 28th day of recovery, the BSI increased to 1.09 but the value was less than the control value (Fig. 33.2). The Hepato Somatic Index (HSI or LSI) remained almost at the same level in the control fish throughout the experimental period. The HSI gradually and significantly decreased with the increase in exposure period. The value depleted from 2.37 to 0.67 on 28th day of exposure (Fig. 33.2). Partial recovery from 0.67 to 1.09 on 14d and to 1.31 on 28d of recovery was noted, when compared to control value. A consistent Kidney Somatic Index (KSI) was recorded in the control fish in the entire period of experimentation. Interestingly and unlike BSI and HSI, the KSI value in the exposed fish when compared to control fish, increased with the increase in exposure period. The KSI value increased from 0.25 to 0.42 on 21st day of exposure and on 28th day of exposure the KSI value decreased from 0.25 to 0.31 (Fig. 33.3). The exposed fish was transferred to toxicant free medium for recovery studies. No recovery was noted. The KSI value increased in all the recovery periods.

The liver somatic index and brain somatic index decreased ($r = -0.986$, $p \leq 0.01$ and $r = -0.931$, $p \leq 0.01$) with the increase in exposure period, when compared to control fish ($r = 0.634$, p = NS and $r = 0.140$, p = NS). The per cent inhibition of BSI initially increased and after 7 days of exposure the BSI and LSI decreased in exposed fish with the increase in exposure period. The control fish remained clinically healthy throughout the period of experiment. The per cent increase in KSI increased with the increase in exposure period and a maximum of 68 per cent increase was recorded on 21st day and 24 per cent increase was noted on 28th day of exposure. The per cent decrease of BSI, increased with the increase in exposure period after 7 days of exposure and a maximum decrease of 22.32 per cent was recorded on 28th day of exposure. When the exposed fish was transferred to toxicant free medium

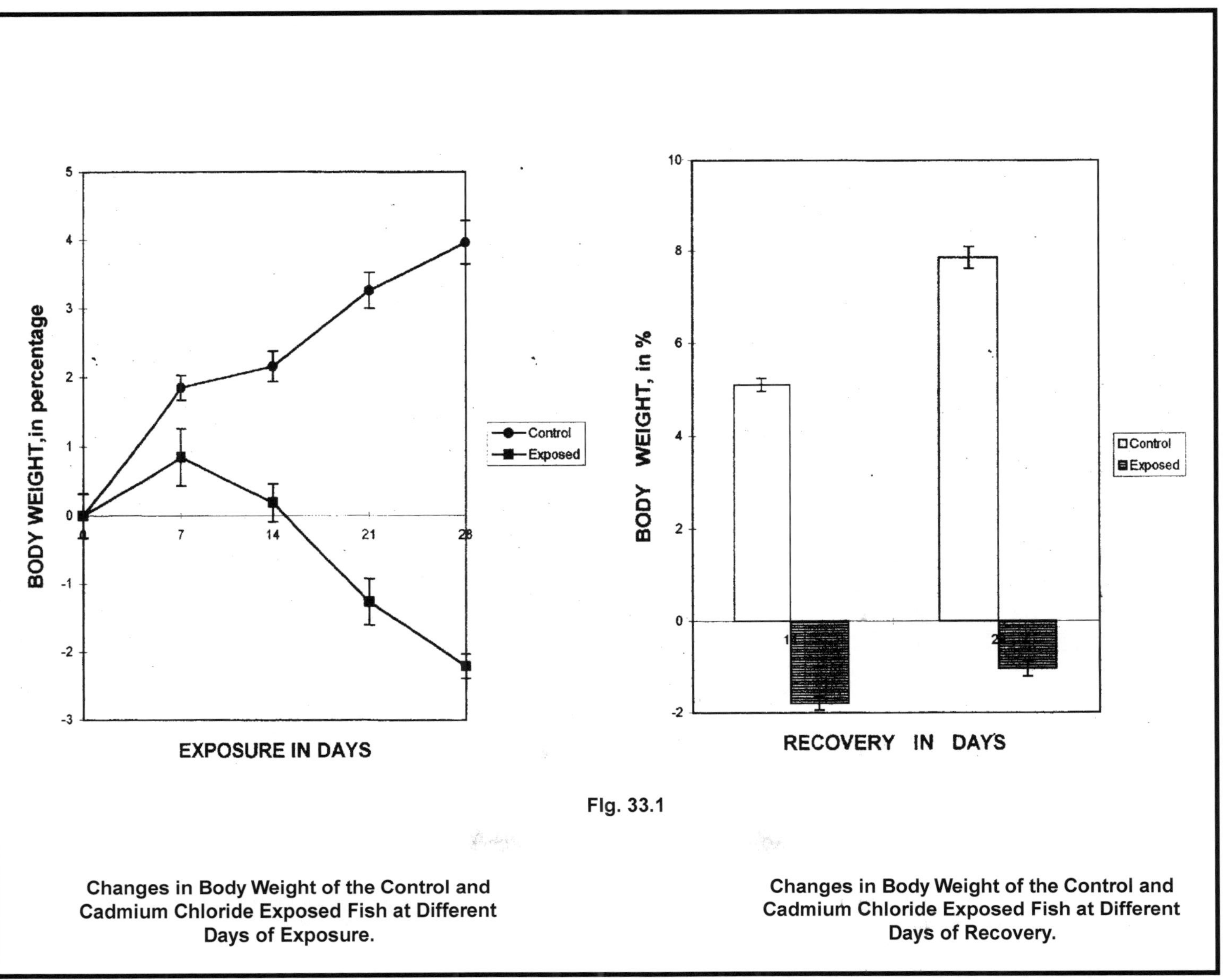

Fig. 33.1

Changes in Body Weight of the Control and Cadmium Chloride Exposed Fish at Different Days of Exposure.

Changes in Body Weight of the Control and Cadmium Chloride Exposed Fish at Different Days of Recovery.

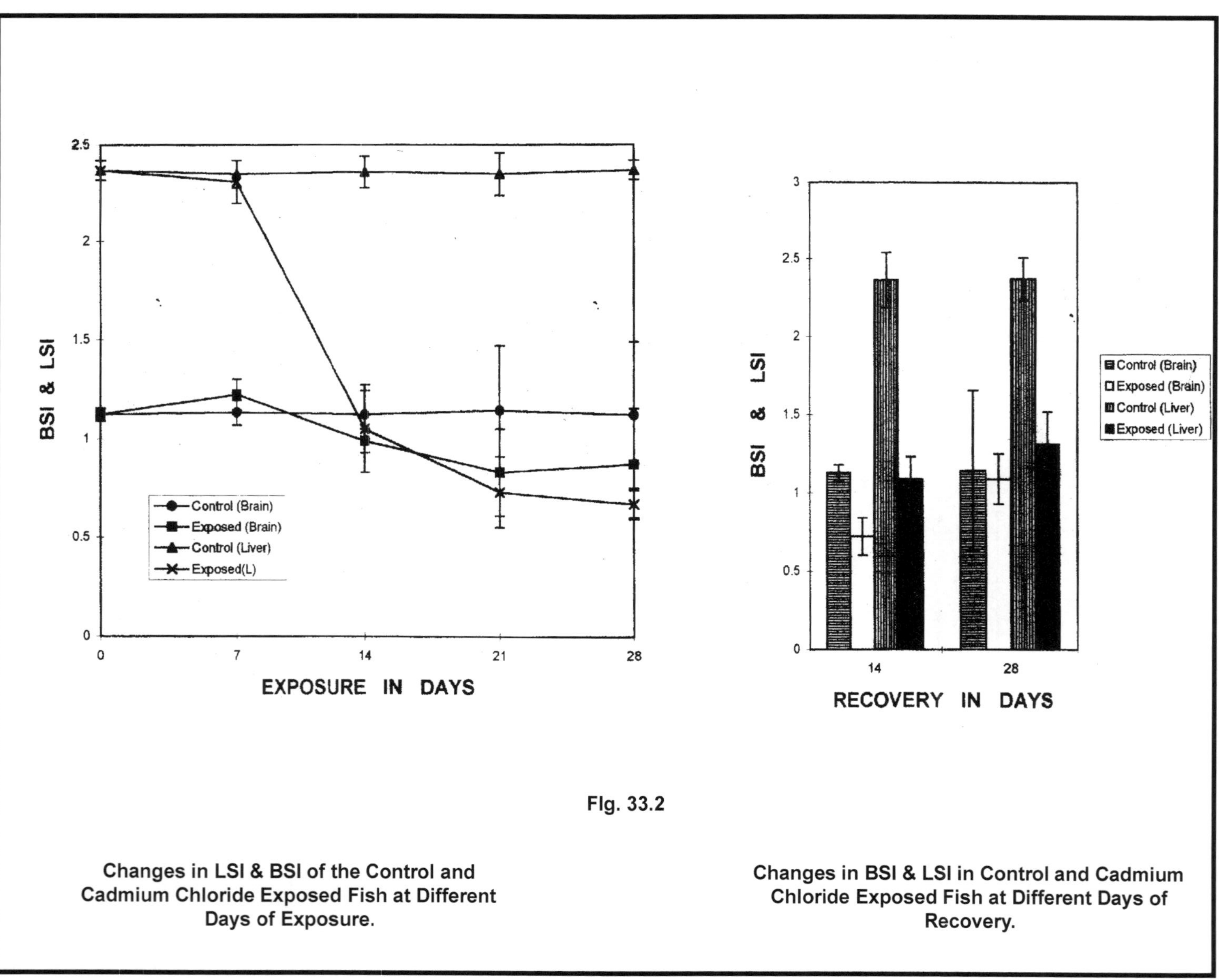

Fig. 33.2

Changes in LSI & BSI of the Control and Cadmium Chloride Exposed Fish at Different Days of Exposure.

Changes in BSI & LSI in Control and Cadmium Chloride Exposed Fish at Different Days of Recovery.

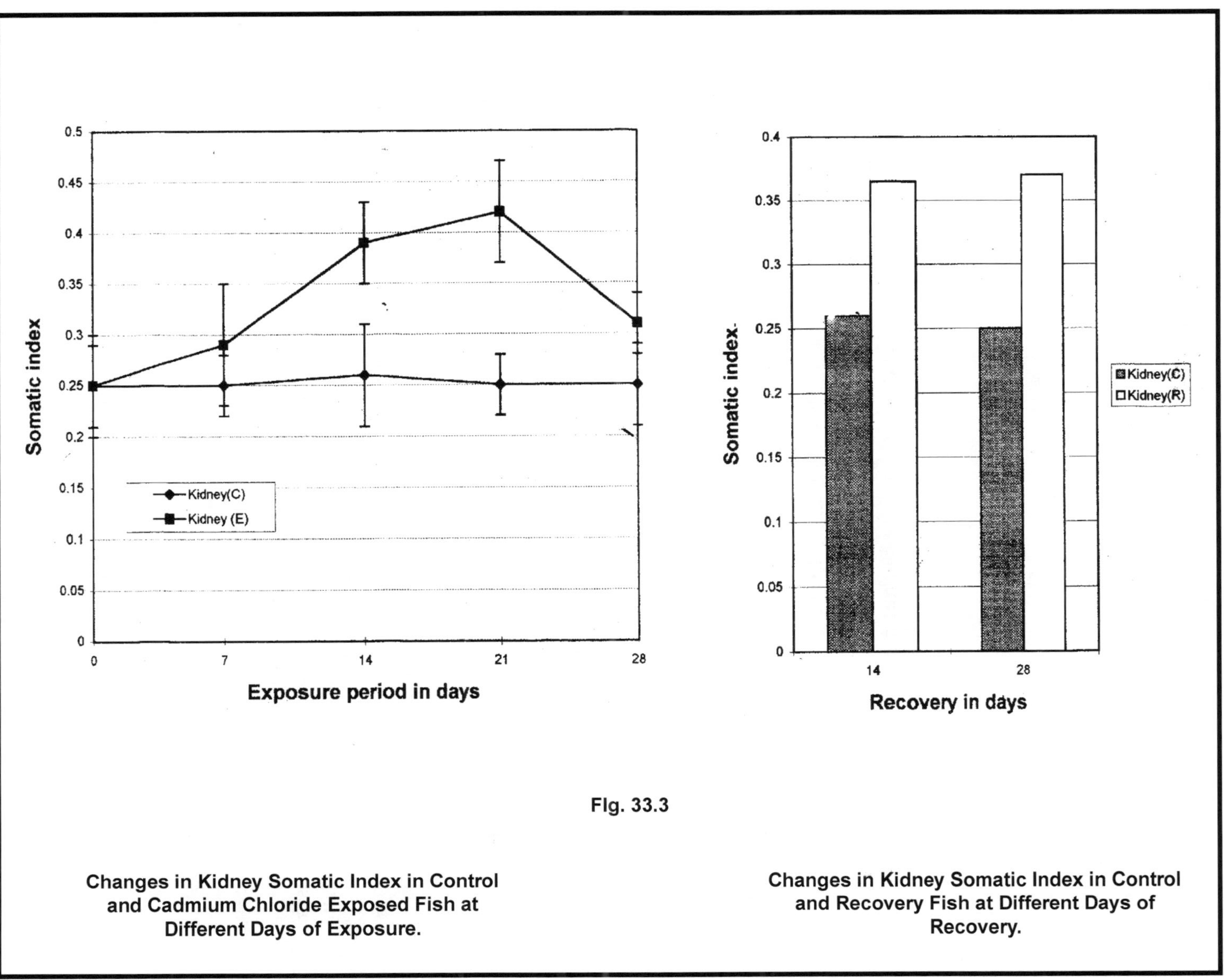

Changes in Kidney Somatic Index in Control and Cadmium Chloride Exposed Fish at Different Days of Exposure.

Changes in Kidney Somatic Index in Control and Recovery Fish at Different Days of Recovery.

Fig. 33.3

17.94 per cent recovery was noted on 28d of recovery in BSI. In case of LSI, a similar trend was observed but here the index decreased maximum by 71.72 per cent, when compared to control fish, after 28 days of exposure. When the fish was transferred to toxicant free medium, partial recovery by 17.91 per cent and 27 per cent was marked after 14 and 28 d of recovery. In case of KSI, the value increased by 24 per cent on 28d of exposure and after 28d of recovery, there KSI value increased by 48 per cent. This value indicated 24 per cent further effect on the exposed fish. The values clearly indicated that during exposure, the kidney was drastically damaged and no recovery was marked. The exposed fish could not recover to its pre-exposure activity.

Discussion

The mortality percentage for any fixed period increased with the increasing concentration of Cadmium chloride. There was an increase in mortality percentage with the increase in exposure period, also. The proportion of these animals that died in a fixed time increased with the increase in the toxicant concentration and the median tolerance limit value decreased with the exposure period (Panigrahi and Misra,1978, 80 and Panigrahi, 1984). The close response approach to assessment of pollutant toxicity involves several assumptions. The route of entry of the toxicant is generally agreed to be via the gills and thus enters directly into the circulatory system. This caused damage to tissues as a result of which there is depression in active metabolism (Macleod and Passah, 1973). Depression in active metabolism that directly reduced scope of activity was demonstrated by Fry (1957). Probably this has relevance with the impaired swimming ability, erratic movements, paralysis, anorexia and ataxia which were observed in all test fish at higher exposure periods.

High percentage mortality of fish, due to the action of Cadmium chloride might be due to the pathological changes, as Mathur (1969) pointed out that fish subjected to insecticides such as lindane, dieldrin, DDT and BHC die due to pathological disorders. During the experimental period temperature, hardness, time of exposure size of the test-fish and all such factors were monitored since the interference of one of these factors will alter the TLM values and MAC values. At present, the sub-lethal concentration selected for further study is 1.20 ppm of cadmium chloride for 28 days, where no death was recorded. The rapid absorption of Cadmium chloride through the gill, skin and gastro-intestinal tract of fish was well evident in the present investigation. The toxicity of Cadmium chloride becomes more apparent in a very shorter period in aquatic animals. Panigrahi and Misra (1978, 1979, 1980) and Panigrahi (1980) documented a detailed behavioral changes in relation to inorganic mercury poisoning. The above authors had observed swelling of eyes and consequent blindness. Such type of symptom was not observed in Cadmium chloride exposed fish, probably such a symptom is unique in mercury contamination. Macleod and Passah (1973) reported that loss of appetite, loss of weight, nervousness, dizziness, loss of equilibrium, erratic swimming and gradual onset of inactivity were the sub-clinical effects of inorganic mercury intoxication. Similar symptoms were marked in cadmium exposure here.

The observed depression in active metabolism in cadmium exposed fish were indicative of damage to nervous tissues, inhibition of enzymes or a vital system, was totally in agreement with Panigrahi (1980). Fry (1957) demonstrated the depression in active metabolism directly reduced "scope for activity". This in turn, may result in decreased growth, impaired swimming ability with erratic out burst at times. Loss in weight due to starvation was noted by Larsen and Lewander (1973). But the loss due to starvation was very less. Panigrahi and Misra (1978, 1980) reported the loss of body weight due to mercury intoxication and confirmed that this loss in body weight was only due to mercury stress. In addition, exposed fish maintained their feeding habit only after a short span of time. Hence the weight loss cannot be correlated with starvation but can only be related with heavy metal intoxication. Whereas, the control fish showed increase in weight. The differences observed between control fish and exposed fish in terms of body weight is only due to the administration of Cadmium chloride in exposed aquarium.

Borg *et al.* (1970) reported inappetance, muscular weakness, ataxia and loss of body weight as the main clinical symptoms of mercury poisoning. On autopsy, he found muscular atrophy, which can be correlated with the weight loss in the goshawks. Hanko *et al.* (1970) reported loss of appetite, weakness of the extremities, excitation in the animal and loss of body weight due to mercury poisoning in chickens. A decline in liver somatic index and brain somatic index in exposed fish could be correlated with the degeneration of cells and decrease in other macromolecular variables. Concisely, at this stage it is not possible to establish the cause of loss in body weight and decrease in liver and brain somatic index, except a generalised comment on the behavior of the exposed fish to heavy metal stress that the observed symptoms were most probably be due to cadmium contamination. The significant increase in KSI in both the exposure period and recovery period indicate positively that cadmium has a significant effect on the kidney and the kidney of the exposed fish was badly damaged by cadmium chloride. The non- recovery of kidney during recovery period also indicate positive permanent damage to the system under study. Hence, with the recognition that all heavy metals are potentially lethal to fish even at relatively low concentrations. It is now a normal practice to test all new chemicals for their toxicity to fish. The LC_{50} and MAC values can be helpful to understand about the amount of heavy metal, cadmium that can be discharged into the environment by any industry, which discharges cadmium in the effluent or in any other waste discharged by the industry.

Acknowledgement

The authors wish to thank the Head, Department of Botany, Berhampur University for library and laboratory facilities.

References

Bhatia, S.C., Sharma, S.C. and Venkitasubramanian, T.A. (1973). Effects of dieldrin on hepatic carbohydrate metabolism and protein biosynthesis in vivo. *Toxicol. Appl. Pharmacol.*, 24: 216.

Borg, K., Wanntorp, H., Erne, K and Hanko, E. (1969). Alkyl mercury poisoning in terrestrial Swedish wild-life. *Viltry*, 6: 301-379.

Finney, D.J. (1971). Estimation of the median effective dose. In: Probit analysis. University Press, Cambridge.

Fry, F.E.J. (1957). The aquatic respiration of fish, pp. 1-63. In: M.E. Brown (ed.). *The Physiology of Fishes. Vol.1.* Academic Press Inc., New York.

Gibilin, F.J. and Massaro, E.J. (1973). Pharmacodynamics of methyl mercury in the rainbow trout (*Salmo gairdnen*): Tissue uptake, distribution and excretion. *Toxicol. Appl. Pharmacol.*, 24: 81-91.

Gouda, R.K., Tripathy, N.K. and Das, C.C. (1981). Toxicity of Dimecron, Sevin and Lindex to *Anabas scandens* and *Heteropneustes fossilis* Comp. *Physiol Ecol.*, 6(3): 170-172.

Hanko, E., Erne, K., Wanntorp, H. and Borg, K. (1970). Poisoning in ferrets by tissues of alkyl mercury fed chickens. *Acta vet. Scand.*, 11: 268-282.

Larsson, A., and Lewander, K. (1973). Metabolic effects of starvation in the eel, *Anguilla anguilla*, L. Comp. *Biochem. Physiol.*, 44: 367-374.

MacLeod, J.C. and Pessah, E. (1973). Temperature effects on mercury accumulation, toxicity, and metabolic rate in Rainbow trout (*Samo gairdnen*) *J. Fish. Res. Board Can.*, 30: 485-492.

McCarty, L.S., Henry, J.A.C., and Houston, A.H. (1978). Toxicity of cadmium to goldfish, *Carassius auratus*, in hard and soft water. *J. Fish. Res. Board Can.*, 35(1): 35-42.

Olson, K.R., Bergman, H.L. and Fromm, P.O. (1973). Uptake of methyl mercuric chloride by Trout: A study of uptake pathways into the whole Animal and uptake by erythrocytes *in vitro*. *J. Fish. Res. Board Can.*, 30: 1293-1299.

Panigrahi, A.K. and Misra, B.N. (1978). Toxicological effects of inorganic mercury on a fresh water fish, *Anabas scandens*, Cuv. and Val and its ecological implications. *Environ. Pollut.*, 16: 31-39.

Panigrahi, A.K. and Misra, B.N. (1979). The effect of mercury on the morphology of erythrocytes in *Anabas scandens*, Cuv. and Val. Bull. *Environ. Contam. Toxicol.*, 23: 784-787.

Panigrahi, A.K. and Misra, B.N. (1980). Toxicological effects of inorganic mercury on a fresh water fish, *Tilapia mossambica*, Peters. *Arch. Toxicol.*, 44: 269-278.

Panigrahi, A.K. (1980). Toxicological effects of mercurial compounds on fresh water fishes and its ecological implications. Ph.D. thesis. Berhampur University, Orissa.

Panigrahy, S.C. (1985). Toxcicological effects of Metacid-50 on a freshwater fish and its ecological implications. Ph.D. thesis. Berhampur University, Orissa.

Rath, S. and Misra, B.N. (1979). Sub-lethal effects of dochlorovos on respiratory metabolism of *Tilapia mossambica* of 3 age groups. *Exp. Geront.*, 14: 37-41.

Rath, S. and Misra, B.N. (1980). Changes in nucleic acids and protein content of *Tilapia mossambica* exposed to dichlorovos. *Indian J. Fish.*, 27(1 and 2): 76-81.

34

Effect of Malathion and Malathion-Temperature Combination of Haematological Parameters of Gobiid Fish, *Glossogobius giuris* (Ham)

G.V. Venkataramana, P.N. Sandhya Rani, M.B. Nadoni and P.S. Murthy

Department of Zoology, Bangalore University, Bangalore - 560 056, Karnataka

ABSTRACT

Temperature variations influence the degree of toxicity of various pesticides in aquatic organisms. *Glossogobius giuris* exposed to various sub-lethal concentrations of malathion for 24, 48, 72, and 96 h. in constant low and high temperature (20, 27 and 30 ± 1°C) altered the blood parameters such as red blood corpuscle (RBC) and white blood corpuscle (WBC) counts, haemoglobin concentration (Hb), mean corpuscular haemoglobin concentration (MCHC), colour index (CI), mean corpuscular haemoglobin (MCH)) mean corpuscular volume (MCV) and haematocrit (Ht). A decreasing tendency in RBC and WBC counts, Hb content, Ht and MCV were noticed in the treated fish with malathion at 30°c and variations in colour index and MCHC concentration. The RBC, Hb, Ht was found to be decreased when exposed for 24 and 48 h with malathion at 20°C. However, WBC, MCV, Cl and MCH increased. The fishes treated with 0.05, 0.25 and 0.5 ppm at 20°c for 72 and 96 h the RBC, Hb, Ht Was found to be increased whereas WBC, MCHC declined with variations in colour index, MCV, MCH values. In the laboratory temperature (27 ± 1°C) the RBC, Ht,. Hb and MCV decreased. On the other hand WBC, MCHC and MCH increased at 24, 48, 72 and 96 h exposure in different sub-lethal concentrations.

Key Words: *G. giuris*, Malathion, Temperature, Heamatology.

Introduction

The rapid use of pesticides to control pests poses serious hazards to non-target organisms including fishes in the aquatic bodies. Pesticides reach aquatic systems by direct application, spray drift from ground or aerial, atmospheric fallout, run-off from agricultural land and discharge of effluents. The concentration of pesticides can produce chronic histopathological changes in aquatic organisms. Survival and dispersal of organisms therefore depend upon their successful adaptation in the changes of environmental conditions in aquatic media Mount and Stephen (1967) and Ramalingam (1982).

These pesticides persist in the environment and are found to be toxic to various aquatic animals including the fishes (Kumar, 1994). The physiological changes occurring as a result of pesticide intoxication may be due to alterations in pH, temperature, dissolved oxygen and hardness of water (Pascoe *et al.*, 1986 and Nemcosk *et al.*, 1987). Haematological analysis have been used for detecting diseases, dietery deficiencies and environments stress (George and Alexender, 1980 and Srivastava, 1989). Such information are important when interpreting the changes which are induced by toxicants and other environmental factors (Basile *et al.*, 1976, Christensen *et al.*, 1978)

Studies on haematology of fishes are of physiological importance. Such study provides valuable information in the health management of fishes. Hence in the present investigation an attempt has been made to study the haematological changes in response to malathion, malathion-temperature combination.

Materials and Methods

The freshwater gobiid fish, *Glossogobius giuris* were collected from Kelaverapalli Dam (located near Bangalore) by using cast and gill nets (mesh: 10 mm), brought alive to the laboratory and were acclimatized for 15 days in 15 liters glass aquarium (60″ × 30″ × 20″) containing aerated tap water prior to use for experiments. They were maintained at 27° ± 1°C temperature and were fed daily with earthworms. Sexually mature fishes were selected for the study.

The pesticide malathion (0.0-dimethyl, S(1-2 dicarbethoxyethyl) phodophorodithioate] was dissolved in acetone and added to the test water to obtain the desired concentrations. The stock solution, 1 mg/litre was prepared separately and desired concentrations were made by adapting the dilution techniques as outlined in APHA, AWWA, WPCF (1985).

The acclimatised fishes were divided into four experimental groups of eight fishes each. The first three groups of fishes were placed in sublethal concentrations (0.05, 025 and 0.5 ppm) of malathion and the fourth group in freshwater served as control. The second set was divided into four groups and were subjected to different concentrations of malathion (0.05, 0.25 and 0.5 ppm) and temperature variation (20°C and 30°C). Eight fishes were used for each concentration in 15 litres glass trough in all experiments. The acclimated fishes were starved for 24 h prior to their use in the experiment and were not fed during the course of experiments (Dalela *et al.*, 1978). The tap water was changed on alternate days and the concentration of pesticide was maintained. In the experimental and control fishes, the blood was collected by severing the caudal peduncle in the intervals of 24, 48, 72 and 96 h, causing minimum stress to the fish, and the blood was drawn with the help of 1 ml graduated syringe fitted with 2.4 Gxl″ needle coated with sodium citrate as anticoagulant.

Total counts of erythrocytes (TEC), leucocytes (TLC), haemoglobin content (Hb), haematocrite (Ht) and blood indices like. Mean corpuscular volume (MCV), Mcan corpuscular haemoglobin (MCH), Mcan corpuscular haemoglobin concentration (MCHC) and Colour index (CI) were determined following the methods of Donald Hunter and Bomford (1963), Blaxhall and Daisley (1973) and Wintrobe

(1977). From the total count of erythrocyte and leucocyte, the erythrocyte and leucocyte ratio (RBC/WBC ratio) was calculated.

Results

The observations revealed a significant reduction in RBC, Hb, PCV, MCV with an increase in WBC, MCH, MCHC. The values of WBC, MCH, MCHC exhibited in inverse proportion with the changes in RBC, Hb and PCV profile.

The decrease in RBC count and Hb concentration were more conspicuous in 0.5 ppm of malathion when compared to control. Similarly a significant decrease of PCV was observed in higher concentration of malathion (0.5 ppm Table 34.1). MCV content was round to have decreased in the higher concentration of malathion (0.5 ppm). On the other hand, the WBC count was found to have increased significantly in the 0.05 ppm of malathion. However in 0.5 ppm the WBC count had increased from 20.58 ± 0.13 × 10^3 cmm to 39.89 ± 0.38 × 10^3 cmm for 24 to 96 h respectively (Table 34.1).

After 24 and 48 h of exposure to sublethal concentrations of malathion (0.05, 0.25 and 0.5 ppm) at 20°c temperature the RBC count significantly decreased, and was more conspicuous in higher concentration (0.5 ppm) (Table 34.2) when compared to control (RBC: 2.85 ± 0.04 × 10^6 cmm, 2.51 ± 0.60 × 10^6 cmm for 24 and 48 h). However at 20°c temperature combination with malathion (0.5 ppm), RBC count significantly increased during 72 and 96 h (RBC: 2.40 ± 0.04 × 10^6 cmm and 2.45 ± 0.08 × 10^6 cmm) when compared to control (RBC: 2.22 ± 0.06 × 10^6 cmm and 1.93 ± 0.13 × 10^6 cmm).

At 30°C temperature with malathion combination a decrease in RBC count was observed in 0.05 ppm of malathion (RBC: 2.07 ± 0.03 × 10^6 cmm, 1.52 ± 0.05 × 10^6 cmm, 1.50 ± 0.10 × 10^6 cmm, 1.46 ± 0.40 × 10^6 cmm) for 24, 48, 72 and 96 h respectively (Table 34.3). The significant decrease was also observed in 0.5 ppm malathion (RBC: 1.66 ± 0.06 × 10^6 cmm for 24 h, 1.43 ± 0.10 × 10^6 cmm for 48 h, 1.36 ± 0.08 × 10^6 cmm for 72 hand 1.37 ± 0.09 × 10^6 cmm for 96 h) when compared to control (30°C: RBC: 2.32 ± 0.04 × 10^6, 1.78 ± 0.05 × 10^6, 1.77 ± 0.07 × 10^6 cmm, 1.69 ± 0.09 × 10^6 cmm) for 24, 48, 72 and 96 h respectively.

The haemoglobin content (Hb) was found to be decreased after exposure to different concentrations of malathion at 20°C for 24 and 48 h (Table 34.2). At 20 °C with higher concentration of malathion (0.5 ppm) the Hb content had further decreased (Hb: 6.52 ± 0.19g per cent and 6.16 ± 0.72g per cent). On the other hand an increase in Hb content in 0.5 ppm malathion at 20°c was recorded (Hb: 7.88 ⊥ 0.17g per cent and 9.20 ⊥ 0.84g per cent) for 72 and 96 h.

At 30°C the Hb concentration had decreased from 24 to 96 h 6.88 ± 0.19g per cent, 6.24 ± 0.17g per cent, 5.96 ± 0.48g per cent and 5.44 ± 0.78g per cent for 24, 48, 72 and 96 h respectively after exposing to 0.05 ppm malathion. The maximum decrease in Hb content was observed at 72 and 96 h of treatment (5.16 ± 1.28g per cent and 3.92 ± 1.23g per cent) with 0.5 ppm of malathion at 30°C (Table 34.3).

A significant decrease of PCV (Ht per cent) was observed in higher concentration (0.5 ppm) of malathion (PCV: 4.68 ± 0.22 per cent for 24 and 4.28 ± 0.23 for 48 h) when compared to control (PCV: 6.16 ± 0.09 and 5.28 ± 0.83 per cent) during 24 and 48 h respectively at 20°C. The significant increase of PCV va1ue was observed at 20°C (PCV: 6.86 ± 0.15 per cent and 8.00 ± 0.51 per cent) for 72 and 96 h of exposure in 0.5 ppm of malathion than the control (5.20 ± 0.77 per cent and 5.94 ± 0.71 per cent for 72 and 96 h) (Table 34.2). The PCV values were found to have significantly declined in the malathion (0.05 ppm) and temperature (30°C) combination, after exposure for 24 to 96 h (PCV: 6.16 ± 1.37 per cent, 4.76 ± 2.00 per cent, 4.00 ± 0.94 per cent and 3.88 ± 0.99 per cent) than the controls. The decrease in the values were more conspicuous in 0.5 ppm (PCV: 5.04 ± 0.97 per cent to 3.32 ± 0.91 per cent) for 24 to 96

Table 34.1: Effect of Sublethal Concentrations of Malathion on Haematological Parameters of *Glossogobius giuris*. (mean ± S.E.)

	Parameters	*RBC* *($x10^6/mm^3$)*	*WBC* *($x10^6/mm^3$)*	*RBC/WBC Ratio*	*Hb* *(g%)*	*Ht* *(%)*	*CI*	*MCV* *(μm^3)*	*MCH* *(pg)*	*MCHC* *(%)*
24 Hours	Control	3.33 ± 0.07	14.08 ± 0.45	23.65 ± 0.38	11.21 ± 0.08	10.0 ± 0.28	0.200 ± 0.01	25.06 ± 0.40	6.01 ± 0.28	24.00 ± 1.22
	0.05	2.44 ± 0.02	16.24 ± 0.56	15.02 ± 0.91	10.53 ± 0.04	7.93 ± 0.08	0.196 ± 0.02	23.06 ± 0.28	5.91 ± 0.10	25.41 ± 0.94
	0.25	2.23 ± 0.03	16.96 ± 0.67	13.14 ± 0.31	10.02 ± 0.07	7.39 ± 0.35	0.186 ± 0.02	15.58 ± 0.95	5.58+0.17	35.82 ± 1.21
	0.5	2.04 ± 0.03	20.58 ± 0.13	10.12 ± 0.48	9.40 ± 0.07	4.26 ± 0.04	0.243 ± 0.01	18.20 ± 0.42	7.30+0.11	40.14 ± 1.32
48 Hours	Control	3.11 ± 0.19	15.08 ± 0.35	20.62 ± 0.68	10.18 ± 0.09	9.43 ± 3.5	0.220 ± 0.09	26.08 ± 0.48	6.21 ± 0.38	25.55 ± 2.11
	0.05	2.76 ± 0.06	18.88 ± 0.22	14.61 ± 1.00	10.42 ± 0.04	7.39 ± 0.13	0.189 ± 0.12	22.53 ± 0.85	5.67 ± 0.03	25.20 ± 1.00
	0.25	2.28 ± 0.05	27.41 ± 0.22	8.31 ± 1.31	9.86 ± 0.07	4.47 ± 0.07	0.200 ± 0.02	16.19 ± 0.41	6.01 ± 0.36	31.13 ± 2.27
	0.5	2.07 ± 0.04	29.33 ± 0.34	7.05 ± 0.97	9.06 ± 0.08	4.06 ± 0.03	0.262 ± 0.02	19.61 ± 0.25	7.87 ± 0.08	40.14 ± 1.31
72 Hours	Control	3.20 ± 0.29	15.43 ± 0.41	20.73 ± 0.40	9.89 ± 0.06	9.38 ± 3.00	0.241 ± 0.02	25.40 ± 0.40	6.91 ± 0.27	25.71 ± 2.66
	0.05	2.45 ± 0.06	27.76 ± 0.36	9.81 ± 1.12	9.48 ± 0.02	6.71 ± 0.07	0.186 ± 0.02	21.78 ± 0.58	5.59 ± 0.07	25.67 ± 0.39
	0.25	2.08 ± 0.90	34.45 ± 0.34	6.03 ± 0.71	8.53 ± 0.01	3.90 ± 0.02	0.212 ± 0.01	15.91 ± 0.73	6.44 ± 0.07	40.51 ± 2.06
	0.5	1.98 ± 0.05	36.58 ± 0.35	5.41 ± 0.41	8.17 ± 0.05	3.80 ± 0.12	0.256 ± 0.02	19.69 ± 0.47	7.70 ± 0.12	40.15 ± 0.01
96 Hours	Control	3.00 ± 0.46	14.48 ± 0.68	20.71 ± 0.65	10.41 ± 0.09	10.00 ± 1.2	0.250 ± 0.02	24.48 ± 0.48	6.48 ± 0.33	26.78 ± 2.06
	0.05	2.38 ± 0.05	30.93 ± 0.05	7.69 ± 0.79	7.70 ± 0.02	5.82 ± 0.08	0.173 ± 0.02	20.93 ± 0.66	5.23 ± 0.19	25.01 ± 0.41
	0.25	2.24 ± 0.09	39.50 ± 0.65	5.67 ± 0.51	8.50 ± 0.04	3.69 ± 0.05	0.189 ± 0.02	16.47 ± 0.52	5.69 ± 0.24	34.57 ± 1.52
	0.5	1.65 ± 0.07	39.89 ± 0.38	4.23 ± 0.38	7.06 ± 0.03	2.58 ± 0.18	0.226 ± 0.01	15.63 ± 0.29	6.42 ± 0.33	41.08 ± 0.98

Table 34.2: Effect of Sublethal Concentrations of Malathion on Haematological Parameters of *Glossogobius giuris* at 20°C (mean ± S.E.)

	Parameters	*RBC* *($x10^6/mm^3$)*	*WBC* *($x10^6/mm^3$)*	*RBC/WBC Ratio*	*Hb* *(g%)*	*Ht* *(%)*	*CI*	*MCV* *(μm^3)*	*MCH* *(pg)*	*MCHC* *(%)*
24 Hours	Control	2.85 ± 0.04	22.40 ± 0.88	12.72 ± 1.83	8.52 ± 0.90	6.16 ± 0.09.	0.996 ± 0.07	21.61 ± 1.78	29.89 ± 0.81	138.31 ± 0.31
	0.05	2.16 ± 0.22	17.90 ± 1.64	12.06 ± 2.40	8.24 ± 0.25	4.20 ± 0.99	1.271 ± 0.09	19.44 ± 2.11	38.14 ± 0.91	196.10 ± 0.41
	0.25	1.72 ± 1.50	27.50 ± 4.1	6.25 ± 1.71	8.00 ± 0.80	4.40 ± 0.75	1.425 ± 0.08	25.58 ± 2.61	46.51 ± 1.20	181.81 ± 0.91
	0.5	1.56 ± 0.60	29.40 ± 1.93	5.30 ± 2.79	6.52 ± 0.19	4.68 ± 0.22	1.392 ± 0.02	30.00 ± 1.99	41.79 ± 0.41	139.31 ± 0.81
48 Hours	Control	2.51 ± 0.60	25.28 ± 1.60	9.92 ± 1.70	8.00 ± 0.06	5.28 ± 0.83	1.062 ± 0.07	21.03 ± 2.41	31.87 ± 0.91	151.51 ± 1.11
	0.05	1.83 ± 0.06	23.72 ± 1.26	7.71 ± 2.00	7.60 ± 0.34	4.08 ± 0.76	1.384 ± 0.05	22.29 ± 3.11	41.53 ± 0.68	186.27 ± 1.33
	0.25	1.53 ± 0.02	30.00 ± 2.00	5.10 ± 3.22	6.52 ± 0.20	4.00 ± 0.40	1.420 ± 0.03	26.14 ± 3.67	42.61 ± 0.81	163.00 ± 0.17
	0.5	1.34 ± 0.06	30.85 ± 0.60	4.34 ± 4.20	6.16 ± 0.72	4.28 ± 0.23	1.532 ± 0.06	31.94 ± 2.66	45.97 ± 1.11	143.92 ± 1.11
72 Hours	Control	2.22 ± 0.06	22.23 ± 0.57	9.98 ± 2.82	7.56 ± 0.73	5.20 ± 0.77	1.198 ± 0.09	23.42 ± 1.77	35.94 ± 0.71	153.46 ± 0.91
	0.05	2.38 ± 0.32	27.87 ± 1.80	8.54 ± 3.33	7.98 ± 0.20	6.40 ± 0.18	1.058 ± 0.04	26.89 ± 2.77	31.76 ± 0.86	118.12 ± 1.60
	0.25	2.24 ± 0.25	22.72 ± 0.41	9.85 ± 3.63	6.94 ± 1.36	5.80 ± 0.26	1.032 ± 0.07	25.89-3.11	30.98 ± 1.12	119.65 ± 1.32
	0.5	2.40 ± 0.04	18.06 ± 0.83	13.28 ± 2.46	7.88 ± 0.17	6.86 ± 0.15	1.063 ± 0.04	27.77 ± 3.00	31.90 ± 1.25	114.86 ± 2.00
96 Hours	Control	1.93 ± 0.13	20.64 ± 0.88	9.35 ± 3.31	6.36 ± 0.73	5.94 ± 0.71	1.098 ± 0.02	30.77 ± 1.97	32.95 ± 1.33	117.07 ± 0.71
	0.05	2.15 ± 0.07	22.72 ± 1.36	9.46 ± 2.10	8.28 ± 0.31	7.28 ± 0.33	1.283 ± 0.04	33.86 ± 1.66	38.51 ± 0.95	113.73 ± 1.21
	0.25	2.20 ± 0.05	18.88 ± 0.76	11.65 ± 3.70	7.96 ± 0.38	6.96 ± 0.44	1.205 ± 0.05	31.63 ± 1.81	36.18 ± 0.71	112.93 ± 1.61
	0.5	2.45 ± 0.08	16.80 ± 1.78	14.58 ± 2.80	9.20 ± 0.84	8.00 ± 0.51	1.251 ± 0.06	32.65 ± 2.22	37.55 ± 1.91	115.00 ± 1.51

Table 34.3: Effect of Sublethal Concentrations of Malathion on Haematological Parameters of *Glossogobius giuris* at 30°C (mean ± S.E.)

	Parameters	*RBC ($x10^6/mm^3$)*	*WBC ($x10^6/mm^3$)*	*RBC/WBC Ratio*	*Hb (g%)*	*Ht (%)*	*CI*	*MCV (μm^3)*	*MCH (pg)*	*MCHC (%)*
24 Hours	Control	2.32 ± 0.04	15.92 ± 0.48	14.57 ± 1.34	8.04 ± 0.17	8.56 ± 1.78	1.150 ± 0.02	36.89 ± 2.16	34.65 ± 0.45	93.92 ± 0.39
	0.05	2.07 ± 0.03	24.08 ± 0.72	8.59 ± 2.08	6.88 ± 0.19	6.16 ± 1.37	1.107 ± 0.07	29.75 ± 2.27	33.23 ± 0.67	111.68 ± 0.86
	0.25	1.82 ± 0.02	24.44 ± 1.42	7.44 ± 1.27	6.40 ± 1.15	5.84 ± 1.49	1.170 ± 0.09	32.08 ± 3.05	35.16 ± 0.29	109.58 ± 0.32
	0.5	1.66 ± 0.06	17.12 ± 1.72	9.69 ± 2.79	5.92 ± 0.23	5.04 ± 0.97	1.188 ± 0.07	30.36 ± 2.86	35.66 ± 1.00	117.46 ± 0.83
48 Hours	Control	1.78 ± 0.05	24.24 ± 2.00	7.34 ± 1.80	6.64 ± 0.21	4.72 ± 1.42	1.243 ± 0.02	26.51 ± 1.86	37.30 ± 0.47	140.67 ± 0.63
	0.05	1.52 ± 0.05	13.68 ± 1.54	11.11 ± 1.70	6.24 ± 0.17	4.76 ± 2.00	1.368 ± 0.22	31.31 ± 2.60	41.05 ± 0.38	131.09 ± 0.26
	0.25	1.48 ± 0.09	12.80 ± 2.40	11.56 ± 3.27	5.84 ± 0.29	3.80 ± 2.11	1.315 ± 0.11	25.67 ± 2.63	39.45 ± 1.31	153.68 ± 0.66
	0.5	1.43 ± 0.10	10.40 ± 3.40	13.75 ± 4.30	4.96 ± 0.49	3.60 ± 2.20	1.155 ± 0.07	25.17 ± 2.65	34.68 ± 0.71	137.77 ± 1.20
72 Hours	Control	1.77 ± 0.07	22.32 ± 2.20	7.93 ± 2.80	6.60 ± 0.40	4.52 ± 1.20	1.242 ± 0.09	25.53 ± 1.66	37.28 ± 0.78	146.01 ± 0.91
	0.05	1.50 ± 0.10	16.56 ± 3.12	9.05 ± 3.86	5.96 ± 0.48	4.00 ± 0.94	1.324 ± 0.19	26.66 ± 2.11	39.73 ± 0.73	149.00 ± 0.94
	0.25	1.38 ± 0.07	14.08 ± 2.40	9.80 ± 3.08	5.24 ± 0.48	3.44 ± 0.96	1.265 ± 0.07	24.92 ± 4.13	37.97 ± 0.92	152.32 ± 0.66
	0.5	1.36 ± 0.08	11.60 ± 0.09	11.72 ± 2.44	5.16 ± 0.28	3.44 ± 1.20	1.264 ± 0.09	25.29 ± 5.55	37.94 ± 0.69	150.00 ± 1.05
96 Hours	Control	1.69 ± 0.09	24.72 ± 0.31	6.83 ± 2.14	5.92 ± 0.88	4.40 ± 1.72	1.167 ± 0.06	26.03 ± 5.93	35.02 ± 0.71	134.54 ± 0.62
	0.05	1.46 ± 0.40	18.00 ± 0.21	8.11 ± 2.72	5.44 ± 0.78	3.88 ± 0.99	1.241 ± 0.05	26.57 ± 6.11	37.26 ± 1.10	140.20 ± 0.72
	0.25	1.40 ± 0.30	14.80 ± 0.31	9.45 ± 1.81	4.28 ± 1.22	3.64 ± 0.71	0.907 ± 0.05	24.97 ± 3.95	29.31 ± 1.31	117.58 ± 1.40
	0.5	1.37 ± 0.09	12.48 ± 0.41	10.97 ± 1.91	3.92 ± 1.23	3.32 ± 0.91	0.953 ± 0.04	24.23 ± 2.85	38.61 ± 0.89	118.07 ± 1.61

h (Table 34.3).

When the fishes were exposed to malathion (0.5 ppm) at 20°C temperature for 24 and 48 h the WBC count was found to have increased (WBC: 29.40 ⊥ 1.93 × 10^3 cmm and 30.85 ⊥ 0.60 × 10^3 cmm) and decreased in the higher concentration (0.5 ppm) of malathion and temperature combination (20°C) for 72 and 96 h (WBC: 18.06 ± 0.83 × 10^3 cmm and 16.80 ± 1.78 × 10^3 cmm) than the control (Table 34.2).

Significant decrease of WBC count was observed in higher concentration of malathion after 24 to 96 h of exposure at 30°C (WBC: 17.12 ± 1.72 × 10^3 cmm to 12.48 ± 0.41 × 10^3 cmm) for 24, 48, 72 and 96 h respectively (Table 34.3).There were variations in values of MCV, MCH and MCHC when the fishes were exposed to sublethal concentrations (0.05, 0.25 and 0.5 ppm) of malathion for 24 to 96 h at 20°C and 30°C (Table 34.2 and 34.3).

Discussion

Blood analysis forms an useful tool for determining pathological conditions of the fishes. The pathological changes were noticed mostly in the peripheral blood of erythrocytes and leucocytes as observed by Reichenbach-Klinke, (1967). In the present study changes in eryhrocyte count was observed after the fish was treated for longer duration. Wlasow and Dabrowska (1989) have reported inhibition of erythroblastosis in circulating blood due to ammonia intoxication in carp. In the present study similar changes of erythrocyte formation was noted in treated fishes. The haematological parameters like Hb content, RBC and WBC count showed significant decrease in treated fishes. The decrease in Hb content and RBC number may result in hypochromic microcytic anaemia, which could be due to deficiency of iron content (Bhai *et al.*, 1971). Since Hb and RBCs are oxygen carrying devices, the quantitative decrease in their levels might have led to the dearrangement of the oxidative metabolism with a concomitant decrease in the tissues of respiratory potential.

G. giuris exposed to sublethal concentrations of malathion resulted in a significant decrease in RBCs count leading to anaemia as a result of inhibition of erythropoiesis, haemosynthesis and increase in the rate of erythrocyte destruction in haemapoietic organs. Similar reports have been made by Goel and Kalpana (1985) and Sampath *et al.* (1998) on the Hb content, RBC count and PCV values resulting in macrocylic anaemia in *Heteropneustes fossilis* and *Oreochromis mossambicus* after exposure to zinc and copper. Srivastava and Shashikala (1979) have shown that exposure to sublethal concentration of Pb produced haemolytic anaemia leading to the lysis of erythrocytes, decrease in Hb content and PCV value in *Colisa fasciatus*. The significant reduction of RBC and Hb content are also reported in fishes exposed to different heavy metals (Goel and Sharma, 1987). Natarajan (1981) reported a reduction in Hb content RBC count and PCV values, resulting in hypochronic anaemia due to deficiency of iron and decreased utilization for Hb synthesis. In the present study on *G. giuris* sublethal concentration of malathion might have impaired the synthesis of iron as a result destruction of mature RBC and reduction in RBC count causing macrocytic anaemia. Sampath *et al.* (1998) have also reported such macrocytic anaemia in *O. mossambicus* after exposure to copper and zinc. Joseph *et al.* (1993) have showed an increase in the relative proportion of immature RBC and decrease maturation of RBC in *Cyprinus carpio* after exposure to sublethal concentration of ammonia. The increased WBC count in treated fishes may be due to the inclusion of thrombocytes in the leucocyte population.

The anaemic condition recorded in the present study could be due to the destruction of mature RBC or inhibition of erythrocyte production. Such a decrease in RHC and anaemic response have been observed by Koundinya and Ramamurthy (1979) in *Sarotherodon massambicus* and *Tilapia mossambicus* after exposure to lethal concentrations of sumithion, Chouhan *et al.* (1983) in *Puntius ticto* treated with herbicide, Lal *et al.* (1986) in *H. fossilis* exposed to malathion, and Sakthivel and Sampath (1989) in *C. carpio* treated with sublethal levels of tannery. Natarajan (1981) reported that impairment in iron synthesizing mechanism corresponds to drastic decrease in haemoglobin content in the fish *Channa striatus* exposed to metasystox. It is evident that the reduction in number of RBC and Hb content and decrease in MCV values might have caused microcytic anaemia in *G. giuris*. Varadharaj *et al.* (1993) have opined that a reduction in the number of RBC and Hb content with all increase in MCV and MCHb values might cause macrocytic anaemic in *O. mossambicus*. Further Pandey *et al.* (1979) and Subramanian *et al.* (1988) have observed a decrease in the rate of oxygen consumption in fishes due to fall in RBC count and Hb content.

An increase in the total WBC count and MCHC content in *G. giuris* after exposure to different sublethal concentrations of malathion coincides with the observations made by Joshi (1987), Radhakrishnan and Prasad (1994). In the present investigation MCV significantly decreased which coincide with the findings of Rai and Quayyam (1981), Dhanekar *et al.* (1985) and Surendra Singh and Bhati (1994). *G. giuris* exposed to different temperatures (20°C and 30°C) and sublethal concentrations (0.05, 0.25 and 0.5 ppm) of malathion the heamatological parameters exhibited variations. Among the various external factors, temperature considered to be as controlling factor and regulating the metabolism (energy utilization) in fishes (Prosser and Brown, 1961 and Kasim, 1979).

The decrease in RBC count, Hb content and PCV in *G. giuris* subjected to temperature (20°C) and malathion toxicity are in agreement with the observations made by Slicher and Pickford (1968) in *Fundulus heteroclitus* and Anthony (1961) in *Carassius auratus*. The results in the present study indicate that the fishes exposed to malathion at 20°C the total number of leucocytes was significantly increased and RBC count, Hb concentration and PCV gradually declined during 24 and 48 h. On the other hand RBC count and PCV increased gradually when the fishes were exposed to malathion with the increased period of exposure at 20°C. However the total WBC, MCHC and MCH significantly decreased when compared to those of controls. The significant decrease in MCV with simultaneous increase in MCHC might have led to microcytosis in *G. giuris* as pointed out by Haws and Goodnight (1962) and Joshi *et al.* (1980).

The fishes exposed to malathion at a temperature of (30°C) and malathion-temperature combination, the haematological parameters except MCHC were found to have decreased with the increase in the exposure period. This phenomenon suggests that a high temperature and malathion toxicity may cause anaemic and macrocytic changes in *G. giuris* when compared to low temperature and malathion stress. Similar results have been recorded by Joshi *et al.* (1980) in *C. batrachus*. In *G. giuris* a decrease in WBC count could have led to a fall in the resistance for malathion and temperature (at 30°C) stress, and this suggests that, malathion in combination with high temperature has an adverse effect resulting in anemic and macrocytic conditions.

Acknowledgements

The author is grateful to Department of Zoology, Bangalore University, Bangalore for providing necessary research facilities and sincere thanks to Dr. S. Krishnan, Professor, Department of Zoology, Bangalore University, Bangalore for their kind help.

References

Anthony, E.H. (1961). The oxygen capacity of gold fish, *Carassius auratus* in relation to the thermal environment. *J. Exp. Biol.*, 38: 93-107.

APHA, AWWA and WPCF. (1985). Standard methods for the examination of water and wastewater,

16th Edition, American Public Health Association, Washington, D.C.

Basile, C., Goldspink, G. Modigh, M. and Tota, B. (1976). Morphological and biochemical characterization of the inner and outer ventricular myocardial layer of adult Tuna fish (*Thymus thymus*) *Biochem. Physio.*, 854: 279-283.

Bhai, L., Nath, N. and Nath, M.C. (1971). Haematological changes in rats injected with acetoacetate, Proc. Soc. Exp. Biol. Med., 138: 597-599.

Blaxhall, P.C. and Daisley, K.W. (1973). Routine haematological methods for use with fish blood. *J. Fish Bioi.*, 5: 771-782.

Chouhan, M.S., Verma, D. and Pandey, A.K. (1983). Herbicide induced haematological changes and their recovery in a fresh water fish, *Puntius ticto* (Ham.), Comp. *Physiol. Ecol.*, 8(4): 249-251.

Christensen, G.M., M.C. Kim, J.M. Brungs, W.A. and Hunt, E.P. (1978). Changes in the blood of the brown bull head, *Ictalurus nebulosus* following short and long term exposure to copper (II), *Toxicol., Appl. Pharmacol.*, 23: 477-482.

Dalela, R.C., Bhatnagar, M.C., Tyagi, A.K. and Verma, S.R. (1978). Adenosine triphosphatase activity in few tissues of a freshwater teleost, *Channa gachua*, following *in vivo* exposure to endosulfan. *Toxicology*, 11: 361-368.

Dhanekar, S., Srivastava, S. and Rao, K.S. (1985). Haematological studies on some fresh water fishes with references to zinc toxicity. *J. Hydrobiol.*, 1(2): 105-107.

George and Alexander, K.M. (1980). Studies of physiological of mammalian cardiac muscle L. A report on differential distribution of some metabolities in myocardia of ox: Boss Sp., Abstr., Proc., 4th All Indian Symposium on comparative physiology and endocrinology. S.V. Unive., Tirupati, Dec.

Goel, K.A. and Kalpana, G. (1985). Haematological characteristics of *Heteropneutes fossilis* under the stress of zinc. *Indian. J. Fish.*, 36: 256-259.

Goel, K.A., and Sharma, S.D. (1987). Some haematological characteristics of *Clarias batrachus* under metallic stress of arsenic. *Comp. Physiol. Ecol.*, 12: 63-66.

Haws, T.G. and Goodnight, C.J. (1962). Some aspects of the haematology of two species of catfish in relation to their habitats. *Physiol. Zool.*, 35: 8-17.

Joseph, K. Manissery and Madhyastha, M.N. (1993). Haematological and histopathological effect of ammonia at sublethal levels of fingerlings of common carp *Cyprinus carpio*. The science of the total environment, supplement, Elsevier Science Publishers B.V. Amsterdam. 913-920.

Joshi, B.D., Chaturvedi, L.D. and Dabral, R. (1980). Some haematologic values of *Clarias batrachus* following its sudden transfer to varying temperature. *Indian J. Exp. Biol.*, 18: 76-77.

Joshi, B.D. (1987). Changes in some blood values of *Clarias batrachus* exposed to lead nitrate. *Ilim. J. Env. Zool.*, 1: 33-36.

Kasim, H.M. (1979). Ecophysiological studies on fry and fingerlings of some freshwater fishes with special reference to temperature tolerance. Ph.D., thesis. Madurai Kamaraj Univ., India.

Koundinya, P.R and Ramamurthy, R. (1979). Effect of Organophosphate sumithion (Fenthion) on some aspects of carbohydrate metabolism in a freshwater fishes *Sarotherodon, mossambicus* and *Tilapia mossambicus* (peters), *J. Expt.*, 15:1632-1633.

Kumar, A.V. (1994). Endosulfan induced biochemical and pathophysiological changes in freshwater fish, *Clarias batractus* (Linn.) Ph.D. Thesis, Osmania University, Hyderabad, India.

Lal, A.S.B., Anitakumari, S. and Sinha, R.N. (1986). Biochemical and haemotological changes following malathion treatment in the freshwater catfish *Heteropneustes fossilis* (Bloch). *Environ. Pollut.*, 42(A): 151-156.

Mount, D.I and Stephan, C.E. (1967). A method for establishing acceptable toxicant limits for fish malathion and the butoxy-ethanol ester of 2,4-D. *Trans. Amer. Fish. Soc.*, 21: 185-193.

Natarajan, G.M. (1981). Changes in the biomodal gas exchange and some blood parameters in the air-breathing fish *Channa striatus* (Bloch) following lethal exposure to metasystox (Dimeton). *Curr. Sci.*, 50: 40-41.

Nemcosk, T., Orban, L., Asztalos, B. and Vig, E. (1987). Accumulation of pesticides in the organs of carp *Cyprinus carpio* L. at 4 and 20°C. *Bull. Environ. Contam. Toxicol.*, 39: 370-378.

Pandey. B.N., Chanchai, A.K., Singh, S.B., Prasad, S. and Singh, M.P. (1979). Effect of some biocides on the blood and oxygen consumption in *Channa punctatus*. The academy of environmental Biology, Muzaffarnagar, 343-348.

Pascoe, D., Evans, S. and Woodworth, J. (1986). Heavy metal toxicity to fish and the influence of water hardness. *Arch. Environ. Contam. Toxicol.*, 15: 481-485.

Prosser, C.L. and Brown, F.W. (1961). Comparative animal physiology (2nd Edn.) Saunders, Philadelphia.

Radhakrishnan, S. and Prasad, G. (1994). A unique haematological response of *Oreochromis mssambicus* to sublethal concentrations of the insecticide, Ekaluk EC-25. Proceedings of the Indian National Science Academy Part B. Biological Sciences 60(6): 499-503:

Rai, R. and Quayyam, M.A. (1981). Haematological studies of mercury intoxicated fleets fish *Catla catla* (Hamilton). *Ind. J. Zool.*, 9: 87-91.

Ramalingam, K. and Ramalingam, K. (1982). Effect of sub lethal levels of DDT, malathion and mercury on tissue proteins of *Sarotherodon mossambicus* (Peters). Proc. Indian. Acad, Sci., (Anim. Sci.) 91: 501-503.

Reichenbach-klinke, H.H. (1967). Untersuchungen uber die Einwirking des Ammoniakgchalts aufden fish organisums, *Arch. Fisch, Wise*, 17: 122-130.

Sakthivel, M. and Sampath, K. (1989). Haematological responses of *Cyprinus carpio* in relation to starvation, *Geobios*, 16: 61-65.

Sampath, K., James, R. and Akbar Ali, K.M. (1998). Effects of copper and zinc on blood parameters and prediction of their recovey in *Oreochromis mossambicus*. *Indian. J. Fish.*, 45(2): 129-139.

Slicher, A.M. and Pickford, G.E. (1968). Temperature controlled stimulation of haemopoiesis in a hypophysectomized cyprinodont fish *Fundulus heteroclitus*. Physiol. Zool., 41: 293-297.

Srivastava, A.K. and Shashikala, M. (1979). Blood dyscrasia in a teleost, *Colisa fasciatus* after acute exposure of sublethal concentration of lead. *J. Fish Biol.*, 14: 199-203.

Srivastava, A.K. (1989).A review of landmarks in teleostean haematology, Proc. Natl., Symp. Emerg. Tr. Anim. Haematol., 165-177.

Subramanian, M.A., Shahul, H. and Veradamj, G. (1988). Haematological changes in the fish *Marcones keletius* (Gunther) after exposure to the tannery effluent. *The Indian J. Zoo.*, 12: 71-74.

Surendra Singh and Bhati, D.P.S. (1994). Effect of zinc chloride on certain morphological parameters of the blood in *Channa punctatus* (Bloch.). *Poll. Res.*, 13(14): 381-384.

Varadharaj, G., M.A. Subramanian and R. Nagarajan (1993). The effect of sublethal concentration of paper and pulp mill effluent on the haematological parameters of *Oreochromis mossambicus* (Peterns). *J. Environ, Biol.*, 14(4): 321-325.

Wintrobe, M.M. (1977). Clinical haematology. Henry Kimpton. London, 448.

Wlasow, T. and Dabrowska, H. (1989). Cellular changes in the blood and haemopoietic tissues of common carp exposed to sublethal concentration of ammonia. *Aquat. Living Resour.*, 2(3): 169-174.

35

Effects of Sublethal Concentrations of Chromium on Gill Tissues of an Edible Fish *Cirrhinus mrigala*

P.L.R.M. Palaniappan, S. Karthikeyan and Selvi Sabhanayakam*

Department of Physics, Annamalai University, Annamalainagar - 608 002
**Department of Zoology, Annamalai University, Annamalainagar - 608 002*

ABSTRACT

Studies on the histopathological effect of the two different sub-lethal concentrations such as lower (1.82 ppm) and higher (6.06 ppm) concentrations of chromium on *Cirrhinus mrigala* fingerlings revealed that the metallic salt is capable of producing severe damages and changes in its cellular levels in gills leading to the death of the fish. The changes in the histology of gills led to the disturbance in basement membrane, disintegrated gill lamellae, reduced inter-lamellar space, fusion of secondary lamellae and disintegrated gill epithelium when fish exposed to higher concentration than lower concentration. All these changes finally caused the failure of the respiratory mechanism resulting in the mortality of the test fish.

Key Words: Chromium, Gill, Disintegrated gill epithelium.

Introduction

Chromium is used extensively in the industry such as pulp and paper, textile, tanning, cement, dyes and ink industries. Chromium is released into aquatic system from industrial sources such as tanning industry (Dad *et al.*, 1980), electroplating processes (Deutch, 1961) and burning of fossil fuels (Torrey, 1978) which affects the aquatic environment. Chromium easily penetrates into biological membranes and irritates cells (Mertz, 1969). Benait (1976), Saxena *et al.* (1980), Stevens and Chapman

(1984) and Abbasi and Soni (1985) have assessed the toxicity of chromium on fish. Srivastava *et al.* (1982, 1991) have observed the histopathological changes and accumulation potential in the fish tissues under chromium stress.

Pollution of aquatic ecosystem by chemicals used in industry and agriculture is increasing day by day. The gills are the most vulnerable organ of fish exposed to the medium containing any type of the toxicants. (Jana and Bandopandhyaya, 1987). The histopathological studies on fish is a noteworthy and promising field to understand the extent to which changes in the structural organization occurring in the organs due to environmental pollutants.

Singh *et al.* (1990) studied about the environmental pollution and its effects on aquatic animals. Heavy metal toxicant leads to many pathological changes in different tissues of fish and has been reported for *Labeo rohita* exposed to mercuric chloride (Jagadeesan, 1999) and *Channa punctatus* exposed to phenyl mercuric acetate (Karuppasamy, 2000).

However, the studies on the effects of heavy metal, chromium on the gills of fingerlings of *Cirrhinus mrigala* are meagre. The present paper deals with the histopathology of gills of *Cirrhinus mrigala* exposed to sub-lethal concentration of chromium treated for 28 days.

Materials and Methods

Cirrhinus mrigala fingerlings of length (6 ± 1 cm and weight 8 ± 1 g) were collected from the freshwater bodies near the local fish farm, Puthur, Tamil Nadu and acclimatized under laboratory conditions (29 ± 1°C). Boiled eggs, rice bran and earthworm pieces were fed on every alternate days.

The LC_{50} for pottassium dichromate for 96 h was found out by using Probit method (Finney, 1971). The $1/10^{th}$ and $1/3^{rd}$ of LC_{50} were taken as lower and higher concentrations, respectively. Control and experimental fishes were maintained in the plastic trough in three groups containing 20 litre capacity of water. Group 1 serves as a control maintained in the water without addition of any toxic chemical. Group 2 and Group 3 were exposed to lower (1.82 ppm) and higher (6.06 ppm) concentration of chromium for a period of 28 days.

At the end of 28 days, the gills of both control, lower and higher concentrations of chromium treated fishes were fixed in Bouin's fluid. The histopathological process and embedding techniques were used according to the method of Gurr (1959). 6 μ sections were stained with haematoxylin and counter stained with aqueous eosin. The slides were examined and microphotographs were taken.

Observations

The microphotographs of gills of *Cirrhinus mrigala* are shown in Figs. 35.1-35.4. It has been observed from the figures that in control animals, the structure of gills are similar to that of other freshwater teleost as described by Laurent (1989) whereas, the gills of *Cirrhinus mrigala* fingerlings exposed to lower and higher concentrations of chromium showed active secretion of mucus depending on the accumulation. At lower concentration of chromium, remarkable changes in the gill structure such as decreased interlamellar space, swollen base of secondary lamellae (Fig. 35.3) are observed. The epithelial layer was detached off and completely destroyed (Fig. 35.4). In few regions, the fusion of secondary lamellae were found to be disappeared (Fig. 35.3) when the test fish is exposed to chromium for 28 days. At higher concentration, changes in its architecture such as reduced interlamellar space, widening of primary gill lamellae and disintegrated primary gill lamellae without any cells in the gill epithelium (Fig. 35.4) are observed.

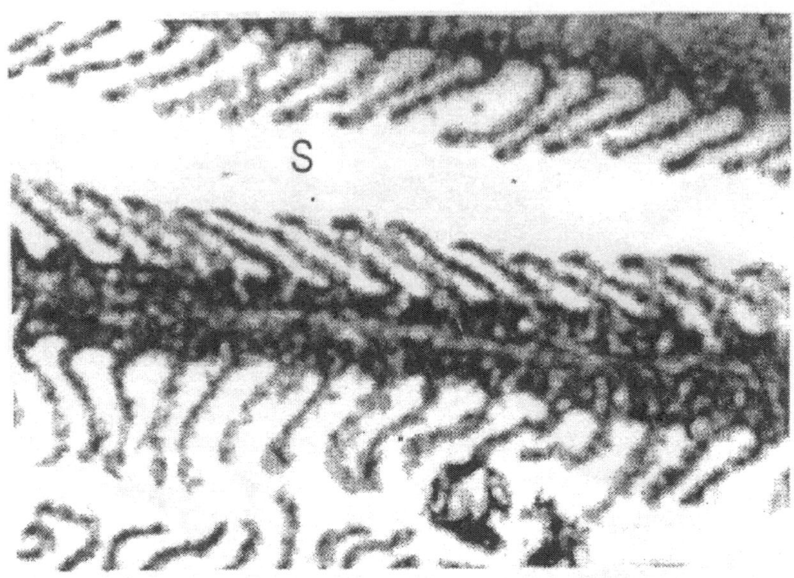

Fig. 35.1: Structure of Gill Under Normal Condition shows Primary and Secondary Gill lamellae, Bouine, 6HE X 100.

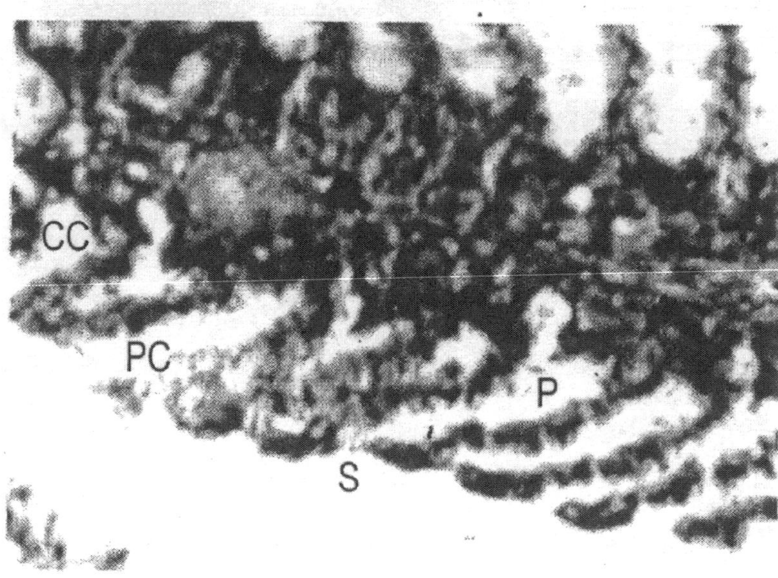

Fig. 35.2: Structure of Gill Under Normal Condition Shows Primary and Secondary Gill lamellae, Bouine, 6HE X 200.
GF: Gill Filament, S: Secondary Gill Lamellae, P: Primary Gill Lamellae.

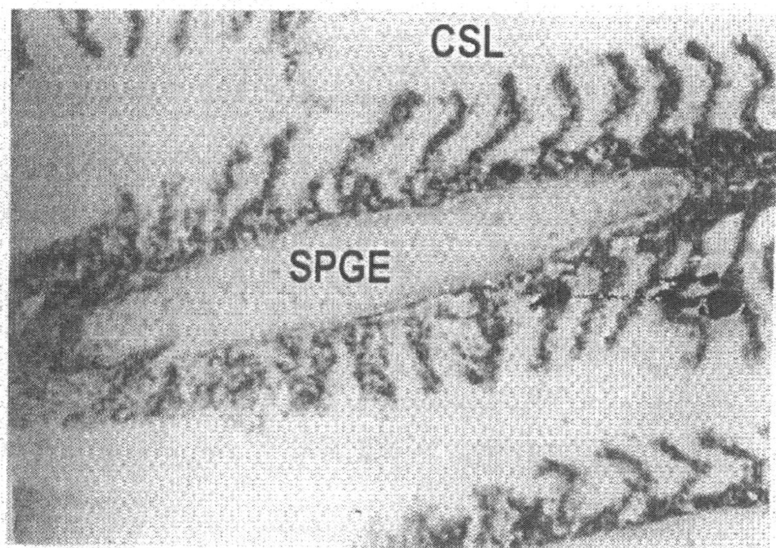

Fig. 35.3: Gill of Fingerlings of *Cirrihinus mrigala* Treated with Lower Concentration of Chromium, Bouine, 6HE X 200.
CSL: Curved Secondary Lamellae, SPGE: Swollen Primary Gill Epthelium.

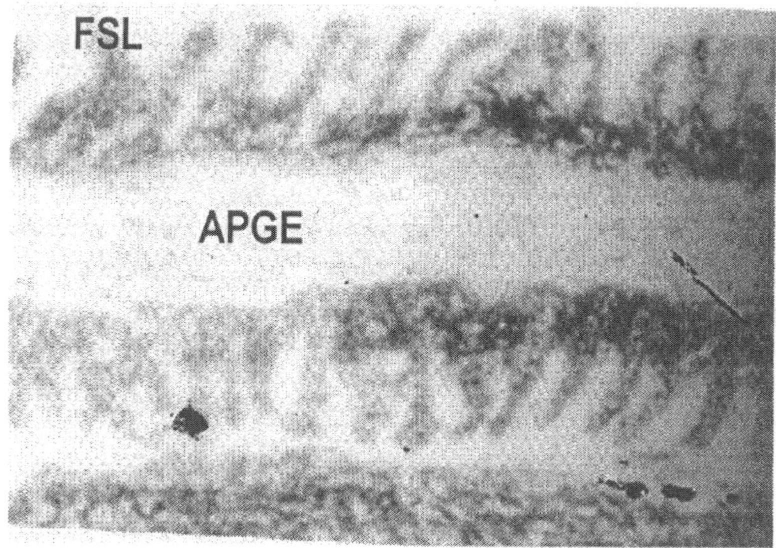

Fig. 35.4: Gill of Fingerlings of *Cirrihinus mrigala* Treated with Higher Concentration of Chromium, Bouine, 6HE X 200.
AGPE: Atrophy of Primary Gill Epithelium, FSL: Fusion of Secondary Lamellae.

Discussion

In the present study, it has been observed that some remarkable histopathological changes have taken place when *Cirrhinus mrigala* was exposed to both lower and higher concentrations of chromium for 28 days. Changes such as disintegrated primary gill lamellae, reduced interlamellar space, necrotic, gill epithelium, swollen base of the secondary lamellae, fusion of secondary lamellae, widening of primary gill lamellae with more space in the gill structure were noticed when the fish was exposed to higher concentration than the lower concentration of chromium. Similar results have also been observed by Skidmore (1970) in rainbow trout when exposed to zinc sulphate.

Histological examinations of the gills of rainbow trout by Lloyd (1965) when exposed to zinc, lead and copper showed separation of the epithelium from the gill lamellae and cells sloughed off into the spaces between the filament. These changes are in agreement with the present observation when the fish is intoxicated with the heavy metal, chromium. This may be attributed due to depletion of oxygen and carbon dioxide accumulation which lead to suffocation and death of the animal.

Further, it has also been observed certain damages to gill filaments at both lower and higher concentrations of chromium on *Cirrhinus mrigala* which leads to decreased efficiency for gas exchange and breakdown of vital functions leading to the death of the fish. The results are in parallel with the works of Kapilamandi *et al.* (1999) who have reported for *Boleopthalmus dumieric* (cuv) exposed to sub-lethal concentration of cadmium.

The present histological observations of gills of *Cirrhinus mrigala* showed that the tested concentrations of chromium have devasting effect on gill tissues which results in the death of the fish.

Acknowledgement

The authors wishes to express their sincere thanks to Dr. S. Krishnan, Professor of Physics for constant encouragement and Dr. S. Lakshmanan, Professor and Head, Department of Anatomy, Rajah Muthiah Medical College, Annamalai University to use microphotograph facilities available in the laboratory.

References

Abbasi, S.A. and Soni, R. (1985). Environment management and treatment level of chromium with respect to impact on common fish. *J. Inst. Eng. Envt, Eng. Divn.* 65: 113-116.

Beliles, R.P. (1975). Toxicology: The basic science of poisons, Ed. Casarett, L.J., Doull, J. Macmillan Publ. Co., Inc., New York.

Benoit, D.A. (1980). Toxic effect of hexavalent chromium to brook trout (*Salvenius fontinalis*) and feather trout (*Salmogairdneri*). *J. Water Res. G.B.* 10: 497-500.

Dad, N.K, Rau, K.S and Qurshi, S.A. (1980). Evaluation of toxicity of leather factory effluents to midge larvae (*Chironomus tentanus*) by bioassay. *Int. J. Environ. Stud.,* 15: 55-56.

Deutch, M. (1961). Incidents of chromium contamination of ground water in Michigan (Proc. Symp. on Groundwater contamination, technical report). Cancinnati OH.

Finney, D.J. (1971). Probit Analysis, 3rd Ed, Cambridge University Press, Cambridge, 333 pp.

Gurr, E. (1959). Methods for analytical histology and histochemistry, Leonard Hill Books Ltd., London.

Jagadeesan, G. (1999). In-vivo recovery of gill tissues of a freshwater fish *Labeo rohita* after exposure to different sublethal concentration of mercury. *Poll. Res.*, 18(3): 289-291.

Jana, S. and Bandopadhyaya, N. (1987). Effect of heavy metals on some biochemical parameters in the freshwater fish *Channa punctatus*. *Environ. Ecol.*, 5(3): 488-493.

Kapilamanoj and Ragothaman, G. (1999). Effect of Sublethal concentrations of cadmium the gills of an estuarine edible fish, *Boleophthalmus dussumieri* (cuv). *Poll. Res.*, 18(2): 145-148.

Karuppasamy, R. (2000). Tissue histopathology of *Channa punctatus* (Block) under phenyl mercuric acetate toxicity. *Bulletin of Pure and Applied Science*, 19A(2): 109-116.

Laurent, P. (1989). Gill structure and function in fishes. In: Comparative Pulmanary Physiology, Current concept series-Lung biology in healthy and diseased fishes -Ed. S. Wood. Marcel and Dekkar Inc. New York, 69-120.

Loyd, R.H. (1965). Factors that effect the tolerance of fish to heavy metal poisoning. Biological problems in the water pollution, 3rd Seminar, U.S. Dept., Health Education and Welfare p. 8.

Mertz, W. (1969). Chromium occurrence and function in biological systems. *Physiol. Rev.*, 49: 163-239.

Singh, N.K, Varma, M.C. and Munshi, J.S.D. (1990). Accumulation of Copper, Zinc, Lead, Iron and Cadmium in certain freshwater fishes of river Subemarekha. *J. Fresh Wat. Biol.*, 2(3): 189-193.

Skidmore, J.F. and Tovell, P.W.A. (1971). Toxic effect of zinc sulphate on the gills of rainbow trout. *Wat. Res.*, 6: 217-230.

Srivastava, V.M.S., Tripathi, R.S. and Saxena, A.K. (1982b). Changes in gill of *Punctius sophore* under chromium stress. *J. Biol. Res.* 2: 85.

Srivastava, V.M.S. and Maurya R.S. (1991). Effect of chromium stress on gill and intestine of *Mystus vittatus* (Bloch): Scaning electron microscopic study. *J. Ecobiol.*, 3: 69-71.

Stevens, D.G. and Chapman, G.A. (1984). Toxicity of trivalent Chromium to early stages of Steel head trout. *Environ. Toxicol. Chem.*, 3: 125-134.

Torrey, S. (1978). Ecological effects, In: Torrey, S. (ed). Trace contaminents from coal. Noyse Data Corp, Trenton. N.J., pp 143-208.

36

Bioaccumulation and Elimination of Cadmium in Freshwater Fingerlings, *Labeo rohita* (Ham.)

PL. RM. Palaniappan, G. Jagadeesan*, P. Venkatachalam, N. Krishnakumar and S. Karthikeyan

Department of Physics, Annamalai University, Annamalainagar - 608 002, Tamil Nadu
**Department of Zoology, Annamalai University, Annamalainagar - 608 002, Tamil Nadu*

ABSTRACT

Bioaccumulation of cadmium is assessed in freshwater fish, *Labeo rohita* after induced acute and chronic exposures. In take, accumulation and elimination of cadmium in different organs of fish have been discussed. The kidney was found to be the prime site of metal binding and brain accumulate least metal concentration. As the exposure concentration was increased, the biological magnification factor reduced and the cadmium concentration in tissues increased by increased exposure period. Also the treatment of chelating agent Calcium disodium ethylene diamine tetra acetic acid ($CaNa_2EDTA$) reduced the metal toxicity significantly and follows the order: Kidney > Liver > Brain > Gill.

Key Words: Bioaccumulation, Biomagnification, Cadmium, $CaNa_2EDTA$, Tissues, *Labeo rohita*.

Introduction

Cadmium is a ubiquitous, non-essential element which possesses high toxicity to both humans and aquatic organisms. According to the hypothesis proposed by Harris and Hohenemser (1978), cadmium is classified as the second most dangerous metal in our environment. In recent years, cadmium and cadmium compounds have been used extensively by various industries such as electroplating, battery manufacture, pigments and plastic production and this has produced sharp increases in contamination of air, water and soil. Population exposure to toxic trace metals are of great concern due to their non-biodegradable nature and long biological half-life for elimination from the body. There are many reports in the literature on metal accumulation in tissues. Usually accumulation differs between tissues and species. For example, in *Hamarus americanus* cadmium accumulated mainly in the digestive gland followed by the gills, whereas in *Callinectes sapidus*, cadmium accumulated mostly in the gills (Darmono, 1990). Cadmium toxicity to aquatic ecothermal animals depends on complex biochemical interactions and a balance between rates of absorption, detoxification and excretion. The main sites of cadmium deposit are the kidney and liver of vertebrate animals (Ferrari *et al.*, 1993). In the present investigation, an attempt has been made to study the accumulation pattern and the efficacy of an antidote $CaNa_2EDTA$ to induce the elimination of cadmium in Cd treated fingerling stage of freshwater fish, *Labeo rohita* (Ham.).

Materials and Methods

Specimens of *Labeo rohita* were collected from the local fish farm and acclimatized to the laboratory conditions for fifteen days. Fishes were fed with oil less groundnut cake twice a day. Fishes ranging from 5-6 cm in length and weighing 8-10 gm were selected for experimental purpose. The physico-chemical parameters of water was estimated according to APHA (1989) and are follows: Dissolved oxygen: 7.2 ppm -7.4 ppm, pH: 7.0 - 7.2, Temperature: 24.0 ± 2.0°C, Salinity: 0.4 ppm - 0.5 ppm and Total hardness: 280 mg/l - 288 mg/l. Preliminary studies were carried out to find the median lethal concentration (LC_{50}) for 96 h by Probit analysis method (Finney, 1971).

The test specimens were divided into seven groups, each consisted of 20 fishes and stocked in 20 litre plastic aquaria, equipped with continuous air supply. The physico-chemical water parameters were measured systematically to maintain its optimum level. The water was changed at regular intervals along with waste feed and faecel materials. Group I was treated as control. Groups (2, 3), (4, 5) and (6, 7) were exposed to cadmium chloride of 2.5 ppm (acute, 1/3 of LC_{50}) for 14 days, 0.75 ppm (chronic, 1/10 of LC_{50}) for 30 days and 60 days respectively. Groups 3, 5 and 7 were again exposed to $CaNa_2EDT$ A of 1: 3 ratio (Cd: $CaNa_2EDTA$) for another 7 days.

At the end of the experimental periods, the fishes were sacrificed, the organs were dissected and kept in frozen at 70°C until used. Before the sample preparation the tissues were dried in an oven at 80°C for 12 hours.

Estimation of Cadmium

Following the method of Topping (1973), dried samples were processed for acid digestion by using perchloric acid and nitric acid in the ratio 1: 3 respectively. Cadmium concentration in the samples were determined by Inductively Coupled Plasma-Atomic Emission Spectrometer (ISA JOBIN YVON 24 Model) installed at Centre for Advanced Study in Marine Biology, Portanova, Annamalai University.

Results and Discussion

Table 36.1 summarize data on cadmium concentration (µg/g) in various tissues of *Labeo rohita*. It is evident from the table that cadmium is taken up by the animals from the surrounding medium and accumulated among various organs. The accumulation of metal in tissues of aquatic animals is dependent upon exposure concentration and period as well as some other factors such as salinity, temperature, interacting agents and metabolic activity of tissue in concern. It is also known that metal accumulation in tissues of fish is dependent upon the rate of uptake, storage and elimination. It is seen from the table that highest cadmium concentration is found in the kidney tissue and least in the brain tissue for both acute and chronic exposures. The accumulation of cadmium in the tissues follows the order: Kidney > Liver > Gill > Muscle > Brain. These results are consistent with the earlier workers (Mount and Stephen, 1967, Kumuda *et al.*, 1972, Andres *et al.*, 2000). It has also been observed that the treatment of chelating agent $CaNa_2EDTA$ has reduced the accumulation of cadmium significantly.

Table 36.1: Organ-wise Average Bioaccumulation (µg/g) of Cadmium in *Labeo rohita* Fingerlings During Acute and Chronic Exposures

Organ	*Control*	*Acute - 14 days - 2.5 ppm*			*Chronic - 30 days - 0.75 ppm*			*Chronic - 60 days - 0.75 ppm*		
		Cd	*B.M.F.*	*Cd+EDTA*	*Cd*	*B.M.F.*	*Cd + EDTA*	*Cd*	*B.M.F.*	*Cd + EDTA*
Kidney	6.142	81.146	32X	31.932	29.412	39X	12.020	75.067	100X	29.031
Liver	3.462	50.238	20X	23.311	22.611	30X	9.640	67.203	90X	37.442
Gill	2.280	32.481	13X	26.427	12.433	17X	9.000	45.612	61X	33.640
Muscle	2.674	30.832	12X	20.332	10.628	14X	6.642	29.524	31X	14.213
Brain	1.026	9.821	4X	5.121	4.514	6X	2.400	11.402	15X	7.631

B.M.F.: Biological Magnification Factor.

Heavy metals primarily affect the kidney, which is involved in the cleaning processes of the body fluids and tissues. In the present study kidney tissue accumulates substantial amount of cadmium (81.146 µg/g, 29.412 µg/g and 75.067 µg/g) for acute and chronic exposures. Since kidney being a major target organ for metals, the perfusion rate is greater compared to other organs (Chris Kent, 1998). Hence the toxic substances present in the blood is delivered to kidneys in large quantities.

Next to kidney, liver accumulates a significant amount of cadmium (50.238 µg/g, 22.611 µg/g & 67.203 µg/g) for acute and chronic exposures. The metal binding protein metal-lothionein is of utmost importance in the accumulation of metals. The liver is the main organ for homeostasis in fish, and is in an advantageous position for clearing the blood substances entering the circulation from the gastro-intestinal tract passes through the liver before reaching the systemic circulation. Therefore, the liver removes the toxicants from the blood, bio-transform them into the bile and prevents distribution to other parts of the body (Coetzee, 1996). Hence, liver, which is a major producer of metal-binding proteins shows high concentrations of heavy metal, cadmium.

Gills are the predominant barriers that control the mechanism of absorption of bioavailable metals from the external medium (direct route of exposure). Hence the gills are one of the target of a direct contamination as they play a significant role in metal uptake, storage and eventually transfer to the internal compartments via blood transport (Andres *et al.*, 2000). In the present study, accumulation of cadmium in the gills are: 32.481 µg/g, 12.433 µg/g and 45.612 µg/g for acute and chronic toxicity respectively.

During acute and chronic exposures, fish accumulated maximum metal in kidney, followed by liver and gill. Muscle and brain were the least cadmium accumulated tissues. According to Bradley and Morris (1986), the muscle tissue is usually not a good indicator of environmental pollution in a freshwater ecosystem, and the fact that the muscle showed low metal concentration is therefore not surprising. Moreover the muscle tissue have larger surface area and the metals are evenly distributed, which leads to low metal significant (30.832 µg/g, 10.628 µg/g and 29.524 µg/g). In the brain, the blood-brain barrier effectively decreases the amount of toxic substances that are transferred into the brain tissues (Chris Kent, 1998). Hence the least metal concentration (9.821 µg/g, 4.514 µg/g and 11.402 µg/g) observed in the brain is in the expected order and are consistent with the observation made by Vincent and Ambrose (1994) for *Catla catla*.

In the present study, the greatest biological magnification occurred at the lowest (0.75 ppm) cadmium exposure level. At this exposure level, the kidney accumulated approximately 39X and 100X the amount of cadmium present in the aquaria water for acute and chronic exposures. As the exposure concentration , increases, the biological magnification reduced. At the highest exposure level (2.5 ppm), the kidney accumulated approximately 32X the amount of cadmium in the aquaria water.

Also, in the liver tissue, greatest biological magnification was found in the lowest (0.75 ppm) cadmium exposure level. At this exposure level, the liver accumulated 30X and 90X the amount of cadmium present in the aquaria water for acute and chronic exposures respectively. At the higher concentration, the biological magnification was reduced by 20X the amount of cadmium present in the aquaria water. Similar trend was observed in the gill, muscle and brain tissues. In the chronic exposure (60 days) the rate of accumulation increases due to the cadmium induced metallothionein synthesis in these tissues, which is an important transport and storage protein for cadmium (Friberg *et al.*, 1992).

Chelating agents are the most versatile and effective antidotes for metal intoxication. Sunda and Zamuda (1982) have demonstrated that organic chelation compounds can form stable complexes which can hardly be accumulated in organisms, thereby reducing the toxicity of the metals to organisms. Of the various chelating agents Calcium disodium ethylene diamine tetra acetic acid ($CaNa_2EDTA$) is an efficient chelator of many divalent and trivalent metals (Ellenhom and Barceloux, 1988) for acute and chronic cadmium exposures. It penetrates cell membranes relatively poorly and chelates extracellular metal ions much more effectively than intracellular ions. The present results suggest that the $CaNa_2EDTA$ reduces the cadmium concentration in the tissues of *Labeo rohita* fingerlings significantly (19 per cent-62 per cent). It is also evident from the present study, that cadmium is released from the organs at different rates (0.54 ppm to 6.57 ppm) and cadmium excretion is invariably found to be higher than the uptake and follows the order: Kidney > Liver > Brain > Gill.

Acknowledgement

The authors express their sincere thanks to Prof. S. Krishnan, Department of Physics, Annamalai University for his constant encouragement and to Prof. L.S. Ranganathan, Head, Department of Zoology, Annamalai University for providing the necessary laboratory facilities to carried out this study.

References

Andres, S., Ribeyre, F., Tourencq, J.N. and Boudou, A. (2000). Interspecific comparison of cadmium and zinc contamination in the organs of four fish species along a polymetallic pollution gradient (Lot River, France). *The Sci. of Total Environ.*, 248: 11-25.

APHA (1989). Standard methods for the examination of water and waste water, 18th Ed. APHA Washington DC.

Bradley, R.W. and Morris, J.R. (1986). Heavy metals in fish from a series of metal contaminated lakes near Sudbury, Ontario. *Water Air Soil Pollut.*, 27: 341-354.

Chris Kent. (1998). Basics of Toxicology. John Wiley and Sons, Inc. New York.

Coetzee, L. (1996). Bioaccumulation of metals in selected fish species and the effect of pH on aluminium toxicity in a cichlid *Oreochromis mossambicus*. M.Sc. Thesis, Rand Afrikaans University, South Africa.

Darmono, D. (1990). Uptake of Cadmium and Nickel in Banana Prawn (*Penaeus merguiensis de man*). *Bull. Environ. Contam. Toxicol.* 45: 320-328.

Ellenhom, M.J. and Barceloux, D.G. (1988). Medical Toxicology: Diagnosis and treatment of human poisoning. Elsevier (Eds.) New York.

Ferrari, L., Alfredo Salibian, and Claudia Muino, V. (1993). Selective protection of temperature against cadmium acute toxicity to *Bufo arenarum Tadpoles*. *Bull. Environ. Contam. Toxicol.*, 50: 212-218.

Finney, D.J. (1971). Probit analysis. Cambridge University Press.

Friberg, L., Elinder, C.G. and T. Kjelltrom (1992). Environmental Health Criteria 134-Cadmium, WHO, Geneva.

Harris, R.C. and Hohenemser, C. (1978). Mercury-Measuring and managing the risk. *Environment*, 20: 25-36.

Kumuda, H., Kimura, S., Yokata, M. and Matida, Y. (1972). Acute and chronic toxicity, uptake and retention of cadmium in freshwater organisms. *Bull. Freshwat. Fish. Res. Lab. Tokyo*. 22: 157-165.

Mount, D.I. and Stephan, C.E. (1967). A method of detecting cadmium poisoning in fish. *J. Wild. Mgnst.*, 31: 168-172.

Sunda, W.G., Zamuda, C.D. (1982). Bioavailability of dissolved copper to the American oyster *Crassostrea virginica. I.* Importance of chemical speciation. *Mar. Biol.*, 66: 77-82.

Topping, G. (1973). Heavy metals from Scottish water. *Aquaculture*. 1: 379-384.

Vincent, S. and Ambrose, T. (1994). Uptake of heavy metals. Cadmium and chromium in tissue of Indian major carp *Catla catla* (Ham.). *Indian J. Environ. Hlth.* 36: 200-204.

37

Herbicides Induced Alterations on Some Haematological Parameters of Freshwater Teleost, *Clarias batrachus (Linn)*

Anju Kumari and K. Pandey

Department of Zoology, M.R.M. College, Darbhanga - 846 004

ABSTRACT

Effects of two herbicides namely Taficide and Butachlor on Haematological Parameters of fresh water fish, *Clarias batrachus* (Linn) exposed to Lc50 at ambient temperature were studied. A significant decrease ($P < 0.001$) in RBC, HB, PVC, MCH and MCHC values except MVC value in male, which remains insignificant ($P > 0.05$) under taficide stress. But in case of Butachlor treatment RBC, Hb content and PVC values significantly decrease ($P < 0.001$) while MCV and MCH values significantly increase. MCHC values insignificantly increase ($P > 0.05$). The ESR values show a highly significant increase ($P > 0.001$) in both the herbicides.

Key Words: Herbicides, Haematological Parameters, *Clarias batrachus*.

Introduction

The extensive use of Pesticides, Insecticides, Herbicides/Seedicides and Fungicides are being promoted by Government of India to enhance the crop production to meet the demand of the growing population. The rationale behind this is the eradication of undesirable insects, pests, weeds and herbs for increased yield but on the other hand, these lead to large scale mortality of one of the most important aquatic fauna, the fishes forming stable food to the mankind. All these in turn cause severe ecological imbalance. Herbicides have been reported to cause undesirable effects (Hodge, 1973, Vardia and Dhurve 1993, Alam, 2000, Dhasarathan, *et al.*, 2000). The herbicides can cause death of fishes either directly or due to starvation by destruction of food chain. Many herbicides have shown to effect the growth and reproduction of fishes with evidence of tissue damage (Paulove, 1980, Jirosek *et al.*, 1981,

Kosanke *et al.*, 1988, Verma and Raji, 2000). Blood is a very important tool for studying the effects of toxicant. Blood is highly susceptible to internal environmental fluctuations. Changes in the Physicomophology of the blood can readily indicate the changes in the quality of the environment. Several reports were available (Hunn, 1967, Blaxhall, 1972, Pandey and Pandey, 2001) which indicates the increasing importance of haematology in health, disease and toxicological screening of fish as test species (Modesley-Thomas, 1971).

Materials and Methods

Adult live specimen of *Clarias batrachus* were procured from the local fish market. They were washed with 0.1 per cent KMno4 solution to get rid of any dermal infection. Only healthy fish of 20-40 cms length and 70-90 gms weight of both the sexes were selected for the experiments. The fish were allowed to adjust laboratory condition in tap water for 10 days under a natural photoperiod and ambient temperature in glass aquarium. The fish were fed with goat liver ad libitum everyday. A batch of twenty fish was transferred each to control and experimental aquaria. The experimental lot of fish was exposed to a LC_{50} Concentration of Taficide and Butachlor for 48 hours. The exposure medium was renewed every alternate day to maintain its effective concentration. Feeding of fish was stopped before they were transferred to experimental jars. Parallel group of control fish were kept in tap water.

Taficide (2, 4-dichlorophenoxy acetic acid) and Butachlor (50 per cent EC) chloro-autanilide herbicides were purchased from local market and solution were prepared in normal tap water. The 48h Lc_{50} value was determined by standard method (APHA, 1985).

The blood sample from both experimental and control group were collected separately at selected period of exposure by severing the Caudal Peduncle region or by direct heart puncture only in the morning hour between 8 to 10 AM to avoid diurnal variations of blood parameters. Sample for fresh blood were collected for Total RBC count (TEC) haemoglobin (Hb), Packet cell volume (PVC), Mean corpuscular haemoglobin (MCH), Mean corpuscular haemoglobin concentration (MCHC), Mean corpuscular volume (MCV) and Erythrocyte Sedimentation rate (ESR) Were calculated employing standard methods and formulae (Sandhu, 1990).

Results and Discussion

Parameters have been shown in Histograms Taficide exposed blood Parameters are shown in Histograms 37.1-37.7 and Table 37.1 reveals that there is a significant decrease ($P < 0.001$) in RBC count Hb content, PVC, MCH, MCHC values in both the saxes except the MCV value in male which remain insignificant ($P > 0.05$). Total count of RBC in male decreases by 1.14 millions/Cmm of blood and in female by 0.81 million/Cmm of blood (Histogram 37.1). The decrease in Hb content in male and female is by 3.42 g/100 ml of the blood and 4.4g/100 ml of the blood respectively (Histogram 37.2). The value of the PVC decreases by 7.4 per cent in male and 7.7°C in female (Histogram 37.3). A respective decrease in MCV, MHC, and MCHC values by 0.91 μm^3, 1.96 pg, 2.72 per cent in male and 5.41 μm^3, 4.64 pg, 4.47 per cent in female respectively (Histogram 37.4-37.6). Butachlor treatment (Histograms 37.1-37.7 and Table 37.2) shows that there is significant decrease ($P < 0.001$) in RBC count, Hb content and PVC values in both the sexes. A significant increase ($P > 0.001$) in MVC and MHC values. While the value of MCHC increases insignificantly ($P > 0.05$) in both the sexes.

The decrease in the erythrocyte number in both male and female sexes are 1.48 million/cmm and 1.87 million/cmm of blood respectively. Hb content by 2.84 g/100 ml and 2.14 g/100 ml of the blood in male and female respectively. The PVC value in male decreases by 6.32 per cent and in female 5.9 per cent. The MCV, MCH and MCHC 8.39 μm^3, 4.36 Pg and 0.55 per cent in male and 14.19 μm^3, 7.69 Pg and 1.85 per cent in female respectively (Table 37.2).

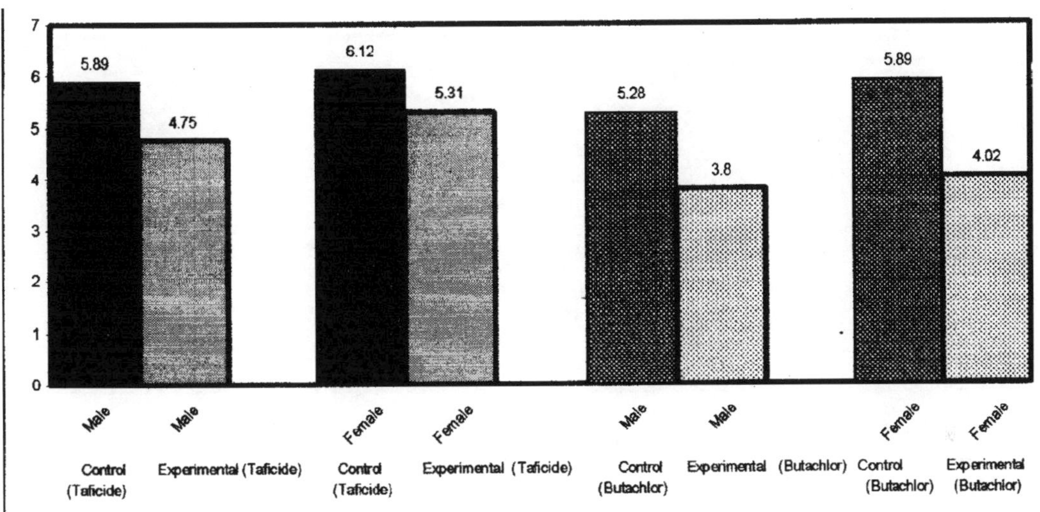

Histogram 37.1: Total Erithrocyte Count (RBC) × 10^6 mm^3.

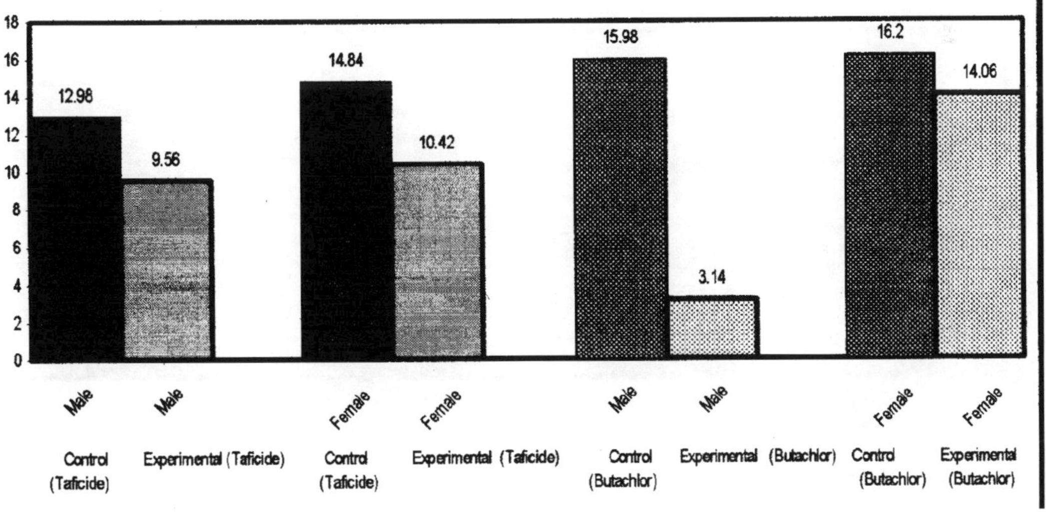

Histogram 37.2: Haemoglobin Content (Hb) (gm/100 ml.).

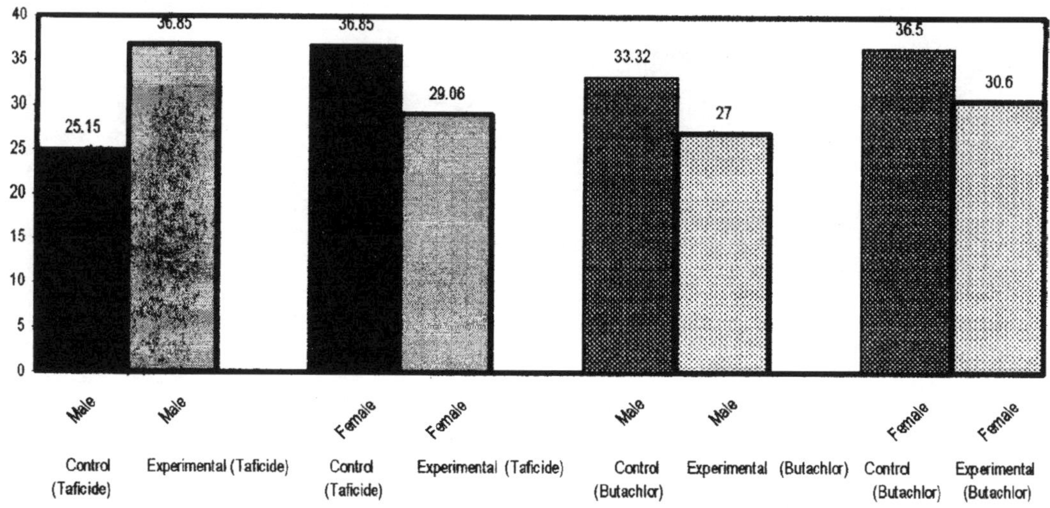

Histogram 37.3: Packed Cell Volume (PVC) (per cent).

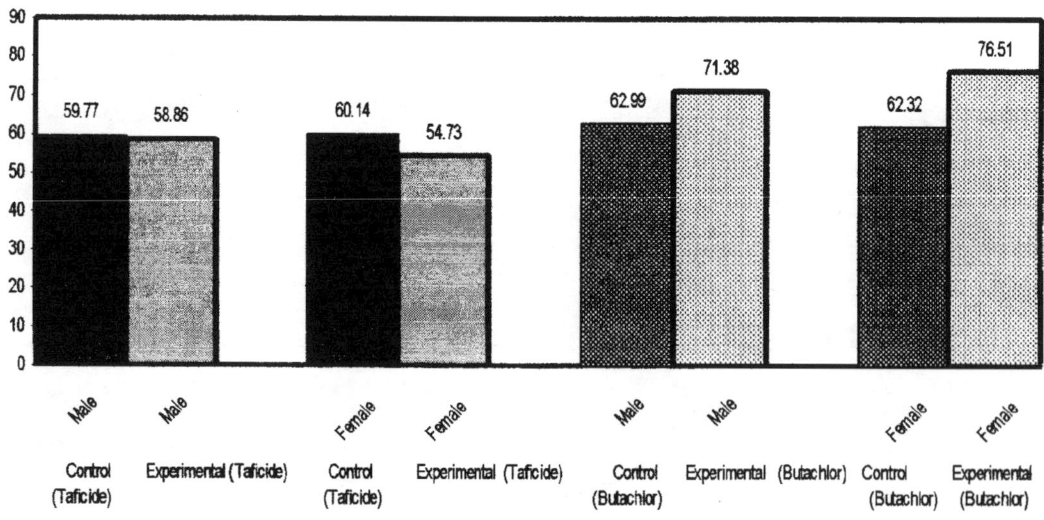

Histogram 37.4: Mean Corpuscular Volume (MCV) (μm³).

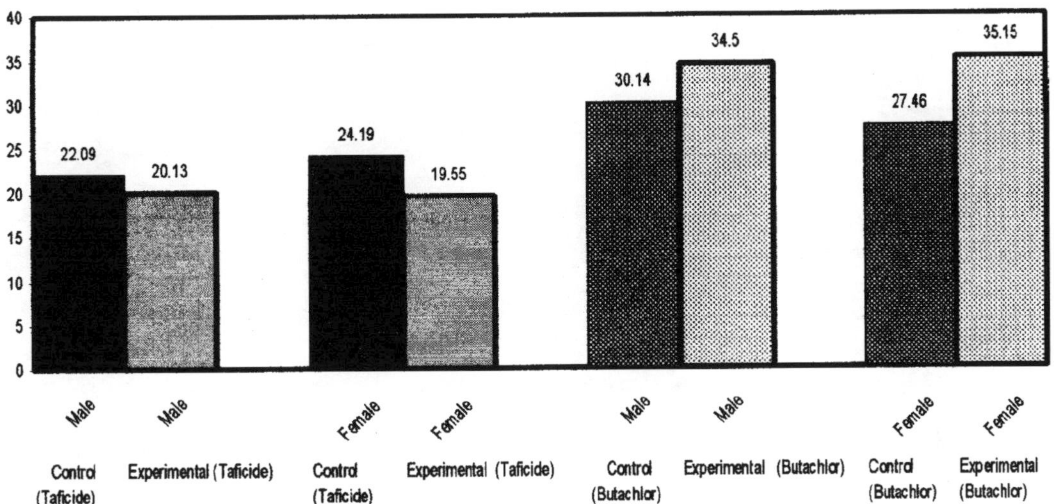

Histogram 37.5: Mean Corpuscular Haemoglobin (MCH) (pg/100 cell).

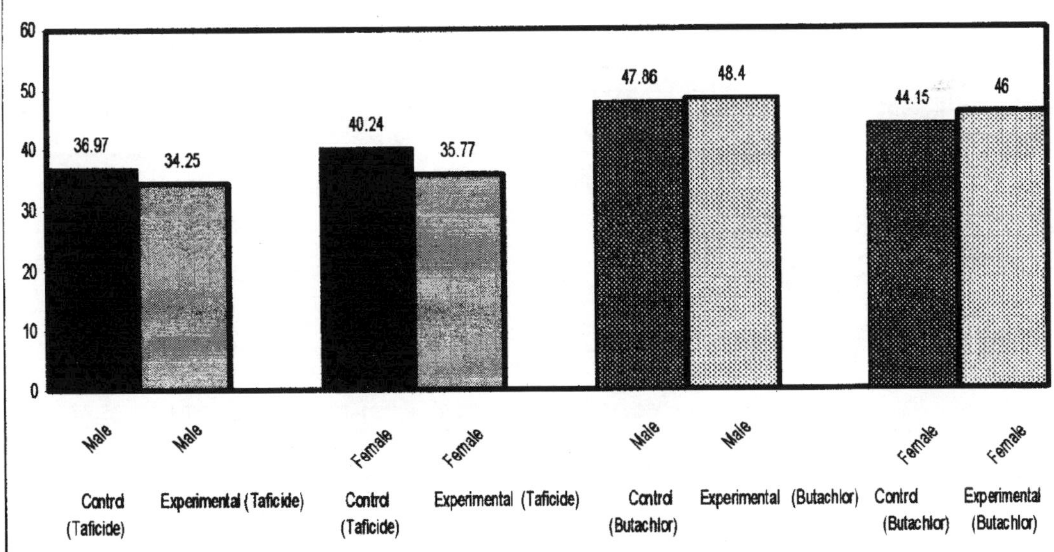

Histogram 37.6: Mean Corpuscular Haemoglobin Concentration (MCHC).

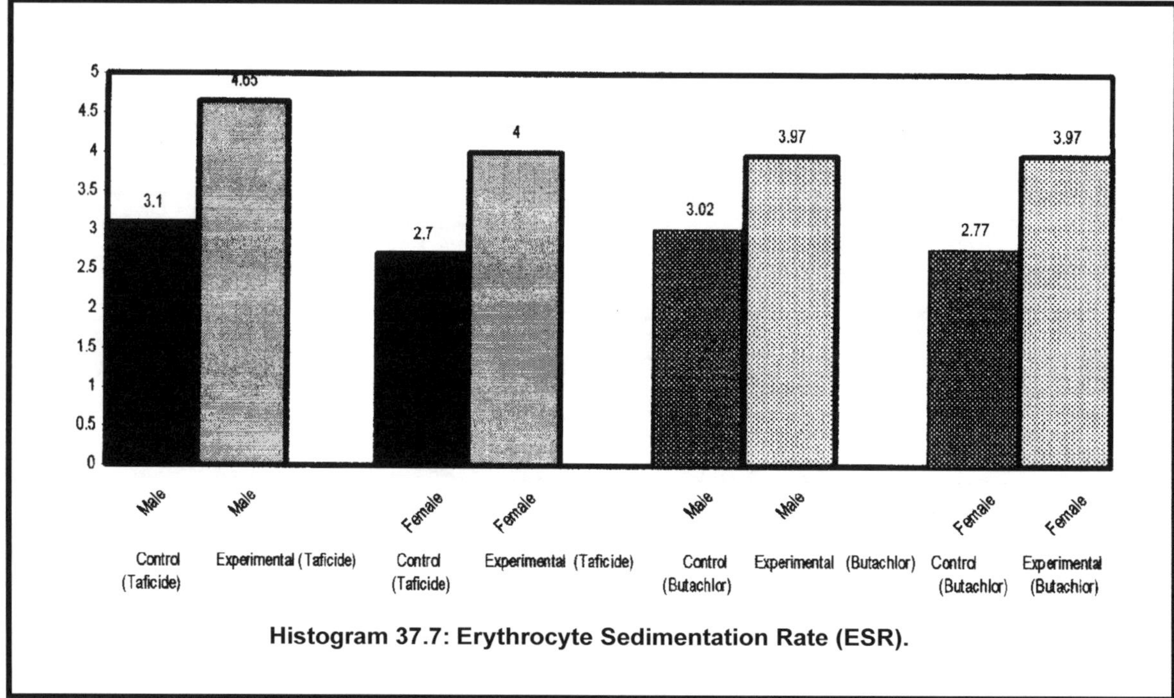

Histogram 37.7: Erythrocyte Sedimentation Rate (ESR).

Table 37.1: Showing Effects of 48-h-Lc50 Concentration of Taficide on Haematological Parameters of *Clarias batrachus*

Parameters	*Control Male*	*Experimental Male*	*Control Female*	*Experimental Female*
Total Erithrocyte Count (RBC) × 10^6 mm^3	5.89 ± 0.54	4.75 ± 0.38***	6.12 ± 0.28	5.31 ± 0.44***
Haemoglobin Content (HB) (gm/100 ml.)	12.98 ± 0.69	9.56 ± 0.84***	14.84 ± 0.65	10.42 ± 1.30***
Packed Cell Volume (PVC) (per cent)	25.15 ± 2.36	36.85 ± 2.06	36.85 ± 2.06	29.06 ± 2.07**
Mean Corpuscular Volume (MCV) (μm^3)	59.77 ± 2.48	58.86 ± 2.89	60.14 ± 1.44	54.73 ± 1.29***
Mean Corpuscular Haemoglobin (MCH) (pg/100 cell)	22.09 ± 1.00	20.13 ± 0.94***	24.19 ± 0.37	19.55 ± 1.23***
Mean Corpuscular Haemoglobin Concentration (MCHC)	36.97 ± 1.22	34.25 ± 1.61***	40.224 ± 0.81	35.77 ± 2.72***
Erythrocyte Sedimentation Rate (ESR) (mm/hr) 4.00 ± 0.57***		3.10 ± 0.42	4.65 ± 0.67***	2.70 ± 0.53

All the values are mean ± standard deviation (SD);

10 observations for each value, (Ranges in parenthesis);

** P < 0.01;

*** P < 0.001 (Student 't' test).

Table 37.2: Showing Effects of 48-h-Lc50 Concentration of Butachlor on Haematological Parameters of *Clarias batrachus*

Parameters	*Control Male*	*Experimental Male*	*Control Female*	*Experimental Female*
Total Erithrocyte Count (RBC) × 10^6 mm^3	5.28 ± 0.22	3.80 ± 0.32***	5.89 ± 0.58	4.02 ± 0.46***
Haemoglobin Content (HB) (gm/100 ml.)	15.98 ± 1.55	13.14 ± 1.32***	16.2 ± 1.84	14.06 ± 1.09**
Packed Cell Volume (PVC) (per cent)	33.32 ± 1.53	27.0 ± 1.68***	36.5 ± 2.38	30.6 ± 3.09***
Mean Corpuscular Volume (MCV) (μm^3)	62.99 ± 1.23	71.38 ± 2.51***	62.32 ± 2.73	76.51 ± 3.25***
Mean Corpuscular Haemoglobin (MCH) (pg/100 cell)	30.14 ± 1.78	34.50 ± 1.24***	27.46 ± 0.74	35.15 ± 2.12***
Mean Corpuscular Haemoglobin Concentration (MCHC)	47.86 ± 2.96	48.41 ± 2.43	44.15 ± 2.25	46.0 ± 1.95
Erythrocyte Sedimentation Rate (ESR)	3.02 ± 0.34	3.97 ± 0.5***	2.77 ± 0.36	3.97 ± 0.51***

All the values are mean ± standard deviation (SD);

10 observations for each value, (Ranges in parenthesis);

** P < 0.01;

*** P < 0.001 (Student 't' test).

The ESR value in Taficide and Butachlor exposed fish shows highly significant increase (P > 0.001) in both the sexes. The ESR value increases by 1.55 mm/hr in male and 1.30 mm/hr in female after treatment with Taficide and 0.95 mm/hr in male and 1.20 mm/hr in female after treatment with Butachlor (Table 37.1 and 37.2).

Similar decrease in RBC, Hb and PVC values have been reported by Eisler and Edmunds (1966) in marine fish treatment with methyl parathion and methoxychlor, Mukhopadhyay and Dehadrai (1980) in *C. batrachus* exposed to malathion observed Erythropenia, which we observe always in the exposed fish in the present study may possibly be due to the pesticidal interference with haemopoiesis causing anaemia (Bansal *et al.*, 1979) which is a ascribed to iron deficiency and subsequent reduction in haemoglobin synthesis Bhai *et al.* (1971) exposed to malathion observed erythropenia, Dieter (1982) in *Herolilapia multispinosa* and *Tilapia leucostic* exposed to labycid. Verma *et al.* (1982) in *Mystus Vittatus* exposed to Thiotox, Dichlorvos, Carbofuran and Mishra (1983) in *H. fossilis* exposed to Fenthion, Dabral and Chaturvedi (1984) in *H. fossilis* exposed to Folidol, Pandey *et al.* (1984) in *Clarias batrachus* exposed to Metacid, Sastry *et al.* (1988) in *Channa punctatus* exposed to Sevin. Singh *et al.* (1991) in *heteropneustes fossilis* exposed to aldrin.

A significant rise in MCV appears, however intriguing when both Total erythrocyte count and PVC decline in exposed fish. This may be possibly due to mobilisation premature erythrocytes (which are usually larger in size) as well as reticulocites from haemopoitic organs of the exposed fish.

Increase in MCH of *Clarias batrachus* under herbicidal stress clearly shows the higher rate of the decline in erythrocyte count than that of the Hb content of the blood, while decrease in MCHC reveals that loss of the Hb is comparatively at higher rate than that of the PVC. Such observations have also been made in other fishes.

Mukhopadhya and Dehadrai find no significant change in ESR in malathion exposed *Clarias batrachus* while enhanced ESR has been reported in fenthion exposed *Heteropneustes fossilis* (Srivastava and Mishra 1983) indicating damaging effect of the pesticides in some of its internal organs. The ESR is negatively correlated with Total erythrocyte count (Srivastava, 1968) *i.e.* lower the TEC higher the ESR. It is found to be true in present investigations. Increased ESR is medically considered as a strong sign of tissue injury. Similar results were also obtained in case of Taficide and Butachlor.

References

Alam, M.N. (2000). Toxicity of metacid 50 to a Paddy field fish *Channa Punctatus* (Bloch). *Indian N Environ and ecoplan.* B: 701-705.

APHA. (1985). *Standard method for the examination of water and waste water*. APHA, AWWA, WPFC. 16th ED. New York, P: 1268.

Bakthavath Salam, R., R. Ramanlingam and A. Ramanwamy (1984). Histopathology of liver, kidney, and Intestine of the fish, *Anabas testudinus* exposed to Furadon. *Environ. Ecol.* 2(4): 243-247.

Bansal, S.K., Gupta, A.K. and Dalela, R.C. (1979). Physiological dysfunction of haemopoietic system in a fresh water toleost, *Labeo rohita* following chronic chlorodane exposure Part 1 Alteration in certain haematological parameters. *Bull. Environ. Contam. Toxicol.* 22: 666-673.

Bhai Nath, N. and Nath, M.C. (1971). Haematological changes in rat injected with acetoacetate 35949, *Proc. Soc. Exp. Bioi. Med.* 138: 597-599.

Blaxhall, P.C. (1972). The haematological assessment of the health of freshwater fish. A review of selected literature. *J. Fish. Biol.* 4: 593-601.

Dabral, R. and L.D. Chaturvedi (1984). Folidol induced changes in some haematological values of *Hetropneustes fossilis* at low temperatures. *J. Adv. Zool.*, 4(2): 92-96.

Dhasarathan, P., Palaniappan, R. and Singh A.J.A.R. (2000). Effect of endosulfan and butachlor on the digestive enzyme an proximate composition of the fish *Cyprinus carpio. Indian J. Environ. and Ecoplan.*, 3(3): 339-446.

Dither, J. (1982). Histopathological and hematological studies on the cichlids (*Heterotilapia multipinsona* and *Tilapia leucostictita*) following organophosphate intoxication with laybycid. *Z. Angew. Zool.* 68 (3): 257-280.

Eisler, R. and Edmunds, P.H. (1996). Effect of methyl parathion and methyl oxychlor on blood and tissue chemistry of a marine fish. *Trans. Am. Fish. Soc.*, 95: 153-159.

Hodges, L. (1973). Environmental Pollution. Holt. Trinchart and Winston, USA.

Hunn, J.B. (1967). Bibliography on the blood chemistry of fishes. Res. U. S. Fish wild. Serv. 72: 1-32.

K. Pandey, R. Kumar, V. Kumari and A.K. Pandey (1996). Effects of some biocides on metabolic rate in Air breathing teleost. Bio-Science Research Bulletin, Vol. 12 (No. 2), pp. 75-80.

Koundinya, P.R. and Rammamurthi, R. (1997). Haematological studies in *Tilapia mossambica* exposed to lethal concentration of sumithion and sevin. *Curr. Sci.*, 48: 877- 879.

Mawdesley-Thomsan, L. E. (1971). Toxic chemicals the risk to fish. New Scientist, 49: 74-75.

M.C. Kim, J.M. and Benoit, D.A. (1971). Effect of long-term exposures to copper survival growth and reproduction of brook trout (*Salvelinus fountinalis*). *J. Fish Res. Board. Can.*, 28: pp. 655-662.

Mukhopadhyay, P.K. and P.V. Dehadrai (1980). Biochemical changes in the air-breathing catfish, *Clarias batrachus* exposed to malathion. *Environ. Pollut. Ser. AE col. Biol.,* 22: 149-158.

Pandey, A.K. Pandey, G.C. (2001). Thiram and Ziram Fungicides induced alterations on some Haematological Parameters alterations on some Haematological Parameters of fresh water catfish, *Heteropneustes fossils. Indian J. Environ & Ecoplan.* 5(3): 437- 442.

Pandey, B.N., S.S. Prasad, A.K. Chanchal and S.B. Singh. (1984). Effect of some pesticides on the oxidative metabolism and haematology of *Clarias batrachus* (Linn.) *Proc. Sem. Eff. Pest. Aq. Face.* 13-18.

R. Kumar, K. Pandey and A.K. Pandey. (1994). Lethal effects of herbicides on fresh water teleost *Anabas testudineus* (Bloch). *Trends in Life Sciences* (India). 9 (2): 79- 84.

Sandu, G.S. (1990). Research *Technique in Biological Sciences,* Anmol Publications, New Delhi, p. 209.

Sastri K.V., Siddique, A.A. and Samuel, M. (1998). Alteration in the intestinal absorption of fructose and typtophan produces by Endosulphan in freshwater teleost *Channa Punctatus. J. Environ, Biol. Supp.* 9: 225-301.

Srivastava, A.K. and J. Mishra (1983). Effect of fenthion on the blood and tissue chemistry of a teleost fish. *Hetropneustes fossils. J. Comp. Pathol.,* 93(1): 27-32.

Vardia, H.K. and V.S. Durve. (1993). Effect of weed controlling agent, 2,4-D on *Rasbora daniconius* (Ham.) *Neilgherrinsis* (Ham.) In: *Advances in Limnology,* H.R. Singh (ed) , Narendra Publishing House, Delhi, pp. 317-322.

Verma, S.R., S. Rani and Dalela (1982). Indicators of stress induced by pesticides in *Mystus vittatus*: Haematological parameters. *India J. Environ. Hlth.,* 24(1): 58-64.

Verma, G.P. and Raji, A.V. (2000). Scanning electron microscopical study of gastric and intestinal mucosa of a dichlorves exposed fresh water teleost, *oreochromis mossambica* (petess). *Proc. Nat. Acad. Sci., India* 70(B): 1.

38

Heavy Metals in Foraminifera from the Tuticorin Coastal Sediments

N. Chandrasekar, Anil Cherian, R. Gopinath, M. Rajamanickam

Chemistry and Marine Geochemistry Research Laboratories,
SPIC Research Centre, V.O.Chidambaram College, Tuticorin - 628 008, Tamil Nadu

ABSTRACT

Sediment samples were collected from off Pandian Tivu and Off Van Tivu with a view to assess the current status of metal concentration in benthic fauna. Foraminiferas were analysed for heavy metals (Cu, Zn, Fe, Mn and Ni) concentrations. Comparison has been done with these two areas levels of metal content. Metal concentrations showed that the sediment off Pandian Tivu is polluted.

Key Words: Sediment, Heavy Metal, Foraminifera, Pollution.

Introduction

Marine organisms have been used successfully as biological indicators of coastal pollution and in the assessment of the influence of waste disposal operations on the environment (Goldberg, 1975, Herut *et al.*, 1996, Hall *et al.*, 1995). Urban communities tend to have distinctive patterns in the trace element content of their soils and benthic fauna. The age of community, population, industrialization and several other factors continue to markedly alter the trace element concentration of soils and benthic fauna in urban area, especially near industrial zone. Monitoring environmental impact through the accumulation of pollution by marine organisms is also applied in a limited number of studies in the sea.

The city of Tuticorin is situated on the east coast of India. Establishment of major industries like copper smelter, petrochemical industry, thermal power station, heavy water plant etc. are posing enormous threat to the quality of environment in Tuticorin. The city is undergoing extensive urbanization and industrialization thereby leading to heavy metal contamination of soils and fauna. Thermal power station and petrochemical industry are supposed to be one of the major source of pollution in Tuticorin. Excess coal fly ash and industrial sludge are disposed into the open sea. These sites and nearby coral islands sediments have been collected to assess the environment impact of wastes on the benthic compartment in particular on foraminefera. Literature on the estimation of heavy metals content in foraminefera is quite meagre. Hence, in this paper attempts have been made for the first time for estimation of heavy metals such as zinc, copper, iron and manganese from foraminifera. This information could serve as a base line data for future investigation.

Materials and Methods

20 sediment samples have been collected and are used for the separation of foraminifera, out of which 12 samples belonged to off Van Tivu Island and 8 samples from an area of fly ash point near off Pandian Tivu (Fig. 38.1). In the laboratory, foraminifera were separated using heavy liquids. A microwave digestion technique was used to achieve total dissolution of the foraminifera. Approximately 0.5 g of foraminifera is weighed into a Teflon digestion vessel and 2 ml aqua region added. (3: 1 Con Hcl and HNO_3). Digestion vessels are then sealed and heated in a 600 w microwave oven for 7 min. When cool 3 ml con. HF (40 per cent) are added, the vessels are then sealed and heated a second time. 20 ml of 0.65 M boric acid are added to complex excess hydrofluoric acid, the final digest (25 ml) is stored in a plastic vial. Metal concentrations in foraminifera is determined by atomic absorption using flame for Cu, Zn, Fe, Ni and Mn (Nakashima *et al.*, 1988). The metal concentrations in sediments of different locations of this region have been reported by Chandrasekar *et al.*, 1998.

Results and Discussion

Foraminifera and heavy metal distributions in the test show well defined distribution patterns with marked areas of different concentration ranges that cause for these patterns can be sought in the transport mechanisms of the fined grained particles delivered into the study area from different sources. Table 38.1 summarizes the heavy metal results in foraminifera. The correlation of metals in the foraminifera is shown in the Table 38.2. There is no significant relationship between size of the organism and metal concentration. A comparison among the sites showed there is significant differences in Cu, Fe, Zn and Ni contents in foraminifera. Positive correlation on Cu, Fe, Zn and Ni is noticed. Cu Vs Zn shows negative correlation. The Cu is released from organic carrier phases in the foram and fixed in anoxic conditions (Calvert and Pedersen 1993). Ni Vs Zn shows high positive correlations. These metals behave as micronutrient being removing quantitatively in surface water by plankton growth and liberated from settling organic debris in the upper test of the foraminifera. We could expect that accumulation of Cu and Zn in the foraminifera is from the detrital sediment input. The concentration of metals are high near the point of coal fly site. In all the stations Zn showed the highest concentration followed by Fe, Mn and Cu. The results of the present study provide evidence that the Pandian Tivu sediments located near the fly ash slurry from the thermal power stations, harbour activity and sewage into the bay. The variation in the concentrations indicated the transport route of heavy metal accumulation may be from industrial sources. The higher content of Fe in the estuary has been reported in the Ennore (Rajathy and Azariah, 1996) which may be due to the anthropogenic input from the industrial sources, urban sewage, fly ash sedimentation from Ennore power station and harbour activity. This can be taken as evidence for the present area. The higher values of copper indicated the biological utilization of copper containing materials which is supported by Lee (1970) who reported that the differential profile of copper, when compared with the other heavy metals, is suggestive of biological fractionation of some selectivity in absorption or due to secondary mobilization of metals by organisms. Some generalizations can be inferred from the mean natural trace metal levels in the studies species. The order of the metal concentrations found in the Pandian Tivu foraminifera is Zn > Fe> Cu > Mn while in the Van Tivu the order is same but the level of concentration is different. We assumed that the values found in the Van Tivu are the natural level in the species.

Table 38.1: Heavy Metal Concentration in Foraminifera Near Coal Fly Ash Sediments and Van Tivu Area (Concentration of Metals 1µg g-1).

Sl. No.	*Location of Sample*	*Ni*	*Cu*	*Zn*	*Fe*	*Mn*
1.	Van Tivu	1.5	9.8	398.0	28.0	8.0
2.	Van Tivu	1.0	6.2	385.0	32.0	6.2
3.	Van Tivu	0.5	5.4	374.0	40.0	1.2
4.	Van Tivu	0.3	3.6	372.0	52.0	1.3
5.	VanTivu	1.0	5.5	387.0	37.0	6.2
6.	Van Tivu	1.5	6.1	395.0	25.0	7.9
7	Van Tivu	1.5	6.1	352.0	29.0	8.3
8.	Van Tivu	1.0	5.5	365.0	31.0	5.7
9.	Van Tivu	0.3	3.7	382.0	54.0	1.0
10.	VanTivu	0.28	3.3	371.0	57.0	0.98
11.	Van Tivu	1.0	6.30	391.0	34.0	1.3
12.	Van Tivu	0.5	10.1	394.0	38.0	1.2
13.	Coal Fly Ash Sediments	1.5	88.0	401.0	90.0	40.6
14.	Coal Fly Ash Sediments	1.0	80.0	385.0	82.0	33.0
15	Coal Fly Ash Sediments	1.2	83.0	362.0	88.0	36.0
16.	Coal Fly Ash Sediments	1.5	88.5	395.0	92.0	38.0
17.	Coal Fly Ash Sediments	0.18	60.1	333.0	52.0	10.5
18.	Coal Fly Ash Sediments	0.17	60.5	325.0	55.0	9.7
19.	Coal Fly Ash Sediments	1.5	87.7	388.0	94.0	38.0
20.	Coal Fly Ash Sediments	1.2	81.0	382.0	89.0	34.0

Table 38.2: Correlation Matrix of Heavy Metals in Foraminifera.

Variables	*Ni*	*Cu*	*Zn*	*Fe*	*Mn*
Ni	1.00	0.31	0.54	0.20	0.57
Cu		1.00	- 0.05	0.90	0.93
Zn			1.00	0.09	0.20
Fe				1.00	0.89
Mn					1.00

Acknowledgement

This work was undertaken as part of the research project funded by NATP-ICAR. Also the financial support by the NATP-ICAR for PSR under grant AED/FLP/P-12 is gratefully acknowledged. We thank Director, CMFRI, Cochin for the logistical support provided to carry out this work.

References

Calvert, S.E and Pedersen, T.F. (1993). Geochemistry of Recent oxic and anoxic marine sediments: Implications for the geologic record. *Marine Geology,* 113: 67- 88.

Chandrasekar, N., Sivakumar, V., Udayanapillai, A.V., Rajamanickam, G.V. (1998). Heavy metal concentrations in salt marsh of Tuticorin, Tamil Nadu. *J. Ind. Assc. of sedimentologists,* 17(2): 245 - 250.

Goldberg, E.D. (1975). The Mussel watch–a first step in global marine monitoring. *Mar. Poll. Bull.* 6: 111.

Hall, L.W., Ziegenfuss, M.C., Anderson, R.D. and Lewis, B.L. (1995). The effect of salinity on the acute toxicity of total and free cadmium to a Chesapeake Bay copepod and fish. *Mar. Poll. Bull.* 30: 376-384.

Herut, B., Homung, H., Kress, N., and Cohew, Y. (1996). Environmental relaxation in response to reduced contamination input: the case of mercury pollution in Haifa Bay, Israel. *Mar. Poll. Bull.* 32: 366-373.

Lee, G.F. (1970). Factors affecting the transfer of materials between water and sediment. Lit. Rev. No.11, Eutrophication information program water Res. Centre, University of Wisconsin, Moddison.

Nakashima, S., Sturgeon, R.E, Whillie, S.N and Berman, S.S. (1988). Acid digestion of marine samples for trace element analysis using microwave heating. *Analyst,* 113: 158-163.

Rajathy, S. and Azariah, J. (1996). Spatial and seasonal variation in heavy metals iron, zinc, manganese and copper in the industrial region of the Ennore Estuary, Madras. *J. Mar. boil. Ass. India,* 38 (1&2): 68-73.

39

Heavy Metals in Cultivated Soils and River Sediments and their Correlation Study

B.H. Patil and V.S. Shrivastava

Centre for P.G. Research in Chemistry, G.T.P. College, Nandurbar - 425 412

ABSTRACT

The soil samples were collected from surface of the soil from different agriculture fields in the Khandesh region where cotton, sugarcane, chilli, banana, jawar crops were cultivated. Tapti river aquatic sediments samples were also collected from five different stations which were 10-15 km away from each other for the similar study. The concentration of heavy metals like Cu, Zn, Cd, Pb, As, Hg, Cr and Fe was estimated by ICP-AES. Besides the metal anlalyses the physico-chemical characteristics of samples have also been evaluated.

Key Words: ICP-AES analysis, Soil, Sediment, Heavy Metals, Tapti river, Khandesh region.

Introduction

The field of heavy metal detection by ICP AES in the industrial waste (IWW) and soil-sediment is still very much in the stage of development (Shrivastava and Choudhary, 2000). Industrial effluents agricultural runoff, transport, burning of fossil fuels, animal and human excretions and geologic weathering and domestic wastes could contribute to the heavy metals in water bodies. The transport of trace elements between sediment and water may be a result of absorption, desertion, precipitation, diffusion, chemical reaction, biological activity and a combination of these phenomenon. The information about the distribution of trace metals in sediments can be of value in assessing the potential impact of sediment resuspension upon water quality. The information about selective distributions of trace metals may also help in evaluating the relative availability of these trace metals in biological communities and their impact on reactions and transformations, (Shreenivasa Rao & Ramamohan Rao, 2001).

Some of industries are discharging their effluents to the open places, city drains, ponds and finally to the Tapti river without taking proper treatment have already been reviewed in several researches (Shrivastava, 1996, Nemade and Shrivastava, 1997 and Michael *et al.*, 1999). This waste water (industrial effluents) causes soil and ground water pollution besides causing a number of adverse effect on agriculture produce animal and health of people living in the neighbouring area since it contains waste chemicals and toxic heavy metal. Hence there is need for the determination of heavy metal in such industrial waste, soil, sediment with the help of advance technique, (Shrivastava and Chaudhari, 2000).

Materials and Methods

The soil and sediment samples were collected from Tapti river basin in Khandesh region, Maharashtra. The 4N HNO_3-HCl extracts were analyzed for the metals with the help of ICP-AES. Soil physico-chemical parameters were determined by following the standard methods. (APHA, 1985, Trivedy and Goel, 1984).

Results and Discussion

The results obtained by ICP-AES analysis of soil and sediment samples are given in Table 39.1 and 39.2. The concentration of heavy metals As and Hg was not to be detected in almost samples. Also (Cr) chromium was not detected in all the sites except in site five whereas in sediment samples the concentration of As ranges from 0.02 to 0.06 µg/gm except in site-2 it was not detected.

Table 39.1: Heavy Metal Concentration in Soil Sample

Sl. No	Site of Soil Sample Collection	Cu	Zn	Cd	Pb	As	Hg	Cr	Fe
1.	Field No. 1 Cultivated by Chilli Crop	1.23	0.64	0.24	0.91	nd	nd	nd	1.50
2.	Field No. 2 Cultivated by Cotton Crop	1.72	0.85	0.21	0.58	nd	nd	nd	185
3.	Field No. 3 Cultivated by Jawar Crop	0.93	0.92	0.28	0.41	nd	nd	nd	262
4.	Field No. 4 Cultivated by Sugarcane Crop	0.90	0.98	0.40	0.58	nd	nd	nd	159
5.	Field No. 5 Cultivated by Banana Crop	1.0	1.32	0.32	1.0	nd	nd	nd	192.4

*All the concentrations are in µg/gm.

Table 39.2: Heavy Metal Concentration in Tapti River Sediments

Sl. No.	Site of Sediment Sample Collection	Cu	Zn	Cd	Pb	As	Hg	Cr	Fe
1.	Station No. 1 at Taloda, Maharashtra	1.26	13.0	0.95	1.6	0.02	1.54	4.0	194
2.	Station No. 2 at Budhaval, Maharashtra	2.10	10.1	0.92	1.64	nd	2.62	2.9	185
3.	Station No. 3 at Kukurmunda, Maharashtra	1.98	12.1	0.98	1.72	0.04	2.14	2.86	207
4.	Station No. 4 at Hatoda, Maharashtra	1.0	20.2	1.0	1.16	0.04	2.86	3.1	218
5.	Station No. 5 at Modalpada, Maharashtra	1.63	40.0	0.58	0.97	0.06	3.0	5.58	249

* All the concentrations are in µg/gm.

The concentration of Hg in sediment samples ranges from 1.54 to 3.0 µg/gm. Concentration of chromium ranges from 2.86 to 5.58 µg/gm. The concentration of Cu was detected in soil and sediment samples. In soil the concentration of Cu ranges from 0.90 to 1.72 µg/gm while in sediment samples it ranges from 1.0 to 2.10 µg/gm. The concentration of Zn in soil samples was found to be in the range of 0.64 to 1.32 µg/gm or ppm while in sediment it ranges from 10.1 to 40.0 µg/gm. The concentration of Cd cadmium is found to be detected in almost all sites of soil and sediments. In soil it ranges from 0.21 to 0.40 µg/gm while in sediment samples it ranges from 0.58 to 1.0 µg/gm or PPM.

Table 39.3: Physico-chemical Parameters of Soil Samples Collected from Cultivated Field Crops

Sl. No.	*Site of Soil Sample Collection from Different Fields*	*pH*	*Ec*	*Cl^-*	*HCO_3^-*	*TA (Total Alkalinity)*	*WES*	*COD mg/kg*
1.	Field No. 1 Cultivated by Chilli Crop	7.8	2500	55	60	1800	1200	2250
2.	Field No. 2 Cultivated by Cotton Crop	8.0	280	77	86.2	2800	7200	215
3.	Field No. 3 Cultivated by Jawar Crop	7.5	250	43.14	75.2	2600	3500	2760
4.	Field No. 4 Cultivated by Sugarcane Crop	8.0	180	55.5	75.2	3000	4500	234
5.	Field No. 5 Cultivated by Banana Crop	8.6	430	67.44	70	2450	2400	230

Concentrations of all the parameters in µg/gm except pH and Ec (umho/cm).

Table 39.4: Physico-chemical Parameters of Sediments of Tapti River

Sl. No.	*Site of Soil Sample Collection*	*pH*	*Ec*	*Cl^-*	*HCO_3^-*	*TA (Total Alkalinity)*	*WES*	*COD mg/kg*
1.	Station No. 1 at Taloda	8.9	3600	326	1200	2300	2100	38
2.	Station No. 2 at Budhaval	8.6	5900	500	9261	1450	2200	52100
3.	Station No. 3 at Kukurmunda	8.1	7200	200	1110	1730	2450	53
4.	Station No. 4 at Hatoda	7.9	7000	140	1011	2350	2650	20
5.	Station No. 5 at Modalpada	7.9	5100	375	910	20	2650	52.4

Concentration of all the parameters in µg/gm Except pH and Ec (umho/cm).

Table 39.5: Correlation Coefficient Among the Metals Present in Soil

COD	*Cu*	*Zn*	*Cd*	*Pb*	*Fe*	
COD	1					
Cu	0.25	1				
Zn	0.59	0.38	1			
Cd	0.025	0.78	0.11	1		
Pb	0.14	0.0079	0.79	0.096	1	
Fe	0.0000012	0.0011	0.25	0.033	0.32	1

The concentration of Pb in soil samples ranges from 0.40 to 1.0 µg/gm while in sediment samples it ranges from 0.97 to 1.72 µg/gm. The concentration of Fe in soil samples ranges from 150 to 262 µg/gm while in sediment samples it ranges from 185.0 to 249.0 µg/gm.

Table 39.6: Correlation Coefficient Among Metals Present in Sediments

	COD	*Cu*	*Zn*	*Cd*	*Pb*	*As*	*Hg*	*Cr*	*Fe*
COD	1								
Cu	0.90	1							
Zn	0.07	0.29	1						
Cd	0.45	0.20	0.35	1					
Pb	0.11	0.49	0.91	0.023	1				
As	0.055	0.30	0.80	0.016	0.032	1			
Hg	0.036	0.041	0.61	0.017	0.72	0.035	1		
Cr	0.66	0.18	0.87	0.032	0.69	0.59	0.23	1	
Fe	0.0033	0.25	0.95	0.10	0.87	0.93	0.60	0.75	1

The correlation coefficient (r) among all the metals is also calculated. The COD-Cu correlation in sediments have large positive correlation (0.90) while COD-Fe correlation have less value among all. In metals the large positive correlation values are found between Zn-Fe (0.95) and Fe-As (0.93).

In case of soil samples the Pb and Zn (Cd-Cu are highly correlated have values of r (0.79) and (0.78) respectively. The very small correlation is found between Fe and COD *i.e.* (0.0000012).

Soil and sediment samples collected from river Tapti contain considerable amount of Cu, Zn, Cd, Pb, As, Hg, Cr and Fe. However metals As, Hg and Cr were not detected in soil samples.

The results show that the sediments of river Tapti were contaminated by the heavy metals. This is because of industrial effluents, domestic waste agricultural runoff, and aquaculture welfare etc. All these activities are gradually increasing the concentrations or heavy metals in sediments and also water to alarming levels, resulting in degrade the quality of water. There is a possibility of bioaccumulation of metals in fish. In such case fish production would decrease and water should not used for drinking purpose and fish is not good for consumption. Hence industrial and aquaculture effluents and domestic waste should be treated thoroughly before leftwing them out into river to avoid the further contamination of heavy metals.

Acknowledgment

The authors are thankful to the Principal, G.T.P. College, Nandurbar for providing laboratory facilities.

References

Shreenivasa, Rao and P. Ramammohan Rao (2001). Heavy metals concentration in the sediments from 'Kolleru Lake", India. *Indian Jr. Environmental Hlth.*, 43(4): 148-153.

Michael, E.R., Panos, H., Kevin, T. and Eric, D.S. (1999). Rapid determination of Cu, Fe, Mg, Mn and Zn in wood pulp by direct sample insertion inducitively coupled plasma atomic emission. Spectrometry using a pyrolytically coated graphite sample probe. *J. Anal Atomic Spectr.*, 14: 1715-1722.

Nemade, P.N. and Shrivastava, V.S. (1997). Detection of metals in pulp and paper mill effluents by ICP-AES and flame photometry and their impact on surrounding environment. *Jr. Industrial Pollution Control*, 13: 143-149.

Shrivastava, V.S. and Choudhari, Ganesh R. (2000). Hazardous heavy metal in and around MIDC Jalgaon by Inductively Coupled Plasma Atomic Emission Spectrophotometer. *Indian Jr. Environment and Ecoplanning,* 3(3): 707-709.

Shrivastava, V.S. (1996). Heavy metals in industrial waste water and sludge by ICP-AES. *Indian Jr. Environmental Prot.,* 16: 328-329.

Trivedy, R.K. and Goel, P.K. (1984). *Chemical and biological methods for water pollution studies*, Environmental Publications, Karad, India.

40

Clinical and Parasitological Studies on Human Fish Borne Trematode Helminthoses with Emphasis on Intestinal Histoenzymatic Changes in Egypt: Comparative Records from India

S.I. Shalaby, Iman B. Mohamed* and Neelima Gupta**

Department of Parasitology and Chairman of Medical Clinic, National Research Centre, Dokki, Cairo, Egypt
**Department of Pathology, Faculty of Veterinary Medicine, Cairo University, Giza, Egypt*
***Department of Animal Science, M.J.P. Rohilkhand University, Bareilly, U.P.*

ABSTRACT

The clinical picture of human fish borne trematode helminthoses are reported including the past history, clinical examination and laboratory investigations. Patients mostly had gastrointestinal symptoms such as indigestion, diarrhoea and abdominal pain or heart problems in the form of arrythmia and ventricular premature beats as shown by ECG. Sometimes obstructive jaundice, complicated by sepsis and dessiminating intravascular coagulopathy-like syndrome was evident. Transhepatic cholangiogram picture and computed tomography scan of the abdomen are described. The histopathology of liver biopsies in these conditions are reported. Enzymatic reactions of the small intestinal mucosa was determined by evaluation of alkaline phosphatase and acid phosphatase activities. Decreased activities of both the enzymes in rats were observed which denotes a decreased absorptive capacity of intestine, while in dogs, mild reduction of the former enzyme and strong reaction of the latter as a function to digest the foreign particles and antagonise any bacterial action was observed.

The Heterophyids reported from India are also incorporated.

Key Words: Heterophyid, Fish Dysentery, Leucocytosis, Eosinophilia, Phosphatases.

Introduction

Fresh water fish is considered to be as one of the important source of parasitic infestations to man and fish-eating mammals particularly after the increased population of rivers and lake. Egypt has become an endemic area of heterophyidae other than *Heterophyes heterophyes* (Mohamed, 1996). Other helminths have been recorded by Shalaby (1982) including other investigators from different parts of Egypt.

Clinically, Byong *et al.* (1980) reported that *Echinostoma cintorchis* in man caused intermittent abdominal pain and nausea. In 1981, Byong *et al.*, found no significant clinical picture or blood chemical changes while eosinophil count was 7 per cent. Line (1981) described abdominal pain, gurgling and loose stools with an eosinophilia of 16 per cent due to the same parasite. Byong *et al.* (1984a) identified 2 cases of human infection by *Stellantchasmus falcatus* (Heterophyidae) in Korea. The first case had gastrointestinal troubles like diarrhoea, abdominal pain and arrhythmia, while the second case complained of indigestion. Stereomicroscopy of diarrheal stools revealed *S. falcatus* in both cases in addition to other 3 heterophyid flukes in the first case. Both patients reported eating mullets which was considered as a source of heterophyid fluke infestation. Byong *et al.* (1984b) found 2 cases of natural human infestation by *Heterophyopsis continua* (Heterophyidae). They reported of clinical complaints of non-specific gastrointestinal symptoms such as epigastric pain, easy fatigability, weakness, indigestion and cardiac arrhythmia which was due to eating raw flesh of several kinds of brackish water fishes. Jong *et al.* (1984) found 2 human cases of *Heterophyes heterophyes* nocens infestation. Case 1 had gastrointestinal symptoms such as epigastric pain and indigestion, while in case 2 there were heart problems like arrythmia and ventricular premature beats in ECG as well as diarrhoea and abdominal pain. Case I was concomitantly infested with *Clonorchis sinensis* and case 2 with three other kinds of heterophyid flukes. The cases were said to have eaten raw brackish water fish such as mullets (*Mugil cephalus*). Byong *et al.* (1985a) found 4 cases of human infestation by *Echinochasmus japonicus*. Patients had gastrointestinal troubles such as indigestion and abdominal discomfort. They had been eating raw flesh of fresh water fishes such as cyprinoids. Byong *et al.* (1985b) reported that the major subjective symptoms in man of *Echinostoma hortense* were abdominal pain and diarrhoea. Jong *et al.* (1985) found 8 cases of human infection by *Heterophyes heterophyes* nocens having general malaise, gastrointestinal troubles and/or heart palpitation. Jong *et al.* (1986) reported vague abdominal discomfort accompanied by diarrhoea and after treatment with a purgative, *H. heterophyes* and *H. dispar* were collected. Jong *et al.* (1988) reported easy fatiguability, general weakness, indigestion and occasional diarrhoea and palpitation in a patient infested with *Stictodora* sp. This patient had a history of eating raw flesh of mullet. Ana *et al.* (1991) found in a patient infected with *Clonorchis sinensis*, shortening of breath. Clinicopathological studies demonstrated elevated bilirubin, alkaline phosphatase and transaminases and 14 per cent eosinophils. The patient developed obstructive jaundice complicated by sepsis and dessiminating intravascular coagulopathy like syndrome with prolonged prothrombin time, partial thromboplastin time and elevated fibrinogen degradation products. Transhepatic cholangiogram demonstrated diffuse irregular dilatation of the biliary tree with abrupt ending of terminal branches. Several small, fusiform filling defects were noticed within the small ducts. Numerous operculated eggs of *Opisthorchis* were found during aspiration of bile. Fernando and Josepth (1991) reported 2 cases of *Clonorchis* associated cholangiocarcinoma. The first case was incidentally noted to have *C. sinensis* in a routine stool examination. Laboratory values demonstrated leucocytosis, polymorphonuclear leucocytosis, monocytosis, eosinophilia, lymphopenia and increased hepatic enzymes. On surgery, the left lobe of liver was fibrotic and atrophic with multiple abscesses. The second case demonstrated jaundice and on computed tomography scan of the abdomen, dilatation of intra and extrahepatic bile ducts with extensive calcification of pancreas and fatty infiltration of the liver. Surgical exploration showed dilatation of the entire biliary tree, enlargement of pancreas, being firm and nodular, the pancreatic tree, enlargement of pancreas, being firm and nodular, the pancreatic duct was markedly dilated containing multiple calculi and the distal common duct was surrounded with dense fibrosis. Rodicheva and Mitasov (1991) reported hepatic cysts caused by *Opisthorchis* infestation. Cysts were predominantly found on liver surface and more frequently in the left than right lobe. They were irregular in shape and their contents varied from transparent and colourless to yellow and turbid with parasites and eggs. Tun and Belobordova (1991) mentioned that on histopathology of 40 liver biopsy of patients with chronic opisthorchiasis, chronic persistant hepatitis was found in 67.5 per cent cases, lobar hepatitis in 22.5 per cent, chronic active hepatitis in 7.5 per cent and cirrhosis in 2.5 per cent. Chieffi *et al.* (1992) found 9 cases infested with *Phagicola* sp. in Brazil as a result of feeding on raw mullet. Out of these patients, 44 per cent demonstrated diarrhoea and 66 per cent showed slight eosinophilia. Jong *et al.* (1993) made an epidemiological study on *Metagonimus* infestation along the upper reaches of Namhan river, with special consideration on the species of worms collected from humans. Eggs were found in 9.7 per cent of the examined people.

In this investigation, the available data on human clinical picture of fish borne trematode helminthoses from our experience is being reported. Moreover, intestinal histoenzymatic changes have been investigated and are being communicated through this piece of work.

Materials and Methods

Sporadic cases suffering from chronic diarrhoea were investigated parasitologically and clinically. Moreover, 5 dogs and 10 rats were experimentally infected with heterophyid encysted metacercariae and 3 others dogs and 5 rats were left as control. After appearance of eggs in faeces of experimental animals, unfixed cryostat (frozen) sections, 8-10 microns in thickness were taken from the intestine for investigation of some enzyme histochemical changes. These included alkaline phosphatase as demonstrated by Gomori's calcium cobalt method (Gomori, 1950a) and acid phosphatase by lead nitrate method. (Gomori, 1950b).

Results

A male patient aged 35 years from Zagazig province was found to be suffering from chronic diarrhoea and loss of weight. Laboratory data demonstrated heterophyid eggs in stools, leucocytosis and eosinophilia. The patient complained of right hypochondrial pain but after treatment with praziquantel 2 tablets twice daily for 3 days, symptoms disappeared and stools proved negative for the parasite. Moreover, 7 cases with chronic dysentery and irritability from Dakahlia province were investigated. Six of them were found to harbour heterophyid eggs in their stools with eosinophilia. The 7th case had eosinophilia and leucocytosis. It is suspected that this case may also be infested with heterophid flukes, as this patient had a history of eating improperly cooked mullets (*Mugil* spp.)

Enzymatic reaction of the small intestinal mucosa was determined by evaluation of alkaline phosphatase and acid phosphatase activities. In experimental rats, alkaline phosphatase activity showed slight depression in the intestinal glands. Also, the villi revealed a marked decrease in the enzyme activity (Fig. 40.1). Whereas acid phosphatase activity in the glands presented a marked diminution and onlya slight depression could be seen in the villi (Fig. 40.2). In experimental dogs, a moderate reaction of alkaline phosphatase was observed in both intestinal villi and glands (Fig. 40.3) while there was intense positive reaction of acid phosphatase along the brush border of the epithelial cells of villi and crypts (Fig. 40.4).

Fig. 40.1: Frozen Section Showing Decreased Acid Phosphatase Activity in the Intestinal Mucosa of Infested Rat. Gomori × 100.

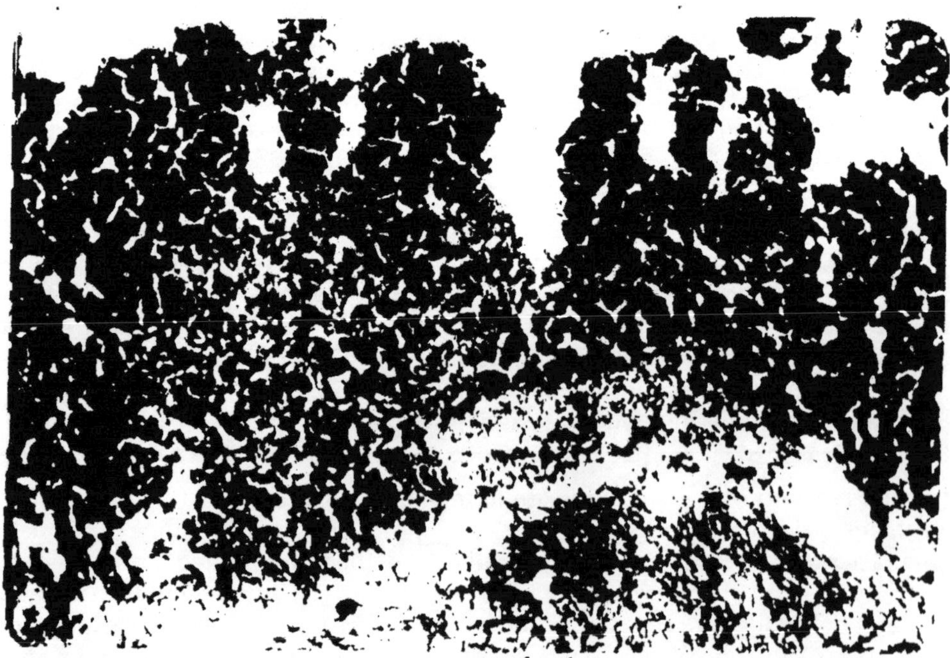

Fig. 40.2: Frozen Section of Intestinal Mucosa of Infested Dog showing Moderate Alkaline Phosphatase Activity. Gomori × 100.

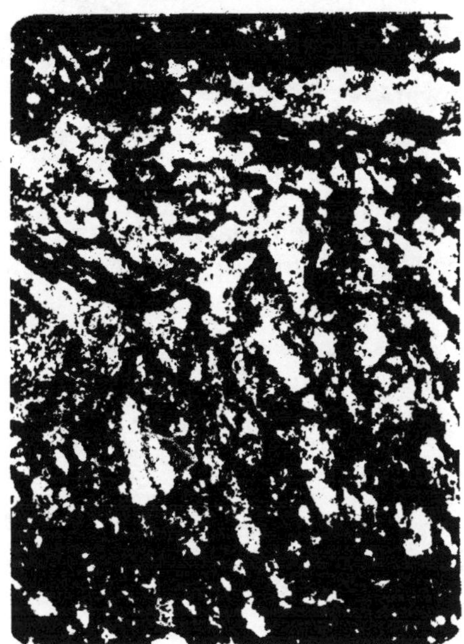

Fig. 40.3: Frozen Section Showing Decreased Alkaline Phosphatase Activity in the Intestinal Mucosa of Infested Rat. Gomori × 100.

Fig. 40.4: Frozen Section of Intestinal Mucosa of Infested Dog Showing Intense Reaction of Acid Phosphatase Enzyme. Gomori × 100.

Discussion

In the present investigation, the clinical picture of fish born trematode helminthoses (heterophiasis) has been represented by diarrhoea and even dysentry. It is proposed to call this form of dysentry that occurs as a result of consumption of improperly cooked fish and which is not associated with fever as fish dysentry. Also, some patients demonstrated irritability which might be related to parasitic toxins circulating all over the body. In all patients, leucocytosis and eosinophilia were detected. The obtained data were in agreement with Line (1981) for infection with *Echinostoma cintorchis,* Byong *et al.* (1984a) for *Stellantchasmus falcatus,* Byong *et al.* (1984b) for *H. continua,* Jong *et al.* (1984) for *H. heterophyes* nocens, Byong *et al.* (1985a) for *E. japonicus*, Byong *et al.* (1985b) for *E. hortense*, Jong *et al.* (1986) for *H. dispar*, Jong *et al.* (1988) for *Stictodora* sp. and Jong *et al.* (1993) for *Metagonimus* sp.

Concerning the histoenzymatic changes in experimental rats and dogs, alkaline phosphatase activity was normally localized in columnar epithelial cells of villi, while moderate reaction was observed in the glands. On the other hand, acid phosphatase enzyme is a hydrolysing enzyme which plays an important role in the process of cell catabolism and in defense against invading organisms as it is associated with lysosomes. Acid phosphatase is normally localized in the villi and mild reaction was observed in apical portion of crypts. In our study, activities of both the enzymes in the small intestine of rats decreased, which denotes a decreased absorptive capacity of the intestine, while in dogs, there was mild reduction of alkaline phosphatase but strong reaction of acid phosphatase. The increased reaction of acid phosphatase could be explained by the increased lysosomal activity as a function to digest the foreign particles and antagonise any bacterial action. These enzymatic changes coincided with that mentioned by Galila *et al.* (1985) and Anwar *et al.* (1990).

Occurrence of Heterophyids from India

Heterophyes heterophyes is an intestinal parasite of dogs, cats, fox, humans and other mammals and is commonly found in Asia and Egypt and has been introduced into Hawaii. It has also been recorded in Palestine and in the Far East. The worm has been held responsible for causing heterophyiasis in humans due to eating of raw or uncooked fish.

The genus *Heterophyes* is not a common parasite in India. It has been recorded in some parts of the country inhabiting mammals. Gill (1972) recorded *H. heterophyes* in 2 out of 88 cats sampled at AIIMS, New Delhi although 84 were infected with helminth parasites. Rajavelu and Raja (1988) surveyed helminth parasites in cats of Madras recording 96 per cent infection and reporting the occurrence of *H. dispar* as the first record from India.

In humans, infection with Heterophyes is rare. The first record of human heterophyid infection was made by Mahanta *et al.* (1995) from Dibrugarh, Assam (India). Only two cases were reported to have passed heterophyid eggs in human stools measuring 31.5 ± 2.0 µm SD × 22.4 ± 1.9 µm SD with well developed miracidium. Due to the paucity of availability of these parasites in India, the pathogenecity of this worm is practically untouched in this part of the continent.

References

Ana, H., Sydney, F. and Benton, R. (1991). A case of Opisthorchiasis diagnosed by cholangiography and bile examination. *Amer. Surgeon,* 57(4): 206-209.

Anwar, A., Nadia, I. and Sohei, M. (1990). Histopathological and histochemical studies on experimentally infected hamsters with *Stictodora tridactyla. J. Egypt. Soc. Parasitol.* 20(2): 653-659.

Byong, S., Sung, Y. and Jong, Y. (1980). Studies on intestinal trematodes in Korea. A human case of *Echinostoma cinetorchis* infection with an epidemiological investigation. *Seoul J. Medicine,* 21(1): 21-29.

Byong, S., Sung, T. and Jong, Y. (1981). Studies on intestinal trematodes in Korea-III-Natural human infections of *Pygidiopsis summa* and *Heterophyes nocens. Seoul J. Medicine,* 22 (2): 228-235.

Byong, S., Soon, H., Jong, Y. and Sung, J. (1984a). Studies on intestinal trematodes in Korea-XII. Two cases of human infection by *Stellantchasmus falcatus. Korean J. Parasitol.,* 22(1): 43-50.

Byong, S., Soon, H., Jong, Y. and Sung, J. (1984b). Studies on intestinal trematodes in Korea III. Two cases of natural human infection by *Heterophyopsis continua* and the status of metacercarial infection in brackish water fish. *Korean J. Parasitol.,* 22(1): 51-60.

Byong, S., Soon, H., Jong, Y. and Sung, J. (1985a). Studies on intestinal trematodes in Korea-XX. Four cases of natural human infection by *Echinochasmus japonicus. Korean J. Parasitol.,* 23(2): 214-220.

Byong, S., Kwang, S., Jong, Y., Sung, J. and Soon, H. (1985b). Studies on intestinal trematode in Korea-XVII. Development and egg laying capacity of *Echinostoma hortense* in albino rats and human experimental infection. *Korean J. Parasitol.,* 23(1): 24-32.

Chieffi, P., Maria, C., Domingas, M., and Rosa, M. (1992). Human infection by *Phagicola* sp. in the municipality of Registro, Sao Paulo State, Brazil. *J. Trop. Med. and Hyg.*, 95: 346-348.

Fernando, V. and Joseph, N. (1991). *Clonorchis* associated cholangiocarcinoma. A report of 2 cases with unusual manifestation. *Gastroenterology*, 101: 831-839.

Galila, A., Samir, M., Hoda, A., Aleya, A. and Soheir, M. (1985). Histochemical studies in experimental *heterophyiasis. J. Egypt Soc. Parasitol.*, 15(1): 109-117.

Gill, H.S. (1972). Incidence of gastro-intestinal helminths in cat (*Felis cotus*) in Delhi. *J. Comm. Dis.*, 4: 2-3, 109-111.

Gomari, G. (1950a). The calcium method for alkaline phosphatases cited by Pearse, AGE (1968): Histochemistry, theoretical and applied, Vol. 1, Churchill, Livingstone, Edinburgh, London and New York.

Gomari, G. (1950b). The lead nitrate method for acid phosphatase. *J. Lab. Clin. Med.*, 35: 802.

Jong Y., Byong, S. and Soon, H. (1984). Studies on intestinal trematodes in Korea. I. Two cases of human infection by *Heterophyes heterophyes nocens. Korean J. Parasitol.*, 22(1): 37-42.

Jong, Y., Byong, S., Soon, H., Sung, J. and Woon, M. (1986). Human infections by *Heterophyes heterophyes* and *H. dispor* imported from Saudi Arabia. *Korean J. Parasitol.*, 24(1): 82-88.

Jong, Y., Sung, H., Hae, R., Jina, K., Kyung, C. and Eun, C. (1993). An epidemiological study of *Metagonimiasis* along the upper reaches of Namhan River. *Korean J. Parasitol.*, 31(2): 99-108.

Jong, Y., Sung, H., Soon, H. and Byong, S. (1988). *Stictodoro* sp. recovered from a man in Korea. *Korean J. Parasitol.*, 26(2): 127-132.

Jong, Y., Sung, J., Woon, M.H. and Byong, S. (1985). Further cases of human *Heterophyes heterophyes nocens* infection in Korea. *Seoul J. Med.*, 26(2): 197-200.

Line, O. (1981). *Echinostoma perfoliatus* in man. A case report. *Natural Medical J. of China*, 61: 249-252.

Mahanta, J., Narain, K. and Srivastava, V.K. (1995). Heterophyid eggs in human stool samples in Assam: First report for India. *J. Comm. Dis.*, 27(3): 142-145.

Mohamed, Iman B. (1996). Pathological changes in some fish eating mammals as a result of consumption of fish infested with encysted metacercariae. Ph.D. (Pathology), Fac. Vet. Med. (Cairo Univ.)

Rajavelu, G. and Raja, E.E. (1988). On helminth parasites in domestic cat in Madras. *Cheiron.*, 17: 1, 11-14.

Rodicheva, E. and Mitasov, C. (1991). Liver-fluke infection as aetiological factor in bile duct carcinoma of man. *Trans. of Royal Soc. Trop. Med. and Hyg.*, 27(4): 787-794.

Shalaby, S.I. (1982). Studies on the role played by some Nile fishes in trasnsmitting trematode worms. M.Sc. (Parasitol), Cario Univ.

Tun, M. and Belobordova, E. (1991). Morphological diagnosis of hepatic lesions during chronic opisthorchiasis. *Meditsinskaya Parazitologiya, Parazitarnye Bolezin*, 3: 19-21.

41

A New Kinetic Spectrophotometric Method for the Determination of Mn (II) in Micrograms in Water

R.D. Kaushik, A.K. Chaubey and Raghvendra P. Singh

Department of Chemistry, Gurukul Kangri University, Haridwar -249 404, Uttaranchal

ABSTRACT

A kinetic-spectrophotometric method based on the periodate oxidation of 2, 4-dimethylaniline in acetone-water medium has been developed for the determination of Mn (II) in water samples in the range 5.0 nanograms/ml to 50.0 nanograms/ml. Validity of Beer's law and effect of a few interferrants has been studied and the best suitable conditions have been worked out in terms of concentration, temperature, pH, dielectric constant etc. Calibration curves have been prepared in terms of absorbance Vs concentration plot, initial rates vs concentration plot and rate constants vs concentration plot. Detection limits, molar absorptivity, correlation coefficients and sensitivity have been worked out. Sandell's sensitivity is 0.001 μg/ cm^2 for the absorbance vs concentration curves. Effect of a few interferrants has also been reported. Reproducibility of the results has been tested. The method is simple with good sensitivity and detection limits.

Key Words: Microgram Determination of Mn (II), Periodate Oxidation of 2, 4- dimethylaniline

Introduction

The commonly used method for the determination of Mn (II) in water is the persulphate method based on spectrophotometry (APHA, 1992). The detection limit for this method is 0.05 mg/lit (Kemmer, 1988). Recently polarographic and differential pulse anodic stripping Voltametric method has been reported (Tamarkar *et al.*, 2001). Besides one report available in literature for the estimation of Mn (II) by applying the kinetic-spectrophotometric technique to the periodate oxidation of p-phenetidine (Dolmanova *et al.*, 1970), a thorough survey of literature reveals that no work has been done on the determination of Mn (II) in aqueous/mixed medium by applying the kinetic-spectrophotometric technique to periodate oxidation of aromatic amines which can be used as model process in this case as these processes are likely to be catalysed even by the traces of Mn (II). Further, Dolmanova *et al.* did not undertake the kinetic studies for developing the reported method. Due to this shortcoming, the sensitivity , (0.01 mg/ml) and detection limits (not reported) were not as good as possible. Also, they have used only absorbance values and not the initial rates or the rate constants for the preparation of the calibration curves (although the word 'kinetic method' was included in the title of the article published by them) that might have improved the sensitivity and the detection limits. Therefore, an attempt was made by us for improving upon the reported method, by blending the kinetic studies with the spectrophotometric technique for the determination of Mn (II) in aqueous medium by using the periodate oxidation of 2, 4-dimethylaniline. A remarkable improvement has been observed and the results are being presented in this article.

Materials and Methods

All chemicals used were of E. Merck/Aldrich (A.R.) grade. Sodium meta periodate and 2, 4-dimethylaniline were used after recrystallization and Zn dust distillation respectively. All solutions and reaction mixtures were prepared by using the doubly distilled water. Thiel, Schultz and Koch buffer (Britton, 1955) was used for maintaining the pH of the reaction mixtures. Absorbance was recorded on Schimadzu double beam spectrophotometer , UV-150-02 with the readability of 0.001 absorbance units. Quartz cells of 1 cm length and 1 cm^2 cross sectional area were used. pH was noted by using Systronics digital pH meter model-335.

The detailed kinetic studies were made separately for the reaction between 2, 4-dimethylaniline (DMA) in acetone-water medium. The reaction was found to be second order being first order in each reactant i.e. DMA and periodate. The other kinetic results *viz.* the effect of pH, dielectric constant, ionic strength, free radical scavengers, and temperature have been reported by us earlier (Kaushik *et al.*, 2000). Negligible effect of ionic strength and no effect of free radical scavengers was observed while the rate-pH profile showed a maxima at pH of 7.5. Rate increased on increasing the dielectric constant of the medium.

The reaction was studied by recording the absorbance at the λ_{max} of the reaction mixture (475 nm) after different time intervals and the best suitable conditions in terms of pH, concentration of the DMA and oxidant, temperature, dielectric constant etc. were worked out. λ_{max} for DMA and sodium meta periodate were found to be in UV region. These were found to differ widely from the λ_{max} of the reaction mixture. The studies were kept restricted only to the period in which the λ_{max} did not change and no precipitate/turbidity appeared. $MnSO_4.2H_2O$ (E. Merck-A.R. grade) was used for the preparation of aqueous solutions of different Mn (II) concentrations (being expressed in micrograms per ml). The dielectric constant of the medium was maintained by keeping the concentration of acetone 5 per cent (v/v). The following values were found to be best suitable for studying the reaction for the purpose of determination of Mn (II) in water medium.

[DMA] = 0.01 M; $[NaIO_4]$ = 0.001 M, Acetone = 5 per cent (v/v); pH = 5.0;

λ_{max} = 475 nm.; Temperature = 35 ± 0.1°C, Dielectric constant = 72.4;

Following procedure was adopted for preparation of the calibration curves:-

Step 1: Noting the Absorbance

A definite volume of stock solution DMA in acetone was mixed with the buffer solution of pH 5.0 and calculated volume of the stock solution of Mn (II) was added. A calculated volume of acetone and DMA stock solution (in acetone) was also added and stirred a little with the help of the pipette.

The mixture was then clamped in a thermostat at 35 ± 0.1°C. Stock solution of sodium meta periodate was also clamped the same thermostat. After 30 minutes, a required amount of the periodate solution was mixed with the mixture to start the reaction. All additions were made according to the quantities calculated for maintaining the concentrations of different reagents as mentioned above. Different sets were prepared in a similar manner by taking the initial concentrations of periodate, DMA and acetone the same, and varying the concentration of Mn (II). Aliquots were withdrawn from the reaction mixture after repeated intervals of 1 minute and the absorbance was recorded.

Slep 2: Evaluation of the Initial Rates, Pseudo First Order Rate Constant (k_1)

Absorbance vs time plots were drawn for different sets. The initial rate of reaction in terms of $(dA/dt)_1$ was found from the slope of the tangent drawn on these curves at one minute. As the pseudo first order conditions were maintained in the sets, the pseudo first order rate constants (k_1) were evaluated by using the Guggenheim's method (Fowler and Guggenheim, 1939). For the Guggenheim's plots (*i.e.* log $[A_t + \Delta t - A_t]$ Vs t plot), the Δt was 4 minutes. Here A_t was the absorbance after a time interval of one minute while Δt was chosen arbitrarily.

Slep 3: Plotting the Calibration Curves

Calibration curves, in the form of straight lines, were obtained in terms of A_8 Vs [Mn (II)], $(dA/dt)_1$ vs [Mn(II)], and k_1 vs [Mn(II)] plots.

Results and Discussion

[Mn(II)] may be determined in aqueous solutions and water samples by mixing the sample with calculated quantity of buffer, DMA and acetone and starting the reaction by adding the $NaIO_4$ and then recording the absorbance as discussed in the step 1 of the procedure given above. After it, the data may be used for finding out the [Mn(II)] by employing different calibration curves already prepared.

Many kinetic runs were carried out in presence of varying [Mn(II)] at the difference of 0.005 µg/ml in the range 0.005 µg/ml to 0.05 µg/ml. The initial rates in terms of $(dA/dt)_1$ obtained at one minute after the start of the reaction, A_8 (*i.e.* the absorbance of the reaction mixture observed after 8 minutes of start of reaction) and the values of pseudo first order rate constants k_1 as evaluated by using the Guggenheim's method in presence of different [Mn (II)], are given in the Table 41.1. The Beer's law was found valid in the concentration range 0.005 µg/ml to 0.05 µg/ml as the absorbance vs concentration plot is a straight line in this concentration range.

Method of Least squares was used for obtaining the calibration curves in terms of A_8 Vs [Mn (II)], (dA/dt)1 Vs [Mn (II)] and k_1 vs [Mn (II)] plots which were straight lines obeying the following relationships:

$$A_8 = 20.84 \times 10^{-2} + 549.45 \times 10^{2} [Mn (II)] \quad (1)$$

$$(dA/dt)_1 = 2.6 \times 10^{-2} + 109.88 \times 10^{2} [Mn (II)] \quad (2)$$

$$k_1 = 1.7 \times 10^{-3} + 116.48 \times 10^{2} [Mn (II)] \quad (3)$$

In these equations, $(dA/dt)_1$ are in min^{-1}, k_1 values are in sec^{-1} and the [Mn (II)] are in $mol.lit^{-1}$. In equation (1), (2) and (3) the slopes are in $lit.mol^{-1}$, $lit.mol^{-1} min^{-1}$ and $lit.mol^{-1}.sec^{-1}$ respectively while the intercepts are in absorbance units, absorbance $unit.min^{-1}$, and sec^{-1} respectively.

549.40×10^2 lit. $mol^{-1}.cm^{-1}$ is the value of molar absorptivity obtained for the A_8 vs [Mn (II)] plot. The Sandell's sensitivity is 0.001 $\mu g/cm^2$ for the same plot. The correlation coefficients for the plots obtained from the equation (1), (2) and (3) are 0.9933, 1.003, and 0.9333 respectively while the coefficient of determination are 0.987, 1.006 and 0.8711 respectively.

Table 41.1: [DMA] = 0.01 M, [$NaIO_4$] = 0.001 M, Acetone = 5% (v/v), pH = 5.0, λ_{max} = 475 nm., Temperature = 35 ± 0.1°C, Dielectric Constant = 72.4

[Mn (II)] (μg/ml)	*Δt (minutes) for the Guggenheim Plots*	*A_8*	*$(dA/dt)_1$ (min^{-1})*	*$K_1 \times 10^3$ (sec^{-1})*
0.005	4	0.213	0.026	1.80
0.010	4	0.218	0.027	1.91
0.015	4	0.223	0.028	2.00
0.020	4	0.228	0.029	2.11
0.025	4	0.233	0.030	2.22
0.030	4	0.238	0.031	2.32
0.035	4	0.243	0.032	2.42
0.040	4	0.248	0.033	2.53
0.045	4	0.253	0.034	2.63
0.050	4	0.258	0.035	2.73

The value of the slope of the calibration curves, molar absorptivity, and Sandell's sensitivity indicate that the sensitivity of the method is very good. A change in absorbance by 0.001 unit is expected on changing the concentration of Mn (II) by 0.001 μg/ml. Further, a change in concentration by 0.005 μg/ml will change the rate of reaction by 0.001 absorbance units/minute. Also, the value of k_1 will change by 0.0064 min^{-1} on changing [Mn (II)] by 0.005 μg/ml *i.e.* a change sufficient enough to be observed easily. The detection limits (5.0 nanograms/ml to 50 nanograms/ml) are also considerably low and these are very good for the trace determination of Mn (II). The correlation coefficient and the coefficient of determination suggests the high precision involved in the determination and good correlation of the data. It should also be noted that the proposed method is far better than that developed by Dolmanova *et al.* as far as the sensitivity and detection limits are concerned. In addition to it, we have used the kinetic technique for employing the reaction rates and the rate constants for getting the calibration curves, which has resulted in the improvement of sensitivity of the method.

[Mn (II)] may be determined in aqueous solutions and water samples by mixing the sample with calculated quantity of buffer, DMA and acetone and starting the reaction by adding the $NaIO_4$ and then recording the absorbance as discussed in the step-1 and II of the procedure given above. After it, the data may be subjected for determination of the [Mn (II)] by making use of different calibration curves already prepared.

The proposed method was tested for many water samples containing known amounts of Mn (II) in the range of the detection limits reported above. The results were found to be reproducible and the maximum relative standard deviation noted was 1.509 per cent.

The effect of various expected interferrants was also studied. The method is not applicable in presence of a few aromatic amines like N-ethylaniline, o-ethylaniline, p-ethylaniline, p-toluidine, o-chloroaniline, p-chloroaniline, p-bromoaniline, p-anisidine, N,N-diethylaniline, N,N- dimethylaniline (when present in the amounts >120 μg/ml), p-phenetidine, 3, 5-dimethylaniline, and 4-chloro-2-methylaniline, as they interfere in this method by getting oxidised by periodate ion and the reaction mixture showing λ_{max} in the range in which it influences the absorbance in the present estimation method. Further, following aromatic amines have been found to have no interference in the proposed method:

Aniline, N-methylaniline, o-toluidine, m-chloroaniline, o-anisidine, m-anisidine, o-phenetidine, m-phenetidine, o-nitroaniline, m-nitroaniline, P- nitroaniline, N, N-dimethylaniline (when present in the amounts < 120 μg/ml), 2,6-dimethylaniline, 2,6-diethylaniline, 3,4-dimethylaniline, 2,5-dimethylaniline, 2,3-dimethylaniline, 2,6-dichloroaniline, 2,5-dichloro-aniline, 3-chloro-2-methylaniline, 3-chloro-4-methylaniline, o-amino-benzoic acid, and p-aminobenzoic acid.

The method may be used in presence of the ions like Na^+, K^+, SO_4^{--}, ClO_4^{--}, NO_3^-, and NO_2^- as they do not interfere in present case. However, the metals like Ag, As, B, Go, Gd, Gr, Gu, Fe, Hg, Mo, Ni, Pb, Sb, Se, U, and Zn are expected to interfere in this method. Therefore, a pretreatment is required for separating precipitating or masking these ions before undertaking the proposed method. For this purpose, H_2S may be passed in presence of 0.3 M H+ solution, followed by filtration and boiling off H_2S. After it, a dilute alkaline solution of α-nitroso-β-naphthol should be added and again the solution should be filtered (Meites, 1963). Thereafter, the solution should be neutralized and the present method be applied. However, the method may be used in presence of the ions like Na^+, K^+, SO_4^{--}, ClO_4^-, NO_3^-, and NO_2^- as they do not interfere in present case. In absence of the interferrants reported above, the proposed method may be successfully used for the determination of Mn (I) in water samples.

As far as the mechanism of the reaction is concerned, it is similar to that reported in our earlier work related to the periodate oxidation of a few aromatic amines (Kaushik *et al.*, 1998, 2000), in which it has been established that a charged intermediate is formed between the active species of periodate (*i.e.* lO_4^-) and aromatic amine. This intermediate is attacked by H_2O molecule and then reacts with another molecule of lO_4^- to form quinoneimine which, subsequently, gets hydrolysed to form the main product of the reaction which, in the present case has been isolated, separated and identified as 3,5-dimethyl-1, 2, -benzoquinone on the basis of the melting point and UV-VIS spectrum which matched with the values reported in literature (Ungnade, 1955, Buckingham, 1982). The structure of this product was supported by the IR and NMR spectral studies, (Kaushik *et al.*, 2000 a). The catalytic effect of Mn^{++} may be attributed to the formation of an intermediate complex between Mn^{++} and DMA. In comparison to DMA itself, this intermediate may be more reactive towards periodate leading to an enhanced reaction rate, probably due to the catalytic effect of Mn^{++} (Dryhurst,1970, Srivastava *et al.*, 1983).

References

APHA, AWWA and WEF. (1992). *Standard methods for the examination of water and wastewater.* 18th edition. APHA, New York.

Britton, H.T.S. (1956). *Hydrogen ions*. D. Von Nostrand Co., 40 p.

Buckingham, J. (1982). *Dictionary of organic compounds*. Chapman and Hall, N.Y. Volume II. 5th edition. p. 2087.

Dolmanova, I.F., Poddubienko, V.P. and Peshkova, V.M. (1970). Determination of manganese by kinetic method using the oxidation of p-phenetidine by KIO_4. *Zh. Anal. Khim*. 25(11): 2146-50.

Dryhurst, G. (1970). *Periodate oxidation of diols and other functional groups, Analytical structural applications*. Pergamon Press.

Fowler, R.H. and Guggenheim, E.A. (1939). *Statistical Thermodynamics*. Cambridge University Press, New York.

Kaushik, R.D. and Joshi, R. (1998). Determination of some aromatic amines in micrograms by kinetic spectrophotometric method based upon periodate oxidation in aqueous/mixed medium. *Asian J. Chem.*, 10(2): 328-332.

Kaushik, R.D., Garg, P.K. and Oswal, S.D. (2000). Kinetics and mechanism of periodate oxidation of 2,4-dimethylaniline in acetone-water medium. *J. Nat. Phys. Sciences,* 14: 129-134.

Kemmer, F.N. (1988). The Nelco water handbook. McGraw-Hill Co., Singapore, 7: 32.

Meites, L. (Ed.) (1963). *Handbook of Analytical Chemistry*. McGraw-Hill book Co., Inc., New York. pp. 3-4 to 3-5.

Srivastava, S.P., Gupta, V.K. and Malik, P. (1983). Kinetics and mechanism of oxidation of o-xylidine by periodate in acetic acid-water medium. *Oxidation Communications*. 5(3-4): 475-487.

Tamarkar, P.K. and Pitre, K.S. (2001). Analysis of trace metals in psilomelane manganese ore. *J. Indian Chem. Soc.,* 78: 482.

Ungnade, H. E. (1953-55). *Organic Electronic Spectral Data*. Interscience Publishers, Inc., New York. Volume II. 136 p.

42

Effect of Lethal Concentration of Metasystox (Oxydemeton-Methyl) on Blood Parameters of an Air Breathing Fish *Heteropneustes fossilis* (Bloch)

Meenaxi Das and Shambhu Prasad

Biotoxin Research Laboratory
University Department of Zoology, L.N. Mithila University, Darbhanga - 846 004

ABSTRACT

The accumulation of pesticide residues in different component of aquatic ecosystem and their detrimental impact is of significant environmental concern. In the present investigation *Heteropneustes fossilis* exposed to lethal concentration of metasystox exhibited striking alteration in blood parameters. The observed effect of exposure to 48h LC_{50} reflected, hyporcromasia, elevated erythrocytic count with increase in size of cremation of cell and their membrane, decrease in total leucocyte count and haemoglobin per cent.

It has been concluded that the present fish in response to metasystox stress faces with a serious physiological crisis due to alteration in several blood parameters that ultimately tells upon the survival of the fish.

Key Words: Metasystox, Blood Profile, *Heteropneustes fossilis*, Pesticides.

Introduction

Pesticides have became essential to modern agriculture but they are also a potential source of chemical degradation of soils. The 1992-93 data show that India used 47.62 million tonnes of insecticide, 19.45 millions tonnes of fungicides 5.79 million tonnes of herbicides, 0.90 million tonnes of rodenticides, and 0.08 million tonnes of fumigants (Employment News 7-13 August 1999). Bulk of pesticides used in India are organochloride compounds. Half life of some of the organochloride pesticides like HCH, chlordane, and heptachlore, in cultivated soils is as high as 80 to 110 days. This means that these pesticides are highly resistant to biological degradation and are potential toxicants. Case studies have indicated that pesticides applied to soil can enter the food chain. Pesticides have undesirable effects on the environment. Some of the effects are excessive mortality and reduced reproductive potential of organism, reduction in the productivity of natural resources and development of resistance to pesticides in target and non-target species.

The increased use of pesticides have helped in controlling insects and pests but have also created undesired environmental deterioration specially hazardous influence in aquatic media causing fatal and long term effects on fishes as well as aquatic fauna (Basok *et al.*, 1980). As a consequences, fishes are being used as indicator of aquatic pollution. Like other tissues, blood is an important physiological reflector of animal body and serves as primary target of many pesticidal action.

There are many reports related to toxicity to chemicals organophosphorus and chlorinated hydrocarbon pesticides on the blood of some Indian fishes (Shrivasatva *et al.*, 1985). But recently some new improved type of systemic pesticides have been developed in the light of growing resistance in the pest of previous insecticides (Sharma *et al.*, 1991).

Metasystox is a systemic pesticide and widely used by the farmers of Bihar to control different crops including paddy, mango, the two main crops. It is already a proved fact that the systemic toxicant have biaccumulative properties and are metobolically degraded through enzymatic action at low pace. This is why an animal is recommended for slaughtering or for milching after a safe period of about 60 days from the time of application or administration of the toxicant.

Indiscriminate use of metasystox may cause greater damage to fresh water fishes as well significant alteration in aquatic ecosystem. The present communication highlights the effect of matasystox on some blood parameters of *H. fossilis.*

Material and Methods

The fish *Heteropneustes fossilis* (28 ± 2g) obtained from local fish market were thoroughly washed in 2 per cent $KMNO_4$ solution and acclimatised for 15 days. Fishes were fed with egg white cut into small pieces once a day. The feeding was stopped one day before the experiment.

48-h LC_{50} value was calculated following recommended procedure (APHA-AWWA and WPCF, 1971). Five groups of twenty fishes were kept in separate aquaria. Group I was treated as control. Four different concentration of metasystox (2, 3, 4, and 5 ml/L.) was used for treating the test animals. LC_{50} was found to be 3 ml/L since 50 per cent population of the fishes died in group II exposed to the said concentration. Surviving fish were used for further studies.

Blood was sampled from the caudal peduncle of fishes using a heparinized tuberculin syringe. Blood film was stained with Leishman's stain. Parameters like total count, differential count, haemoglobin per cent were studied following routine techniques. Cells were measured by 'Sico' make micrometer.

Results and Discussion

The present effects of metasystox 3ml/L. on *H. fossilis* included cremation, hypochromasia in erythrocytes and significant increase in size of erythrocyte including nucleus. Total erythrocyte count and haemoglobin percentage revealed a decreasing trend. Similar observations of hypochromais, a cremation and shrinkage of cell membrane were observed in Anabus exposed to Rogor (Singh, 1990). Chakarbarty and Banarjee, (1988) also reported simulating findings in *Channa punctatus* exposed to lethal exposure of organophosphate pesticides and gradual increase in toxicity of Mahua oil cake respectively. Chauhan *et al.* (1983) reported hypochromasia and vaculation of erythrocytes in *Puntiusticto* exposed to herbicide but in present case inspite of hypochromasia no vacuolar degeneration could be marked.

Increase in erythrocyte surface area finds support with the findings of Chakrabarty and Banerjee (1988) in *Channa punctatus* exposed to thimet 10G and Elsan 2 per cent dust.

Increase in total erythrocyte count (TEC) and Haemoglobin percentage HB%) as found in present case have also been reported earlier by Mahajan and Juneja (1979) in *Channa punctatus*, Dhilon and Gupta (1983) in *Clarias batrachus* and Ghosh and Chatterjee (1986) in *Sarotherodoth-mossambicus* due to Aldrin exposure.

In contrast to the finding of Kumari *et al.* (1989) who reported significant increase in TLC (Total Leucocyte count) in *Clarias batrachus* exposed to organochlorine pesticide (DDT and Lindane). The present pesticide (Metasystox) causes significant decrease in TLC. Mishra and Shrivastava (1985) have also reported increase in TLC of *H. fossilis*. The fish were treated with dichlorous and lindane. *Clarias batrachus* exposed to the same pesticide also reflected increase in TLC (Natrajan, 1981).

In lethal exposure of metasystox (5 ml/L. for 48 h) *Channa straitus* reflected decreased TEC, Haemoglobin percentage and TLC whereas when the same fish were exposed to 48 h LC_{50} of Thimet 10G and Banzanon 5G the same fish reflected significant increase in TEC but a fall in haemoglobin level (Natrajanm 1981).

Thus metasystox induces significant changes in TLC, TEC and Hb% reflecting its toxic effects in *Heteropneustes fossilis*.

Table 42.1: Blood Parameter of *H. fossilis* Following 48 h LC50 exposure to, Metasystox

Parameters	*Control*	*Treated*
Erythrocyte Length (μm)L	9.12 ± 0.01	9.17 ± 0.03
Erythrocyte Breadth (μm) B	8.5 ± 0.01	8.72 ± 0.04
Nucleus Length (μm) L'	3.42 ± 0.01	4.08 ± 0.02
Nucleus Breadth (μm) B'	3.28 ± 0.01	3.58 ± 0.01
L/B (Cell)	1.07: 1	1.05: 1
L'/B' (Nucleus)	1.03: 1	1.1: 1
Hb%	63% to 76%	60% to 62%
Small Lymphocyte	12%	10%
Large Lymphocyte	18%	20%
Monocyte %	00	00
Neutrophil%	06	08
Eosinophil%	10	1
Basophil%	04	00
TLC	1632 to 1964 mm^3	1600 to 1820 mm^3
TEC	169 × 10^3 mm^3 to 189 × 10^3 mm^3	143 × 10^3 mm^3 to 161 × 10^3 mm^3

TLC = Total Lecocyte Count
TEC = Total Erythrocyte Count

Hb% = Haemoglobin Percentage.

Acknowledgements

The authors are grateful to Dr. N.K. Dubey, Prof. and Head for provided laboratory facilities during the investigation and Dr. Shishir K. Verma, Principal Investigator, I.C.A.R Project, University Department of Zoology, L.N. Mithila University, Darbhanga for having critically gone through the manuscript and suggesting improvements.

References

APHA-AWWA and WPCF (1971). *Standard Methods for the Examination of Water and Waste Water*, American Public Health Association. Washington D.C.

Basak, P.K. and Konar, S.K. (1980). Proc. Nat. Acad. Sci. India 50(B): 218.

Chauhan, M.S., Verma, D. and Pandey, A.K. (1983). Herbicide Induced haematobiogical changes and their recovery in fresh water fish, *Puntius ticto* (Ham.) *Camp. Physial. Ecol.*, 8(A): 249-251.

Chakrabarty, P. and Banerjee, V. (1988). Effect of organophosphorus pesticides on the peripheral haemogram of the fish, *Channa punctatus. Environ and Ecol.*, 6(2): 390-394.

Dhilon, S.S. and Gupta, A.K. (1983). A clinical approach to study the pollutants intoxication in fresh water *Clarias batrachus. Water Air and Soil Pollution*, 20: 63-68.

Employment News. Vigyan Prasar, 7-13 Aug. 1999.

Ghosh, T.K. and Chatterjee, S.K. (1986). Effect of Aldrin on some haematological variables of a fresh water fish *Sarotheryodon massambicus. Environ, Ecol.*, 4(3): 427-429.

Hoemechaudhuri, S., Pandit, T., Poddar, S. and Banerjee, S. (1986). *Proc. Ind. Acad. Sci.* 9: 617.

Kumari, M., Kumari, G. and Yadav, S. C. (1989). Perils of environmental pollution by pesticides, an Assessment through leucocytic and haemostatic response of a fresh water fish *Clarias batrachus* (L). *Him. J. Env. Zool.*, (3): 36-43.

Mahajan, C.L. and Juneja, C.S. (1979). Effects of Aldrin on peripheral blood of a fish *Channa punctatus. Indian J. Environment Health*, 21: 169-172.

Mishra, J. and Srivastva, A.K. (1985). Effect of dichlorous on the blood and tissue chemistry of the Indian catfish. *Proc. Nat. Symp. Industry, Pollu. Pest.*, Bhopal Tragedy, Gorakhpur.

Natarajan, G.M. (1981). Changes in the bimodal gas exchange and some blood parameters in the air breathing fish *Channa Striatus* (Blacker) following lethal (LC_{50} /46 hr) exposure to metasystox (emeten). *Curr. Sci.*, 50(1): 40-41.

Sharma, G.D. (1991). Toxicity of BHC on the gills of *H. fossilis. Proceeding of the 78th session of Indian Science Congress*, Indore.

Singh, S.K.P. (1990). Changes in Haematological parameters and haemopoetic tissue of *Anabas testidineus* (Bloch) under influence of stress. Ph.D. Thesis, Patna University, Patna, India.

Srivastava, A.K. and Mishra, J. (1985). Lindene induced haematological changes in catfish. *Nat. Acad. Sci. Letter*, 8: 391-394.

43

Acute Toxicity of Fish Fingerlings in Paper Board Mill Effluent

M. Elayarajan and P. Jothimani*

Department of Soil Science and Agricultural Chemistry,
Tamil Nadu Agricultural University, Coimbatore
** Department of Environmental Sciences,*
Tamil Nadu Agricultural University, Coimbatore

ABSTRACT

Both urban and industrial complexes grow rapidly in the neighbourhood of rivers and lakes, as water is the basic raw material for their survival. In ancient times, conservation of water resources was considered as a basic law of self preservation, whereas, now sanitation of water resources is going from bad to worse due to population explosion, urbanization. The total supply of water is limited and geographically irregular through seasons and years. The problem is more serious for India, since we have to support about 15 per cent of the world population in only 6 per cent of water resources on 2.5 per cent of land area. Bioassay experiments for determining acute toxicity were carried out according to the methods recommended in American Public Health Association (APHA, 1980). To study the effect of effluent at different concentrations *viz.* 0, 25, 50, 75 and 100 per cent on survival per cent of fingerlings for 96 hours. Healthy fingerlings *viz., Catla catla, Cyprinus rogu, Mrigal* of about 4-5 cm in length and 10 g in weight were used and acclimatized in dechlorinated tap water. The fingerlings were fed with minced meat on alternate days. Excess food and water were cleared from the bottom of the aquarium periodically. Results obtained from the present work are quite interesting. Marked behavioural changes were observed in effluent treated fingerlings, as they showed restless condition, erratic and jerky movements, loss of balance, jumping, breathing rapidly and keeping the fin stretched laterally. When the period of restlessness subsided in 4 to 6 hours it settled quietly on at the bottom. Bioassay experiment showed that the survival per cent of fingerlings was increased in *Cyprinus rogu* with increasing concentrations of treated effluent and decreased in *Catla catla. Mrigal* was found very sensitive to treated effluent.

Introduction

Water pollution is a worldwide phenomenon and is one of the most serious problems confronting mankind today. The problem is accentuated by rapid urbanization and industrialization, which are fast transferring air, water and soil into natural reservoirs of pollutants. Both urban and industrial complexes grow rapidly in the neighbourhood of rivers and lakes, as water is the basic raw material for their survival. In ancient times, conservation of water resources was considered as a basic law of self preservation, whereas, now sanitation of water resources is going from bad to worse due to population explosion, urbanization. The total supply of water is limited and geographically irregular through seasons and years. The problem is more serious for India, since we have to support about 15 per cent of the world population in only 6 per cent of water resources on 2.5 per cent of land area. Sometimes, industries produce highly toxic effluents, which can neither be thrown into water bodies nor used for agricultural purpose as the toxic elements may enter the food chain through plants, animals and fish. However, effluents of some industries have useful characteristics and have the potential to improve soil productivity (Kansal, 1994). This effective management of wastes brings economic benefits and protects fragile ecosystems from degradation.

Materials and Methods

Bioassay experiments for determining acute toxicity were carried out according to the methods recommended in American Public Health Association (APHA, 1980). To study the effect of effluent at different concentrations *viz.* 0, 25, 50, 75 and 100 per cent on survival per cent of fingerlings for 96 hours.

Healthy fingerlings *viz., Catla catla, Cyprinus rogu, Mrigal* of about 4-5 cm in length and 10 g in weight were brought from Department of Fisheries, Bhavanisagar, and acclimatized in dechlorinated tap water. The fingerlings were fed with minced meat on alternate days. Excess food and water were cleared from the bottom of the aquarium periodically.

Results and Discussion

Impact of Effluent on Survival of Fingerlings (Bio-assay Study)

The effect of treated effluent on survival percentage of *Catla catla, Cyprinus rogu, Mrigal* fingerlings was studied at different effluent concentrations (0, 25, 50, 75 and 100 per cent) under laboratory conditions. The tap water was used as control. The results are presented in Table 43.1, Fig. 43.1.

Table 43.1: Impact of Effluent on Survival Per Cent of Fingerlings at Different Dilutions

Effluent Concentrations	*Catla catla*	*Cyprinus rogu*	*Mrigal*
0%	100	90	0
25%	90	90	60
50%	90	100	40
75%	80	100	50
100%	80	100	20

Results obtained from the present work are quite interesting. Marked behavioural changes were observed in effluent treated fingerlings, as they showed restless condition, erratic and jerky movements, loss of balance, jumping, breathing rapidly and keeping the fin-stretched laterally. When the period of restlessness subsided in 4 to 6 hours it settled quietly on at the bottom.

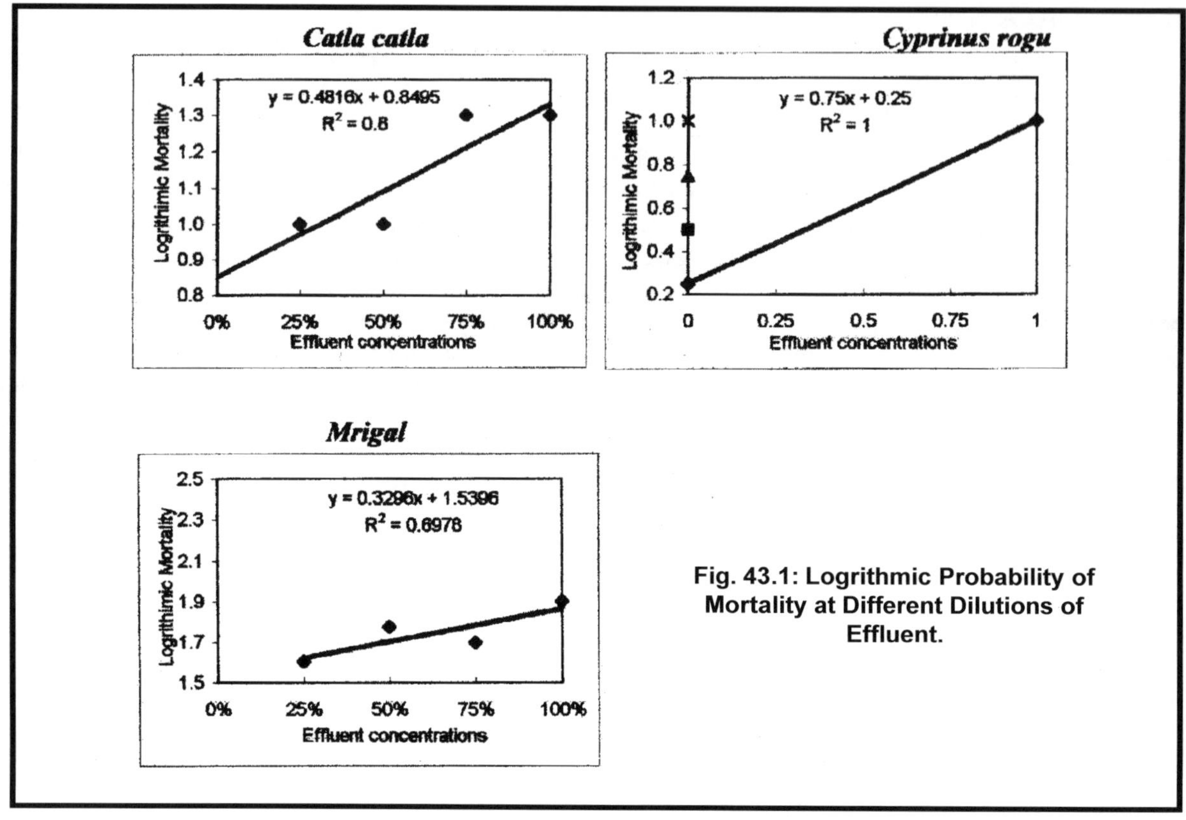

Fig. 43.1: Logrithmic Probability of Mortality at Different Dilutions of Effluent.

Catla catla

Among the treatments, 0 per cent treated effluent (tap water) was found to be the best recording 100 per cent survival and 75 and 100 per cent treated effluent recorded the poorest survival (80 per cent). As the treated effluent concentration increases, the survival per cent was found to be reduced.

Cyprinus rogu

The survival rate of *Cyprinus rogu* was increased with the increasing concentration of effluent and the maximum survival per cent was recorded in 50, 75 and 100 per cent effluent concentration.

Mrigal

In tap water (0 per cent effluent) all the fingerlings died within 96 hours. Fourty per cent survival was observed in the 50 per cent treated effluent. The survival per cent was maximum at 25 per cent effluent concentration, which decreased drastically due to increasing concentration of effluent in case of *Mrigal.*

In the present study, it was found that the treated effluent affected survival per cent of *Catla catla* which is relatively sensitive to water quality. The reduction in survival per cent from 100 to 80 at higher concentration might be due to the reduced level of DO in the tested effluent. Jyoti *et al.* (1989) have also observed the same negative correlation between survival per cent and effluent concentration in the same species. In contrast, the survival per cent of *Cyprinus rogu* was increased from 90 to 100 with higher effluent concentrations. The increase in survival per cent might be due to its capacity to survive even under reduced levels of DO besides effective assimilation of available nutrients at higher concentration of effluent by this species. Whereas in *Mrigal* the survival per cent decreased from 60 to 20 with increasing concentration of effluent. *Mrigal* was very sensitive to the treated effluent.

Conclusion

Bioassay experiment showed that the survival per cent of fingerlings was increased in *Cyprinus rogu* with increasing concentrations of treated effluent and decreased in *Catla catla. Mrigal* was found very sensitive to treated effluent.

References

APHA. (1980). *Standard methods for examination of water and waste water*. American Public Health Association. Washington D.C. (15 ed).

Jyoti, A., A.A. Khan, N. Haque, M. Fatima and S.I. Barbhuyan. (1989). Studies on the effect of Malathion on a freshwater fish *Channa punctotus* (Bloch). *J. Environ. Biol.*, 10(3): 251-257.

Kallas, J. and R. Munter. (1994). Post-treatment of pulp and paper industry waste waters using oxidation and adsorption processes. *Wat. Sci. Techno.*, 29: 259-272.

44

Sublethal Effects of BHC on Oxygen Consumption by Fish *Cyprinus carpio* and Role of an Aquaphyte (*Chara*) as Detoxicant

R. Punitha, N. Krishnan and G.N. Subbiah*

Department of Zoology, Madura College, Madurai - 625 011
**Department of Zoology, Thiagarajar College, Madurai - 625 009*

ABSTRACT

Exposure of carp, *Cyprinus carpio* to a sublethal concentration of an organochlorine insecticide, BHC for 30 days under laboratory conditions decreased the rate of oxygen consumption by fish. However addition of an aquatic plant, *Chara* (Family: Charophyceae) to intoxicated test medium (100 g/10 litres) enhanced the rate of oxygen consumption by fish. It is interpreted that *Chara* could be tested further as phytodetoxicant in reducing the insecticidal toxicity to cultivable fishes.

Key Words: *Cyprinus carpio*, BHC, Oxygen Consumption, *Chara*, Detoxicant.

Introduction

Environmental pollution in freshwater ecosystems is caused by a variety of pollutants. In an agricultural country like India, insecticides constitute the major components of aquatic chemical pollutants. Laboratory investigations in fish models have documented a variety of physiological, bio-chemical, haematological and behavioural responses of fishes under toxic stress of insecticides (Review:

Pritchard, 1993). Oxygen consumption by fishes under toxic stress of various aquatic pollutants including insecticides was estimated quantitatively by several workers in the past (Jeyachandran and Chockalingam, 1986, Roy and Datta Munshi, 1988, Baskaran *et al.*, 1989, Khillare, 1990, Jayasuriya *et al.*, 1991). Use of aquatic plants as phytodetoxicants or biofilters their by reducing the toxic stress on fishes is one of the known method in the control of insecticidal pollution in aquatic ecosystems (Subbiah and Suryaprabha, 1986, Jampani, 1987, Chandrasekaran *et al.*, 1995). In this background, the present study ascertains the valve of aquatic algae, *Chara* as a detoxicant in sharing the toxic stress of BHC by *Cyprinus carpio* in relation to its oxygen consumption.

Materials and Methods

Healthy fingerlings of *C. carpio* (4 + 0.5 g/live weight) were obtained from local fish farm and acclimatized to the laboratory conditions at 29°C for 7 days before experimentation. They were fed daily with chopped goat liver but food was withheld 24 hrs before bioassay test. The water was changed daily with an intermittent aeration, *chara* (Family: Charophyceae), commonly known as stonewart was selected as phytodetoxicant. It was collected from the college pond, washed with tap water and stored in aquarium. For the present study BHC, an organochlorine insecticide was selected as experimental toxicant. The commercial grade BHC (50 per cent wp, trade name Hexidole) is obtained from Rallis India Ltd. Its stock solution in distilled water (1 g in 500 ml) is prepared and then diluted to required test concentrations (8-15 ppm). The acclimatized fishes, ten in each group were exposed to test concentrations along with suitable control. Three replicates were maintained for each treatment. Water and concentration of BHC in aquaria were renewed every 24 hr. Mortality rate was observed for 96 hr in each test concentration. LC_{50} for 96 hr was computed by plotting the percentage mortality against the concentration of BHC (Abbott,1925). The 96 hr LC_{50} was found to be 10 ppm (Punitha, 1990). Oxygen consumption by fish was studied under sublethal concentration (5 ppm) by closed-chamber method. Fishes were exposed to 10 litres of insecticide treated water with adequate replicates for 30 days. Simultaneously another set was maintained along with *Chara* (100 g/10 litres). Oxygen consumption was determined at weekly interval in both groups by Winkler's method. The results are expressed in mg/g wet weight/hr.

Results and Discussion

In fishes treated with BHC but without *Chara*, the rate of oxygen consumption decreased from an initial value of 7.457 in first week to 3.442 mg/g wet weight/hr. The percentage reduction data in every week of exposure supports this conclusion (Table 44.1). Similarly earlier works have estimated a significant reduction in oxygen consumption by many fish species under the toxic stress of aquatic pollutants such as insecticides (Roy and Datta Munshi, 1988, Khillare, 1990), detergents (Jayasuriya *et el.*, 1991), textile dye effluent (Baskaran *et al.*, 1989) and tannery effluent (Jeyachandran and Chokalingam, 1986). The gills of fishes are in direct contact with the toxic test medium. Histopathological studies have revealed marked alternations in gill structure of the fishes exposed to organochlorine insecticides (Sahai, 1991). For instance, BHC induces an acute inflammation, loss of epithelial layer and fusion of gill lammellae in *Rosbora daniconius* (Singh and Sahai, 1984). It is presumed that BHC may induce similar alternations in gills of *C. carpio* and this could be the reason for the drop in oxygen consumption.

In constast, fishes exposed to BHC with *Chara*, it was dropped from 7.681 to 7.001 mg/g wet weight/hr. Eventhough there were no significant differences between the weeks in mean valves when compared to groups treated without *Chara*, the difference between the overall mean values is significant at $P < 0.05$. It showed that addition of *Chara* to the test medium enhances the rate of oxygen consumption by fish. This algae may act as detoxicant or biofilter possibly accumulating a considerable quantity of BHC. This presumption needs further study. For instance, Chandrasekaran *et al.* (1995) have shown that *Chara* accumulated endosulfan, an organochlorine insecticide from the test medium significantly their by reducing its accumulation in kidney, liver and gills of *Macrones keletius*. Similarly aquaphytes like *Hydrilla verticilata* and *Pistia stratiotes* have been shown to be detoxicant of endosulfan (Subbiah and Suryaprabha, 1986). Detoxification of pesticides such as phosphomidon and monocrotophos by the blue green algae, *Nostac colci* was reported by Jampani (1987). Further investigations are needed to confirm such BHC accumulation in *Chara* and degree of protection by this algae to other species of fishes.

Table 44.1: Oxygen Consumption by *C. Carpio* (in mg/g Wet Weight/hr) Exposed to Sublethal Concentration of BHC (5 ppm)

	Exposure Duration (in Weeks)				
Fish Group	*I*	*II*	*III*	*IV*	*Grand Means*
Without *Chara*	7.457	6.310	4.302	3.442	5.4
	(1.088)	(1.060)	(0.620)	(0.103)	(1.581)
%Reduction	34.82	29.33	19.99	16.01	
With *Chara*	7.681	7.465	7.101	7.001	7.3
	(2.002)	(0.620)	(0.098)	(0.860)	(0.082)
% Reduction	26.62	25.52	24.27	23.93	

Values are Mean (SD)

n = 3 Difference Between Grand Means is Significant of $p < 0.05$ by t Test.

Acknowledgements

This work was carried out by RP at Thiagarajar College under the guidance of Prof. GNS. We are grateful to Board of Madura College, HOD, Principal and all faculty members for encouragement. NK gratefully acknowledge Prof. Dr. Arvind Kumar, Editor for his constant support and encouragement.

References

Abbott, W.S. (1925). A method of computing the effectiveness of an insecticide. *J. Enta,* 18: 263-267.

Baskaran, B., Palanichamy, S. and Arunachalam, S. (1989). Effects of textile dye effluent on feeding energitics, body composition and oxygen consumption of the fish *Oreochromis mossambicus. J. Ecobiol.,* 1: 203-207.

Chandrasekaran, N., Murugavel, P. and Krishnan, N. (1995). Control of endosulfan residue accumulation in *Macrones keletius* with *Chara* (Charophyceae). *J. Appl. Zool. Res.,* 6: 193-194.

Jampani, C.S.R. (1987). Detoxification of insecticides phosphomidon and monocrotophos by the blue green algae *Nostoc colci. Natl. Acad. Sci. Lett.* 10: 381-382.

Jayasuriya, S., Subramanian, M.A. and Varadaraj, G. (1991). Effect of detergents on the oxygen consumption of the cat fish *Mystus vittatus. J. Ecobiol.,* 3: 217-220.

Jeyachandran, K.P.S. and Chokalingam, S. (1986). Effects of tannery effluent on respiratory parameters and biochemical constituents of fish *Channa punctatus. Comp. Physiol. Ecol.* 12: 197-200.

Khillare, Y.K. (1990). Effects of pesticide on fresh water fish, *Puntius stigma*. In: *Trends in Ecotoxicology*.

Academy of Environmental Biology, Karad. pp.185-188.

Pritchard, J.B. (1993). Aquatic toxicology: past, present and prospects. *Environ. Health Persp.,* 100: 249-257.

Punitha, R. (1990). The role of the aquaphyte *Chara* as a detoxicant. M. Phil. Thesis, Madurai Kamaraj University, India.

Roy, P.K. and Datta Munshi, J.S. (1988). Oxygen consumption and ventilation rate of major carp *Cirrhinus mrigala* in malathion treated water. *J. Environ. Biol.,* 9: 5-13.

Subbiah, G.N. and Suryaprabha, R. (1986). A study on the control of insecticide residue accumulation in a freshwater fish *Macrones keletius* with aquaphyte. *Proc. Natl. Symp. Fish. Environ.* Madras, pp. 19-23.

Sahai, S. (1990). Pesticide pollution and its impact on some fish tissues: a review. In: *Trends in Ecotoxicology*. Academy of Environmental Biology, Karad, pp. 63-72.

Singh, S. and Sahai, S. (1984). Histopathological changes in the gills of *Rasbora daniconius* induced by BHC. *J. Environ. Biol.,* 6: 65-69.

45

"Ochratoxin A": A Naturally Occurring Mycotoxin Found in Fish Feed Samples

Shambhu Prasad and Shishir K. Verma

Biotoxin Research Laboratory, University Department of Zoology, L.N. Mithila University, Darbhanga (Bihar)

ABSTRACT

"Ochratoxin A" is naturally occurring mycotoxin produced by some strains of *Aspergillus* and *penicillium* genera as their secondary metabolites. A comprehensive survey of fish farmer storehouse and retail shops in different block of Darbhanga district was carried out to determine incidence of *A. ochraceous* group of fungi. Altogether 868 fish feed samples collected during October 2000 to September 2001, were screened. *A. Flavus, A. ochraceous, A. niger, A. versicolor, A. parasiticus, A. nidulanse, A. tamari, Altermanria* sps, *Curvularia* sps. and *Penicillium* genera appeared as frequent invaders with variable degree of occurrence. Degree of *A. ochraceous* infestation on these fish feed samples varied between 13.13-22.91 per cent being maximum during pre-winter season (22.91 per cent) and minimum during winter (13.33 per cent). During monsoon 18.47 per cent samples were found to the contaminated. However, *A. flavus* was recorded to be the most dominant mycoflora.

Among the agricultural commodities oil cakes (Mustard, sunflower, linseed), wheat bran, Rice flour and corn flour appeared to be substrates of preference and the climatic condition prevalent during pre-monsoon and pre-winter months were observed to be congenial for growth of *A. ochraceous* group of fungi.

Key Words: Ochratoxin A, Mycotoxin, Fish Feed, Mold Spoilage.

Introduction

Sustainable and economically viable culture of fish requires supplementation of artificial feed. Generally two types of fish feeds are used for fish culture: commercially available formulations and raw agricultural products and their byproducts. Most of the fish farmer of Darbhanga district generally use locally available agricultural byproducts *viz.* Mustard Oil Cake (MOC) Linseed Oil Cake (LOC), Sunflower Oil Cake (SOC), Soyabeen Cake (SC), Rice Bran (RB), Rice Flour (RF), Wheat Bran (WB), Wheat flour (WF), Corn Flour (CF) etc. Use of commercial rations has yet to gain popularity in this locality.

Since the outbreak of rainbow trout hepatoma epidemic during 1960s in USA and Great Britain (Rucker *et al.*, 1961) food/feeds all through the world have been subject of investigations for mycobial infestation. The studies till date reveal that almost all dietary items of either plant or animal origins are prone to mold infestation. Many of these molds are capable of elaborating potentially lethal toxins as their secondary toxic metabolites. It has been further suggested (WHO, 1979) that these toxic fungi are ubiquitous in occurrence and their magnitude of invasiveness vary with the climatic factors. Hot and humid climate are often congenial for their growth and subsequent metabolic activities.

Out of a long list of mycotoxins (Verma, 2001) a few considered detrimental to aquaculture practices are Aflatoxins, Ochratoxins, zeralenone, sterigmatoeystin and Tricothecene. Ochratoxins are secondary metabolites of some toxigenic strains of *Aspergillus ochroceous* and *Penicillum viridicotum*. Ochratoxin A (OTA) contamination of food and feeds of both plant and animal origin in various parts of the world has been reviewed by Merwe *et al.*, (1965a,b), Marraufi *et al.*, (1995) and Stoev (1999). It has been suggested that cereals and derived products are assumed to be the major dietary source of OTA. In addition, other products of vegetable origin as nuts, beans, coffee, cocoa, spices, dried fruits and beer may contain OTA. The intake of OTA through contaminated feed may lead to residues in the blood, the kidney and the liver of pigs and of poultry and to a lesser extent in muscle adipose tissue and eggs (Krogh, 1987).

The general toxicity of OTA has been reviewed by Dirheimer and Creppy (1991), Pohland *et al.* (1992), Stormer (1992) and by Morquart and Frohlich (1992). OTA causes proximal tubule necrosis and renal degeneration in all species (Pigs, dog, rat, rabbit, mouse, avians) tested so far (Kuiper-Goodman *et al.*, 1989). In addition it has been shown to produce renal tumors in male mice and both sexes of rat (Boorman, 1989) and has been reclassified as a possible human carcinogen by the International Agency for Research on Cancer. OTA is a highly suspected etiologic agent in the development of Balkan Endemic Nephropathy and urinary tract tumors. In toxic potency it is estimated to have an oral LD_{50} of about 150 µg per ducking, one-tenth to the toxicity of aflatoxin (Merwe, *et al.*, 1965).

Materials and Methods

General survey of the fish feeds for the incidence of toxic fungi was conducted during pre-winter (October 2000), winter (November 2000 to January 2001) Pre-summer (February 2001), Summer (March 2001 to May 2001), Pre-monsoon (June 2001), and monsoon (July 2001 to Sept. 2001). The samples of MOC, SOC, SC, LOC, RB, RF, WB, WF and CF were obtained/purchased directly from retail shops and fish farmer storehouses from different blocks of Darbhanga district, following the guideline suggested by U.S. food and drug administration and Tropical Products Institute, London.

The samples (minimum 500 gm) and maximum depending upon the size of the lot (upto 2 kg in 500 gm units) were physically verified, serially numbered and brought to the laboratory in sterilized polythene tape woven bags, The samples before screening were stored in ambient condition. Following physical verification the apparently fresh looking samples were designed as normal (N) and so on like, slightly molded (SM), and Highly molded (HM) depending upon the intensity of mold infestation.

Representative working sub samples of 50 g were prepared by quartering the bulk samples following Dickens and Whitkar (1982). A few pinch of each sub sample were plated on sterilized Czapek's Agar Medium (30.0 gm $NaNo_3$, 1.0 gm, K_2HPO_4, 0.5 gm KCl, 0.01 gm $FeSO_4$, 30.0 gm Sucrose, 15.0 gm agar, 1000 ml distilled water) following Raper *et al.*, (1965). The cultures were incubated for 7-10 days in BOD, at 28 ± 2°C. All the fungi appearing in culture were purified by streaking. Repeated sub culturing was done to maintain the pure culture of different isolates. Identification of fungal species was done by studying their micromorphological characters with reference to Moreau (1979), Wyllie and Morehouse (1977). Frequency of occurrence of various forms at the end of incubation period was determined in terms of percentage incidence and percentage abundance (Verma 1989).

Results

During the period of one year (October 2000 to September 2001) a total number of 868 samples were collected from different areas of Darbhanga districts (Table 45.1). Record of natural infestation of the present substrates by different fungal forms has been presented (Table 45.2).The fungi recorded on all the nine types of fish feeds ingredients round the year were *Rhizopus Aspergillus flavus, A. versicolor, A. parasiticus. A. niger, A. nidulanse, A. ruber, A. ochraceous,* and *Penicillium* species. Table 45.1 also indicates about the samples that appeared to be fresh during sampling but developed fungal colonies following incubation under optimal climatic condition. However, such incidence was variable in different seasons as well as type of the substrate (Table 45.2). Analysis of the data reveal that different type of oil cakes (Mustard, Sunflower and linseed) and wheat bran are highly susceptible to mold invasion showing incidence of molds above 70 per cent followed by rice bran (Table 45.2).

Table 45.1: Detail of the Samples Collected in Different Seasons

Item	*Pre-winter*	*Winter*	*Pre-summer*	*Summer*	*Pre-monsoon*	*Monsoon*	*Total*
MOC	16	36	12	40	20	40	164
SOC	08	16	10	20	04	16	74
SC	08	16	12	20	04	12	72
LOC	08	16	06	20	04	16	70
RB	12	20	16	20	12	20	100
RF	12	20	16	24	08	20	100
WB	16	20	20	28	16	24	124
WF	08	14	08	28	08	24	100
CF	08	12	08	20	04	12	64
Total	96	180	108	220	80	184	868

MOC = Mustard Oil Cake, SOC = Sunflower Oil Cake, LOC = Linseed Oil Cake, SC = Soyabeen Cake, RB = Rice Bran, WB = Wheat Bran, RF = Rice Flour, WF= Wheat Flour, CF= Corn Flour.

Incidence of *A. ochraceous* on these substrates in different climatic conditions has been presented (Table 45.3). Significantly, it was observed that the present substrates, almost invariably, were contaminated with more than one fungal strain. In particular *A. flavus* and *A. niger*, the two common genera of the group Aspergillus, were found as most common invaders along with *A. ochraceous*. A seasonal record of the two strains competing with *A. ochraceous* has also been shown (Table 45.4).

Table 45.2: Mold Contamination of Fish Feeds Collected from Different Sources in Darbhanga

Items	*Total No. of sample*	*Natural*	*Contamination*	*Total*	*Source*

	Physical condition (N/SM/HM)	infestation % (A)	appearing after incubation % (B)	Contamination % (A+B)	Farmer Store ~~House~~	Retail Shop
MOC	164 100/40/24	39.02	36.58	75.60	84	80
SOC	74 40/20/14	45.94	32.43	78.37	40	34
LOC	70 44/20/10	42.85	28.57	71.77	44	26
SC	72 50/16/06	30.55	22.22	52.77	22	50
RB	100 60/24/16	40.00	20.00	60.00	60	40
RF	100 70/20/10	30.00	16.00	46.00	70	30
WB	124 80/20/24	35.48	40.32	75.8	60	64
WF	100 84/08/08	16.00	40.00	56.00	60	40
CF	64 50/10/04	21.87	31.25	53.12	40	24

N = Normal, SM = Slightly Moldy, HM = Highly Moldy.

Table 45.3: *A. ochraceous* Positive Sample (%) in Different Seasons

Substrate → *Season ↓*	*MOC*	*SOC*	*SC*	*LOC*	*RB*	*RF*	*WB*	*WF*	*CF*
Pre-winter	37.5	37.5	0	11.11	8.34	16.66	25.0	37.5	12.5
Winter	25.0	12.5	6.25	18.75	5.0	15.0	10.0	8.33	8.33
Pre-summer	33.33	30.0	0	33.33	0	6.25	30.0	25.0	12.5
Summer	30.0	30.0	0	25	5.0	8.33	14.28	18.7	5.0
Pre-monsoon	30.0	25.0	0	50.0	0	12.5	31.25	25.0	25.0
Monsoon	30.0	12.5	8.33	25.0	5.0	15.0	20.83	20.83	8.33
Total	29.87	22.97	2.77	25.71	4.0	12.0	20.96	19.0	9.37

Abbreviations as in Table 45.1.

Table 45.4: Seasonal Incidence of *A. flavus, A. niger* and *A. ochraceous*

Strain	*Season*					
	Pre-winter	*Winter*	*Pre-summer*	*Summer*	*Pre-monsoon*	*Monsoon*
A. flavus	21.04%	17.77%	14.81%	15.90%	20.0%	23.91%
A. niger	10.60%	10.36%	8.41%	7,14%	10.0%	11.4%
A. ochraceous	22.91%	13.33%	17.59%	15.45%	22.5%	18.47%

Attempts were made to establish impact of the prevalent climatic factors like temperature, relative humidity and water content of the substrate (Table 45.5). Data suggest that pre-monsoon and pre-winter months were most congenial for the growth of *A. ochraceous* preferably on the substrates like oil cakes and wheat bran. However, the *A. flavus* maintains its score of being most dominant mycoflora except during pre-summer and pre-monsoon month. *A. niger* finds edge over *A. ochraceous* only during winter months when atmospheric temperature is relatively low and water content of the substrate is little below its maximum as observed during monsoon (Table 45.5).

Table 45.5: Climatic Factors and Water Content of the Substrate Facilitating Growth of Some Fungi of Aspergillus Group

Season	*Range of Temperature (°C)*	*Range of RH (%)*	*Corresponding Water Content of the Substrate (%)*	*Degree of Spoilage by A. ochraceous (%)*	*Degree of Spoilage of A. flavus (%)*	*Degree of Spoilage by A. niger (%)*
Pre-winter	20-32	60-83	12.0-13.8	22.91	21.04	10.60
Winter	4-18	68-80	12.6-14.2	13.33	17.77	10.36
Pre-summer	14-32	62-81	9.8-10.0	17.59	14.81	8.41
Summer	30-42	59-74	9.3-10.0	15.45	15.90	7.14
Pre-monsoon	27-36	66-90	10.3-11.2	22.5	20.0	10.0
Monsoon	26-40	80-96	15.6-18.0	18.47	23.91	11.4

Discussion

Mycobial spoilage of agricultural byproducts incorporated in fish/animal feeds appear to be significantly high being 57.6 per cent in comparison to the average national record of 26 per cent as mentioned by Anonymous, RRL. Hyderabad (1967) Neelkanthan (1979). The reason may primarily be attributed to the prevalent climatic condition, which is relatively hot and humid. It is the climatic factors that make the tropical and subtropical countries a congenial ground for mycobial spoilage (WHO 1979). Besides, improper sanitation, storage and handling of the food/feed stuffs also contribute significantly in providing opportunity for the mycobial straines to contaminate and propagate over the substrates (Schroeder *et al.*, 1961).

From mold contamination point of view the storage period has been suggested to be most critical (Moreau, 1979). The problem is further compounded if the prescribed protocol for storage and preservation are not followed. Earlier works suggest that lack of ventilation, moist ceiling and floor, use of packaging materials that do not effectively prevents penetration of moisture, increase the risk of mold spoilage. It has also been emphasized that use of Jute bags for storing the grains are excellent source of mold contamination (Fusy *et al.*, 1951).The agricultural commodities from sowing to storage are always at risk of mold contamination since fungal spores are air borne (Verma and Mukherjee, 2001). Besides, the insects also play vital role as a carrier for spores.

Records of the fungal strains suggest that most of them belong to the group recognised as "Storage fungi" by Christensan and Kaufman (1965), since these fungi usually contaminate the substrates during storage, However, Lillehoz *et al.*, (1976) extends ample evidence in favour of pre-harvest infestion of the substrates by these fungi specially those belonging to Aspergillus group. Thus, once a substrate has been contaminated the infesting spore awaits congenial climatie to grow over the substrates. Though, the factors operating into the microenvironment facilitating mold propagation are multiple *viz.* nutrient climatic, physical conditions of the substrate etc., but the later two plays significant role. The climate, if provides ambient temperature and favourable relative humidity, the infesting spore get opportunity to appear on the substrates. At this stage, water content of the substrate and heating of substrates during storage become supplementary or often primary factors. If the agricultural commodities have been subjected to storage without proper drying *i.e.*, with high moisture content and have been stacked in a manner that causes heating during storage the risk of mold spoilage increases manyfold even if the climatic factors are not supportive. Effect of water content of the substrates has been worked out by Mukharjee and Verma (2001).

The present substrates may be broadly categorized in three groups *viz.* Oil Cakes (MOC, SOC, LOC, and SC), bran (RB, and WB) and flour (RF, WF, CF). Of these the oil cakes appear to be most favoured substrate for the toxic fungal strains. Since, altogether 69.66 per cent samples of the oil cakes were contaminated. Though per cent contamination on Soya cake (SC) was relatively low (Table 45.2). The oil cakes in general are left out byproduct after extraction of oil. The fresh samples usually contain 7-8 per cent of water that may increase upto 18 per cent during storage if the moisture is made available by any means. The contributing factors are high RH and moisture available from other sources like storage in direct contact of soil etc. Thus, high water content of the substrates added with hot climate accelerates mold spoilage of oil cakes. Though appearance of mold on these substrates are little delayed, probably due to leftout amount of oil that delays the metabolic interaction (Verma and Mukherjee, 2001). It has been further suggested that the left out amount of oil also accounts for enhanced carcinogenecity of the secondary toxic metabolites like aflatoxins. Synergistic effect of cyclopropenoid fatty acids on carcinogenic potentials of alfatoxins have been reported (Halver, 1969).

The samples of brans (RB and WB) harboured a range of toxigenic strains. Though the degree of spoilage was little below oil cakes being 60.5 per cent. In case of wheat bran, like oil cakes, left out amount of wheat serves as a good carbohydrate source facilitating growth of toxic fungi. Besides, the bran in general, sustain water for longer period due to biochemical affinity with seed coat components. During low moisture condition the xerophytic strains of Aspergillus group finds dominance. The present substrates also harboured many of the soil fungi like *Rhizopus* and *curvularia*. The reason may again be attributed to improper storage in direct contact of soil.

Unlike the other commodities under investigation the flours are mainly marketed for human consumption. The reason of mold invasion on these substrates may specifically be attributed to the rich bio-chemical profile of the wheat and corn. Water content of the parent products (Wheat and corn) are relatively low and their processing involves thorough sun drying. However, it has been suggested that the dried products gain moisture more readily than the others. In retail shops these products are often kept open for sale. This makes them prone to high atmospheric RH.

Table 45.5 summarises some of the critical findings of the present study. Earlier works suggest that the fungal species *viz. A. flavus, A. Parasiticus* and *A. niger* congenially grow at 87 per cent RH (Diener and Davis, 1970) while *A. ochraceous*, is somewhat xerophytic in nature and grow well at 80-82 per cent RH with 6 per cent water content of the substrate (Christenson, 1966). The maximum and minimum temperature for sporulation and growth of *A. ochraceous* has been reported to be 10 and 40°C (Sansing *et al.*, 1973). Similarly, *A. flavus* and *A. niger* require humidities in excess of 85 per cent RH corresponding to the substrate water content of 10-12 per cent. It has been suggested that 32°C climatic temperature is optimum for *A. flavus* and 35°C for *A. niger*. However, the later may grow conveniently even at low RH of 77 per cent (Ayrest, 1969). Jackson (1965) suggested that *A. flavus* is capable of growth over the temperature range of 6-54°C. While the optimum growth occurs at 30-35°C.

Thus, the present findings of seasonal dominance of *A. ochraceous, A. flavus* and *A. niger* more or less are in agreement with the earlier reports. (Table 45.5). Dominance of *A. ochraceous* during pre-winter, pre-summer and pre-monsoon months clearly Indicate that the climatic factors during these months do not favour the most dominant species, the *A. flavus* and *A. ochraceous* finds opportune moment to dominate over the substrates. Almost comparable percentage occurrence of these two strains (*A. flavus* and *A. ochraceous*) during summer (Table 45.4) indicate that infested substrate must be gaining moisture from any other source but has not attained saturation otherwise *A. ochraceous* would have gone out of the scene. Because except atmospheric temperature neither water content of the substrate nor RH exhibit congenial range for the growth of *A. ochraceous*. The findings suggest that these agricultural commodities if prior to storage are properly dried to bring the water content level at its minimum and the factors that increase the water content of the substrate are carefully prevented, the risk of mold spoilage may be effectively avoided.

Acknowledgments

The first author is grateful to Dr. N.K. Dubey Prof. and Head, University Department of Zoology, L.N. Mithila University, Darbhanga (Bihar) for providing the necessary facilities.

References

Ayerst, G. (1969). The effects of moisture and temperature on growth and spore germination in some fungi. *J. Stored Prod. Res.*, 5: 127-141.

Anonymous (1967). Annual Report, 1965-1967, Regional Research Laboratory. Hyderabad.

Boorman, G. (1989). *Toxicology and carcinogensis studies of ochratxin* A.U.S. National Institute of Health Publication, 89-2813, Washington. p.142.

Christensen, C. M. (1966). Invasion of stored wheat by *Aspergillus ochraceus. Cereal Chem.*, 39: 100-106.

Creppy, E.E. and Dirheimer, G.I (1991). Human ochratoxicosis in France. Mycotoxins. Endemic nephropathy and urinary tract tumors. Lyon. IARC. Scientific Publication no. 115. p. 145-151.

Diener, U.L. and Davis, N.D. (1970). Environmental factors affecting the production of aflatoxin. In: proceeding of the first US-Japan conference on toxic Microorganism: Washington, ed. Heriberg, P., p. 43-47.

Dickens and Whitaker (1982). Sampling and sample preparation. In: Experimental carcinogenesis. Selected Methods of Analysis. eds. Egan, H, *IACE, Monograph*, 5: 17-32.

Fusy, P. and Nicot, J. (1951). *Quclquer examples degradation paries champignons*. La Mycotheque. cataloque collection museum, Paris, ler suppl., pp. 53-59

Halver, J. E. (1969). Aflatoxicosis and trout hepatoma. In: *Aflatoxins and Implications*. Goldblatt. L.A. (ed.), Academic Press. New York, pp. 265-304.

Jackson, C.R. (1965). Peanut Pod mycoflora and Kernel infection. *Plant and Soil,* 23: 203-212.

Krogh, P. (1987). *Mycotoxins in Food*. Academic Press, London, p. 97.

Kuiper, T., Goodman, R.M. Scott (1989). Biomed. Environ. *Science.* p. 179.

Lillehoj, E.B. Knolek, W.F., Peterson, R.E., Shotwell, O.L., and Hesseltine. C.W. (1976). Alfatoxin contamination, fluorescence and insect damage in corn infected with *Aspergillus flavus* before harvest. *Cereal Chem.,* 53: 505-512.

Maaroufi, K. *et al.* (1995). Ochratoxin A in human serum in relation to nephropathy in Tunisia. *Human and Exp., Toxical.,* 14: 609-615.

Marquardt, R.R. and Frolich, A.A. (1992). A review of recent advances in understanding ochratoxins. *J. Animal Science,* 70: 3968.

Moreau, C. (1979). *Moulds, Toxins and Food*. John Wiley and Sons, Chichester, New York, Brisbane, Toronto.

Neelkanthan, S. (1979). *Final report project no. HCS/DST/17/76.* Department of Science and Technology. Tamil Nadu Agriculture University, Coimbatore.

Pohland. A.E., Nesheim's, Friedman, L. (1992). *Pure Appl. Chem.,* 64: 1029.

Raper, K.B. and Fennell, D.I. (1965). The genus *Aspergillus.* Williams and Wikins. Baltimore Maryland.

Rucker, R.R., Yasutake, W.T. and Wolf, H. (1961). Trout hepatoma: A Preliminary report. *Prog. Fish Cult.,* 23: 3-7.

Sansing, G.A., Davis, N.D. and Diener, U.L. (1973). Effect of time and temperature on ochratoxin A production by *Aspergillus ochraceous. Can. J. Microbiol.,* 19: 1259-1263.

Schroeder, H.W. and Sorensen, J.W. (1961). Mould development in rough rice as affected by aeration during storage. *Rice J.T.* 65: 8-23.

Schindler, A.F., et Nesheims. (1970). Effect of moisture and incubation time on ochratoxin a production by an isolate of *Aspergillus ochraceus. J. Ass. of F. Anal, Chem.,* 53: 89-91.

Stormer, F.C. (1992). *Ochratoxin A–a mycotoxin of concern. Handbook of applied mycology* Vol. 5, mycotoxins in ecological system Marcel Dekker, N.Y. pp. 403-432.

Stoev, S.D. (1999). Historical Data, spreading, A etiology and Pidemiology of Mycotoxic Nephropathy (Ochratoxicosis) in Pigs (revieo). *Bulguarian J. of Agri. Sc.,* 50: 515-524.

Van der Merwe, K. J., Steyn, P.S., fouire L., Scott, de B. and Thersen, J.J. (1965a). Ochratoxin A., A toxic metabolite produced by *A. Ochraceous nature,* 205: 1112.

Van der Merwe, K.J., Steyn, P.S. and faurie, L. (1965b). Mycotoxins II, The constitution of ochratoxin A, B, and C metabolites at *A. ochraceous. J. Chem. Soc.,* 7083.

Verma, S.K. and Mukherjee, S.C. (2001). Mycotoxin: Histopathological Manifestations in fishes and possibilities of Human uptake. Annual report ICAR Project, L. N. Mithila University Darbhaga, Bihar, India.

Verma, S.K. (1989). Ph.D. Thesis L.N. Mithila University, Darbhanga, Bihar.

WHO (1979). Collective views of an international group of experts in Environmental health Criteria II, "Mycotoxins".

Wyllie, T.D. and Morehous, L.G. (1977). *Mycotoxic fungi, Mycotoxins, Mycotoxicosis,* Vol. 1, Marcel Dekker Inc. New York.

46

FT-IR Study of the Effect of Nickel on the Tissue Proteins of an Edible Fish *Cirrhinius mrigala*

PL. R.M. Palaniappan, Mrs. Selvisabhanayakam*, S. Karthikeyan, N. Krishnakumar and P. Venkatachalam

Department of Physics, Annamalai University, Annamalainagar - 608 002
**Department of Zoology, Annamalai University, Annamalainagar - 608 002*

ABSTRACT

Fourier transform infrared spectra of tissues isolated from the fingerlings of an edible fish *Cirrhinius mrigala* treated with two different concentrations of Nickel have been investigated in the region 2000-1000 cm^{-1}. The characteristic spectral bands were compared with those of control group to confirm the effect of the heavy metal, Nickel on tissues. FTIR spectra revealed significant differences in band position and absorbance intensities between control and treated tissues. A decrease in protein and lipid level is observed as compared to control and there is a gain in these level during the period of recovery.

Key Words: FTIR Specta, Protein, Nickel, Fish Tissue, *Cirrhinus mrigala*.

Introduction

Nickel is an essential metal for animal nutrition and to man (WHO, 1984). Nickel is extensively used in electroplating in the manufacture of steel and other alloys, and in the manufacture of electronic devices (Venugopal and Luckey, 1978) and is well recognized carcinogen in man and animals (IARC, 1990). Nickel concentration is found to be greater in freshwater than in sea water, particularly in the vicinity of industrialised area (Nriagu, 1980). It leads to harmful effects to the survival and productivity of the freshwater fauna, especially fish and shellfish, thereby disturbing the normal ecosystem and its food chain (Moore and Ramamoorthy, 1984). Fish, a common source of protein contains greater quantity of protein content than any other living organisms, contributing roughly about 75 per cent by weight of fish. Protein content was experimentally verified for the muscle sample of catfish, *Claria batrachus* (Claude, 1996) and was found to be 70 per cent. Protein plays a major part in all biological system with a wide variety of structural and functional roles.

Infrared spectroscopy is a powerful method for the study of molecular structure and intermolecular interaction in biological tissues and cells (Patrick *et al.*, 1993). Several authors have studied infrared spectroscopy on biological substance like muscle, liver etc. Luis *et al.* (2000) has studied infrared spectra of normal and cancer liver tissues such as glycogen, DNA and RNA. Galina *et al.* (2000) have studied the changes in primary, secondary and tertiary structure of nucleic acid of RNA in rats exposed to gamma radiations and Patric *et al.* (1993) have studied on human colon tissues at molecular level from normal epithelium to malignant tumour.

FT-IR spectroscopy is one of the techniques that can be applied to identify micro-organisms and to obtain information on different aspects of protein structure in a wide range of environments because of the fingerprinting capabilities of the techniques. The ability to characterise structure of protein in a wide range of environments makes FT-IR spectroscopy an important analytical tools (Parvez I Haris and Feride Severcan, 1999). Methods for quantitative estimation of secondary structure using this techniques are progressively improving. The present work is aimed to get insights into the composition of tissue proteins from Nickel treated fish as compared with control fish using FT-IR spectroscopy.

Materials and Methods

Cirrhinius mrigala fingerlings of length (6 ± 1 cm and weight 8 ± 1 g) were collected from the freshwater bodies near the local fish farm, Puthur, Tamil Nadu and acclimatized under Laboratory conditions (29 ± 1°C). Boiled eggs, rice bran and earthworm pieces were fed on every alternate days. The LC_{50} for Nickel chloride for 96 h was found out by using probit method (Finney, 1971). The 1/10th and 1/3rd of LC_{50} were taken as lower (1.083 ppm) and higher (3.61 ppm) concentrations respectively. The experimental animals were exposed to these two sublethal concentrations for a period of 28 days. Simultaneously, control was run in a pure water without any treatment. After the period of exposure one group of fish was allowed to remain in a pure water for 28 days taken as a recovery period for the elimination of Nickel. At the end of the periods fishes were sacrificed and muscle tissues were taken out and kept in a freezer-drier to remove water from the tissues. The dried sample was then powdered using agate mortar and made pellet using KBr technique (Wang xiulin, 1992). Spectra were recorded in the range 2000-1000 cm^{-1} using Nicolet Avatar-360 FT-IR spectrometer equipped with KBr beamsplitter and a DTGS detector, at Centralised Instrumentation and Services Laboratory (CISL), Annamalai University.

Results and Discussion

FT-IR spectra of control, lower and higher Nickel treated samples are presented in Figs. 46.1-46.3 shows spectra for Nickel treated samples along with recovery for lower and higher concentrations respectively. Considerable changes are observed in the intensities of the bands. The tentative vibrational frequency assignment of absorption spectra are presented in Table 46.1. The modes most widely used in protein structural studies are amide I, amide II and amide III. The amide I band arises principally from the C = O stretching vibrations of the peptide group. The amide II band primarily N-H bending with a contribution from C-N stretching vibrations are found around 1540 cm^{-1}. The amide III arises primarily from C-H/N-H deformation and is normally very weak and occurs around 1250 cm^{-1}. The overall proteins found inside the tissues are dominated by the C = O (amide I) band at ~ 1650 cm^{-1}. In the present study C = O (amide I) band has been observed at 1657 cm^{-1}, 1659 cm^{-1} and 1657 cm^{-1} for the control, lower and higher concentrations of nickel treated medium, respectively. This amide I band are primarily associated with the stretching motion of the C = O group. This C = O band is sensitive to the environments of the peptide linkage and also depends on the protein's overall secondary structure (Max Diem *et al.*, 1999). The absorption intensities of amide I are found to be 0.8 for control, 0.42 for lower Nickel treated and 0.3 for higher Nickel treated samples. For amide II band, the absorption intensities are 0.61 for control, 0.32 and 0.24 for lower and higher concentrations, respectively. In the present work the amide II bands have been observed at 1545 cm^{-1} for control, 1546 cm^{-1} and 1545 cm^{-1} for lower and higher concentrations of Nickel treated samples. The amide III band has been observed at 1240 cm^{-1}, 1238 cm^{-1} and 1238 cm^{-1} for control and Nickel treated samples, respectively.

Table 46.1: The Frequency Assignment of FT-IR Absorption Spectra for Control and Nickel Treated Tissues of *Cirrhinus mrigala*

Sl. No.	*Control*	*Nickel Treated*		*Recovery*		*Frequency Assignment*
		Lower Concentration	*Higher Concentration*	*Lower Concentration*	*Higher Concentration*	
1.	1657 (vs)	1659 (vs)	1657 (m)	1653 (vs)	1653 (m)	C = O Stretching Amide I
2.	1545 (vs)	1546 (m)	1545 (w)	1542 (vs)	1541 (m)	C-N Stretching N-H bending, Amide II
3.	1240 (w)	1238 (w)	1238 (vw)	1240 (w)	1238 (vw)	CH/NH Deformation Amide III
4.	1446 (s)	1445 (w)	1444 (vw)	1450 (s)	1448 (vw)	CH_3 Scissoring Lipids

vs = Very Strong, s = Strong, m = Medium, w = Weak, vw = Very Weak.

The presence of amide I (~ 1650 cm^{-1}) and amide II (~ 1550 cm^{-1}) band are very sensitive to the conformational substructure in tissue proteins (Patric *et al.*, 1993) and weaker protein vibrations include the amide III at ~ 1240 cm^{-1} (Max Diem *et al.*, 1999). It is observed from the present study that intensity of absorption of protein (due to amide I, amide II and amide III) decreases by about 45.25 per cent and 53.31 per cent for lower and higher Nickel treatment, respectively. Interestingly, it has also been observed that there is a regain in absorption intensity of about 52.18 per cent and 5.28 per cent for lower and higher treatment of Nickel for the fingerling of *Cirrhinius mrigala*.

The band observed at ~ 1450 cm^{-1} is identified as CH_3 scissoring and is mainly due to lipids. In the present study, this bands are observed at 1444-1450 cm^{-1} in all the samples (Table 46.1). The absorption intensity of lipid bands decreases by about 34.37 per cent and 43.75 per cent for lower and higher treatment of Nickel. There is also gain in the intensity of absorption of lipids of 43.24 per cent for lower treatment and 11.11 per cent for higher treatment respectively. Experimental studies on proteins of known structure shows that α-helical conformation gives rise to infrared absorption in the range of 1650-1658 cm^{-1} (Parvez I. Haris and Feride Severcan, 1999). The amide I maximum observed at 1652-1659 in the present work could therefore be assigned to α-helical structure.

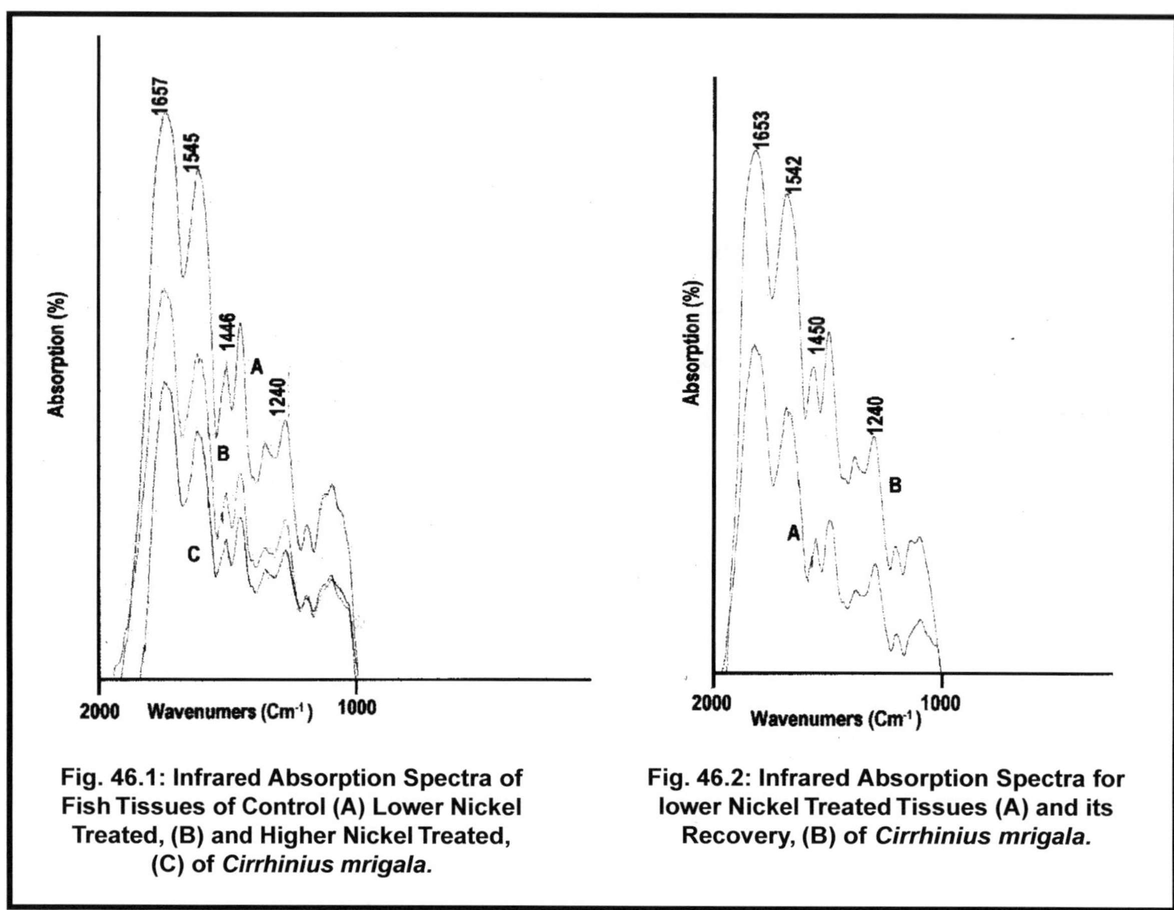

Fig. 46.1: Infrared Absorption Spectra of Fish Tissues of Control (A) Lower Nickel Treated, (B) and Higher Nickel Treated, (C) of *Cirrhinius mrigala*.

Fig. 46.2: Infrared Absorption Spectra for lower Nickel Treated Tissues (A) and its Recovery, (B) of *Cirrhinius mrigala*.

In the light of the present observation in infrared that there is a decrease in the overall proteins and lipids due to toxicity of Nickel on fish muscle. The depletion of the protein may be due to the effect of Nickel that affect the protein content of tissue by inhibition of RNA synthesis and protein synthesis due to anerobiosis which depletes ATP molecules or increase the rate 1 of degradation of protein in the cells (Rao, 1989). The decrease in lipid content of muscle may be due to the conversion of fatty acid composition of total lipids to form the total amino acids due to toxic conditions (Villalan *et al.*, 1988). Also there has been a gain in protein and lipids during the period of recovery. The substantial gain can be attributed to the elimination of Nickel from muscle tissues through excretion or depuration of metal which results in recovery of proteins and lipids. Hence, it could be concluded that the treatment of nickel results in changes in protein and lipid contents and these may be easily detected using the FT-IR technique. Also, FT-IR spectroscopy can be used to differentiate nickel treated tissues from healthy ones at the molecular levels.

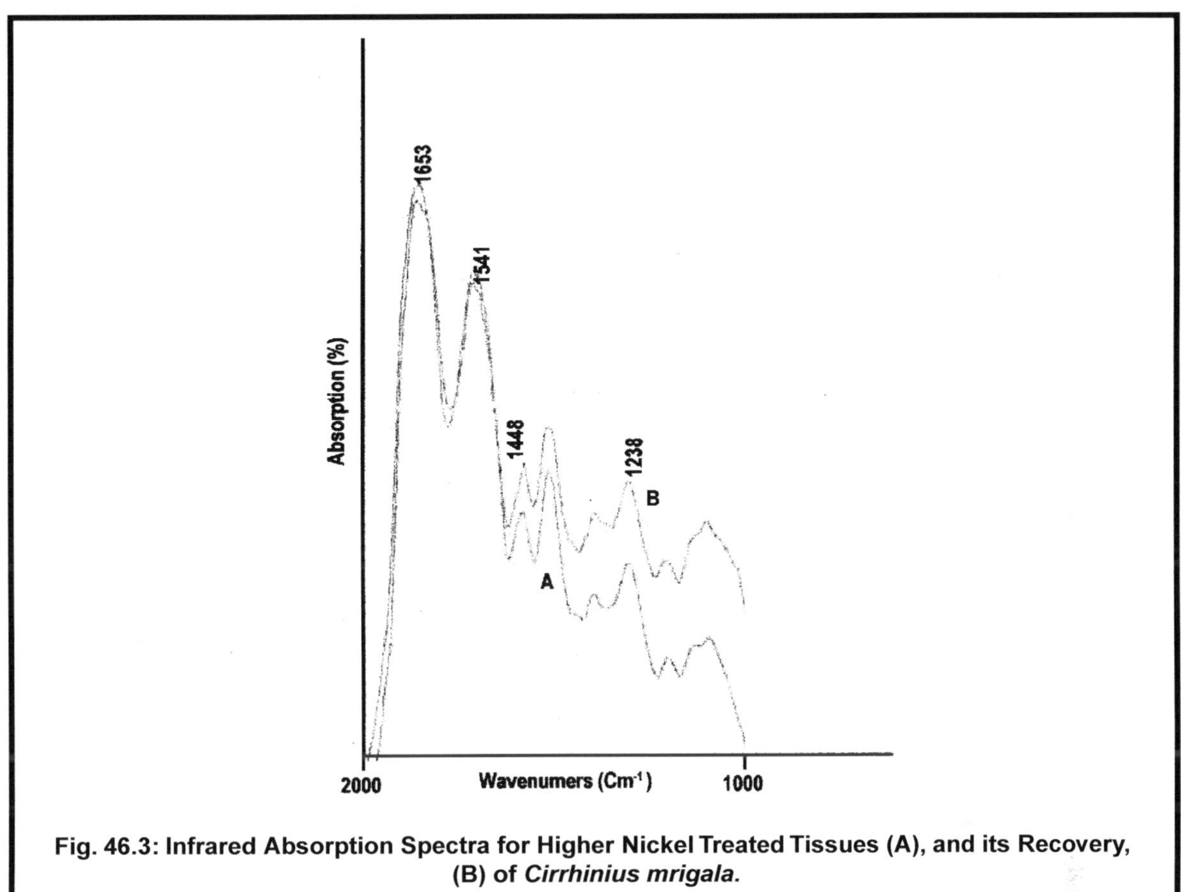

Fig. 46.3: Infrared Absorption Spectra for Higher Nickel Treated Tissues (A), and its Recovery, (B) of *Cirrhinius mrigala*.

Acknowledgement

The author wishes to express their sincere thanks to Dr. A.N. Kannappan, Professor and Head, Department of Physics, for his constant encouragement to carry out this work successfully.

References

Claude, A. (1996). Transport of charge carriers in fish proteins, M.Phil Thesis, Loyola College, Madras.

Feride Severcan, Neslihand Toyran, Nese Kaptan and Belma Turan (2000). Fourier Transform infrared study of the effect of diabetes on rat liver and heart tissues in the C-H region, *Talanta*, 53: 55-59.

Finney, D.J. (1971). Probit Analysis, 3rd Ed, Cambridge University Press, Cambridge, pp. 333.

Galina, I. Doubeshko, Nina Ya. Gridina, Elena B. Kruglova and Olena P. Pashchuk. (2000). FTIR spectoscopy studies of nucleic acids damage, *Talanta*, 53: 233-246.

IARC (1990). Monographs on the Evaluation of carcinogenic Risks to Human, Vol. 49, Chromium, Nickel and Welding, International Agency for Research on Cancer, Lyon, pp. 257- 455.

Luis Chiriboga, Herman Yee and Max Diem (2000). Infrared spectroscopy of Human Cells and Tissue: Part VI: A Comparative Study of Histopathology and Infrared Micro Spectroscopy of Normal, Cirrhotic and Cancerous Liver Tissue, *Applied Spectroscopy*, 54: pp. 1 -8.

Max Diem, Susie Boydston White and Lius Chinboga (1999). "Infrared spectroscopy of cells and Tissues: shining light onto a Novel subjects", *Applied Spectroscopy*, 53: pp. 148 -161.

Moore, J.W. and Ramamoorthy, S. (1984). Nickel in heavy metals in industrial waters. In: *Applied Monitoring and Impact Assessment*. R.S. Desano (ed), Springer Verlag, New York, pp. 161-173.

Nriagu, J.O. (1980). Global Cycle and Properties of Nickel. In: *Nickel in the Environment*, Wiley, New York, pp. 1-26.

Parvez, I. Haris and Feride Severcan (1999). FTIR spectroscopic characterization of protein structure in aqueous and non-aqueous media. *Journal of molecular catalysis B: Enzymatic*, 7: 207-221.

Patrick, T.T. Wong, Suzanne Lacelle and Hossein M. Yazdi (1993). Normal and Malignant Human colonic Tissues investigation by pressure tunning FT-IR spectroscopy, *Applied spectroscopy*, 47: 1830 -1836.

Rao, K.S.P. (1989). Studies on some aspects of metabolic changes with emphasis on carbohydrate utilization in cell free systems of the freshwater teleost, *T. Mossoambica* subjected to methyl parathion exposure, Ph.D. thesis, S.V. University, India.

Venugopal, D. and Luckey, T.D. (1978). *Metal Toxicity in Mammals*: 2 (chemical Toxicity of metals and metalloids). Plenum Press, New York, pp. 289-297.

Villalan, P., K.R. Narayanan, S. Ajmalkhan and R. Natarajan (1988). Proximate composition of muscle, hepatopancreas and gill in the copper exposed estuarine crab *Thalamita crenata* (*Latr eille*), Proceeding of the second national symposium on ecotoxicology, Department of Zoology, Annamalai University, India.

Wang Xiulin (1992). Fourier transform infrared spectroscopic studies on the interaction between Copper, Amino acids and Marine solid particles, *Analyst*, 117: 165-171.

WHO (1984). *Guidelines for Drinking Water Quality*, Vol. 2 Health Criteria and other supporting information, World Health Organization, Geneva.

47

Metal Concentrations in Water Bodies of Schirmacher Oasis Antarctica: An Assessment

A.N. Sahi*, R.K. Gupta and S. Kaushik

Department of Botany, Government P.G. College, Rishikesh, Dehradun - 249 201, Uttaranchal
**Amity Institute of Biotechnology, Sector 125, Noida - 201 303, Uttar Pradesh*

Introduction

Antarctic Treaty Consultative Meeting (ATCM) held in Gorizia, Italy in 1993 emphasized the necessity of long term monitoring programmes to verify the impacts of human activities such as tourism and scientific research on the physico-chemical attributes of antarctic ecosystems. The SCAR/COMNAP document on the environmental monitoring (1992) also referred close monitoring of the chemical compounds (*e.g.* Heavy metals and/or selected anions in waste water, soil, snow etc.) alongwith the biological indices of the antarctic ecosystems (SCAR report, No. 12, October, 1996). Monitoring of the concentrations of metal ions (including heavy metals) is an important prerequisite for the assessment of the aesthetic value of the lake water of Antarctica. The toxic impacts of heavy metals on plants and human beings are well established therefore, a sufficient bulk of baseline data regarding the concentrations of heavy metals in lake water as well as time bound variations in the concentrations of these metals is essential, for the timely implementation of remediation processes.

Present communication describes the results obtained by the analysis of the samples collected during the austral summer (1997-98) of the 17th Indian Antarctic Expedition. Concentrations of twelve metals were determined in the water samples collected from different sites around the permanent Indian station 'Maitri' situated in Schirmacher Oasis.

Materials and Methods

The objective of the water quality monitoring of the lakes around 'Maitri' (70°, 45′ 52" S: 11°, 44′ 03" E) in the Schirmacher Oasis, Antarctica was to (i) determine the quality of water in its natural state which might be available for the future needs and to assess, (ii) the impact of activities by human and scientific operations upon the quality of water in its suitability for required uses.

Study Sites

Water samples were collected from 40 different locations (Fig. 47.1) at a particular time and represents only the composition of the source at that time.

Control

This comprises of lakes where no human activity was found in the vicinity *e.g.* DG lake and Epsilon or Long lake in the Central Schirmacher, west of 'Maitri' and CL1, CL2 and CL3 lakes as shown in Fig. 47.1. The observations of these lakes have been pooled to represent the undisturbed control sites (Table 47.1).

Table 47.1: Metal Concentrations (µg/ml) in Water Bodies Around Permanent Indian Station "Maitri"

Metal	*Control*	*Priyadarshini*	*WDP*	*BAG*	*Pipeline Sites*	*LSD*
Cadmium	0.0	0.047 ± 0.03	0.050 ± 0.003	0.020 ± 0.005	0.029 ± 0.018	0.046
Copper	0.0	0.055 ± 0.035	0.073 ± 0.024	0.313 ± 0.156	0.130 ± 0.045	0.297
Chromium	0.017 ± 0.002	0.078 ± 0.032	0.010 ± 0.004	0.020 ± 0.005	0.005 ± 0.002	0.427
Nickel	0.008 ± 0.005	0.200 ± 0.009	0.029 ± 0.013	0.015 ± 0.007	0.022 ± 0.011	0.028
Lead	0.016 ± 0.002	0.055 ± 0.009	0.042 ± 0.006	0.048 ± 0.015	0.030 ± 0.015	0.034
Calcium	0.040 ± 0.022	2.483 ± 0.688	3.380 ± 0.675	4.980 ± 0.529	4.271 ± 1.036	1.932
Iron	1.036 ± 0.792	0.382 ± 0.031	0.657 ± 0.163	2.902 ± 0.226	0.690 ± 0.189	1.23
Potassium	1.250 ± 0.250	1.250 ± 0.250	2.830 ± 0.210	7.833 ± 0.0641	3.333 ± l.062	1.711
Magnesium	0.320 ± 0.020	0.885 ± 0.217	1.202 ± 0.184	3.073 ± 0.079	1.520 ± 0.367	0.614
Manganese	0.012 ± 0.003	0.050 ± 0.011	0.050 ± 0.04	0.147 ± 0.07	0.175 ± 0.082	0.109
Sodium	3.500 ± 0.000	10.167 ± 2.076	16.333 ± 2.905	36.166 ± 0.601	16.833 ± 4.233	7.239
Zinc	0.075 ± 0.054	0.051 ± 0.005	0.064 ± 0.078	0.056 ± 0.02	0.056 ± 0.073	0.039

Table 47.2: Consequences of Metal Concentrations Above the Permissible Limits in Drinking Water

Metal	*Permissible Limit in Drinking Water*	*Consequences of Above Permissible Limit Concentrations*
Cadmium	0.01	Renal dysfunction, Respiratory diseases
Copper	0.05	Eye inflammation, hepatic and renal damage
Chromium	0.05	Liver necrosis, nephritis, irritation, gastrointestinal mucosa
Nickel	0.05	Embryo toxic and nephrotoxic effects, contact dermatitis
Lead	0.10	Reduced hemoglobin
Iron	0.03	Peripheral nerve dysfunction
Manganese	0.10	Decrease in systolic BP, disturbed excretion of 17 ketosteroid

Priyadarshini

Sampling from M1 to M6 and M11 to MI9 has been presented in the Fig. 47.1.

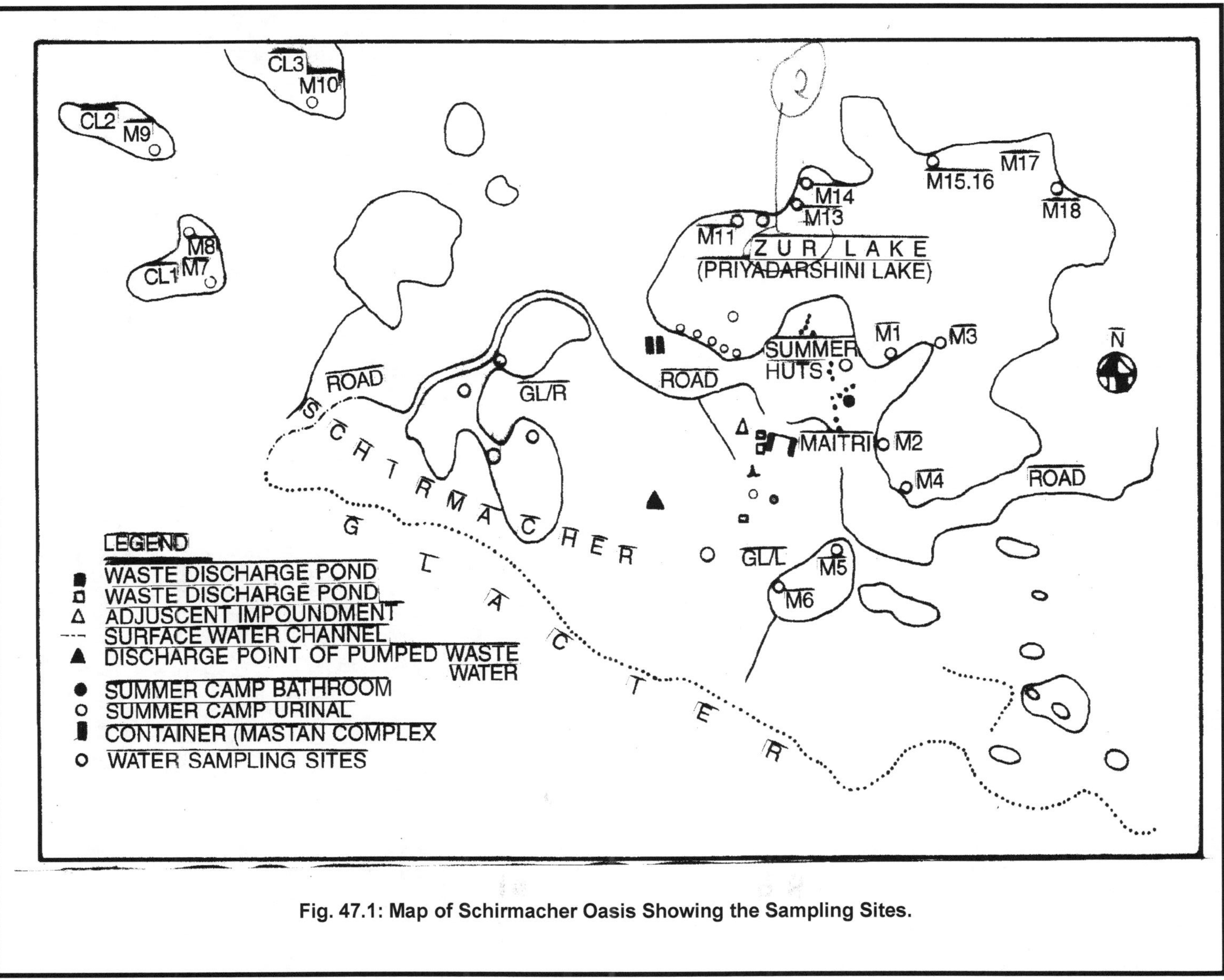

Fig. 47.1: Map of Schirmacher Oasis Showing the Sampling Sites.

Waste Disposal Pond (WDP)

The waste water is collected from the kitchen, urinal and wash room in the waste discharge pond. Periodically it is pumped out and discharged few hundred meters away as shown in Fig. 47.1. From the discharge point of the pumped waste water, the water trickles down to the waste disposal pond (GL/L). Water renewal is a constant feature in this pond due to the snow melts of the Glacier in the South. This causes rapid dilution of the metals in the summers (see sampling points WDP1 to WDP4 and Li1 to Li2 in Fig. 47.1).

Behind Aditya Generator Complex (BAG)

Sampling was done in the small stagnant pool of water behind the generator complex expressed as BAG1 to BAG4 and GL/R (Fig. 47.1 and Table 47.1).

Pipeline Sites

This comprises of water samples just beside the drinking water pipeline in the Priyadarshini-lake (P2 to P6) upto the container complex and P1 from the pump house (Fig. 47.1 and Table 47.1).

Sampling Procedure

Samples were collected in 1 litre bottles for the study or Acid-Extractable Metals. At the time of collection the entire sample was acidified with 5 ml conc HNO_3/L sample. Before analysis, the sample was well mixed and transferred to 100 ml flask and added 5 ml 1 + 1 high purity HCI. The same was heated for 15 min on a steam bath and filtered through membrane filter and the filtrate volume was adjusted to 100 ml for further analysis. Metal analysis was carried out by Atomic Absorption Spectrophotometer were as the analysis of Sodium and Potassium were estimated by flame photometer (Greenberg *et al.*, 1992).

Results and Discussion

Cadmium is regarded as one of the most toxic elements in the environment. Its presistence in the environment, rapid uptake and accumulation in the food-chain contributes to its potential hazards. Recommended level of cadmium in drinking water is 0.01 µg/ml (US and Indian Standards) and 0.005 µg/ml according to WHO guideline values. Maximum concentration of 0.05 µg/ml Cd was recorded in waste disposal pond which was slightly lowered in Priyadarshini (0.047 µg/ml). Taking into account the concentration of Cd adjacent to the pipeline sites (0.03 µg/ml), it is assumed that per day intake of 2 liters of water and only 6 per cent absorption in the gut, the daily intakes of cadmium is 3.6 µg which is very high. (WHO 1984, Murti CRK and Vishwanathan, P., 1991). Most of the Cd absorbed is deposited in the liver, binding to a low molecular weight protein-metallothionin which has a high binding capacity for Cd, Zn and Cu (Table. 47.2). Later on Cd is transferred from the liver to the kidney and eventually accumulates in the kidney cortex. In long term, low level exposure, the kidney is regarded as the critical organ and it has been estimated that renal dysfunction may appear when the Cd concentration in the renal cortex is around 200 mg/kg wet weight (Friberg *et al.*, 1974 and Roels *et al.*, 1983). Toxicity of Cd to plants and microorganisms has been assumed in the presence of divalent cations (eg. Zn, Mg, Mn, Ca and Fe). Cd has been shown to inhibit enzyme activity, oxidative phosphorylation and photosynthetic rates, to alter cell membrane permeability and integrity, to interfere with RNA and protein synthesis and to complex with DNA. By this multipronged influence on the cell, it is not surprising that Cd may affect several aspects of microbial growth (Babich and Stotzky, 1978).

Chromium is ubiquitous in nature most commonly found in the trivalent state, but hexavalent compounds are also in small quantities (Langard and Norseth, 1978). Chromium is an essential micronutrient in some mammalian diet including humans (Mertz, 1969, Schroeder *et al.*, 1962) but not essential to plants. After uptakes Cr accumulates in the roots and only a small amount reaches the leaves (Wallace *et al.*, 1976) which produced Chlorosis in tomato and potato (Hewitt, 1953) and a number of deleterious effects *viz.* Chromosome aberrations and mutations in *Allium cepa* (Sahi *et al.*, 1998). Similar results are expected in the mosses growing in close contact of the water having high chromium content *e.g.* Priyadarshini lake (0.078 μg/ml) where chromium content was considerably high above the prescribed limit of 0.05 μg/ml by the WHO, US and Indian standards.

Copper is frequently found in surface water and is an essential micronutrient for plants and animals. Copper deficiency as an outcome of the disturbance of trace metal metabolism in the human body may occur *e.g.* Genu Valgum syndrome, a new manifestation of classical flurosis. Copper concentration in the water samples around 'Maitri' was found to be appreciably higher than the prescribed limit of 0.05 μg/ml in drinking water except the undisturbed sites designated as control.

There is evidence that Manganese occurs in surface water, both in suspension in the quadrivalent state and in the trivalent state in a relatively stable, soluble complex. It is rarely present beyond 1 mg/L. But at few sites *e.g.*, BAG and Pipeline sites the level of Mn was appreciably higher than the prescribed limit of 0.01 mg/L. Similarly the concentration of Iron also exceeded the limit of 0.03 mg/L (WHO, 1971) in drinking water at almost every site and ranged between 0.38 μg/ml to 1.03 μg/ml in Priyadarshini Lake and control, respectively.

The present study will serve as the baseline data for future environment impact assessment programme and in predicting a better environment management strategy.

Acknowledgment

The authors wish to thank the Department of Ocean Development, New Delhi for providing the opportunity to work in Antarctica, One of the author (ANS) is thankful to Mr. K.R. Sivan (Leader, XVII IAE) and Secretary, D.O.D., New Delhi, for the facilities extended by them during the expedition and thanks are also due to the Head, Department of Botany, B.H.U., Varanasi for laboratory facilities. ANS and SPS are thankful to CSIR, New Delhi for financial assistance in the form of Post-doctoral fellowship.

References

Babich, H. and Stotzky, G. (1978). Effect of cadmium on the biota: Influence of environmental factors. *Adv. Appl. Microbiol.*, 23: 55-117.

Fostel, P.L. (1982). Species association and metal contents of algae from rivers polluted by heavy metals. *Fresh Water Biol.*, 12: 17-39.

Greenberg, A.E., Clesceri, L.C. and Eaton, A.D. (1992). Standard Methods for the Examination of Water and Waste Water 18th Edition. American Public Health Association, Washington D.C.

Hewitt, E.J. (1953). Metal interrelationship in plant nutrition: Effect of some metal toxicities on sugar bash, tomato, oat, potato and marrowstem vale grown in sand culture. *J. Exp. Bot.*, 4: 59-64.

Langard, S. and Norseth, T. (1978). *Handbook on the Toxicology of Metals*. Ed. L. Friberg. Elsevier-North Holland, Amsterdam.

Mertz, W. (1969). Chromium occurrence and function in biological systems. *Physiol. Rev.*, 49: 163-239.

Murti, C.R.K. and Vishwanathan, P. (1991). *Toxic Metals in Indian Environment*. Tata McGraw-Hill Publishing Company Ltd., New Delhi.

Sahi, A.N., Singh, S.K., Sen, P.K. and Singh, R.N. (1998). Cytogenetic response of hexavalent chromium-induced somatic cell abnormalities in *Allium cepa. Cytobios* Vol. 96: 71-79.

Schrocder, H.A., Balassa, J.J. and Tupton, I.H. (1962). Abnormal trace metals in man. *Chromium. J. Chron. Dis.* 15: 941-964.

Wallace, A., Soufi, S.M., Cha, J.W. and Romney, E.M. (1976). Some effects of chromium toxicity on bush bean plants grown in soil. *Plant Soil,* 44: 471-473.

W.H.O. (1971). *International Standards for Drinking Water*, 3rd Edition, WHO, Geneva.

W.H.O. (1984). *Guidelines for Drinking Water Quality*, Vol. 1 "Recommendations" 6 WHO, Geneva.

W.H.O. (1984). *Guidelines for Drinking Water Quality*, Vol. 2: Health Criterion and other Supporting Information, WHO, Geneva.

48

Toxic Effects of Chromium Sulphate on the Indian Catfish *Heteropheustes fossilis* (Bloch) in Short-term and Long-term Exposure

D.N. Roy and N.K. Dubey**

Lecturer in Zoology, MLSM College, Darbangha

***Professor & Head, University Department of Zoology, L.N. Mithila University, Darbhanga - 846 004*

ABSTRACT

The effect of invivo exposure to a sub-lethal concentration of trivalent chromium (9.945 mg/g/L chromium sulphate) for 96 hrs, 15 days (short-term exposure) 30 days and 45 days (long-term exposure) on blood glucose, plasma cholesterol, total serum and plasma potassium has been studied. Short-term exposure (0 to 96 hr and 96 hr to 15 days) of Cr^{+3} to Heteropneustes fossilis exhibited elevated levels of blood glucose (23.7 per cent and 17.01 per cent), elevated levels of plasma cholesterol (35.25 per cent and 12.61 per cent) where as 9.30 per cent rise in TSP followed by a decline of 32.14 per cent in 15 days. However, levels of plasma potassium was intact upto 0 to 96 hr of exposure which elevated 0.7 per cent in 15 days of exposure.

In long-term exposure (30 days and 45 days) the results showed a definite pattern. Blood glucose and plasma cholesterol were found elevated (glucose 83.65 per cent and 102.65 per cent, cholesterol levels were 54.59 per cent and 59.24 per cent elevated) where as total plasma protein showed depletion (23.7 per cent and 43.56 per cent). The levels of plasma potassium were also found elevated (0.8 per cent and 3.79 per cent respectively).

Elevated levels of blood glucose, plasma cholesterol and total plasma protein points towards increasing toxicity with duration of exposure. Possible reasons behind such blood biochemical alteration have been discussed.

Key words: Cr^{+3} toxicity, stress Hyperglycemia, Hyper cholesterolemia, Hypoproteinemia Hyper Kalaemia, Acidosis.

Introduction

Environmental pollution by compounds of heavy metals is increasing with extensive industrial developments. Biological interest in chromium arises from its prominent role in industrial pollution and its toxicity to microbes, plants and animals (Royle 1975, Baetjer *et al.*, 1974, Langard 1980). Large amounts of chromium are introduced into the environment through sewage water, sludge, electroplating, textiles and tannery effluents, thus made available to plants, animals and human. In the river Ganga, within the Rishikesh-Diamond Harbour Stretch, the highest concentration of metals have been observed at Kanpur. In the tannery effluents in this area, the concentrations recorded during 1987's were Cr 0.200 ppm and Cd 0.014 ppm (Samanta, 1999). Exposure to chromium at the working place is another mode of introduction to human system (Langard, 1980).

Trace amount of chromium is considered to be essential for normal metabolic process (Mertz, 1969), however intakes at higher levels have been found to be toxic mainly to kidney and liver (Berndt, 1976, Tandon *et al.*, 1978). Laborda *et al.* (1986) have reported about the hepato and nephrotoxic effects of chromium in rats.

Olson and Foster (1956) studied the effects of chromium on growth and early life history stages of Chinook salmon and Rainbow trout. The toxicity of chromium to Brook trout and Rainbow trout has been examined by Benoit (1976). Chromium has been reported to effect salinity tolerance (Sugatt, 1980). However, very little information is available on the toxic effects of chromium to Indian teleosts.

The present study was undertaken to examine the short-term and long-term toxicity of this heavy metal to Indian teleost *Heteroporeustes fossilis*, a fresh water cat fish and deals with alterations in blood glucose, plasma cholesterol total serum protein and plasma potassium.

Materials and Methods

Living specimens of *Heteropneustes fossilis* of 12 to 14 cm length and 16 to 18 g weight were procured from local fresh water sources and also from fish market. They were, maintained in laboratory glass aquaria in clean tap water for 3 to 4 days prior to experimentation, water used in the aquaria had a pH 7.6 ± 0.2 temperature 23 ± 3°C, dissolved oxygen 6.8 ± 1.2 mg/L, free CO_2 1.3 ± 1.7 mg/L, total hardness (as $CaCO_3$) 185 ± 15 mg/l and total alkalinity (as $CaCO_3$) 135 ± 2.5 mg/L. Fish were fed with chopped chicken intestine obtained from local poultry shop.

Feeding was, however, stopped 24 hrs before the start of the experiment. A static bioassay test was performed to determine the 24 hr, 48 hr, 72 and 96 hr. LC_{50} of chromium sulphate (Cr^{+3}) to *Heteropneutes fossilis* following the methods of APHA, AWWA and WPCF (16th ed). The sublethal concentration was obtained by employing the formula of Hart *et al.* (1945) using 0.3 as application factor. The test chemical used in this study was the chromium sulphate powder $Cr_2(SO_4)_3 . 4H_2O$ of Burgyne Chemical Limited, Kolkata. Test concentrations were prepared in tap water.

In all 80 fish were selected for biochemical study placed in 8 glass aquaria, each with 10 L water and 10 acclimated fish. In aquania ABCD each provided with a control group of 10 fish, were exposed to sublethal concentrated of Cr^{+3} (9.94 mg/g/L) for a period of 96 hrs, 15 days 30 days and 45 days. Exposure medium was renewed every 24 hrs to maintain the effective concentrations of the heavy metal. In long-term exposure, fish were given equal amount of feed at similar time interval.

At the end of each exposure period, fish were anaesthesized by MS 222 (Tricane Methane Sulphonate, Sandoz) for two min. The blood samples were collected in heparinized plastic vials with the help of 1 ml disposable syringe equipped with 22-gauge hypodermic needle by caudal peduncle cut. Blood obtained from individual fish was centrifuged at 3,500 rpm for 10 min, Serum was collected in different plastic vials and stored at 20°C until analysed.

Plasma glucose was determined colorimetrically by Anthrone method (Seifter *et al.*, 1950). Total plasma protein was determined colorimetrically by Biuret method (Gornall *et al.*, 1949). Plasma cholesterol was determined according to Webester (1962).

Plasma potassium was measured by Flame emission photometry (after Wotten, 1977) using flame - photometer.

Results and Discussion

The results are shown in Table 48.1.

A close perusal of pertinent literature reveals that Cr^{+3} is considered as a substance of average acute toxicity (Peres and Brichon, 1980). The most commonly described sublethal effects in fish concern the inhibition of the Na^{+}/K^{+} dependent ATP-ase activity of gills, intestine or kidney (Sastry and Sunita, 1983, Vanderputte *et al.*, 1982, Kunhert *et al.*, 1976). A decrease in the alkaline activity has been shown in dichromate intoxicated trout and bass (Cr_2O_7) (Boge *et al.*, 1981). Chromium is also a compound of biological interest of which has a role in glucose and lipid metabolism (Mertz 1965, Venugopal and Luckey, 1977).

Vincet *et al.* (1996) studied the impact of heavy metal Cr on bioenergetics of the Indian Major Carp, *Catla catla* (Ham). Bioenergetics comprises energy transformation in living organism and growth, is a broad index of various physiological functions. To understand the action of xenobiotics, qualitative and quantitative changes in metabolism must be studied at the level of organisation (Metelvet *et al.*, 1983).

Biologically, stress is a stimulus or succession of stimuli of such magnitude as to tend to disrupt the homeostasis of the organism. (Homeostasis-maintenance of internal constancy and an independence of the environment). The factors responsible for development of stress in the organisms of an ecosystem varied greatly with different magnitude of impact. Some of the stress factors may change, the biochemical reactions of different enzyme and the organism can adapt to the changed situation, while in other case the organism may fail to adopt and may gradually disappear from the ecosystem which may even disrupt the existing food chain. Stress response and aspect of energy drain in fish and higher vertebrates are characterised by several metabolic alterations (Larson *et al.* 1985).

The onset of stress in vertebrates causes an increased concentration of cortisol in plasma which in turn increases the concentration of glucose. Glucose is the major carbohydrate reservoir utilized during cortisol elevation (Leach and Taylor, 1982).

In the present study Cr^{+3} exposed experimental fish *H. fossilis* exhibits a hyperglycemic response both in short-term and long-term exposure. The significant increase in the blood glucose in experimental fish *H. fossilis* indicates an imbalance of glucose homeostasis which may be due to decreased utilization of glucose (glycolysis) or increased glucose formation (gluconeogenesis). Change in blood glucose concentration is the most widely recognised and consistent response to stressors.

When mullet were exposed acutely and chronically to cadmium, a dose related increase in glucose concentration in plasma was observed. Cadmium exposure caused a reduced capacity of mullet to produce and secrete insulin (Thomas, 1976).

Decreases serum insulin concentration in fish exposed to zinc were concomitant with increased

Table 48.1: Blood and Plasma Parameters of Control and Experimental Fish *Heteropneustes Fossilis* Treated with Cr^{+3} Compound for 45 days.

Parameters	*Short-term exposure*				*Long-term exposure*			
	96 Hrs		*15 Days*		*30 Days*		*45 Days*	
	Control	*Treated*	*Control*	*Treated*	*Control*	*Treated*	*Control*	*Treated*
1. Blood Glucose (Plasmaglucose) mg/100 ml	62.7 ± 0.02	77.56 ± 0.135 (-23.7%)	70.52 ± 0.35	82.52 ± 3.364 (-17.016%)	68.64 ± 1.852	126.06 ± 1.717 (-83.653%)	67.86 ± 2.237	137.52 ± 0.585 (-102.652%)
2. Plasma Cholesterol mg/100 ml	218 .4 ± 0.707	295.4 ± 1.938 (-35.256%)	219.08 ± 0.281	246.72 ± 0.230 (-12.616%)	222.6 ± 0.925	284.42 ± 0.286 (-27.77%)	223.26 ± 0.884	320.52 ± 0.618 (-43.563)
3. Plasma Protein mg/100 ml	3.44 ± 0.05	3.76 ± 0.05 (-9.3%)	3.364 ± 0.03	2.28 ± 0.09 (+32.14%)	3.7 ± 0.08	1.68 ± 0.245 (+54.59%)	3.46 ± 0.105	1.41 ± 0.157 (+59.24%)
4. Plasma Potassium mEq/L	2.11 mEq/L	2.11 mEq/L	2.11	2.12 (0.07%)	2.11 mEq/L	2.12 mEq/L (0.8%)	2.11 mEq/L	2.19 (-73.79%)

Values are Mean X ± SE of 5 Animals
For Plasma Potassium n=3

concentration of glucose in plasma and decreased concentration of glycogen in liver (Wagner and Mc. Koeown, 1982).

The effects may also be related to the suppression of insulin release and also hyper glycemic action of epinephrine.

Ghafghzi *et al.*, (1979) reported the hyper glycemic effect of acute chromium treatment in rats which was nullified by adrenelectroy. It is also reported that dichronate feeding in rabbits decreased insulin level due to B. cell damage. (Kucher *et al.*, 1967).

A similar hormonal imbalance may have resulted in the observed elevation of the blood glucose in Cr^{+3} treated fish *H. fossilis.*

Like blood glucose, the cholesterol level of blood of *H. fossilis* also shows higher values both in short-term and long-term exposure to Cr^{+3}, in experimental fish *H. fossilis.*

Depletion in the cholesterol content during stress and metal exposure has been observed by Borek (1958), Idler and Bitners (1958) and Lau *et al.* (1978). Increase in cholesterol value have been reported by Sastry and Sharma (1980), Mukhopadhyay and Dehadrai (1980) and Bano (1981, 1982) following heavy metal treatement.

Elevated levels of lipids or its fractions such as cholesterol is suggestive of increased lipid mobilization. Hexavalent choromium has no proven biological function and Cr^{+3} on the other hand is assumed to be the biologically active form of Cr in nature. Cr^{-} has been shown to be capable of increasing the activity of certain enzymes e. g. "Succiniccytochrome c Dehydrogenase" or satisfying the requirement of metal co-factors in other enzymes e.g. -phosphoglucomutase (Mertz, 1969).

Cr^{-} invitro has been shown to be active as the metal co-factor in cholesterol synthesis and maintaining the activity of "Trypsin and Renin" the two protein splitting enzyme (Mertz, 1969). It is therefore possible, that in Cr^{+3} toxicity in fish *H. fossilis* failure of insulin to induce glucose utilization by tissues, causes diversion of insulin towards stimulating cholesterol synthesis. It has also been reported that serum cholesterol and phospholipids rise together during liver damage (Sherlock, 1968) and Chromium treatement has been shown to increase cholesterol synthesis (Curran, 1944).

Exposure of Cr^{+3} salt to *H. fossilis* causes a gradual fall generally in the level of total serum protein besides a 9.3 per cent elevation in 96 hr exposure. This elevation in TSP may be due to haemo-concentration. Bilgrami *et al.* (1978) reported fall in serum protein of *Clarais batrachus* following mercury exposure. Bano *et al.* (1981) and Bano *et al.* (1982) observed lower value of plasma protein subjected to compounds containing heavy metals. The fall in plasma protein might be due to adrenocorticoid hyper activity causing increased protein catabolism leading to glconeogenesis (Madhurima and Pandey, 1991).

Oxidation of amino acids for energy might be the cause of depletion of plasma protein in Cr^{+3} exposed *H. fossilis.* During toxicant induced stress, protein catabolism increases and increased proteolysis to meet the energy demands of stress might be one of the cause of depletion of plasma protein. Depending upon length of stressfull condition glycogen and/or lipids are generally utilized first and proteins becomes the last and ultimate source of energy in morbid fish in long-term exposure.

Thus, in nutshell, it can only be emphasized at present that Cr^{+3} exposure to *H. fossilis* both in short-term and long-term perhaps affects B. cells first which causes deminished synthesis of insulin. This causes multiple effects on general metabolism. Carbohydrate metabolism is affected by glucose in-tolerance by the tissues, increased glycogenolysis and gluconeogenesis. Lipid metabolism is affected due to increased liposysis and cholesterol synthesis where as protein metabolism is affected by decreased protein systhesis, splitting of proteins, increased oxidation of amino acids and also by enhanced protein catabolism for different purposes.

This situation adversely affects plasma potassium level, pH of blood and water balance of fish. Enhanced potassium level observed in the present study is suggestive of decreased entry of K into the cells. Potassium is an important plasma cation which reduces excitability and hastens fatigue. For this ion, potassium, the range of variability is very high. This may in part be due to a leakage of potassium ions from the cells affected by low oxygen or raised level of pco2, or otherwise disturbed (Fange and Lidman, 1976).

Hypoxia might have contributed to a rise of plasma potassium, since potassium tends to remain mostly within the cells, breakdown of cells causes its rise in blood and urine. Therefore, monitoring of potassium in blood plasma is a good guide as to degree of disintegrating cells of the body. A rise in plasma potassium level indicates tissue damage and lowering of pH (acidosis). For every 0.1 pH unit decrease in serum potassium will rise approximately 0.6 m Eq/L and conversely (Schrier, R.W. 1976). The kidney is the main channel of excretion of H^+ ion from the body and its failure leads to acidosis.

Acknowledgement

The first author is grateful to the Head, PG Dept. of Zoology, L.N. Mithila University, Darbhanga for providing facilities.

References

APHA, AWWA and WPCF. (1985). *Standard methods for the examination of water and waste water,* American Health Association, American Water Works Association and Water Pollution Control Federation, 16th edition, American Public Health Association, Washington, DC.

Baetjer, A.M., Bermingham, D.J., Enterlino, P.E., Mertz, W. and Pierce II. J. (1974). Effects of Chromium Compounds on Human Health. In: *Chromium*, National Academy of Sciences, Washington, 42-73 (1974).

Bano, Y. (1982). Effect of Aldrin on Serum and Liver Concentrations of Fresh Water Catfish Clarias batrachus (L). *Proc. Ind. Acad. Sci (Anim. Sci).* 91(1): 27-32.

Bano, Y. Sheikh, A.A. and Hameed, J. (1981). Effect of Sub-Lethal Concentration of DDT on Muscle Constituents of an Air Breathing Catfish Clarias Batrachus (Lim). *Proc. Ind. Acad. Sci (Anim. Sci.)* 90(1): 33-37.

Becker, A.D. and T.O. Thatcher. (1973). *Toxicity of Power Plant Chemicals to Aquatic Life*, Wash-124, U.S. Atomic Energy Commission Washington, DC.

Benoit, D.A. (1976). Toxic effects of hexavalent chromium on brook trout (Salvelinus Fontinalis) and rainbow trout (Salmogairdneri). *Water. Res.* (10): 497.

Berndt, W.D. (1976). Renal Chromium Accumulation and its Relationship to Chromium Induced Nephortoxicity. *J. Toxicol Environ Health,* I: 449-459.

Bilgrami, Salma, R. and Quayyum. (1978): Effect of Mercury Intoxication on Blood Serum Protein Fraction of Clarias Batrachus (Linn). *Ind. J. Zool.,* 6(1): 8-15.

Boge, G., M. Leydet and D. Hauvet (1992). The effects of hexavalent chromium on the activity of alkaline phosphatase in the intestine of rainbow trout (Oncorhynchus Mykiss). *Aquatic Toxicology.*

23: 247-260.

Borck, Z. (1958). The contents of Lipids and Other Components in the Crucian's Body during Hibernation and Experimental Starvation. Polskie, Archwan, *Hydrobil*. 5:65-91.

Curran, G.L. (1944). Effect of certain transition group elements on hepatic synthesis of cholesterol in the Rat. J. Biol. Chem. 210: 765.

Ghfghazi, T., Maghbarch. A., and Bernell. R. (1979). Chromium induced hyperglycemia in the Rat. *Toxicology*. 12(1): 47.

Hart, W.B., Doudoroff, P. and Green Bank J. (1945). *The evaluation of toxicity of industrial wastes, chemicals and other substances to fresh water fishes*. Atlantic Refining Company, Philadelphia. pp. 317-326.

Idler, D.R and Britners, I. (1958). Biochemical study on Sockeys Salmon during Spawning Migration Cholesterol, Fat, Protein and Water. *Can. J. Biochem. Physiol.*, 36: 793-798.

Kucher, I.M. and Sahbanow, A.M. (1976). Histochemical Investigation of the Pancreatic Islets in Potassium-dichromate poisoning. Gistokhim, Norm Patol. Morfol. p. 353. Ed. Subbotin. M. Yaizd. Navka Sib otd. (1967) CA. 411. 27b/vol.72.

Kunhert, P.M., B.R. Kunhert and R.M. Stokes (1976). The effects of invivo chromium exposure Na-K and Mg-ATPase activity in several tissues of rainbow trout (Salmo gairdelneri) *Bull. Environ. Contam. Toxicol.*, 15: 383-390.

Laborda, R, Mayans, D.J. and Nunex. A. (1986). Nephrotoxic and Hepatotoxic effects of Chromium Compounds in Rats. *Bull. Environ. Contam. Toxicol.*, 36: 332-336.

Langard, S. (1980). Chromium. In: *Metals in the Environment*, H.A. Waldron (ed), Academic Press, London, pp. 111-116.

Larson, A., Haux. C and Sjobeck, M.L. (1985). Fish Physiology and Metal Pollution. Results and Experiences from Laboratory and Field Studies. *Ecotoxicol Env. Saf.* 9: 250-281.

Leach, G.J. and M.H. Taylor (1982). Effect of Cortisol Treatement on Carbohydrate and Protein Metabolism of Fundulus Heteroclitus. *Gen. Comp. Endocrinol.*, 41. 328 p.

Mertz, W. (1969). Chromium Occurrence and Function in Biological System. Physiol. Rev. 49: 163-239.

Metelev. V.V., Karsev, A.K. and Szasokhova N.G. (1983). *Water Toxicology*. Amerind, New Delhi, 216.

Mukhopadhyay, P.K. and Dehadri, P.V. (1980). Biochemical Changes in the Air Breathing Catfish Clarias batrachus (L) Exposed to Malathione. *Environ Pollu.*, 22: 149-158.

Olson, P.A. and R.F. Foster (1956). Effect of Chronic Exposure to Sodium Dichromate on Young Chinook Salmon and Rainbow Trout, Hangorod Biol. Res. Annu. Rep. H.W. 41500 (1956): 35.

Peres, G. and M.H. Brichon. (1980). Etude Comparative an Moyen D'um Test Stalique (AFNOR Sur 24) de la Sensibilite Relative De Plusieurs Especes De Poissous on Bichromate Depotassium. *Report Dusymposium International Desophia-Antipolic*, pp. 320-328.

Ragnar Fange and Ulf Lidman. (1976). Comparative Studies in Inorganic Substances in the Blood of Fishes from the Scagerac Sea. *J. Fish Biol.* (1976): 8, 441-448.

Royle, H. (1975). Toxicity of Chromic Acid in Chromium Plating Industry. *J. Environ. Res.* 10: 39-53.

Samanta, S. (1999): Heavy Metals as an Indicator of Stress Condition of the Ecosystem. Short Course Training on Aquatic Environment Impact Assessment. *CICFRI (ICAR) Bull No-848.* January 1999.

Sastry, K.V. and K. Sunita (1983). Enzymological and Biochemical Changes Produced by Chronic Chromium Exposure in a Teleost Fish. (Channa panctatus). *Toxicol Lett.*, 16-15.

Sastry, K.V. and Sharma, K. (1980). Mercury Induced Haematological and Biochemical Anomalies in Ophiocaphalus Punctatus. *Toxicol. Lett.* (AMST). 5 (3/4): 245-250.

Schrier, R.W. (1976). *Renal and Electrotyle Disorders.* Little Brown and Company, Philadelphia.

Sherlock, S. (1968). *Diseases of the Liver and Biliary System.* Blackwell Scientific Publication, Oxford, p. 13.

Singh, Madhurima and S. Pandey (1991). Study of Toxic Effects of Manganese Sulphate on Haematology of *Heteropneustes fossilis. Biojournal,* 3(1): 177-184.

Sugatt, R.H. (1980). Effect of Sodium Dichromate Exposure in Freshwater on the Salinity Tolerance and Serum Osmolority of Juvenile Cohosalmon Onchorhynchus Kisutchin Sea-Water. *Arch. Environ. Contam. Toxicol.,* 9 (1980): 41.

Tandon, S.K., Saxena, D.K., Gaur, J.S. and Chandra, S.V. (1978). Comparative Toxicity of Trivalent and Hexavalent Chromium Alteration in Blood and Liver. *Environ. Res.* 15: 90-99.

Thomas, P. (1982): Effect of Cadmium Exposure on Plasma Cortisol Levels and Carbohydrate Metabolism in Mullet (Mugil Cephalus). *J. Endocrinol. Supp.* 94: 35 p.

Vander Putte. I, W. Vander Galien and J.J.T.W.A. Strik. (1982). Effects of Hexavalent Chromium in Rainbow Trout (Salmo gairdneri) after Prolonged Exposure at Two Different pH Levels. Ecotoxicol. *Environ. Safety.* 6: 246-257.

Venugopal, B. and T.D. Luckey (1977): Chemical Toxicity of Metals and Metalloids. In: *Metal Toxicity in Mammals.* Vol 2. Plenum Press, New York and London.

Vincet, S., L. Cyril, Arun Kumar and T. Ambrose (1996). Impact of Heavy Metal Chromium on Bio-Energetics of the Indian Major Carp, *Catla-catla* (Ham). *Poll. Res.* 15(3): 273-275.

49

A Composite Approach for Evaluation of the Effect of Malathion on Gobiid Fish *Glossogobius giuris* (Ham.)

M. Ramachandra Mohan

Department of Zoology, Fishery Research Lab., Bangalaore University, Bangalore - 560 056

ABSTRACT

Indiscriminate use of pesticides, careless handling, accidental spillage, and discharges of untreated effluent into natural waterways have resulted in the harmful effects on the fish population. According to Haider and Inbaraj (1988), insecticides used for pest control, are also toxic to non-target organisms in the aquatic ecosystem. Indiscriminate use of pesticides can be considered one of the factors, which changes the environment causing several imbalances in the ecosystem, especially denizens of the aquatic environment (Reddy and Rao, 1990).

Malathion has low toxicity for mammals but in relatively highly toxic for fish Mount and Stephen (1967), Datta (1995) and Mohan (2000). In mammals it is hydrolyzed rapidly and becomes inactive, whereas such hydrolysis does not occur in insects and proceeds slowly in fish. This is due to the lack of hydrolytic enzymes in insects and fish. (Arecchon and Plumb, 1990). Oxygen analogues of malathion (Malaxon) appear to be the active part which binds vigorously with acetylcholine esterase (O'Brien *et al.*, 1974).

Introduction

Toxicological effect of malathion has been observed in the hypophysis of the fish *Glossogobius giuris* while estimating the degree of damage in the prolactin secreting cells and gonadotrophins.

Malathion used in agriculture, finds its way with run off water interfering with all the metabolic processes and getting accumulated in vital organs of the body thereby affecting the functional activity of both endocrine and exocrine system of non-target aquatic organisms including fishes (Sahai, 1987). Even though it is certain that pesticides affect reproduction in fishes, their mechanism of action is not clearly understood in many cases particularly from fisheries development point of view. Therefore, several attempts have been made to study the histopathological changes in the ovary of a commonly available and edible fish, *G. giuris* in response to sublethal concentration of malathion.

Regarding the insecticides induced physiological disorders in the organism, very little is known about the effects on the fish intestine which is believed to be an important route of insecticides (Johnson, 1968). Intestine plays vital role in digestion and absorption of food and any interference by toxic substances may cause direct/indirect effect on there vitality of organism. When a fish habitat is exposed to toxic chemicals, its food also gets contaminated causing damage to the intestine.

Most of the toxicological studies are restricted to the mortality and growth rate of fishes and only few studies have been focussed on the organ system. Hence, study have been carried out to find the effect of malathion on gastro-intestinal system of freshwater gobiid fish *G. giuris* which will throw light on alterations in the above tissue and damages due to these toxicants. Such study may help in the better management of the water bodies, and in improving the production of economically important fish like *G. giuris*.

The Thyroid is an important endocrine gland in different groups of fishes and is present in the pharyngeal region of all vertebrates. It is well known that thyroid hormone plays an important role in metabolism of vertebrates. From the available literature it appears that in fishes pesticides affect the thyroid gland and in turn it alters the metabolism of the fish, thereby causing many physiological imbalances including regression of gonads. Histopathological studies on other teleostean species after exposing them to organophosphorus pesticide in general, malathion in particular are lacking. Hence, in the present study histopathological changes of the thyroid of *G. giuris* have been made during non-breeding and breeding phases under laboratory condition after exposing fish to different concentrations of malathion.

Gills play a vital role in cosmo regulation and gaseous exchange. Any interference by toxic substance may cause direct or indirect effects on gills. When a fish exposed to pesticides, they cause damage to the gills by absorbing toxic chemicals. Since gill covers considerable 60 per cent surface area of the fish, and its external location as well as intimate contact with water makes it the most vulnerable target organ for the pollutants. Roberts (1989) made an attempt to record the branchial histopathology in response to the sublethal mercuric chloride treatment.

The gill is the primary corridor for molecular exchange between the internal milieu of a fish and its environments. Most of the toxicological studies are restricted to the mortality and growth rate of fishes and only few organ systems. Hence, the present study has been taken out to find out the effect of malathion on internal gills of freshwater gobiid fish *G. giuris* to study in depth of its histopathlogical alterations after exposing them to various concentration of malathion at different intervals.

Methodology

The result obtained from the multiple measures (five aspects) used for assessing the effects of organophosphorus pesticide (Malathion) on freshwater fish *G. giuris* (an Indian gobiid fish) are discussed in this section. This section comprises five parts: they cover histomorphological and histopathological aspects of hypophysis, ovary, intestine, thyroid, and internal gills.

Histopathology

Histopathology appears to be very sensitive parameter and is crucial in determining cellular changes that may occur in the target organs such as hypophysis, ovary, intestine, thyroid and gills.

Hypophysis

The hypophysis of mature *G. giuris* is presented in (Fig. 49.1). It is composed of a nervous part-the neurohypophysis, and glandular part - adenohypophysis. Histologically, the adenohypophysis is divided into a pro- (rostral pars distalis-RPD), meso-(proximal pars distalis-PPD) and meta-(Pars intermedia-PI) adenohypophysis. Though the hypophysis is divided into three different glandular regions, there are no distinct boundaries between them. Hence, there is a possibility of mixing up of the various cell types of adjoining regions during sexual maturity and spawning periods.

In this study, the gonadotrophic hormone (GTH) and prolactin secreting cells have been identified by their cytomorphological changes during the exposure of fishes to malathion treatment of 0.05, 0.25 and 0.5 ppm for 24, 48, 72 and 96 hrs duration.

On treatment of the fish with malathion for 24 hrs of exposure to 0.05 ppm concentration, the GTH cells were affected. In the treated fish, the promixal pars distalis (PPD) increased in size, since, the cells in this region were loosely arranged. The PPD region contained granulated cells situated towards the peripheral region. The degranulated cells and a few vacuoles were found towards the central region. Intercellular spaces became conspicuous and cellular disturbances were also noticed in the PPD region. When they were kept in 0.05 ppm malathion solution for the same period the GTH cells exhibited hypertrophy as evidenced by degranulation and decrease in size of the cells and nuclei. The lightly stained cells, though degranulation, had become chromophobic. In addition to these degranulated cells, a few vacuoles and empty spaces were seen in the intercellular spaces, which might be due to cytoplasmic disintegration. In fishes examined 48 and 72 hrs after malathion treatment (0.5 ppm), the GTH cells showed further decrease in diameter. Few small vacuoles were found on the peripheral and central regions of the gland. After 96 hrs of malathion treatment the GTH cells appeared granulated. The cell boundaries were inconspicuous. In this concentration the PPD decreased in size gradually, and the cells of this region were loosely arranged. The cytoplasm on the cell was densely stained while in others was with partial degranulation which were formed towards the periphery. Few vacuoles were seen in some areas of the PPD, and were highly vascularised. Intercellular spaces had increased towards the central region of the gland. (Fig. 49.2)

The cytoplasm of prolactin cells of central was mostly filled with granules and was positive to erythrosin or phloxine. Along with the granulated cells, a few granulated over with large vacoules were also seen. The vacuoles contained vesicular fluid which stained feebly with erythosin (Fig. 49.3).

On treatment of the fish with malathion for 24 hrs the eta cells showed variation. In the treated fish of different concentrations of malathion (0.05-0.5 ppm), these cells showed marked cytological changes in the RPD. Histological observation revealed that the prolactin cells showed signs of degranulation and vacuolization with the increase in the concentration of malathion. Intercellular spaces became conspicuous and cellular disturbances were noticed in the RPD (Fig. 49.4). Most of the cells had undergone degranulation and showed vacuoles. In the treated fish there was an increase in the quantity of granulated eta cells than those of the control. (Fig. 49.5) The cell boundary was not clear and as such they appeared to be closely packed.

Ovary

In the experimental fish exposed to sublethal concentrations of (0.05. 0.25 and 0.5 ppm) of malathion for 24, 48, 72 and 96 hrs, the ovarian growth was inhibited significantly with the presence

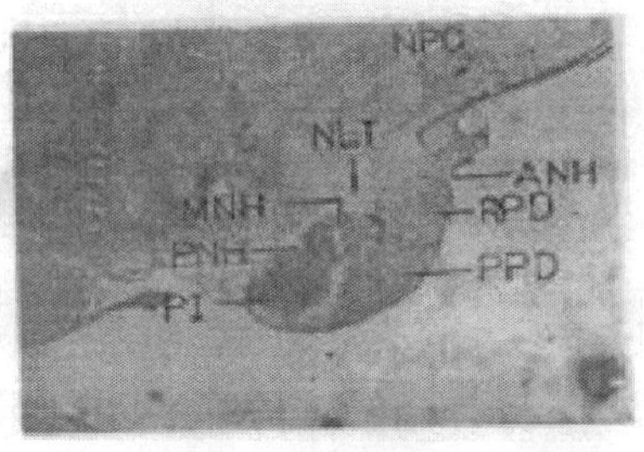

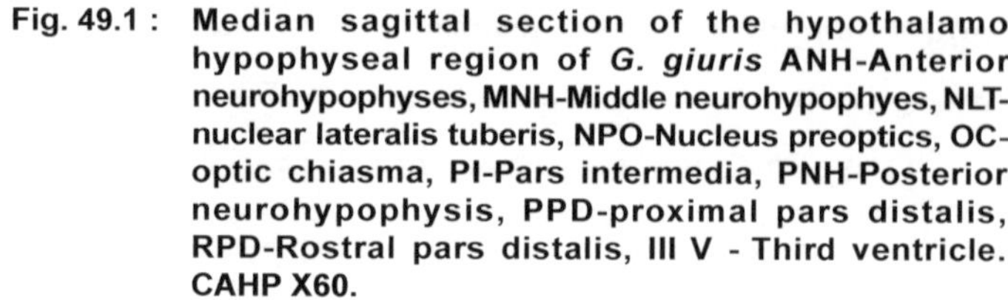
Fig. 49.1 : Median sagittal section of the hypothalamo hypophyseal region of *G. giuris* ANH-Anterior neurohypophyses, MNH-Middle neurohypophyes, NLT-nuclear lateralis tuberis, NPO-Nucleus preoptics, OC-optic chiasma, PI-Pars intermedia, PNH-Posterior neurohypophysis, PPD-proximal pars distalis, RPD-Rostral pars distalis, III V - Third ventricle. CAHP X60.

Fig. 49.2 : Median sagittal section of PPD, after 96 hrs of 0.5 ppm malathion treatment indicating increase in degranulated GTH cells with distinct nuclei. BV-Blood vessel; DGC-degranulated cell; GC-granulated cell; N-Nucleus; V-Vacoule. CAHPX 650.

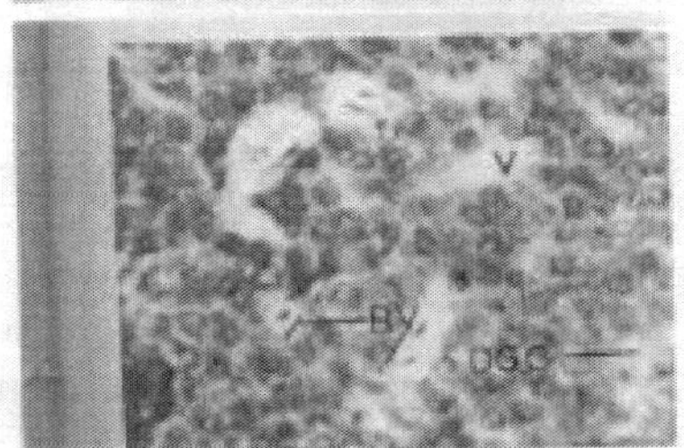

Fig. 49.3 : Medium sagittal section of the PPD region, after 24 hrs of 0.5 ppm malathion treatment indicating various stages of degranulation of GTH cells. CAHPX440.

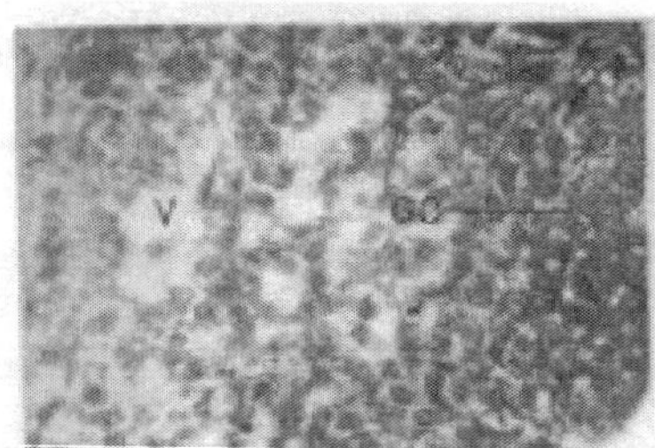

Fig. 49.4 : A portion of RPD, after 24 hrs of 0.5 ppm malathion treatment indicating granulated eta cells towards periphery. Note a large number of vacoules in the ventral region. CAHPX750

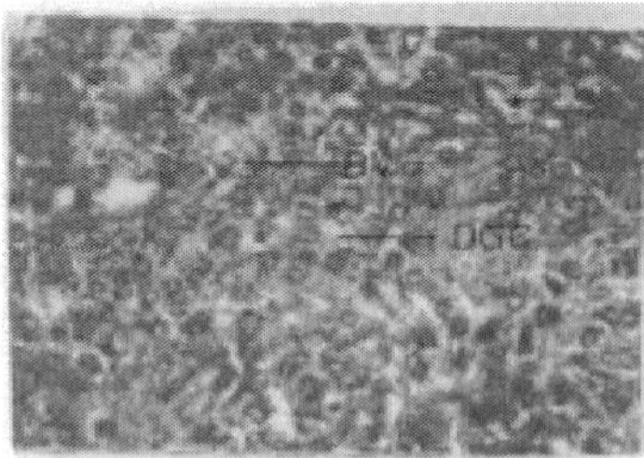

Fig. 49.5 : A portion of the RPD, after 96 hrs of 0.5 ppm malathion treatment indicating, accumulation, increase in number of degranulated prolactin cells. Note a few vacoules and blood vessels. CAHPX650.

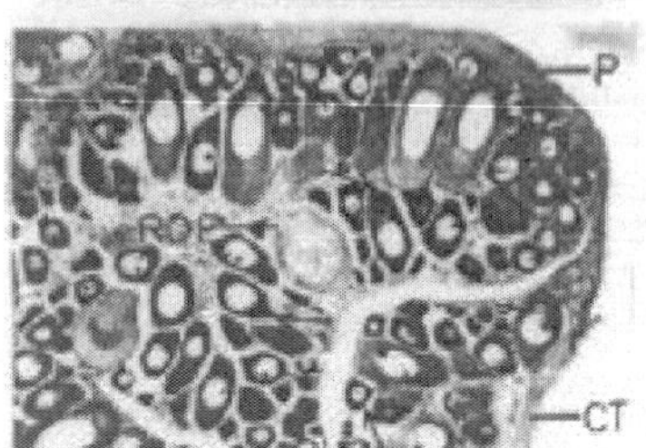

Fig. 49.6 : T.S. of ovary of *G. giuris* (Control) indicating different stages developing oocytes of variable sizes (arrows) Ehrhlick's haematoxylene and cosin X200.

of large numbers of non-vitellogenic stage-I and very few stage-II and stage - III oocytes. Some of the stage I and II oocytes in 0.05 and 0.5 ppm for 24 hrs, exhibited nuclear degeneration characterized by the destruction of nuclear membrane. A thin layer of connective tissue, tunica albugenia, enclosed the ovaries of control fishes. The ovaries exhibited a thick germinal epithelial layer and distinct ovocoel (Fig. 49.6). These were considered as the organizing ovaries. Among the oocytes, a few had ruptured follicles, which showed signs of disintegration. Other oocytes, which were not extruded during the previous season slowly, got reabsorbed in the ovary. In addition to these oocytes the ovaries exhibited different stages of developing oocytes of various sizes. The smaller oocytes were spherical or ovoid in outline and were found on either side of the ovigerous lamellae. In addition to the immature oocytes, a few large unspawned oocytes that were not extruded during the previous season were also present.

The ovaries of 24 hrs, treated fishes with 0.05 ppm of malathion were enclosed by a thin layer of tunica albugenia. The germinal epithelium was thick followed by a larger layer of connective tissue. The ovigerous lamellae were prominent and their number had increased in comparison to that of control group. In most of the ovaries the developing oocytes showed degeneration with an abundance of stage I oocytes (Fig. 49.7a). There was an increase in the number of artetic follicles also. The increase in percentage of stage I oocytes and GSI indicated that the rate of oogonial proliferation was very high at this time. The percentage of stages I and II oocytes and atretic follicles was 67.0 per cent, 15 per cent and 3 per cent respectively. In the experimental fish exposed to 0.5 ppm of malathion for 24 hrs, the ovarian growth was characterized by the presence of large number of atretic follicles of stages I and few stage II and III oocytes. (Fig. 49.7b). The occurrence of atretic oocytes was characterized by the shrinkage of the oocytes, accompanied by the disorganization of ooplasmic and nuclear components. Some of the stage II oocytes exhibited nuclear degeneration through the breakdown of nuclear membrane.

The ovaries of 96 hrs, treated fishes (with 0.5 ppm of malathion) were enclosed by a thin layer of peritoneum. The germinal epithelium became very thick and the ovigerous lamellae were quite prominent. The ovaries were in the immature condition having a large number of stage II, III and fewer stage I oocytes. In this concentrations the retardation of ovarian growth is reflected on the gonosomatic index. This is further evidenced by the presence of loosely arranged follicles in the ovaries, the percentage of oocytes in stages I, II and III oocytes and atretic follicles being 40 per cent, 25 per cent, 15 per cent and 20 per cent respectively. The results indicate that the malathion induced the growing oocyte to undergo higher development resulting in early atresia. The histology of the malathion treated fish was quite different. The oogonia showed cytoplasmic clumping, degeneration and deformaties in shape. The nuclei were in degenerating condition. Due to toxic stress wide spaces were found between the oocytes. A significant decline in gonosomatic index, diameter and number of the different oocytes in the treated fish were also observed in fishes treated at 0.5 ppm for 96 hrs. (Fig. 49.8).

Intestine

Histology: The foregut of *G. giuris* shows four layers of tissues namely, serosa, muscularis, submucosa and mucosa (Fig. 49.9). The outermost serosa is followed by a well developed muscularis (longitudinal and circular muscle) embedded in loose connective tissue richly supplied with blood capillaries. It merges with tunica propria of the underlying mucosal coat. The mucosa is raised into several longitudinal folds and the epithelial lining of mucosa consists of absorptive cells. The intestinal mucosal cells are oval in shape with serrated margin having goblet cells in the interspaces. The hindgut resembles the foregut with a slight reduction in size and thickness. (Fig. 49.10).

Histopathology

In *G.giuris* malathion induced alterations both in the foregut and hindgut after exposing to each of 0.05, 0.25 and 0.5 ppm malathion for 24, 48, 72 and 96 hrs of intervals.

Foregut: Treatment with (0.05. 0.25 and 0.5 ppm) malathion for 24, 48, 72 and 96 hrs resulted in the breakdown and fragmentation of the serosal layer. In muscularis the longitudinal muscle layer greatly reduced and circular muscle got thickened with an increase in the number of blood vessels. At the highest concentration (0.5 ppm) the longitudinal and circular muscles were completely damaged.

In the submucosa the connective tissue got detached with an increase in empty blood vessels. In the mucosal region the columnar epithelial cells became atrophic and vacuolated and lost their texture. There was a reduction in the size of the villi with more interspaces in the tunical propria and the villi become highly fringed.

Hindgut: After treatment with (0.05. 0.25, 0.5 ppm) malathion, serosa was seen to be degenerating. In the muscularis longitudinal muscle was loosely arranged after treating with 0.05 ppm for 24 hrs. The circular muscle got reduced and the fibres were broken down at many places (Fig. 49.11). In the submucosal region large vacoules appeared and the villi got deepened into the submucosa resulting in its reduction. Few cells showed necrosis and degranulation. In the mucosal region the villi became flattened and got reduced in size, number and height. Tunica propria was reduced in thickness and got detached from the columnar epithelium. The mucous layer increased enormously with degranulation in the absorptive cells.

Thyroid

Histology: Histologically, the thyroid gland is composed of a large number of follicles, each of which is in the form, of hollow ball consisting of a single layer of epithelial cells enclosing fluid filled space. These follicles vary in shape and size and are bound together by connective tissue. The gland is highly vascular and is generally well supplied with blood. The epithelium surrounding the follicle may by thick or thin and the height of the cells depends upon its secretory activity. The less active follicles generally show a thin epithelium and a few collapsed ones may also observed in a section. The epithelium is mainly composed of two kinds of cells, the chief cells are cuboidal cells contain droplets of secretory material. The lumen of each follicles is full of colloid which may be basophilic or acidophilic depending upon the secretory activity of the follicle and may show vacuoles in it (Fig. 49.12).

Histopathology: The thyroid follicles of *G. giuris* showed lesser damages in its surface, after exposure to lower concentrations (0.05 ppm) of malathion for shorter duration (24 hrs). On the other hand its impact was more at higher concentrations in longer duration.

The follicles of thyroid, after treatment with 0.05 ppm of malathion for 24 hrs (Fig. 49.13) exhibited marked cytological changes. The mean follicular diameter, epithelial cell height and height of colloid were found to have decreased when compared to that of controls. On treatment with 0.05 ppm of malathion, there was further decrease in mean follicular diameter, height of colloid and epithelial cell height Cellular hypertrophy and follicular hyperplasia accompany these changes. The epithelial cells were found to be cuboidal with dense colloid, thus many vacuoles are frequently recognized in the colloid adjacent to the epithelium. Most of the follicles were devoid of colloid.

After treatment with 0.05 ppm malathion for 96 hrs, the follicles of thyroid exhibited a decrease in epithelial cell height, follicular diameter and height of colloid when compared to that of controls. In fishes treated with 0.05 ppm of malathion there was decrease in epithelial cell height and increase in follicular diameter and height of colloid consequently the follicular lumen was obliterated due to

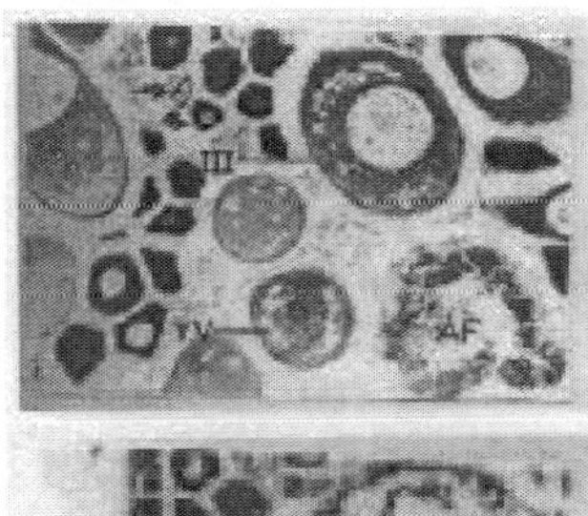

Fig. 49.7a : T.S. of the ovary, after 24 hrs of 0.5 ppm malathion treatment indicating a large number of atretic follicles of stage I and few stages II and III oocytes. Ehrilch's haematoxylin and eosin X160.

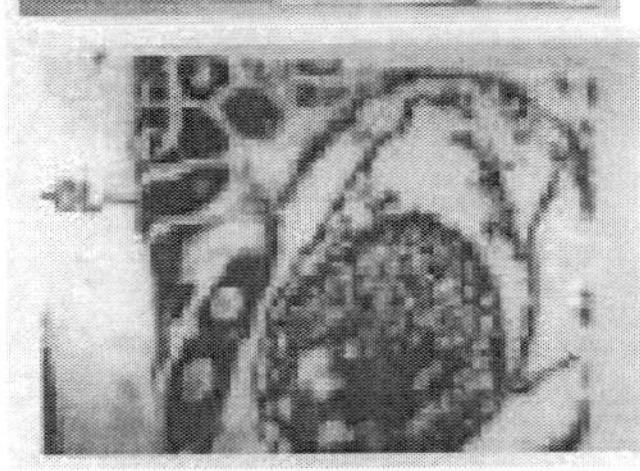

Fig. 49.7b: A portion of T.S. of ovary, after 24 hrs of 0.5 ppm, malathion treatment indicating shrunken, degenerated atretic phagocytic oocyte (arrow). Note the ovigerous lamallae (OL) Echrilch's haematoxylin and eosin X256.

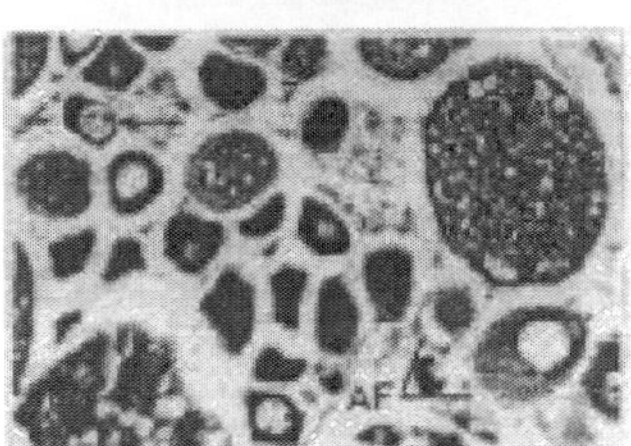

Fig. 49.8 : T.S. of ovary, after 96 hrs of 0.5 ppm malathion treatment indicating nuclei degeneration of stages II and III atretic oocytes (arrow). Ehrlich's haematoxylin and eosin X250.

Fig. 49.9 : T.S. Foregut during preparatory period (control) showing serosa (S) longitudinal muscle (LM) circular muscle (CM), submucosa (SM) mucosal cells (MC) and columnar epithelial cells. Mallory's tripple X63.

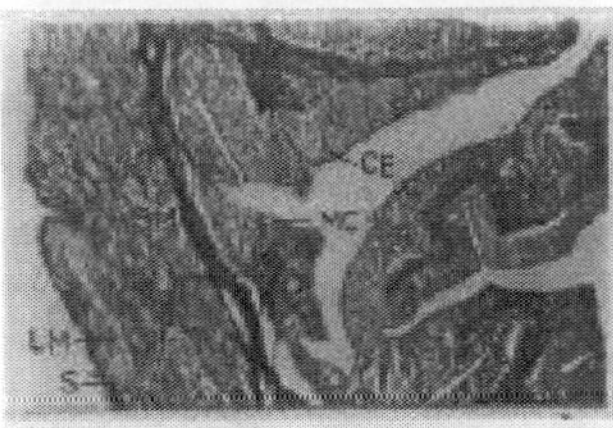

Fig. 49.10 : T.S. of Hindgut (Control) showing serosa (S), Longitudinal muscle (LM), Circular muscle (CM), Submucosa (SM), Blood capillaries, Tunica Propria, Mucosal cells (MC) and Columnar epithelial cells. Mallory's tripple X160.

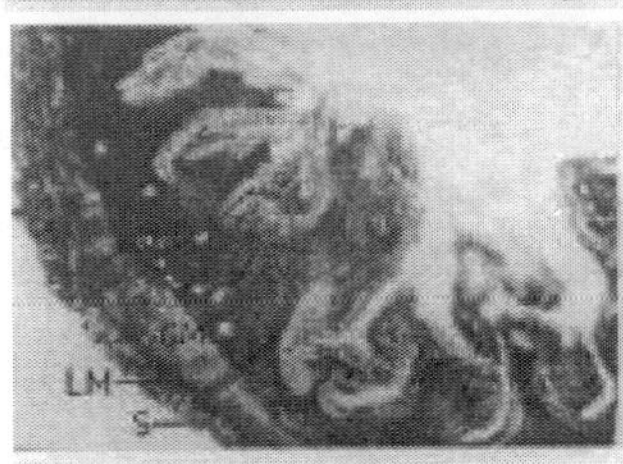

Fig. 49.11: T.S. hindgut exposed to 0.05 ppm for 24 hrs in malathion solution shows the broken serosa (S), the detachment of tunica propria from the columnar epithelium (arrow) Mallory's tripple X100.

Fig. 49.12 : T.S. of the thyroid gland of *G. giuris* (Control) showing thyroid acinus composed of specialised thyroid epithelium (TE) resting on a basement membrane (BM). These epithelial cells enclose a lumen filled with thyroid colloid (TC) and are surrounded by a fine network of capillaries associated with thin fibrous septa. Haematoxylin-eosin X1000.

hyperplasia and hypertrophy of epithelial cells, in some follicles the follicular lumen was totally obliterated with concomitant total loss of colloid, few follicles with little amount of colloid were also noticed (Fig. 49.14).

Gills

Histology of Gills: The gills of the fresh water gobiid, *G. giuris* consists of laterally compressed leaf like gill filaments (Primary gill lamellae) arranged alternately on either side of the interbranchial septum. Each primary filament bears a row of secondary lamellae on both sides perpendicular to its long axis comprising of a central core of cartilaginous rod with lining epithelial cells and blood vessels. The secondary lamellae consist of a flattened epithelial cell layer attached to the basement membrane contractile pillar cell system and few blood spaces or sinusoids (Fig. 49.15). Mucous cells are scattered over the arch support epithelium, filament and lamellae. Chloride cells are found on the epithelium, lamellar portion of the filament and medial margin of the lamellae. These cells are most frequent on the filament surface between the lamellae (inter lamellar filament epithelium) and around the afferent - trailing edge with respect to water flow.

Histopathology

The gill filament of *G. giuris* showed lesser damages in its surface, after exposure to lower concentrations of malathion for shorter duration. On the other hand its impact was more in longer duration at higher concentrations.

The gills of *G. giuris,* after treatment with 0.05 ppm of malathion for 24 hrs, exhibited marked cytological changes for the mean cell diameters of treated and controls. The mean cell diameters of chloride and mucous cells were found to have deceased when compared to that of the controls. On the other hand the pillarcell diameter was almost equal to that of the control. On treatment with 0.5 ppm of malathion the chloride and pillar cell diameter were found to have increased, while that of the mucous cell got reduced. These changes are accompanied by necrotic changes like oedematous separation of primary lamellae from the basement membrane. Shortening of the height of the primary lamellae, vacuolization at the basal region of primary lamellae, and the lamellar talengectases at the secondary lamellae. (Fig. 49.16).

After treatment with 0.05 ppm malathion for 96 hrs, the gills of *G. giuris* exhibited a decrease in chloride, mucous and pillar cell diameters when compared to that of controls. In fishes treated with 0.5 ppm of malathion, there was a decrease in chloride and mucous cell diameter while there was an increase in pillar cell diameter. Severe oedema, fusion of secondary lamellae with damaged blood vascular system. (Fig. 49.17) was also observed.

Discussion

Mohan (1991) have shown that during preparatory period gonadotropic contents of pituitary gland decreased significantly when exposed to pesticides. The histological changes in the gonadotrophs of malathion exposed *G. giuris* are comparable to the changes described in the gonadotrophs of teleost *H. fossilis.* (Jagadeesh and Sahai, 1989). In the present study on *G. giuris* it was observed that after 24 to 96 hrs of treatment with malathion, the GTH cells showed degranulations and hypotrophy of cells and nuclei. The cells depicted deformed shapes and vacuolization in the cytoplasm along with intercellular spaces, which might be due to the cytoplasmic disintegration. Further, these intercellular spaces increased with an increase in the concentration of malathion. Such abnormalities in the PPD have been recorded in other fishes also (Vajkpai and Mathur, 1986, Jagadeesh

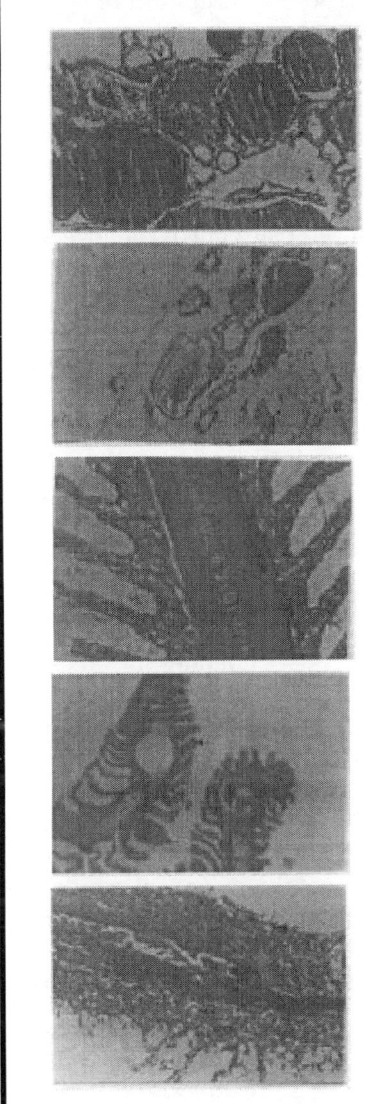

Fig. 49.13 : T.S. thyroid of *G. giuris* treated with 0.5 ppm of malathion for 24hrs showing extensive hyperplasia of thyroid follicles with dense colloid (C) surrounded by cuboidal epithelium (EP) and vacoules (V). Haematoxylin-eosin X400.

Fig. 49.14 : T.S. of thyroid of *G. giuris* after treatment with 0.05 ppm of malathion for 96 hrs showing activation of thyroid follicles by indicating cuboidal epithelium (EP) and concomitant loss of colloid (C) in few follicles (arrows) Haematoxylene eosin X160.

Fig. 49.15 : T.S. of the gill filament of *G. giuris* (Control) showing primary lamella (P) secondar lamella (S) and central cartilaginous supporting axis (CSA) rod (C) Haematoxylin-eosin X400.

Fig. 49.16 : T.S. of the gill filament of *G. giuris* treated with 0.05 ppm of malathion for 96 hrs showing hypertrophied epithelial cells along with lamellar talengectases (arrows). Haematoxylin-eosin X400.

Fig. 49.17 : T.S. of the gill filament of *G. giuris* treated with 0.05 ppm of malathion for 96 hrs showing fusion of secondary lamellae, note the hypertrophied Chloride (CC), Mucous (MC) and Pillar cells (PC) (arrows). Haematoxylin-eosin X160.

and Sahai, 1986 and 1988 and Sahai, 1989). Ram and Sathyanesan (1983) also have made similar observations in *C. punctatus* following the mercuric chloride treatment.

The decreasing trend of gonadotropic level in response to different concentrations of pesticide was found to be more conspicuous during the spawning phases.

In *G. giuris,* during spawning phase, malathion exposure resulted in a significant reduction in the content of gonadotrophin in PPD. The short term effects of malathion in *G. giuris* during spawning phase was similar to those observed in the spawning females of *H. fossilis* (Singh and Singh, 1988). These changes may be the result of an altered or impaired synthesis of gonadotropin in response to the malathon treatment. Thus in *G. giuris,* exposed to malathion for 24 to 96 hrs, inhibition of ovarian growth and comparable changes in the pituitary gonadotropins were noticed indicating a possible impairment of pituitary gonadal axis.

During preparatory period of ovary in *G. giuris* the most important indication of the ovarian damage due to exposure to malathion is the reduction in the weight of the ovaries. Similar observations have been made in a few species of teleost in relation to different concentrations of pesticide (Saxena and Garg, 1978, Kaur and Virk, 1983 and Shukla, *et al.,* 1984). In *G. giuris* it was observed that the malathion greatly affect the secondary growth of primary oocytes due to the impairment of vitellogenesis. Such abnormalities in the ovary have also been recorded in other fishes also (Sahai, 1987, Koppar and Kulkarni, 1987 and Singh and Singh, 1987). Further Saxena and Agarwal (1986) have shown that cadmium chloride blocked all the oogonial activity at the vitellogenic growth phase in *Clarius batrachus* and suggested that it might be due to the retardation of oocyte proliferation, growth of oocytes and increase in the number of atretic follicles. The present study on *G. giuris* indicated that malathion caused similar nature of ovarian structure during preparatory period. This may be due to an altered synthesis and release of gonadotropin as a result of malathion treatment. Ram and Sathyanesan (1983) have pointed an inhibition of gonadal growth and cytomorphological changes in the pituitary gonadotrophs of *C. punclatus* after exposure to mercury. This indicates the impairment of pituitary-gonadal axis without any obvious sign of gonadal degeneration. However, in *G. giuris* it was observed that the rate of oocyte degeneration in the ovary was more in higher concentration of malathion (0.05 to 0.5 ppm) when exposed for 24 to 96 hrs.

During the spawning phase in *G. giuris* malathion induced degenerative changes in stages I, II and VII oocytes suggest its effect on the ovary indirectly through the hypothalamus by inhibiting the secretion of GnRH. This in turn might have decreased the synthesis and release of gonadotrophs from the pituitary followed by reduced ovarian activity. Singh and Singh (1983) have also made similar observation on *H. fossilis* following the malathion treatment. Saxena and Garg (1978) emphasized that the arrested ovarian recrudescence in *C. punctatus* occurred due to its exposure to fenthion and carbaryl and is brought about the inhibited gonadotropin release. Similar phenomena were observed in *G. giuris* exposed to higher concentration of malathion. The inhibition of gonadal growth is also reflected on the GSI. Ram and Sathyanesan (1983) also have suggested that pesticides cause reduction in ovarian growth of the teleost *C. punctatus.* In light of these findings it is presumed that in *G. giuris* the malathion interferes with gonadotropic secretion in the hypophysis resulting in a decreased ovarian activity during all phases of annual reproduction cycle.

In the present study the intestine of *G. giuris* showed histopathological changes in response to different concentrations of malathion ranging from 0.05, 0.25, to 0.5 ppm for durations of 24, 48, 72 and 96 hrs. Such changes include degeneration of the serosal and muscular layer and the presence of large vacuoles in the submucosal region. The villi became highly fringed, reduced in size and submerged into the submucosal region. Tunica propria was detached from columnar epithelium and epithelial cells showed necrosis and degranulation. Similar observations have been made by Mathur (1959) in *Ophiocephalus punctatus.* Anees (1978) has shown the effect of BHC and endrin on different organs of *Ophiocephalus punctatus, Heteropneustes fossilis, Trichogaster* Sp. and *Barbus stigma.*

The result of the effects of malathion on the gastrointestinal system *G. giuris* clearly show that malathion excerts toxic effects on the different layers of intestine which in turn might have impaired the different growth rate of fishes. Hence, it is suggested that spraying of pesticides in higher dosage will be lethal to fish population in general and economic important fish like *G. giuris* in particular.

In the present investigation, the toxic effect of different concentrations of organophosphorous pesticide malathion were made in the thyroid gland of *G. giuris* after exposing for a period of 24 - 96 hrs. Treatment with sublethal levels of malathion (0.05 and 0.5 ppm) for 24 hrs on *G. giuris* elicited several changes in the thyroid including cellular hypertrophy and follicular hyperplasia. The epithelial cells were found to be cuboidal with dense colloid, having many vacuoles with a decrease in follicular diameter and epithelial cell height. Similar responses to the various pesticide pollutants have been observed by Wani (1998) after thiourea treatment on the thyroid gland of a freshwater cyprinid fish, *Garra mullya,* by Prasad Rao and Mukherjii (1972) in *Heteropneustes fossilis* after thiourea administration, Leatherland and Sonstegard (1980) in *Oncorhynchus kisutch,* Srivastava and Sathyanesan (1971) in *Mystus vittatus* after treatment with antithyroid drug, thiourea, Singh *et al.* (1974) in the air breathing fish *Heteropneustes fossilis*, and Van Overbeeke and Mc Bride (1971) in gonadectomised sockey salmon *Oncorhynchus nerka.* After 96 hr of 0.05 and 0.5 ppm of malathion. The follicular lumen was obliterated due to hyperplasia and hypertrophy of epithelial cells. In some follicles, the follicular lumen was totally obliterated with the loss of colloid. There were some follicles with little amount of colloid. Van Overbeeke and Mc Bride (1981) made similar observations in gonadeatomised sockey salmon *Onchorhynchus nerka.* Baker-cohen (1961) working on the thyroid follicles during breeding period of few teleost have also reported by Singh *et al.* (1974). In the present study on *G giuris* it was observed that malathion treatment produced a hyperthyroid condition and this finding coincides with that of Rangeker and Latey (1977) who observed a similar change in *Tilapia mossambica.*

In the light of present findings of thyroid on the gobiid fish, *G. giuris* malathion seems to cause damage on the structure of the follicle, amount of colloid, epithelial cell height and function of the cell.

The aquatic toxicology test considers fish as a representative of the biota, and is also considered as screening test. It determines the presence of absence of harmful wastes, which are ascertained by survival of fish in the sample water.

The present study clearly demonstrates that even the sublethal concentrations (0.05. 0.25 and 0.5 ppm) of malathion induces pathological lesions in gill of which covers more than 60 per cent surface area of the fish and its external location renders it the most vulnerable target organ for pollutants (Roberts, 1989). Histopathological changes were noticed in the gill of *G. giuris* after exposing them to malathion. Similar branchial responses to the various pesticide pollutants have also observed by Ramamurthy *et al.* (1987), Roberts (1989) and Pandey *et al.* (1993). The proliferate thickening (hyperplasia) of gill epithelium appears to be a general safety measure against irritation by the environmental toxicants as reported by Roberts, (1989) and Pandey *et al.* (1993).

Kumar and Pant (1984) remarked that histopathological studies are useful in evaluating the pollution due to pesticides, since trace levels of these chemicals, which did not bring about animal mortality over a given period, were capable of producing considerable organal damage.

Severe oedema, increased interlamellar space and fusion of secondary lamellae after prolonged treatment (96 hrs) of malathion in *G. giuris* occurred as a result of malathion treatment. Similar observations on gill have been noticed by Kapilamanoj and Ragothaman (1999) in the esturine fish *Boleopthalmus dussumieri* (Cuv) after exposing them to sublethal concentrations of cadmium, by Kiemer and Black (1997) in *Salmo salar* (L) after exposing them to hydrogen peroxide, Kumaraguru *et al.* (1992) in rainbow trout *Salmo gairdneri,* with permethrin, and by Pandey *et al.* (1997) in *Liza parsia* induced by sublethal exposure to BHC.

The present observations in *G. giuris* clearly indicate that gills are affected most severely in response to malathion exposure.

References

Arecchon and Plumb, J.A. (1990). Sublethal effects of Malathion on Channel catfish *Ictalurus punctatus.* Bull. Environ. *Contam. Toxicol.* 44: 435-442.

Anees, M.A. (1975). Acute toxicity of four organophosphorous insecticides to a freshwater Teleost *Channa punctatus* (Bloch) Pakistan L. Zool. 7: 135-141.

Baker-Cohen K.F. (1961). *The role of the thyroid in the development of platyfish*, Zoological. New York, 46: 181-222.

Cassano, G.B., Amaducci, L. and Viola, P.L. (1966). *Rev. pat. Nerv. Ment.*, 87: 214.

Datta Munshi, J.S. Hiran M. Dutta. (1995). *Fish morphology*, horizon of new research.

Haiders, and Inbaraj, R.M. (1988). In vitro effect of Malathion and endosulfan on the LH-induced oocyte maturation on the common carp, *Cyprinus carpio* (L). *Water, Air and Soil Poll.* 39: 27-31.

Jagadeesh, C.G. and Sahai, S. (1986). Effect of malathion on the pituitary gland by *Mystus vittatus* (Bloch). *Abstract, National conference on water pollution - A potential threat to mankind*, Sagar: p. 23.

Johnson, D.W. (1968). *Pesticide and Fishes - A review of selected literature*, Trans. Amer Fish., Soc., 97: 398-424.

Kapila Manoj and Ragothaman. (1999). Effect of sublethal concentrations of cadmium of the gills of an estuarine edible fish *Boleohthalmus dussumieri* (CUV). *Poll Res.* 18(2): 145-148.

Kenneth, R. Olson. (1995). *Scanning electron microscopy of the fish gill, fish morphology*. Horizon of New Research.

Kiemer and Block. (1997). The effects of hydrogen peroxide on the gills tissues of Atlantic Salmon, *Salmo salar. Aquaculture.* 153, 181-189.

Kumars, and Pant, S.C. (1984).Orgonal damage caused by alsicarb to a freshwater teleost, *Barbus conchonius* Hamilton, *Bull. Environ, Conta, Tolicok*, 33: 50-55.

Kumaraguru, A.K. Beamsih, F.W.H. and Ferguson H.W. (1982). Direct and circulatory paths of permethrin (NRDC 143) causing histopathlogical changes on the gills of rainbow trout, *Salmo gairdneri* Richardson. *J. Fish Biol.*, 20, 87-91.

Lameperti, A.A. and Niewenhuis, R. (1976). *Cell Tissue Res.*, 170: 315.

Leatherland J.F. and Sonstegard R.A., (1980). Seasonal changes on thyroid hyperplasia, serum thyroid hormone and lipid concentrations and pituitary gland structure in lake ontario coho salmon, *Oncorhynchus Kisutch* Walbaum and a Comparison with Coho salmon from lakes Michigan and Erie. *T. Fish. Biol.*, 16: 539-562.

Mathur, D.S. (1959). Toxicity on organic insecticides to certain fish. *Proc. Nat. Acad. Sci. Ind.* 29: 380-382.

Mohan, M. (1991). Experimental studies on the Hypothalamo hypophyseal ovarian system of fresh water gobiid fish, *Glossogobius Giuris* (Ham). Ph.D. thesis, Bangalore University, Bangalore.

Mohan, M. (2000). Malathion induced changes in the ovary of freshwater fish, *Glossogobius giuris* (HAM) *Poll. Res.*, 19(1): 45-47.

Mount, D.I. and Stephan, C.E. (1967). A method for establishing acceptable toxicant limits for fish-malathion and the butoxy-ethanol ester of 2, 4-D. *Trans. Amer. Fish. Soc.* 21: 135-193.

'O' Brien, R.D., Hetnarkshi, B, Tripathi. R.K, and Hart. G.J (1944). Recent studies on acetylcholine estarase inhibition. In: G.K. Kohn (ed), *Mechanisms of Pesticide Action*. American Chemical Society, Washington, D.C.

Pandey, A.K., Mohamed M.P., George, K.C. and Shyamla, (1993). Histopathological changes in the gill kidney and liver of an estuarine mullet, *Liza parsia,* induced by sublethal exposure to DDT. *J. Indian Fish Assoc.,* 23: 55-53.

Pandey, A.K. George, K.C. and Peer Mohamed, M. (1997). Histopathological alterations in the gill and kidney of an estuarine mullet, *Liza parsia* (Hamilton-Buchanan), caused by sublethal exposure to lead. *Indian. J. Fish,* 44 (2): 171-180.

Prasad Rao, P.D. and Mukherjee, A., (1972). Effect of thiourea on the thyroid gland and gonads of the Catfish, *Heteropneustes fossilis* (Bloch), *Z. Mikrosk-Anat. Forsch.* 85: 187-198.

Ram, R.N. and Sathyanesan, A.G. (1983). Effect of mercuric chloride on the reproduction cycle of the teleost fish, *Channa punctatus. Bull Environ. Contam. Toxicol.,* 30: 24-27.

Ramamurthy, Nagarathnamma, B. Jayasunderamma and Rama Rao, P. (1987). Histopathological lesions in the gills of fresh water teleost, *Cyprinus carpio,* induced by methyl parathion. *Matsya.* 13: 144-147.

Rangneker, P.S. and Latey., A.M., (1977). Influence of the thyroid gland on gonadal activity in the teleost, *Tilapia massambica* (Peters). *J. Anim. Morpho. Physiol.,* 24(2): 344-352.

Reddy, M.S. and Rao, K.V.R.(1990). Influence of salinity on the toxicity of phosphamiden to the estuarine crab, *Cylla serrata* (Forskal). *Bull. Environ. Contam. Toxicol.,* 44: 859-864.

Roberts, R.J. (1989). Fish pathology, 2nd Edition, Bailliere, Tindall, London.

Sahai, (1987). Toxicological effects of some pesticides on the ovaries of *punctius ticto* (Teleostei). *Proc. VIII A.E.B SES and Symp.,* pp. 53-54.

Saxena, D.N. and Agarwal, A. (1986). Effect of mercuric chloride intoxication on ovarian activity of a teleost fish, *Channa punctatus* (Bloch). *Int. J.Acad. Ecthyol.,* 7(1), 1-6.

Saxena, P.K. and Garg, M. (1978). Effect of insecticidal pollution on Ovarian recrudescence in the fresh teleost, *Channa punctatus. Indian J. Exp. Biol.,* 16: 689-691.

Shukla, L., Srivastava, A., Merwani, D. and Pandey, A.K. (1984). Effect of sublethal malathion on ovarian histopathology in *Sarotherodan mosambicus. Comp. Physiol. Ecol.,* 9: 13-17.

Singh, and Singh T.P. (1983). Pesticide induced modulations of endocrine physiology of reproduction in the catfish, *Heteropneustes fossilis. Proc. Inst. Int. Symp. Life Sci.,* 138-148.

Singh, S.P. and Singh, T.P. (1987). Impact of malathion and hexachlorcyclohexane on plasma profiles of three sex hormones during different phases of reproductive cycle in *Clarius batrachus. Memoire,* 42: 57-62.

Singh, B.R., Thakur, R.N. and Yadav, B.N. (1974). The relationship between the change in the inter renal, Gonadal and thyroidal tissue of the air breathing fish *Heteropneustes fossilis* (Bloch) at different period of the breeding *Cycle. J. Endocr.,* 61: 309-316.

Srivastava, S.S. and Sathyanesan, A.G. (1971). A comparative study on the effect of Thiourea and large doses of radioiodine on the thyroid and thyrotrophs of *Mystus vittatus* (Bloch). *Acta anal.* 79: 556-

569.

Van Overbeeke, A.P. and M.C. Bride, J.R. (1971). Histological effects of 11-Ketotestosterone 17 Methyl testosterone, Estradiol, Estradiol Cypionate, and cortisol on the inter renal tissue, thyroid gland, and pituitary gland of gonadectomised sockety Salmon (*Oncorhynchus nerka*). *Journal Fisheries Research Board of Canada*, 28(4).

Wani (1998). Effect of thiourea treatment on the thyroid gland, ovary and liver of a freshwater cyprinid fish, *Garra mullya* (Sykes), *Proc. 4th National Seminar on fish and their environment*, IAES (Abstract), 30.

50

Acute Toxicity of Carbaryl and Methyl Parathion on Survival of *Rana tigrina* Tadpoles

K. Sampath, I.J.J. Kennedy and R. James

Department of Zoology, V.O.C. College, Tuticorin - 628 008, Tamil Nadu

ABSTRACT

Toxic effects of carbaryl and methyl parathion on survival were studied in the common frog *Rana tigrina* tadpoles as a function of different time intervals. The mortality of tadpoles was time and concentrations of pesticide dependent. The LC_{50} values of tadpoles exposed to carbaryl were 9.40, 7.98, 6.71 and 5.68 ppm and these values were 6.28, 5.15, 4.63 and 4.36 ppm for methyl parathion exposed tadpole at 24, 48, 72 and 96 h respectively. During the long-term study, no mortality was observed in the control tadpoles at different development stages. However, both the pesticides caused the mortality of tadpoles in fourth stage and methyl parathion executed mortality in third stage itself and it indicates the severe adverse effect of methyl parathion over carbaryl even before metamorphosis. Besides, tadpoles exposed to maximum concentration of methyl parathion caused malformation like scoliosis and failure of forelimb emergence during metamorphosis. Methyl parathion about 30 per cent more toxic than carbaryl.

Key words: Methyl parathion, carbaryl, survival, tadpoles, *Rana tigrina*.

Introduction

Application of pesticides has become invitable to protect the crop plants from pest and diseases. The agricultural loss due to pests and diseases was estimated around 50 per cent of total production in developing tropical countries (FAO, 1975). Pesticides reach water bodies either by direct application or indirectly. The indirect sources include run-off from agricultural fields, spray drifts, rain water, sewage and effluent from industries which are manufacturing the pesticides or using them in their processes (Bhaskaran, 1980). The effect of pesticides on aquatic life is often acute resulting in mass mortality of animals or chronic changes in behaviour and/or reduction in the rates of survival, growth and reproduction. Chronic effects of pesticides may induce histological and physiological changes in aquatic animals (Konar, 1981). Pesticide pollution also poses a constant threat to frog population in India. The ecologists warn that the diminishing frog population can upset the ecological balance, resulting in the abundance of vector and pests (Abdul Ali, 1985). *Rana tigrina* is commonly seen in paddy fields. Water bodies adjacent to paddy fields are directly exposed to the different kinds of pesticides and hence a study on *R. tigrina* tadpole with reference to adundantly used pesticides like carbamate (carbaryl) and organophosphate (methyl parathion) has become imperative. The present paper reports on the comparative study of toxic effects of carbaryl and methyl parathion on survival in the common frog, *Rana tigrina*.

Materials and Methods

Rana tigrina tadpoles were produced in the laboratory following the technique of hypophysation. Tadpoles were fed with pellet feed (40 per cent animal protein) and feeding of tadpoles commenced 2 days after hatching (Pandian and Marian 1986). Two series of experiments were conducted in the present study.

Series 1: Short-term study

In the first series of experiment, static, renewable bioassay test was conducted to determine the 96 h LC_{50} value of tadpole against carbaryl (CA) methyl parathion (MP). Three days old 10 healthy tadpoles (15 ± 2 mg) were separately exposed to different concentrations of CA and MP. A control was also maintained in pesticide-free water. The experiment was conducted in circular plastic trough containing 61 of test media. Test animals were not fed during the bioassay test. Fresh concentrations were prepared daily to maintain the constancy of the pesticide in the test media. Mortality of the tadpoles was observed at different time intervals (24, 48, 72 and 96 h). Regression equation was obtained following the least square method (Zar, 1974).

Series 2: Long-term study

A long-term study was carried out after determining 96 h LC_{50}. Five levels of 96 h LC_{50} value of CA and MP, 10, 20, 30, 40, and 50 per cent were taken as the test concentrations. A control was also maintained. Ten tadpoles (15 ± 2 mg) were reared in each concentration and control group until they reach froglet stage. The experiment was conducted in circular plastic trough containing 3 1 of test media and triplicates were maintained for each group. Tadpoles undergo distinct morphological changes during metamorphosis (growth). In the present study, tadpoles were classified into four types based on their morphological indices following the method of Gosner (1960). They are (i) appearance of hind limb, (ii) disappearance of external gill, (iii) appearance of fore limb, and (iv) disappearance of tail. The experimental tadpoles were fed *ad libitum* with pellet feed and uneaten feed was removed after 2 h. Strings of faecal matter and the left out faecal matter were pipetted out and filtered out while changing the water on alternate days.

Dead tadpoles were removed daily just prior to water change. Death was confirmed by turning the tadpoles by a glass rod upside down, tadpoles which failed to return to the normal position were considered as dead and removed. Statistical analyses were carried out to determine LC_{50} as per standard method (Finney, 1971). Toxicological statistics (Busvine, 1971) was followed and computerized (see

Syed Othuman, 1994). The LC_{50} values, fiducial limits and regression equation for CA and MP on the survival of *R. tigrina* tadpoles at 24, 48, 72 and 96 h durations were calculated. The LC_{50} for the entire larval period was arrived from the percentage mortality observed during the fourth stage.

Results

Mortality of tadpoles in any fixed period increased with increasing concentrations of CA and MP and in any fixed concentration. The mortality rate increased with the extension of exposure period. The LC_{50} values of tadpoles exposed to CA were 9.40, 7.98, 6.71 and 5.68 ppm and these value for MP were 6.28, 5.15, 4.63 and 4.36 ppm at 24, 48, 72 and 96 h respectively (Table 50.1, Fig. 50.1 and 50.2). The mortality rate of tadpoles was higher in MP than CA. Cent per cent survival was observed in the control as well as tadpoles exposed to 20 and 1.5 ppm of CA and MP respectively. The upper (not CA) and lower 95 per cent fiducial limits showed decreasing trend with the extension of exposure period (Table 50.1).

In the long-term study, no mortality was observed in the control tadpoles at different development stages. The sublethal concentrations of CA did not show any lethal effect upto III stage, however, 10, 20, 40, 50 and 70 per cent mortality was observed in IV stage of tadpoles exposed to 0.6, 1.2, 1.8, 2.4 and 3.0 ppm CA respectively (Fig. 50.2). In MP treated group, mortality occurred in third itself, 10, 20, 30 and 40 per cent mortality was observed in 0.8, 1.2, 1.6 and 2.0 ppm respectively at III stage and 10, 30, 50, 60 and 70 per cent mortality was recorded in 0.4, 0.8, 1.2, 1.6 and 2.0 ppm respectively in IV stage (Fig. 50.2). The LC_{50} values arrived from the corresponding log dose were 2.2 and 1.3 ppm for CA and MP respectively. The relative potency of MP on mortality was low (0.59) as compared to carbaryl (1.00) (Table 50.2).

Table 50.1: Acute Toxicity Values of Carbaryl and Methyl Parathion to *Rana tigrina* Tadpoles Exposed for 96 h

Exposure Duration (h)	LC_{50} (ppm)	95% fiducial limit (ppm)		y = (bx - a)	Variance	Relative potency of the pesticide
		Upper	Lower			
		Carbaryl				
24	9.40	8.39	10.52	20.19x - 14.64	0.001	1.00
48	7.98	6.02	10.57	16.33x - 9.72	0.004	0.90
72	6.71	3.76	11.96	9.66x - 2.98	0.016	0.84
96	5.68	4.46	7.23	8.95x - 1.75	0.003	0.61
		Methyl parathion				
24	6.28	4.95	7.96	9.93x - 2.92	0.003	1.00
48	5.15	3.96	6.71	7.85x - 0.59	0.003	0.82
72	4.63	3.30	6.47	8.70x - 0.79	0.006	0.72
96	4.36	3.30	5.69	9.03x - 0.77	0.004	0.69

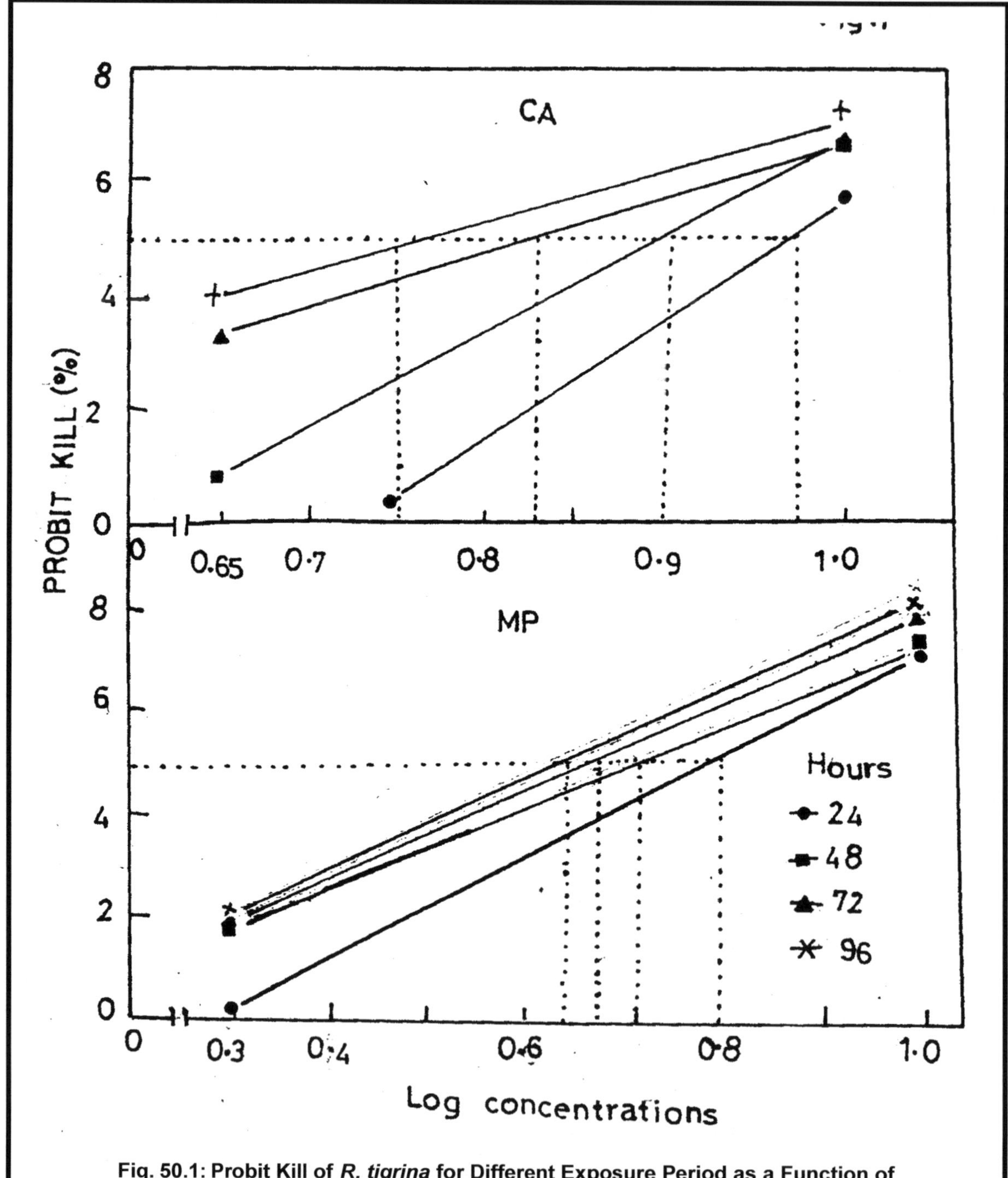

Fig. 50.1: Probit Kill of *R. tigrina* for Different Exposure Period as a Function of Log Concentrations of Carbaryl (CA) and Methyl Parathion (MP).

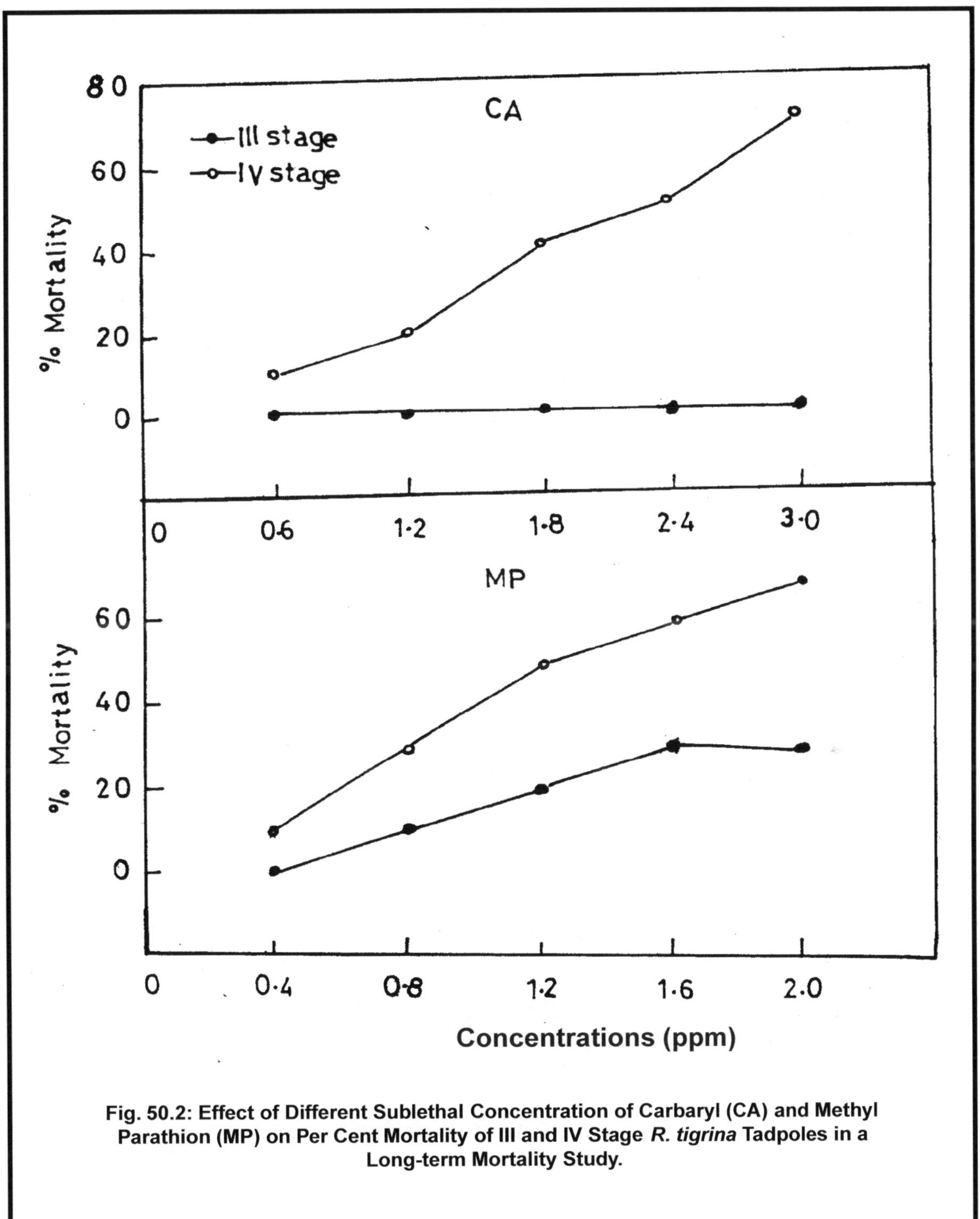

Fig. 50.2: Effect of Different Sublethal Concentration of Carbaryl (CA) and Methyl Parathion (MP) on Per Cent Mortality of III and IV Stage *R. tigrina* Tadpoles in a Long-term Mortality Study.

Table 50.2: Acute toxicity values of carbaryl and methyl parathion to Rana tigrina tadpoles in IV stage during long-term study

Pesticide	*LC_{50} (ppm)*	*95% fiducial limit (ppm)*		*y = (bx - a)*	*Variance*	*Relative potency of the pesticide*
		Upper	*Lower*			
Carbaryl	2.2	1.54	3.20	2.50x + 1.64	0.007	1.00
Methyl parathion	1.3	0.89	1.75	2.58x + 2.17	0.006	0.59

Discussion

The bioassay results revealed that the mortality of tadpoles was time and concentrations of CA and MP dependent. Pollutants cause disruption in enzyme activities leading to subsequent death of the animals (Agarwal, 1991). He also reported time and concentration-dependent mortality in *Channa punctatus* exposed to mercuric chloride. The LC_{50} value for CA fall between 5.68 and 9.40 ppm and for MP between 4.36 and 6.28 ppm. This suggests that MP was more toxic than CA. Due to the lack of pertinent literature for pesticides on anurans, these values could be compared with those compared for fish. The 24 h LC_{50} value of carbaryl for *Saccobranchus fossilis* was 22.44 ppm and it reduced to 16.67 ppm, when the exposure period was extended to 96 h (Verma *et al.*, 1979). The LC_{50} value of carbaryl for adult *R. tigrina* was reported to be 640 mg/kg body weight (Sampath *et al.*, 1992). The 96 h LC_{50} value reported for MP was 5.9 ppm in *Mystus cavasius* (Murty *et al.*, 1984) and this value lies very close to the value observed in the present study (4.36 ppm). The 96 h LC_{50} value of MP reported for *R. tigrina* in the present study was 30 per cent higher than CA. Methyl parathion has been proved to be more toxic than carbaryl in catfishes (Sukumar and Balaparameswara Rao, 1985, Arunachalam *et al.*, 1990).

The 95 per cent fiducial limits of each response curve for different exposure periods showed a wide range of fluctuations for carbaryl and a narrow range for MP treated tadpoles (Table 50.1). It indicates the effective sharp response of tadpoles to the MP. The decreasing trend of fiducial limits due to increase in exposure period was reported by Mohapatra and Noble (1991) for the mullet, *Liza parsia* exposed to DDVP, an organophosphate. Relative toxicity revealed a time dependent increase of toxicity for CA and MP in the case of *R. tigrina* tadpole which is more sensitive to the above pesticides.

In the long-term study, both the pesticides cause mortality in IV stage and MP executed mortality in III stage itself. It reveals the overriding adverse effect of this organophosphate over carbamate even before metamorphosis. Evidently it has been observed that MP causes malforamtion like failure of forelimb emergence during III stage and scoliosis (lateral spinal flexures) during development particularly in tadpoles exposed to 30 per cent MP. The incidence of scoliosis has been reported in *Channa striatus* exposed to methyl parathion (Bhaskaran, 1980). Symptoms of vertebral deformation in amphibian tadpoles exposed to MP has been reported for the first time. Toxicants like lead acts as teratogenic agent on *Bufo arenarum* embryos (Perez-Coll and Herkovits, 1990). In both the long-term and short-term studies, methyl parathion affects the survival of the tadpoles more than that of carbaryl.

As mortality was observed during the larval period in sublethal levels separate LC_{50} values were derived for CA and MP from the per cent mortality observed in IV stage, which would be considered as the LC_{50} for the whole larval period.

References

Abdul Ali, H. (1985). On the export of frog legs from India. *J. Bombay Nat. Hist. Soc.*, 18: 347-375.

Agarwal, S.K. (1991). Bioassay evaluation of acute toxicity levels of mercuric chloride to an air-breathing fish *Channa punctatus* (Bloch): Mortality and behavioural study. *J. Environ. Biol.*, 12: 99-106.

Arunachalam, S. Palanichamy, S. Vasanthi, M. and Bhaskaran. P. (1990). The impact of pesticides on the feeding energetics and body composition in the freshwater catfish *Mystus vittatus.* In: The Second Asian Fisheries Forum, Hirano. R. and Hayna, I. (Eds.). Asian Fisheries Society, Manila, Philippines: pp. 939-941.

Bhaskaran, R. (1980). Biological studies on chosen thermoconformers (*Channa striatus*). Ph.D. Thesis, Madurai Kamaraj University, Madurai, India, p. 135.

Busvine, J.R. (1971). A critical review of the techniques for testing insecticides. Commonwealth Agricultural Bureau, England: pp. 267-282.

FAO. (1975). *The monographs, FAO, Rome, Proceeds Publ. AGP/1973/M/491.*

Finney, D.J. (1971). *Probit analysis*. 3rd Edn. Cambridge, Cambridge University Press: p. 333.

Gosner, K.L. (1960). A simplified table for staging anuran embryos and larvae with notes on identification. *Herpetologica,* 16: 183-190.

Konar, S.K. (1981). Pollution of water by pesticides and its influence on aquatic ecosystem. *Indian Rew. Life Sci.*, 1: 139-165.

Mohapatra, B.C. and Noble, A. (1991). Static bioassay with *Liza parsia* exposed to DDVP, an organophosphate. *Indian J. Fish.*, 38: 194-197.

Murty, A.S. Ramani, A.V. Christopher, K. and Rajabhusanam, B.R. (1984). Effect of methyl parathion and fensulforthion to the fish, *Mystus cavasius. Environ. Pollut.*, 34: 37-46.

Pandian, T.J. and Marian, M.P. Prediction of absorption efficiency from food nitrogen in amphibians. *Proc. Indian Acad. Sci.*, 95: 387-395.

Perez-Coll, C.S. and Herkovits. (1990). Stage dependent susceptibility to lead in *Bufo arenarium* embryos. *Environ. Pollut.*, 63: 239-245.

Sampath, K., Elango, P. and Thanalakshmi, S. (1992). Effect of carbaryl (sevin) on the carbohydrate metabolism of the common frog, *Rana tigrina. Environ. Ecol.*, 10: 278-281.

Sukumar, R.V. and Balaparameswara Rao, M. (1985). Toxicity of γ-HeH-methyl parathion and carbryl to two varieties of a tropical freshwater gastropod *Bellamya bengalensis* (Lamarck) (Gastropoda: Viviparidae). *Proc. Symp. Assess. Environ. Pollut.* (Eds. Dalela. R.C. and Mane. M.H.). The Academy of Environmental Biology, India, pp. 101-106.

Syed Othuman, H. (1994). Toxicity of distillery effluent on a chosen freshwater fish. Ph.D. Thesis to Madurai Kamaraj University, Madurai, India, p. 107.

Verma, S.R., Bansal, S.K., Gupta, A.K., Pal, N., Tyagi, A.C., Bhatnagar, M.C., Kumar, K. and Dalela, R.C. (1979). Acute toxicity of three pesticides to a freshwater teleost, *Saccobranchus fossilis.* Proc. Symp. Environ. Biol. (Eds. Verma. S.R.: Tyagi. A.K. and Bansal, S.K.). The Academy of Environmental Biology, Muzaffarnagar, India. pp. 481-497.

Zar, J.E. (1974). *Biostatistical Analysis*. Prentice-Hall, Engelwood, New Jersey, p. 620.

Index

B

C

D

E

F

G

H

M

N

O

P

Q

R

S

T

U

V

W

X

Z